Evolution of Plant–Pollinator Relationships

What are the evolutionary mechanisms and ecological implications behind a pollinator choosing its favorite flower? Sixty-five million years of evolution have created the complex and integrated system which we see today, and understanding the interactions involved is key to environmental sustainability.

Examining pollination relationships from an evolutionary perspective, this book covers both botanical and zoological aspects. It addresses the puzzling question of co-speciation and co-evolution and the complexity of the relationships between plant and pollinator, the development of which is examined through the fossil record. Additional chapters are dedicated to the evolution of floral displays and signaling, as well as their role in pollination syndromes and the building of pollination networks. Wide ranging in its coverage, the book outlines current knowledge and complex emerging topics, demonstrating how advances in research methods are applied to pollination biology.

SÉBASTIEN PATINY is a scientific collaborator in the Laboratory of Zoology, Université de Mons, Belgium. A large part of his research focuses on desert species of bees, their distribution, and the importance of biogeographical features in some species-level radiations. He is currently developing a series of papers dedicated to the inference of large phylogenetic topologies.

The Systematics Association Special Volume Series

DAVID J. GOWER

Department of Zoology, The Natural History Museum, London, UK

The Systematics Association promotes all aspects of systematic biology by organizing conferences and workshops on key themes in systematics, running annual lecture series, publishing books and a newsletter, and awarding grants in support of systematics research. Membership of the Association is open globally to professionals and amateurs with an interest in any branch of biology, including palaeobiology. Members are entitled to attend conferences at discounted rates, to apply for grants and to receive the newsletter and mailed information; they also receive a generous discount on the purchase of all volumes produced by the Association.

The first of the Systematics Association's publications *The New Systematics* (1940) was a classic work edited by its then-president Sir Julian Huxley. Since then, more than 80 volumes have been published, often in rapidly expanding areas of science where a modern synthesis is required.

The Association encourages researchers to organize symposia that result in multiauthored volumes. In 1997, the Association organized the first of its international Biennial Conferences. This and subsequent Biennial Conferences, which are designed to provide for systematists of all kinds, included themed symposia that resulted in further publications. The Association also publishes volumes that are not specifically linked to meetings, and encourages new publications (including textbooks) in a broad range of systematics topics.

More information about the Systematics Association and its publications can be found at our website: www.systass.org

Previous Systematics Association publications are listed after the index for this volume.

Systematics Association Special Volumes published by Cambridge University Press:

78. *Climate Change, Ecology and Systematics*
   Trevor Hodkinson, Michael Jones, Stephen Waldren, and John Parnell
79. *Biogeography of Microscopic Organisms: Is Everything Small Everywhere?*
   Diego Fontaneto
80. *Flowers on the Tree of Life*
   Livia Wanntorp and Louis Ronse De Craene

THE SYSTEMATICS ASSOCIATION SPECIAL

VOLUME 81

Evolution of Plant–Pollinator Relationships

EDITED BY

SÉBASTIEN PATINY

University of Mons, Belgium

THE
Systematics
ASSOCIATION

CAMBRIDGE
UNIVERSITY PRESS

CAMBRIDGE
UNIVERSITY PRESS

University Printing House, Cambridge CB2 8BS, United Kingdom

Cambridge University Press is part of the University of Cambridge.

It furthers the University's mission by disseminating knowledge in the pursuit of
education, learning and research at the highest international levels of excellence.

www.cambridge.org
Information on this title: www.cambridge.org/9780521198929

First published 2012
Reprinted 2013

A catalogue record for this publication is available from the British Library

Library of Congress Cataloguing in Publication data
Evolution of plant–pollinator relationships / [edited by] Sébastien Patiny.
      p.   cm. – (Systematics association special volume series)
    Includes bibliographical references and index.
    ISBN 978-0-521-19892-9 (hardback)
    1. Pollination by insects.   2. Pollination by animals   3. Plants – Evolution.
    4. Pollinators – Evolution.   I. Patiny, Sébastien.
    QK926.E96   2011
    576.8'75–dc23   2011032468

ISBN 978-0-521-19892-9 Hardback

Additional resources for this publication at www.cambridge.org/9780521198929

Contents

Color plate section appears between pages 248 and 249

Contributors

NADIR ALVAREZ Department of Ecology and Evolution, University of Lausanne, Switzerland

BRUCE ANDERSON Botany and Zoology Department, Stellenbosch University, South Africa

W. SCOTT ARMBRUSTER School of Biological Sciences, University of Portsmouth, UK

JAMES COOK Philip Lyle Building, School of Biological Sciences, University of Reading, UK

ASTRID CRUAUD INRA-UMR Centre de Biologie et de Gestion des Populations, CBGP (INRA/IRD/CIRAD/Montpellier SupAgro), Campus international de Baillarguet, France

AMOTS DAFNI Laboratory of Pollination Ecology, Institute of Evolution, Haifa University, Israel

YANG DA-RONG Key Laboratory of Tropical Forest Ecology, Xishuangbanna Tropical Botanical Garden, Chinese Academy of Sciences, China

DAVID L. DILCHER Department of Biology, Indiana University, USA

ACHIK DORCHIN Laboratory of Pollination Ecology, Institute of Evolution, Haifa University, Israel

ANNA DORNHAUS Center for Insect Science / Department of Ecology and Evolutionary Biology, University of Arizona, USA

STEFAN DÖTTERL Department of Plant Systematics, University of Bayreuth, Germany

YOKO L. DUPONT Department of Biological Sciences, Aarhus University, Denmark

CONNAL D. EARDLEY Agricultural Research Council, Plant Protection Research Institute, School of Biology and Conservation Science, University of KwaZulu–Natal, South Africa

ALLAN G. ELLIS Botany and Zoology Department, Stellenbosch University, South Africa

MICHAEL S. ENGEL University of Kansas, Division of Entomology and Department of Ecology and Evolutionary Biology, USA

GWENAËLLE GENSON INRA-UMR Centre de Biologie et de Gestion des Populations, CBGP (INRA/IRD/CIRAD/Montpellier SupAgro), Campus international de Baillarguet, France

ANTOINE GUISAN Department of Ecology and Evolution, University of Lausanne, Switzerland

MELANIE HAGEN Department of Biological Sciences, Aarhus University, Denmark

SHUSHENG HU Division of Paleobotany, Peabody Museum of Natural History, Yale University, USA

ROULA JABBOUR-ZAHAB INRA-UMR Centre de Biologie et de Gestion des Populations, CBGP (INRA/IRD/CIRAD/Montpellier SupAgro), Campus international de Baillarguet, France

TOM J. DE JONG Plant Ecology, Leiden University, The Netherlands

ANDREAS JÜRGENS School of Biological and Conservation Sciences, University of KwaZulu-Natal, South Africa

H. ELIZABETH KIRKPATRICK Biology Department, University of Puget Sound, USA

FINN KJELLBERG CNRS – UMR Centre d'Ecologie Fonctionnelle et Evolutive, CEFE, France

MICHAEL KUHLMANN Department of Entomology, The Natural History Museum, UK

ANNE S. LEONARD Center for Insect Science / Department of Ecology and Evolutionary Biology, University of Arizona, USA

YAEL MANDELIK Department of Entomology, The Robert H. Smith Faculty of Agriculture, Food and Environment, The Hebrew University of Jerusalem, Israel

TALYA MAROM-LEVY Laboratory of Pollination Ecology, Institute of Evolution, Haifa University, Israel

DENIS MICHEZ University of Mons, Laboratory of Zoology, Belgium

SIMON VAN NOORT Natural History Division, South African Museum, Iziko Museums of Cape Town, South Africa

JENS M. OLESEN Department of Biological Sciences Aarhus University Denmark

DANIEL R. PAPAJ Center for Insect Science / Department of Ecology and Evolutionary Biology, University of Arizona, USA

SÉBASTIEN PATINY University of Mons, Laboratory of Zoology, Belgium

LOÏC PELLISSIER Department of Ecology and Evolution, University of Lausanne, Switzerland

RODRIGO AUGUSTO SANTINELO PEREIRA Depto de Biologia/FFCLRP-USP, Brazil

GIDEON PISANTY Department of Entomology, The Robert H. Smith Faculty of Agriculture, Food and Environment, The Hebrew University of Jerusalem, Israel

CLAUS RASMUSSEN Department of Biological Sciences, Aarhus University, Denmark

JEAN-YVES RASPLUS INRA-UMR Centre de Biologie et de Gestion des Populations, CBGP (INRA/IRD/CIRAD/Montpellier SupAgro), Campus international de Baillarguet, France

NINA RØNSTED Department of Medicinal Chemistry, Faculty of Pharmaceutical Sciences, University of Copenhagen, Denmark

OTILENE SANTOS-MATTOS Instituto Nacional de Pesquisa da Amazônia, Brazil

VINCENT SAVOLAINEN Imperial College London, UK

FLORIAN P. SCHIESTL Institute of Systematic Botany, University of Zürich, Switzerland

BORIS O. SCHLUMPBERGER University of Munich, LMU, Germany (Current address: Herrenhausen Gardens, Hannover, Germany)

YUVAL SHIMRAT Laboratory of Pollination Ecology, Institute of Evolution, Haifa University, Israel

DAVID WINSHIP Taylor Department of Biology, Indiana University Southeast, USA

KRISTIAN TRØJELSGAARD Department of Biological Sciences, Aarhus University, Denmark

ROSICHON UBAIDILLAH Entomology Laboratory, Zoology Division (Museum Zoologicum Bogoriense), Center Research for Biology, LIPI, Indonesia

MARYSE VANDERPLANCK University of Mons, Laboratory of Zoology, Belgium

JANA C. VAMOSI Department of Biological Sciences, University of Calgary, Canada

STEVEN M. VAMOSI Department of Biological Sciences, University of Calgary, Canada

PAUL WILSON Department of Biology, California State University, Northridge, USA

TAINA WITT School of Biological and Conservation Sciences, University of KwaZulu-Natal, South Africa

PENG YAN-QIONG Key Laboratory of Tropical Forest Ecology, Xishuangbanna Tropical Botanical Garden, Chinese Academy of Sciences, China

Preface

Pollination has been a source of questioning and fascination as long as there have been naturalists. Aristotle and Herodotus before him were already according a specific interest to the topics of fig and palm pollination. In the eighteenth and nineteenth centuries – adopting a more adaptationist point of view – Kölreuter (1761), Sprengel (1793), and Darwin (1862) authored books that constitute the early stepping stones in development of modern pollination biology. From the time Darwin published the *On the Origin of Species* (1859), the functional relationships of plant and pollinator were cast in evolutionary scenarios in which plants adapted to pollinators and pollinators adapted to flowers.

Over the last decade, many edited volumes have been published on pollination-related topics (Proctor et al. 1997; Chittka and Thomson 2001; Dafni et al. 2005; Waser and Ollerton 2006; Harder and Barrett 2007). The present volume originated in a symposium dedicated to the evolution of plant–pollinators relationships (EPPR), which was organized in the framework of the *SYSTEMATICS 2009* meeting in Leiden in the Netherlands. Given the intense scientific activity in pollination biology (Mitchell et al. 2009), the idea behind this symposium was to provide a forum for authors to pull together recent advances in pollination in the context of systematics. The present book constitutes an outcome of this symposium along with its natural prolongation. It has been developed with an explicit sensitivity for the evolutionary aspects of pollination. It includes contributions from the participants in the symposium and additional authors who joined the book project to round out its evolutionary coverage.

The attention currently given to pollination can be considered as the consequence of the combined importance of the pollination ecological service, plus threats weighing on a continually increasing number of pollinator populations worldwide, and interest in how plant–pollinator relations evolve in the face of environmental change.

(1) The economic value of pollination has focused the interests of numerous research groups (Aizen et al. 2009; Allsopp et al. 2008; Buchmann and

Nabhan, 1996; Gallai et al. 2009). Two very recent reports underscore the economic importance of pollination. Gallai et al. (2009) estimated about €150 billion per year are contributed by insect pollination to crops worldwide. Allsopp et al. (2008) showed that, despite the deep divergences in the methods used and the results tabulated, the varied studies that have been done converge in concluding that pollination constitutes a key component of the world economy.

(2) Echoing the economic importance of pollination services, the observation of continued population regressions and diversity erosions (e.g. Biesmeijer et al. 2006; Kluser and Peduzzi 2007; Potts et al. 2010) is increasing the urgency for better conservation of pollinators. Conservation of pollinators, in turn, demands the development of better supporting science.

(3) In addition to these econocentric and conservation interests, pollination systems emerge as wonderful models for the study of adaptation, cophylogeny and speciation, topics in which a wealth of questions are puzzling the scientific community. This last point constitutes the main focus of the present book, and in the following pages expert authors discuss in detail varied aspects of the evolution of pollinators, pollinated plants, and pollination systems.

Considering the above points, the improvement of the understanding of the evolution of interactions between pollinators and pollinated plants within their ecological webs is highly desirable. Likewise, renewed models of the evolution of pollination in space and time are needed.

Nowadays, understanding of the evolutionary dance between pollinators and pollinated plants remains quite fragmentary. The simple coevolutionary model – envisioned as specialized forms adapting reciprocally to one another – and the basic picture of progress to specialization have been questioned (e.g. Danforth et al. 2006a, 2006b; Cruaud et al. Chapter 4). The scale of pollination processes range from the molecular to the community level, but studies at the various scales seem not to have settled into a coherent model of evolution. The aim of the present book is to embrace an evolutionary point of view, bringing together the contributions from a large panel of research groups that have explored pollination with various approaches. The following chapters address a series of domains within the biology of pollination:

(1) Evolutionary biology of pollination integrating phylogenetic thinking

(2) Evolution of pollination syndromes, floral displays and rewards

(3) Evolution of feature of pollination networks

The contributions in these sections outline both the state of the knowledge in the three domains and novel aspects under development.

Phylogenetics are the new toolkit in studying all aspects of the diversification of life. This is, of course, true when studying examples of coevolution involving distinct groups of living forms, as for instance in the evolution of pollination. The methodological and analytical opportunities to map phylogenies onto one another, to date clades using molecular clocks, and to trace evolution of characters on trees are the operational promises of phylogenetics as applied to pollination biology. However, despite this promise of phylogenetics, very few specific empirical results have been produced so far for the study of pollination as has been the case in biogeography (e.g. Ree and Sanmartin 2009; Salvo et al. 2010). Evolutionary pollination biology has so far mostly benefited from the general progress made in phylogenetics. For example, phylogenetics allow us to recast our conceptual understanding of macroevolution. The first chapters of this book present some of the key phylogenetic insights into pollination biology.

For a long time, attention has been paid to the concordant evolution of plants and their pollinators, notably the evolutionary strategies developed by plants to increase pollination by their best pollinators and the senses used by pollinators to identify and locate food. The conceptual framework of pollination syndromes developed from these two topics (Faegri and van der Pijl 1979; Proctor et al. 1997). With increased ease over the next few years, genomics, transcriptomes, and floral physiology point to a wealth of new directions for investigation. This will open new vistas in pollination biology. Our understanding of the ways in which pollinators perceive flowers and are rewarded by flowers seems to be constantly improving. While all these avenues of research are of interest in and of themselves, they also are providing a fountain of data of a new kind to be used in deciphering the evolutionary relationships between plants and pollinators. Aspects of these topics are developed in the last chapters dedicated to the evolution of pollination syndromes.

Moving out in scale, to the level of communities, we must now recognize that pollination webs evolve. This topic is directly related to conservation, as well as evolution. Pollination webs are fundamental to understanding the general patterns in community context in which the evolution of the pollination systems occurs. Studying pollination webs means considering the interactions of multiple species with distinct levels of respective knowledge and interacting partners that are, at some level, competing for niches and resources within niches. The study of pollination webs is, *par nature*, integrative. The chapters dealing with pollination webs discuss both the theoretical aspects of pollination evolution and several study cases. A particular focus has been set on evolution of the plant sex systems, emergence of unusual floral rewards, and relationships between herbivory and pollination.

This book should become a reference for questions related to the evolution of pollination systems in varied contexts. Secondly, it documents the ways in which the complexity of pollination ramifies into many areas within biology. Finally, it serves as an example of new research methods applied to pollination systems that give us the opportunity to revisit old problems such as the usefulness of varied species concepts.

References

Aizen, M. A., Garibaldi, L. A., Cunningham, S. A. and Klein, A. M. (2009). How much does agriculture depend on pollinators? Lessons from long-term trends in crop production. *Annals of Botany*, **103**, 1579-88.

Allsopp, M. H., de Lange, W. J. and Veldtman, R. (2008). Valuing insect pollination services with cost of replacement. *PLoS ONE*, **3**, e3128.

Biesmeijer, J. C., Roberts, S. P. M., Reemer, M., Ohlemüller, R., Edwards, M., Peeters, T., Schaffers, A. P., Potts, S. G., Kleukers, R., Thomas, C. D., Settele, J. and Kunin, W. E. (2006). Parallel declines in pollinators and insect-pollinated plants in Britain and the Netherlands. *Science*, **313**, 351-4.

Buchmann, S. L. and Nabhan, G. P. (1996). *The Forgotten Pollinators*. Washington, DC: Island Press.

Chittka, L. and Thomson, J. D. (eds.). (2001). *Cognitive Ecology of Pollination. Animal Behaviour and Floral Evolution*. Cambridge, UK: Cambridge University Press.

Dafni, A., Kevan, P.G. and Husband, B. C. (eds.) (2005). *Practical Pollination Biology*. Cambridge, Canada: Enviroques,.

Danforth, B. N., Sipes, S. D., Fang, J. and Brady, S. G. (2006a). The history of early bee diversification based on give genes plus morphology. *Proceedings of the National Academy of Sciences USA*, **103**, 15118-23.

Danforth, B. N., Fang, J. and Sipes, S. D. (2006b). Analysis of family-level relationships in bees (Hymenoptera:Apiformes) using 28S and two previously unexplored nuclear genes: CAD and RNA polymerase II. *Molecular Phylogenetics and Evolution*, **39**, 358-72.

Darwin, C. D. (1859). *On the Origin of Species by Means of Natural Selection, or the Preservation of Favoured Races in the Struggle for Life*. 1st edn. London, UK: John Murray.

Darwin, C. D. (1862). *On the Various Contrivances by which British and Foreign Orchids are Fertilised by Insects*. London, UK: John Murray.

Fægri, K. and van der Pijl, L. (1979). *The Principles of Pollination Ecology*. Oxford, UK: Pergamon Press.

Gallai, N., Salles, J. M., Settele, J. and Vaissière, B. E. (2009). Economic valuation of the vulnerability of world agriculture confronted with pollinator decline. *Ecological Economics*, **68**, 810-21.

Harder, L. D. and Barrett, S. C. H. (eds.) (2007). *Ecology and Evolution of Flowers*. New York, NY: Oxford University Press.

Kölreuter, J. G. (1761). *Vorläufige Nachricht von einigen, das Geschlecht der Pflanzen betreffenden Versuchen und Beobachtungen*. Leipzig, Germany: In der Gleditschischen Handlung.

Kluser, S. and Peduzzi, P. (2007). Global pollinator decline: a literature review. UNEP/GRIDEurope.

Mitchell, R. J., Flanagan, R. J., Brown, B. J., Waser, N. M. and Karron, J. D. (2009). New frontiers in competition for pollination. *Annals of Botany*, **103**, 1403–13.

Potts, S. G., Biesmeijer, J. C., Kremen, C., Neumann, P., Schweiger, O. and Kunin, W. E. (2010). Global pollinator declines: trends, impacts and drivers. *Trends in Ecology and Evolution*, **25**, 345–53.

Proctor, M., Lack, A. and Yeo, P. (1997). *The Natural History of Pollination*. London, UK: Harper Collins.

Ree, R. H. and Sanmartín, I. (2009). Prospects and challenges for parametric models in historical biogeographical inference. *Journal of Biogeography*, **36**, 1211–20.

Salvo, G., Ho, S. Y. W., Rosenbaum, G., Ree, R. and Conti, E. (2010). Tracing the temporal and spatial origins of island endemics in the Mediterranean region: a case study from the *Citrus* family (Ruta L., Rutaceae). *Systematic Biology*, **59**, 1–18.

Sprengel, C. K. (1793). Das entdeckte Geheimniss der Natur im Bau und in der Befruchtung der Blumen. Berlin, Germany: Bei Friedrich Vieweg dem aeltern.

Waser, N. M. and Ollerton, J. (2006). *Plant–Pollinator Interactions: from Specialization to Generalization*. Chicago, IL: The University of Chicago Press.

1

Macroevolution for plant reproductive biologists

PAUL WILSON

1.1 From micro- to macroevolution

Just as there is a microevolutionary process that explains organismal adaptations, so is there a macroevolutionary process that explains biological diversity. Consider western North America's wildflowers. How is it that there are 246 penstemons that are hymenopteran pollinated, and 40 penstemons that have taken on hummingbirds, but no penstemon species has adapted to fly or butterfly or beetle pollination? How is it that there are 60 kinds of dudleyas, all with ranges emanating from the coastal mountains? And how about mariposa lilies, a group of 35 species varying in flower colors and petal hairs yet all pollinated by both beetles and bees via a highly generalized floral mechanism? The amounts of diversity and the patterns in which they are arranged are the products of a macroevolutionary process.

The microevolutionary process is more familiar. Mutations occur from time to time. They are undirected. Many are deleterious to the functioning of the organism in its environment. For a while they contribute to the genetic load, then eventually they are lost due to natural selection. Many other mutations are neutral or nearly neutral given the environment where the organism lives and the genetic state of the organism at other loci. Neutral alleles change in frequency due to genetic drift. A few new mutations are beneficial to the individuals that carry them, or to

Evolution of Plant–Pollinator Relationships, ed S. Patiny. Published by Cambridge University Press. © The Systematics Association 2012.

their close relatives, and these are selected up in frequency. The beneficence of these alleles may depend on the outside environment, for example on the kinds of animals that are pollinating those plants in their local population. Likewise, whether or not an allele is beneficial may depend on the genetic state of the rest of the organism. If the outside environment or the genetic background change, then what was once deleterious or neutral may become beneficial.

This dynamic that happens at the microevolutionary scale has an analogue at the macroevolutionary scale, what Stephen Jay Gould (2002) called a "grand analogy." The analogue to selection among individuals within a population is selection among clades in a biota. Clades with certain character states diversify more. For example, flowers with nectar spurs have higher rates of diversification than flowers without nectar spurs (Hodges 1997a). The analogue of mutation is the punctuation in punctuated equilibrium, the shift to a new adaptive state, such as when an isolated population shifts to a different pollinator. The analogue of genetic drift in allele frequencies at a locus is clade drift in the frequency of species having a particular trait in a region's biota. In Gould's hierarchical process, characters come to be fixed in a lineage through organismal adaptation, and then those fixed differences among lineages become the criteria for selection at a higher level.

Individual selection along with some other microevolutionary ingredients such as mutation and drift are mainly what is responsible for the adaptations of organisms: how a bee-pollinated penstemon has come to have purple vestibular flowers that make nectar of a certain sort and have a staminode for levering the anthers and stigma onto the bee's back in a certain way, etc. But there is more to explain about life than just the adaptations of this or that flower: there is the amount and pattern of biodiversity. Clade selection and other macroevolutionary ingredients are responsible for the diversity of organisms: how many species of penstemons there are, how many are specialized for pollination by bees versus birds, the size of penstemon geographic ranges, the way they remain clustered into groups nested within the larger penstemon clade, how each of those smaller groups is characterized, the disparity of specializations within the groups, etc. (Wilson et al. 2006). Microevolution and macroevolution work together and end up affecting one another. Together they constitute one unified machine that generates order out of history.[1]

---

[1] Those who are reading for pleasure should ignore my footnotes. My chapter is an introduction to hierarchical evolution aimed at people interested in pollination. As such, I have refrained from reviewing many philosophical distinctions and historical debates. For instance, I do not review the claims of Gould and associates circa 1980 and the criticisms of those early attempts. An improved and less controversial version of how hierarchical evolution works followed from a change in definitions announced in Gould and Lloyd (1999). Refinements to the grand analogy beyond Gould (2002) are continuing, and I here add some of my own. True, the logic of hierarchical evolution could use some

1.2 Four forms of clade selection

A key innovation is a derived feature of a lineage that leads to greater diversity than would otherwise arise. Typically, this is detected by finding more species in the clade with the innovation than in a sister clade without the innovation (Kay et al. 2006). Nectar spurs seem to have led the groups that possess them to be species-rich compared to sister clades. Such a key innovation can work by either increasing the rate of speciation or decreasing the rate of extinction, and new statistics are starting to allow people to tease apart the two (FitzJohn et al. 2009). Evolutionary biologists have gotten used to thinking of speciation and extinction, but I shall ease into my developing argument by using slightly different language. I invoke *cladogenesis* (which is like speciation without focusing on the point when reproductive barriers become permanent) and *persistence* (which describes a lineage before its extinction). There are two causal paths for an innovation to be favored: it may be favored via increasing the likelihood of cladogenesis or increasing the likelihood of a clade persisting. Innovations may also be disfavored via lowering the likelihood of subsequent cladogenesis, or more generally, lowering the time that lineages are likely to persist. Consider all four cases (Fig 1.1).

(1) An innovation that favors cladogenesis is bilateral symmetry in flowers. Sargent (2004) found that groups with bilaterally symmetric flowers had more species in them than sister groups. Kay et al. (2006) worry that six of 22 sister groups show the reverse pattern, with radially symmetric flowers being more species rich. Nevertheless, the pattern seen in most cases is that bilateral symmetry increases the rate of cladogenesis. How would this work? Flowers that guide their pollinators to visit in a receiving line place pollen on the pollinator more accurately (Armbruster et al. 2009b). That tends to promote speciation in the form of a reproductive isolating barrier whereby different kinds of pollinators are employed or different areas of the pollinator's body are used by different plants. In addition to presenting numerous species to pollinators, having high rates of cladogenesis might multiply the chances that a sub-lineage of a

scholarly help, but that should be done elsewhere. I add only a very few footnotes to help readers who are of a more critical mind.

The overarching semantic debate would be whether it is better to draw a grand analogy or to use verbiage at the level of clades distinct from the verbiage of microevolution. Many would prefer to not use loaded terms like clade selection as an analogue to individual selection because by their definitions there is only one kind of selection. They use other terminology to write about the phenomenon of some clades being more successful than others because of the traits of those clades. I choose to put as much as possible into a theory of hierarchical evolution. I do, however, believe the hierarchy should be presented with some exploration of how the two levels are not parallel.

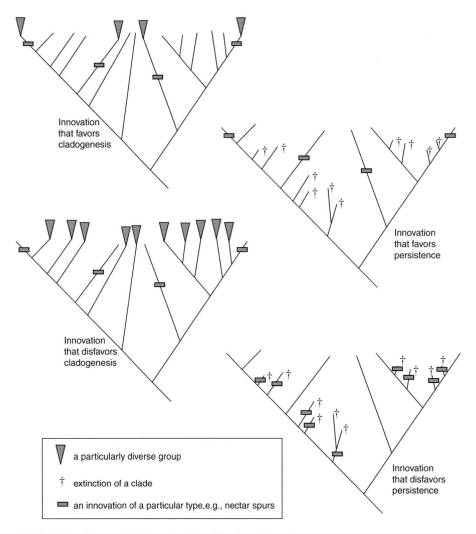

Innovation
that favors
cladogenesis

Innovation
that favors
persistence

Innovation
that disfavors
cladogenesis

Innovation
that disfavors
persistence

▼ a particularly diverse group

† extinction of a clade

▭ an innovation of a particular type, e.g., nectar spurs

Fig 1.1 Four forms of clade selection affecting diversity.

clade will survive through catastrophes. Thus, an innovation that favors cladogenesis has two effects: in a snapshot in time, the groups with the innovation have many species; and in the long term, as ecological divergence proceeds, the clade as an aggregate is likely to have varied chances of surviving.

(2) An innovation that favors long persistence by means other than the multiplication of its clades might be the evolution of seed dormancy. Seed dormancy allows seeds to survive in a seed bank for longer than the seeds of cousins that lack seed dormancy. I know of no phylogenetic analysis that shows this pattern, but a bit of inspiration can be drawn from work done on an ecological time scale. Kalisz et al. (1997) have used population data on blue-eyed

mary to parameterize a model showing how a seed bank buffers a population against the vagaries of bad years. Stöcklin and Fischer (1999), reporting on a grassland community, found that species with seeds that live for more than five years are less likely to go to extirpation than species with short-lived seeds. On a longer time scale, clade selection in favor of seed dormancy seems likely. If one had quantitative measures of seed dormancy for a group of species and a phylogeny relating the species, one could test for a phylogenetic effect of seed longevity on how deeply rooted the dormant clades are. It would also be worthwhile to see if the effect was contingent on the life history of the plants involved. The clade selection might be stronger in annuals than in perennials. It might also be stronger in biomes with highly stochastic weather than in biomes where rainfall is relatively constant. Comparative tests of hypotheses about seed dormancy will surely be complicated (Baskin and Baskin 1998).

(3) An innovation that disfavors cladogenesis is the shift to abiotic pollination (Dodd et al. 1999). Most major lineages of flowering plants were once animal pollinated and those that are still animal pollinated have high rates of cladogenesis, but those that have gone over to wind or water pollination have lower rates of cladogenesis. Why? First, animal pollinators tend to be picky about the appearance of the flowers they visit. For example, individual bees are prone to becoming temporarily constant to a particular color or appearance of flower (Gegear and Laverty 2005). In a local community, flowers that evolve to be distinctive in appearance compared to co-flowering species encourage such constancy and thereby have their pollen moved with less wastage and purer delivery (Wilson and Stine 1996). This may even cause ecological sorting allowing species with distinctive colors to become abundant in their community (McEwen and Vamosi 2010). In addition to appearance, the mechanical fit of flowers around pollinator bodies is probably selected to be as efficient as possible (Castellanos et al. 2003). Second, pollinators differ discontinuously in physical dimensions, so flowers pollinated by different types of animals could be experiencing diversifying selection (Wilson and Thomson 1996). Finally, aside from being an organ of local differentiation (Johnson 2006), when divergent flowers come back together in sympatry, the functional variation is grist for positive assortative mating whereby similar flowers mate with one another. The assortative mating is caused by animals having a behavioral tendency to categorize (Jones 2001), and the assortative mating maintains or even adds to genetic correlations. Genetic correlations in turn predispose lineages to evolve reproductive isolating barriers (Kondrashov and Shpak 1998). When a clade changes from animal pollination to pollination by wind or water, there is then a relaxation of the tendency towards subsequent cladogenesis.

(4) An innovation that disfavors persistence might be the evolution of separate sexes. Dioecious clades have been found to have fewer species than their sister clades that are co-sexual (Heilbuth 2000), and they tend to have more endangered and threatened species (Vamosi and Vamosi 2005). Compared to hermaphroditic lineages, dioecious plants are likely to be inferior at establishing new sub-populations in the meta-population dynamic. Also, dioecious plants have more of a seed-shadow handicap whereby seedlings are clumped around mother plants and compete with each other to the detriment of the population. Finally, dioecious plants have a stronger reliance upon pollinators in the face of stochastic variation in pollinator services. Not only is self-pollination impossible, but there is more of a chance that neighboring plants will be of the same gender. The phylogenetic patterns could be because dioecious clades have low rates of cladogenesis, but it seems more likely that they have a higher rate of extinction than co-sexual clades. In other words, dioecy evolves from time to time but tends to be an evolutionary dead end.[2]

1.3 Other macroevolutionary ingredients

Just as clade selection is an analog of allelic selection, so there is a process of clade drift that is analogous to genetic drift. Gould suggested that at the macroevolutionary level drift might be more important than it is at the level of sexual individuals adapting to their surroundings. More generally, there is the possibility that ingredients that are most important for macroevolution might not be parallel to ingredients that are most important for microevolution. At any rate, several ingredients other than selection need to be introduced as I proceed to layer my argument for recognizing a hierarchical evolutionary process.

---

[2] I choose to use clade selection generally and species selection as a special case. Gould used the word "species" a great deal, as in species selection, species drift, and directional speciation. He defended punctuated equilibrium at the species level; at levels above the species level, he would say the dynamic was punctuational. I am reluctant to extend this usage (Mishler 2010). It seems particularly odd to speak of species selection resulting from characters acquired in a lineage making it more prone to subsequent cladogenesis and through that proliferation to the clade's extended life than if the characters were otherwise. I am more comfortable speaking of species selection when its mechanism is to delay extinction by some means other than favoring additional cladogenesis, but the term clade selection works in all cases. I probably picked up my usage by taking a class from George Williams, who considered it a fallacy to tie much of anything to the taxonomic species level (1992 starting on p. 118). "Clade" is actually also objectionable because, unlike an individual, a clade includes its descendants (Okasha 2003). A clade is like a family, so the strict analog of clade selection would be clone selection.

As a way of appreciating clade drift, consider the role of founder effects in the Hawaiian Islands. The original colonizers dispersed a fantastic distance and were not absolute outcrossers, yet aside from these traits (which need not have been innovations), once a founding colonist got to Hawaii and established a population, it had a very good chance of undergoing an adaptive radiation. Ricklefs and Renner (2000) sought explanations for Hawaii's radiations and found little more than the usual tendency for animal-pollinated groups to diversify. Thus, it would seem that local populations have adapted, each guided by individual selection, while at the level of the larger clades, those clades that diversified and those that did not have been largely drawn at random with respect to their traits. An appreciation for both levels in the hierarchy greatly aids understanding of the process.

Founder effects are not the only form of clade drift. The frequency of a trait among species in a biota on a phylogeny naturally staggers up and down even without the trait of interest causing the changes in its frequency. Differences might *seem* big in terms of numbers of species, but from a purely statistical standpoint, all possible partitions of species richness into two groups are equally likely (Slowinski and Guyer 1993). Maybe the number of petals has experienced clade drift. Vast swaths of angiosperm diversity have five-parted flowers; fewer, such as mustards, have four-parted flowers. The monocots are the main group that has three-parted flowers. Even if the number of parts has not causally affected the diversification process, the frequencies of five-, four- and three-parted flowers would still have changed as the tree of life has grown. So far as I know, clade drift could have accounted for the way in which biogeographic provinces around the world have different proportions of their floras that are five-, four- and three-parted.

Another macroevolutionary ingredient is clade hitchhiking. If you were taught evolution by focusing on quantitative traits being optimized in sexually reproducing organisms with selection among individuals, then you are not likely to think hitchhiking is a very important feature of evolution. If you were taught evolution by comparing DNA sequences for a particular gene where there is little chance of recombination, you are more likely to be familiar with the idea. Hitchhiking is when one character's frequency is dragged quickly up or down by linkage to another character that is undergoing a selective sweep (Barton 2000). Orchids have inferior ovaries, and there are many orchids, so maybe inferior ovaries favor cladogenesis. But maybe not. For orchids, it is possible that having inferior ovaries is merely coincidental with other traits that favor diversification: like having bilaterally symmetric flowers, stigma and anthers on a rigid column, pollen dispersed in pollinia, tiny seeds, ectomycorrhizae, the proclivity to grow as epiphytes, etc. The inferior ovary might have been dragged to high frequency by clade hitchhiking. Because there is usually no recombination between clades, clade selection is similar to selection in a strictly asexual organism, or selection on a stretch of DNA that does not recombine, and hitchhiking is to be expected.

Examples of clade drift and clade hitchhiking bring up the possibility that non-deterministic factors can explain many macroevolutionary patterns. A particular ancestor was at the right place at the right time and so it gave rise to a diverse lineage. It had a peculiar feature, not generally superior to alternative character states, and that feature was lucky to get to go along for the ride in a group that for other reasons diversified. Gould gave the name "contingency" to the way such arbitrariness can be propagated, and he suspected that many of the great successes and failures in the pageant of life were contingent turns of history, rather than determined by the features of the clades.

Let's say, the contingent bit of luck is dispersal to a different biogeographic province. That dispersal event might have the effect of favoring cladogenesis, favoring clade persistence, disfavoring cladogenesis, and/or disfavoring clade persistence. Moreover, luck and innovation could interact: the value of an innovation could be contingent upon the dispersal event. Moore and Donoghue (2007) considered both dispersal and innovation, looking to see if either or both might affect diversification in the Adoxaceae and Valerianaceae. They looked for changes in diversification rate anywhere in their phylogenies, and then they tested for associations with characters and with dispersal events. Of seven phylogenetic segments where diversification rates shifted into high gear, three were associated with dispersal to a new province. One of those also might have been associated with a decrease in stamen number. No specific reason was found for the remaining four increases in diversification rate.

The final evolutionary ingredient to be introduced early in my chapter is less whimsical and more tractable than drift or hitchhiking. It is transitional drive. Transitional drive corresponds to what is called mutational drive at the level of the gene. Mutational drive is an inequality in the direction of mutations. For example, if mutations from allele *Violet* to allele *White* are very frequent but mutations from allele *White* to allele *Violet* are nearly impossible, then over time a population of violet flowers will be converted to a population of mostly white flowers (if selection is negligible). At the level of clades, transitional drive is an inequality in shifts from adaptive mode *A* to *B* versus from adaptive mode *B* to *A*. An example of transitional drive is found among penstemons. Hummingbird pollination has evolved from hymenopteran pollination many times, and there is no evidence for any reversals, nor have there been shifts to any other pollination syndrome (Wilson et al. 2007).

Transitional drive may figure largely in macroevolution, whereas at the level of organismal evolution it is rarely treated as a very strong ingredient. Within populations, mutation–selection balance on flower color seems to keep albino-flowered individuals very rare despite mutational drive that eliminates floral pigments (Waser and Price 1981). At the macroevolutionary level, eudicots started out having purplish anthocyanin-pigmented flowers, but a great many lineages have transitioned to warmer or paler colors and few have transitioned back to purple

(Rausher 2006, 2008). Perhaps color has not had a consistent effect on cladogenesis or clade persistence, so transitional drive would seem to greatly explain why now there are untold numbers of eudicot species with flowers that are red, orange, yellow, or white.[3]

1.4 A full-blown example: the evolution of selfing

With these evolutionary ingredients in mind, contemplate an extended example: the rise and fall of self-pollination in the flora of a region. Self-pollination is when pollen on an individual plant ends up siring seeds on the same individual. It is not asexual: meiosis and fertilization remain part of the genetic lifecycle. However, as for asexuality, some of the inefficiency of outcrossing and the cost of male function can be saved by selfing. For example, a population of selfers would be expected to have a growth rate higher than a population of outcrossers if all other things were equal.

But all other things are not equal. There are often fitness advantages to outcrossing. If the particular species has been outcrossing for a large number of generations, then deleterious recessive alleles will have built up in the gene pool. This is the dominance genetic load that is carried around by diploid organisms such as poppies and people. If, from this state, a plant self-pollinates, homozygocity will increase. Selfing brings together recessive deleterious alleles, and this makes for seeds and seedlings plagued with genetic disease. Across much of angiosperm diversity, outcrossing has been maintained as the norm (Stebbins 1957).

Nevertheless and despite the norm, selfing has arisen over and over in monkeyflowers and lupines, gilias and lotuses, and collinsias and drabas. In scattered species, selfing becomes habitual (although not necessarily obligate). Anthers and stigmas evolve to mature at the same moment, to have no positional separation, and to be self-compatible. Nectar evolves to nothing. Petals become diminutive. Pollen production declines. All these traits tend to evolve together or as a close cascade (Cruden 1977; Aarssen 2000).

Near the microevolutionary scale, there are many circumstances that can favor selfing. Perhaps a population finds itself in a situation where pollinators are scarce. Perhaps selfing allows the plants to set seed quickly over a growing season that has

---

[3] Vrba and Gould (1986) distinguish upward versus downward causation in the hierarchical evolutionary process. Transitional drive upwardly causes patterns among clades. Selection downwardly causes patterns among gene frequencies. In this passage, I presume flower color is acted on by individual selection and is adaptive at the level of the organism. This translates into drive among clades. I entertain the possibility that although flower color is selected upon at the individual level, at the clade level its frequency distribution might be determined by transitional drive (caused by selection at a lower level) plus clade drift, and not by clade selection.

become compressed compared to the growing season of ancestors. Perhaps the costs of putting on a show for pollinators and rewarding them with nectar have become exorbitant because the plants, though they once lived in good conditions, are now living in harsh conditions. All these reasons can be considered aspects of the "efficiency of selfing."

Another class of reason for the evolution of selfing is often called "the two-fold advantage." This is not a way in which selfing is advantageous to the health of the organism, rather it is a way in which selfish alleles tend to spread in the population. If, in a population of outcrossers, a mutant arises that makes its bearers put their pollen on their own stigmas and still present about as much pollen to pollinators for outcrossing, then that allele will nearly double its success via male function. There are, however, complications that might make the two-fold advantage less than two-fold.

(1) The mechanism by which the flowers self-pollinate might be that pollination happens quickly and pre-empts outcrossing. In this case, as the population becomes dominated by plants that never present fresh stigmas, which are available to receive outcrossed pollen, the two-fold advantage goes to zero.

(2) By selfing, the plant may use up some of the pollen that would have been available for outcrossing. If so, then the two-fold advantage would be reduced by pollen discounting (Holsinger and Thomson 1994; Harder and Wilson 1998).

Presumably because of the efficiency of selfing and the up-to-two-fold allelic advantage, there are many species that have transitioned to become selfers, but it is hypothesized that selfing clades tend to be dead ends (Stebbins 1957). The dead-end-selfers hypothesis asserts first, that there is transitional drive from outcrossing to selfing, and second, that it is counteracted by clade selection favoring outcrossing clades.

The first assertion, that of transitional drive, is expected since, as selfing becomes the norm for a species, deleterious recessive alleles ought to be purged, inbreeding depression ought to be lessened, and there would then be less of an immediate microevolutionary selective reason for outcrossing (Lande and Schemske 1985). This diminution of selection-for-outcrossing ought to remain even after the species enters better environments where selfing would not have been favored in the first place. The reasons why a species evolved from outcrossing to habitual selfing do not work in reverse.

The second assertion, that of clade selection favoring outcrossing, is suggested by the facts that most selfing species are closely related to outcrossing species and few large genera consist only of selfers. Eventual extinction of selfing lines compared to outcrossing clades can proceed by either of two mechanisms.

(1) Selfing species may ratchet themselves into poor genetic health (Lynch et al. 1995). They have thousands of loci that can mutate and that, because of selfing, have a high chance of becoming homozygous for the deleterious allele. To the degree that the population size is small, mutation-free genotypes will be lost due to drift, and without outcrossing, they will not be reinvented. Hence there is a ratcheting down of viability. If a local population were to self-pollinate exclusively for hundreds of generations, it would be expected to have a mutational meltdown (Lynch et al. 1995), although a small amount of outcrossing would delay this fate (Charlesworth et al. 1993). Extinction is not inevitable as long as outcrossing occurs occasionally, but still mutational meltdown puts lineage persistence at risk.

(2) Selfing species have less opportunity to recombine genetic variation, so they are less adaptable to changing conditions. They would be dependent on occasional outcrossing to bring together beneficial alleles at different loci for polygenic adaptation. They might go extinct just because they cannot keep up with environmental changes, or they might eventually be displaced by outcrossing species that have become generally superior.

In principle, the dead-end-selfers hypothesis can be tested with phylogenies. It predicts more transitions to selfing than away from it, and that selfing lineages should be shallowly rooted in the phylogeny. Takebayashi and Morrell (2001) attempted such a test, but they failed to confirm the expected patterns with confidence, and they worried their phylogenies lacked enough resolution. More progress has been made on the related topic of transitions from self-incompatibility to self-compatibility. Igic et al. (2006) have presented evidence stemming from ancient polymorphism for many losses of self-incompatibility without reversal in the Solanaceae.

The rise and fall of selfing offers an example of evolution involving ingredients from several levels. There is mutational drive that is creating the deleterious alleles, selection "for" selfish alleles via male function to "cheat" on the social contract of outcrossing, selection at the level of the individual to be efficient in reproducing, transitional drive, and there is also clade selection that keeps selfers at the tips of phylogenies.

1.5 Phylogenetic conservatism is like heritability

At the species level, lack of change is called stasis (Eldredge et al. 2005). When the same species occurs essentially unchanged through many strata in the fossil record, it is said to be static. Stasis may also be inferred from a species having a large geographic range, if the range is thought to be ancient (Levin 2000). Above the species level, lack of change in a character or complex of characters is called phylogenetic conservatism. Phylogenetic conservatism is lack of change despite divergence in other (often subsidiary) characters. Conservatism is probably caused by stabilizing selection and similar, but more complex, forms of past selection.

The morphology of a larkspur flower with its characteristic nectar spur is conserved among 340 species of *Delphinium* + *Consolida*, and in a larger clade that also includes *Aconitum* the arrangement of having nectariferous petals tucked inside the dorsal sepal and of being bilaterally symmetric is conserved more broadly. The overwhelming experience of systematists is that taxa can be characterized morphologically, developmentally, functionally, and ecologically. Experience with the fossil record and with comparative developmental genetics further confirms the impression that conservatism is the rule, and evolutionary change very often (but not always) is restricted to new kinds of divergence nested within old norms. The diversity of flowers often appears as variations on themes that are themselves variations nested within the themes of more inclusive groups (Endress 1996).

In macroevolution, conservatism is the analog to what quantitative geneticists call heritability. You might be familiar with the function $R = S \times h^2$, or in words, response to selection follows from selection times heritability. In an artificial selection experiment in which a quantitative trait (like floral tube length) is selected upon, if heritability is zero, then response to selection is zero (tube length does not evolve). By analogy, species stasis or phylogenetic conservatism is what makes clade selection amount to something that is especially hierarchical.

Clade selection would have an effect even without conservatism, just with ordinary heritability, but it has a different sort of effect when there is conservatism among clades. Consider a character that, after being fixed in a species, is irreversible. Imagine total loss of nectaries is irreversible. Species that have lost their nectaries will then go extinct or thrive, and their lack of nectaries might affect the outcome. If on the other hand, it is a character that continues to be evolutionarily labile – imagine the amount of nectar produced – then a prolonged wrestling match will ensue between levels of selection. Sexual selection might drive flowers to produce more nectar, but species selection might act against species that invest heavily in nectar. Perhaps the average amount of nectar in most flowers would be moderate with occasional local populations and neoendemic species producing copious nectar because they have yet to be eliminated by species selection. (For birds, Doherty et al. 2003 find a similar pattern in color differences between the sexes.)

Clade selection acting above individual selection is one way in which the evolutionary process is hierarchical; phylogenetic conservatism further contributes to the hierarchical nature of the evolutionary process. In Gould's view, punctuated equilibrium is the analog to mutation creating alleles that are not subsequently fluid. Punctuated equilibrium creates species, and then species selection sorts those static entities. The punctuations would generally be caused by a local population adapting to new conditions via ordinary selection among organisms. Rephrasing so as not to dwell on the species level, clade selection affects patterns of diversity by sorting alternative adaptive states that evolved once upon a time but that are not constantly continuing to change at lower levels.[4]

There is an abundance of monophyletic clades and paraphyletic grades at every taxonomic rank that display considerable floral conservatism. Many of them are or once were named sections, genera, tribes, families, etc. As an example, consider the ceropegias (Ollerton et al. 2009). Ceropegia flowers are flasks that temporarily trap medium-sized flies (Fig 1.2). The flask shape, mechanical function, and taxonomic order of pollinators are all highly conserved, but within that morphological architecture, the flowers vary in their decorations. There are 180 species. The ceropegia morphology is a conserved evolutionary formation centered on an adaptive mode of pollination. There are some other groups, such as *Aristolochia*, that have converged on having flask-shaped temporary flytraps, but the adaptive mode would be obvious without any convergence.

Floral conservatism ought to be seen as a largely nested hierarchy of morphological ground plans. Ceropegia flowers have the diagnostic morphology of the erstwhile genus *Ceropegia* (now thought to be paraphyletic). The flowers also have features of the broader subfamily Asclepiadoideae, including the packaging of pollen in pollinia. Going farther back in time and outward in the taxonomic hierarchy, one can see characters of the asterid clade, such as the fusion of petals to one another and of stamens to petals. The ancestor's tale continues as we recognize the consolidation of the core eudicot flower into four whorls (not spirals) of organs with the number of sepals and petals fixed at five. We can go farther back to when angiosperms settled on closed carpels with a style and stigma, and yet farther back to when the seed–plant lifecycle was established. This nested conservatism has accumulated in the ceropegia flower.

---

[4] The word sorting is used inconsistently. Ecologists typically are talking about selection, not drift. Vrba and Gould (1986) attempted to establish sorting as an umbrella term for sorting due to selection or drive or drift with no specification of the level of causation. I sense they failed to get those speaking about community assembly to follow their usage, even those who comprehend it (e.g. Hererra 1992; Ackerly 2003). The way ecologists speak is often the opposite of what Vrba and Gould recommended. To me, ecological sorting is to clade selection as phenotypic sorting is to allelic selection. Williams (1992) taught discernment of the material domain versus the codical domain. Ecological sorting to me is the same as Janzen's (1985) "ecological fitting." It allows for species stasis while the species' range waxes and wanes (Levin 2000). While I am reciting analogies, I'll add that geographic-community ecology is to macroevolution as within-population ecology is to organismal evolution.

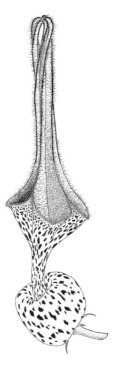

Fig 1.2 A ceropegia flower. Flies are attracted by scent and get trapped in the flask where they often get a foot caught in the pollinium apparatus before they escape. There are 180 species of ceropegias, united by their conserved pollination system and diverse in the coloring, hairs, and dimensions of the flask.

Admittedly, phylogenetically nested conservatism is often not absolute. In the case of ceropegia, the group is paraphyletic because out of it sprung two other groups that do not have the flask-shaped flowers, *Brachystelma* and the Stapeliads. Sometimes new lineages break with the past, but the breaks are not so frequent as to obliterate what we see as the morphological hierarchy. In fact, even for characters that are not particularly diagnostic of a taxon, and thus are far from absolutely conserved, statistics quantifying phylogenetic conservatism still find that related species tend to be more similar than one would expect of evolution by Brownian motion (Ackerly 2009 gives statistical options).

The reasons for species stasis and phylogenetic conservatism have been written about for two hundred years, starting at least with Georges Cuvier. Futuyma (2010) reviews recent explanations. Here I give only a taste.

In the case of the ceropegia flower, I suppose that the various characters all work well together to allow for pollination by flies that are temporarily trapped, get a foot caught in the gynostegium, and eventually escape, and that deviations from this ground plan are generally maladaptive. By this view, stabilizing selection maintains those aspects of the flower that are integral to its mode of pollination. Other aspects of the flower – in the case of ceropegia, color patterning, hairs, and proportions of the flask – are free to vary, and so the 180 species are diverse in these subsidiary features. Taking the Asclepiadoideae as a whole, I would suppose the gynostegium is integral to the functioning of Asclepiadoideae flowers, including ceropegia flowers.

Explanations for conservatism can be divided into externally enforced reasons and internal genetic reasons. External reasons for conservatism include "habitat tracking:" species or more inclusive groups that are adapted to one niche do not survive in other habitats for long enough to permit directional selection to change the norm (Eldredge 2003). In other words, ecological sorting puts organisms in situations where they experience stabilizing selection (Ackerley 2003). Internal reasons for conservatism could have the same systems-structural organization but instead of each character being maintained because the whole flower works as an integrated functional module, a network of genes is maintained because they collectively work in an integrated way, and those same genes are responsible for the characters we notice as conserved. When the integration is via developmental genetics instead of external coordinated function, the conserved characters might be as inscrutably connected as, say, stamen shape and mode of photosynthesis. According to the systems view by which conservatism results from integration, the internal and external reasons amount to the same thing, either co-adapted gene complexes or co-adapted aspects of ecological function.

It is helpful to here mention the writings of Rupert Riedl (Wagner and Laubichler 2004). Riedl held that the characteristics of major taxonomic groups that are conserved are traits that have become "burdened" with other traits built upon them, making them centrally connected (Riedl 1975, 1977). Among animals, the body plans have become the most burdened, the features that distinguish orders of insects have become subsequently burdened in a subsidiary manner, and so on. Within a lineage's ground plan other traits are relatively free to vary; they have not become burdened and might never become burdened. Riedl (1975) illustrated the labellae of orchids as an organ that freely varies and is associated with varied pollinators.

Innovations do not start out burdened when they originate. It is just that the innovations we notice as being useful at a high taxonomic level are the ones that became burdened. Burden increases presumably because of selection, perhaps even stabilizing selection on the trait itself, though not necessarily. It could alternatively increase because of selection for subsequent adaptations that are built on part of the trait's established developmental genetic system. The selection that originally burdens a ground plan need not continue to be ever-present for the ground plan to be conservative. It becomes internalized so that the conserved traits as we see them are not the target of on-going stabilizing selection. Thus, as the tree of life has grown, different branches have become conservative in different ways, and subsidiary branches have themselves become conservative each in their own way.

1.6 The evolution of evolutionary ingredients

In Riedl's view, the nested hierarchy of conserved traits not only marks the course of evolution, but also represents evolutionary change in adaptation ability along the branches of the tree. A fern need not have the same ability to adapt as an orchid. An orchid need not have the same ability to adapt as an iris. We might suspect that the adaptation ability of orchids and irises is more similar than either is to the adaptation ability of ferns. Riedl believed that the macroevolutionary fixation of traits potentially has consequences for subsequent evolutionary dynamics. I wish to extend this argument to evolutionary ingredients other than the ability to adapt. Not all systems have the same proclivity to specialize, speciate, expand geographically, persist, or undergo subsequent cladogenesis (Levin 2000). The capacities for evolution themselves evolve along the branches of the tree of life. Roy et al. (2009) have shown phylogenetic conservatism in rates of extinction, and it would be expected in other macroevolutionary ingredients.

The establishment of one set of characters may spawn taxon-specific diversity in other subsidiary aspects of the organism (Riedl 1975 starting on p. 157). Establishment of the pea-type flowers in the lineage leading up to the Papilionoideae might have spawned diversification in the placement of pollen on specific parts of the bodies of varied pollinators, and this may well have permitted high species richness (Leppik 1966). A corollary would be that different groups, such as different subfamilies, may diversify (or not) in their own specific way. In contrast to the Papilionoideae, the Mimosoideae has diversified in colors and rewards while being conservative in its mess-and-soil pollination mechanics. It also seems less prone to speciation, probably because of the difference in flower morphology, along with the greater tendency of the Papilionoideae to be herbaceous and to have shorter generations times.

Explanations for high species richness in species-rich groups might generally be posed in the form of a lucky-streaks account. The dynamics that lead to high diversification depend on a long and winding series of contingencies that act together to foster cladogenesis. We should not attribute the high species richness of columbines only to their nectar spurs, but to their nectar spurs in the context of being perennials that tolerate the accentuated cold season of high mountains, and in the broader context of being animal pollinated, and in the even broader context of being herbaceous while having limited seed dispersal (Hodges 1997b).

There is an analogy to how we study organismal adaptations. In Gould's 2002 book he gave an expansive alternative to the cartoon adaptationist's program (Gould and Lewonton 1979), explaining how we ought to consider the possibility that traits once adapted for one purpose have been co-opted for another. By analogy, we can use the same logic as an alternative to a cartoon version of the key-innovationist's program given near the beginning of my chapter. A character that favors diversity may have favored diversification from the start or it may have been long established in its lineage and then later came to be important. This dissection of possibilities is analogous to the distinction between an adaptation and an exaptation (Gould and Vrba 1982; Armbruster et al. 2009a). Plausibly, closed carpels was not an innovation of any consequence when it came into being, but later after pollen-style interactions became more complex, closed carpels may have become an "exovation" favoring angiosperm diversification (Mulcahy 1979).

At the risk of taking the grand analogy too far, consider the following algorithm of propositions:

(1) More local populations are produced than can possibly survive. To a progressively lesser extent, this is also true of "species" and "genera," whatever they are. All these entities can be called incipient clades.

(2) To some extent the differences between incipient clades affect the likelihood that they will survive and not merge back into the mother species, find a niche in a community that sustains the population demographically, and expand to have a geographic range that allows the clade to persist. To some extent the differences will affect the tendency of the clade to undergo subsequent cladogenesis, producing subsidiary clades ("grand-daughter species").

(3) To some extent the clade differences become conservative tending to become synapomorphies among the subsidiary clades that are produced. Thus there ensues a struggle for existence that occurs among clades. The effect of the characters on the likelihood of success may be very small but nonetheless cumulatively telling. The clade heritability of the traits may be far from absolute but nonetheless come to be consolidated as the process is extended. So, in various branches of the tree of life, the various capacities to evolve have accumulated in differing manifestations. In some branches extreme diversity has evolved. In other branches extreme clade longevity has evolved. In many clades a mixed strategy has evolved.

This grand analogy to the process of organismal adaptation should be recited only with some criticism. The action of clade selection is analogous to that of ordinary natural selection, but it is not identical. It might be more similar to the action of evolution on DNA sequences that are not recombining than to the action of evolution on sexual organisms. Aside from the defining difference of success being measured at a higher level (Gould and Lloyd 1999), clade selection, unlike ordinary natural selection, does not involve the recombination that occurs through sex, i.e. through meiosis and fertilization.

Presumably sex itself was consolidated in the core eukaryotes via a process of clade selection (Nunney 1989). Most but not all other adaptations of organisms are probably mostly caused by selection among individuals, family groups, and groups of cooperating friends, whereas clade selection and its attendant dynamics probably mostly affect features of biodiversity, such as species richness, morphological disparity, the geographic ranges of clades, and the breadth of their specialties. Many features of the tree of life could be outcomes of clade selection, although this is far from proven since we do not know the relative importance of transitional drive, clade hitchhiking, and clade drift.

Also complicating matters, the environments that have formed the context for the diversification process have changed at the same time that the process has been producing patterns of diversity. This is also true of the evolution of adaptation in the sense that an organism's adaptations have arisen in the long and winding tale of its ancestors who lived through changing environments that accreted one adaptation and its byproducts after another. In the end, the story of diversification, like the story of adaptation, is a singular pageant colored by streaks of contingency. Nevertheless, it is imaginable that the ancestor's tale for diversification involves less discovery-of-reasons and more invention-of-idiosyncrasies than the ancestor's tale of adaptation.

Another way in which clade selection is probably unlike organismal selection is that typically incipient clades are allopatric and arise in different environments whereas individual organisms often importantly vie with each other within a population (Brandon 1990). Species vie with each other at a site within a community, but the species as represented at the site are generally not closely related. Ecological sorting leads to community succession, but I presume communities have low heritability, so they tend not to accumulate much evolutionary organization (Dawkins 2004). Phylogenies have astoundingly high conservatism, albeit the sorting tends to be what a gambler would call "each player against the house" rather than "all players facing off."

In organismal adaptation, components of fitness trade-off against one another; how about in diversification? The analogy is only suggestive. A dynamic is sometimes established whereby certain clades are prone to give rise to varied specialists that tend to go extinct. The flywheel of specialization-followed-by-extinction seems evident in leaf-eating insects and various other parasites (Kelley and Farrell 1998). The host-plants available to leaf-eating insects differ from one another discontinuously reflecting the categories of plant taxonomy. There is selective pressure to specialize, and speciation seems to often involve host-shifts or shifts in degree of specialization. Leaf-eating insects, then, are species rich, but they are not as species rich as they would be if specialization did not also come with a shortened period of persistence-to-extinction.

What about floral specialization on pollinators? Flowers are not usually so specialized, so we might doubt that there is much of an extinction effect (Waser et al. 1996). We might also doubt that floral adaptation to different principal pollinators directly causes reproductive isolation (Waser 2001). However, in orchids, genera with species that have few pollinators are more species rich than genera with species that are less specialized (Schiestl and Schlüter 2009). Finally, because floral divergence is often caused partially by sexual selection rather than only by survival selection, the extinction effect ought to be less pronounced. I am not sure of the net effect of the mix of ingredients involved in pollination. Some groups of flowers do not diversify. Others do. I could believe specialization in these flowers is relatively decoupled from extinction.

1.7 Pollination leads to everything

Pollination biology can be made more profound by considering hierarchical evolution, but then consideration of hierarchical evolution leads beyond pollination to a more complete interest in the organism's reproductive biology.

It has been suggested that while pollination characters often diverge during cladogenesis, seed dispersal characters are relatively static (Schaefer et al. 2004). Animal-pollinated flowers often seem to be selected to be different from coflowering species, but animal-dispersed fruits do not seem to be under selection to look different from cofruiting species. We might even expect fruits to mimic each other thereby cooperatively using animal dispersers. Also floral characters might be freer to vary because floral flags are often different organs than floral rewards, whereas fruits are generally both the signal and the reward to dispersers.

To partially test the theory that flowers are more distinct than fruits, Whitney (2009) studied the animal-pollinated fleshy-fruited species of three regional floras: the Great Plains, Hawaii, and the Virgin Islands. He scored the size and color of flowers and fruits. In each flora, most-similar species were more similar for fruit characters than for flower characters. It would be interesting to do a similar study but comparing closest relatives in a flora (Grant 1949). The predictions would stem from a theory of which types of plants would involve pollination traits in cladogenesis (perhaps those in groups that are more specialized) and which types of plants would involve dispersal characters in cladogenesis (possibly in groups that use ant dispersal Lengyel et al. 2009).

Although dispersal characters might not so often be the characters that diverge in a vigorously diversifying group, surely dispersal traits play a big role in allowing for cladogenesis. Homosporous pteridophytes tend to have species with large ranges compared to heterosporous plants (Moran 2004 starting on p. 178). In many floras, the proportion of endemics and the smallness of their ranges is much more dramatic for flowering plants than for ferns. An exception that proves the rule is that the ranges of species in *Selaginella*, a heterosporous pteridophyte, tend to be much smaller and more allopatric (Valdespino 1993). This can be blamed on the limits of megaspore dispersal: the wind-born spores of a normal fern prevent cladogenesis, whereas large megaspores or seeds predispose the groups that have them to diversify in other characters.

Nested within the seed plants, there would be additional consequences of the size of seeds, their means of dispersal, and their longevity. All of these features of dispersal ought to affect the likelihood of allopatry and thus biological speciation. Orchids have tiny seeds that allow them to have widely scattered populations. How could this not affect the dynamics of diversification? Similarly, all other things being equal, when a clade shifts from dumping its seeds on the ground to having birds disperse its seeds, all subsequent species will have a much greater ability to expand their geographic ranges. Bird-dispersed genera seem to have lower rates of endemism than genera with seeds that have no special means of dispersal.

Good dispersal allows a species to expand its range, and the range size of a species surely ought to affect its ability to elude extinction (Payne and Finnegan 2007, cf. Lester et al. 2007). What then are the correlates of geographic range size? Lavergne et al. (2004) surveyed 20 pairs of endemic and widespread species from the Mediterranean. Endemic and widespread species did not differ significantly in leaf traits. Endemics did make significantly fewer seeds than widespread congeners and had a number of characters associated with greater inbreeding (smaller flowers, lower pollen:ovule ratios, less stigma–anther separation). Finally, endemics tended to be in steeper, rockier, and less vegetated habitats than closely related widespread species. I suppose that clades of plants that are good competitors with habitual outcrossing and good dispersal ought to have long persistence times.

I cannot stop at going beyond floral characters to dispersal characters. Just as I had to mention dispersal characters because I think they allow floral characters to become diverse, I have to at least touch on the rest of the plant's lifecycle because it is likely to be tied into increasing or decreasing the diversification rate of a lineage. The many ingredients that affect diversification have evolved up and down along the various branches of the tree of life. It is the accumulated biology that determines the capacity of, for example, a genus to disperse and diversify (Fig 1.3).

For example, whether dispersal by animals increases the diversification rate or decreases it apparently depends on whether the plant is woody or not. Woodiness makes fleshiness increase diversification; herbaceousness does not (de Quieroz 2002). Tiffney and Mazer's (1995) explanation for this pattern is that woodiness is a character related to having an ecology in which plants are seriously limited by competition. Fleshiness would then allow the plants to escape from competition.

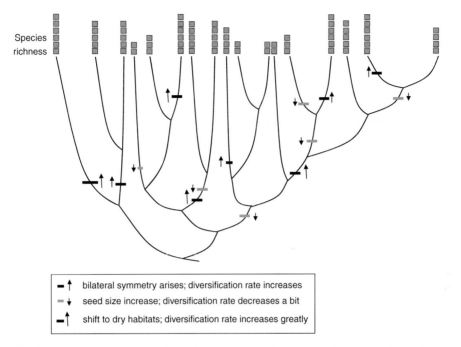

Fig 1.3 As one traces a path from the root to any tip, various characters affect the capacity for diversification, positively or negatively. With a much more extensive phylogeny, such as the phylogeny of flowering plants resolved to the genus level, one could do an analysis in the spirit of multiple regression to see how several traits cumulatively affect the species richnesses of the genera.

Herbaceous plants differ ecologically, being tolerant of overstory trees or occupying sites early in succession or living in ecosystems where water is only seasonally available. Bolmgren and Eriksson (2005) add to the explanation by documenting that fleshiness seems to evolve with shifts to closed-canopy ecosystems that have spatially less predictable disturbances such as tree-fall gaps and trampling by herds of mammals.

Not just woodiness versus herbaceousness, but more generally a lineage's schedule of life history events is going to frame the context for its macroevolutionary dynamics. Annuals evolve differently than perennials. Even within a group of generally similar plants, subtle changes in life history are likely to imply differences in components of species success.

Consider evolutionary changes in life history in the genus *Dudleya*, a group of succulents that has formed many neoendemic species in the coastal mountains of California and Baja (Dorsey and Wilson 2011). Species vary in their life histories, with some rushing to reproduce early and nearly ceasing growth, while other species delay reproduction and become larger and tougher. Correlated with position along this life-history trade-off envelope is geographic range size. The species that rush to reproduce have not expanded out of the tiny ranges where they originated, whereas the species that delay reproduction have considerably broader geographic ranges. The former are government-listed as threatened; the latter are not. It is easy to understand how other aspects of life history would similarly affect components of clade success. Imagine evolution along the trade-off envelope defined by allocation to larger seed size versus greater seed number, or the investment in excessive flowers versus fruits.

Ecological niches are another aspect of conservatism that colors macroevolution. Related species tend to have similar requirements and tolerances for wetness of soil, tend to grow at similar elevations, and so on (Prinzing et al. 2001). Furthermore, these ecological traits are associated with such issues as the likelihood of speciation. Kimball et al. (2004) compiled data on the ecological differences and geographic ranges of plants in a landscape on the eastern side of California's Sierra Nevada. The plants that grow in wet sites have broad geographic ranges, whereas the endemic species are almost all plants of dry rocky sites. Stebbins (1952) argued that groups of plants that live at intermediate levels of aridity are more prone to diversification than groups that live in mesic or desert ecosystems. Living on ecotones spurs on diversity in combination with other aspects of the plant's heritage.

The various factors I have mentioned as being the basis for clade selection and transitional drive might be only the tip of the iceberg on features that affect macroevolution. It is possible that the body of the iceberg is hidden in the dark waters of the operational systems inside the skin of the organism. To add just one other suggestion, perhaps the rate at which chromosomes diverge in their ability to recognize homologues at meiosis would have an important effect on the tendency for speciation in a group.

Acknowledgements

Carrie Wessinger, Kathleen Kay, Michael Mesler, Jocelyn Holt, Lena Coleman, Katherine Gould, and Nickte Mendez reacted to previous drafts.

References

Aarssen, L. W. (2000). Why are most selfers annuals? A new hypothesis for the fitness benefit of selfing. *Oikos*, **89**, 606–12.

Ackerly, D. D. (2003). Community assembly, niche conservatism, and adaptive evolution in changing environments. *International Journal of Plant Sciences*, **164**, S165–S184.

Ackerly, D. D. (2009). Conservatism and diversification of plant functional traits: Evolutionary rates versus phylogenetic signal. *Proceedings of the National Academy of Sciences USA*, **106**, 19699–706.

Armbruster, W. S., Lee, J. and Baldwin, B. G. (2009a). Macroevolutionary patterns of defense and pollination in *Dalechampia* vines: Adaptation, exaptation, and evolutionary novelty. *Proceedings of the National Academy of Sciences USA*, **106**, 18085–90.

Armbruster, W. S., Hanen, T. F., Pélabon, C., Pérez-Barrales, R. and Maad, J. (2009b). The adaptive accuracy of flowers: Measurements and microevolutionary patterns. *Annals of Botany*, **103**, 1529–45.

Barton, N. H. (2000). Genetic hitchhiking. *Philosophical Transactions of the Royal Society B*, **355**, 1553–62.

Baskin, C. C. and Baskin, J. M. (1998). *Seeds: Ecology, Biogeography, and Evolution of Dormancy and Germination*. San Diego, CA: Academic Press.

Bolmgren, K. and Eriksson, O. (2005). Fleshy fruits – origins, niche shifts, and diversification. *Oikos*, **109**, 255–72.

Brandon, R. N. (1990). *Adaptation and Environment*. Princeton, NJ: Princeton University Press.

Castellanos, M. C., Wilson, P. and Thomson, J. D. (2003). Pollen transfer by hummingbirds and bumblebees, and the divergence of pollination modes in *Penstemon*. *Evolution*, **57**, 2742–52.

Charlesworth, D., Morgan, M. T. and Charlesworth, B. (1993). Mutation accumulation in finite outbreeding and inbreeding populations. *Genetical Research*, **61**, 39–56.

Cruden, R. W. (1977). Pollen–ovule ratios: A conservative indicator of breeding system in flowering plants. *Evolution*, **31**, 32–46.

Dawkins, R. (2004). Extended phenotype – but not *too* extended. A reply to Laland, Turner and Jablonka. *Biology and Philosophy*, **19**, 377–96.

de Quieroz, A. (2002). Contingent predictability in evolution: Key traits and diversification. *Systematic Biology*, **51**, 917–29.

Dodd, M. E., Silvertown, J. and Chase, M. W. (1999). Phylogenetic analysis of trait evolution and species diversity variation among angiosperm families. *Evolution*, **53**, 732–44.

Doherty, P. F., Sorci, G., Royle, J. A., Hines, J. E., Nichols, J. D. and Boulinier, T. (2003). Sexual selection affects local exinction and turnover in bird communities. *Proceedings of the National Academy of Sciences USA*, **100**, 5858–62.

Dorsey, A. and Wilson, P. (2011). Rarity as a life-history correlate in *Dudleya* (Crassulaceae). *American Journal of Botany*, **98**, 1104–12.

Eldredge, N. (2003). The sloshing bucket: How the physical realm controls evolution. In *Evolutionary Dynamics*, ed. J. P Crutchfield and P. Schuster. Oxford, UK: Oxford University Press, pp. 3–32.

Eldredge, N., Thompson, J. N., Brakefield, P. M., Gavrilets, S., Jablonski, D., Jackson, J. B. C., Lenski, R. E., Lieberman, B. S., McPeek, M. A. and Miller III, W. (2005). The dynamics of evolutionary stasis. *Paleobiology*, **31**, 133–45.

Endress, P. K. (1996). *Diversity and Evolutionary Biology of Tropical Flowers*. Cambridge, UK: Cambridge University Press.

FitzJohn, R. G., Maddison, W. P. and Otto, S. P. (2009). Estimating trait-dependent speciation and extinction rates from incompletely resolved phylogenies. *Systematic Biology*, **58**, 595–611.

Futuyma, D. J. (2010). Evolutionary constraint and ecological consequences. *Evolution*, **64**, 1865–84.

Gegear, R. J. and Laverty, T. M. (2005). Flower constancy in bumblebees: A test of the trait variability hypothesis. *Animal Behavior*, **69**, 939–49.

Gould, S. J. (2002). *The Structure of Evolutionary Theory*. Cambridge, MA: Harvard University Press.

Gould, S. J. and Lewontin, R. C. (1979). The spandrels of San Marco and the Panglossian paradigm: A critique of the adaptationist programme. *Proceedings of the Royal Society London B*, **205**, 581–98.

Gould, S. J. and Lloyd, E. A. (1999). Individuality and adaptation across levels of selection: How shall we name and generalize the unit of Darwinism? *Proceedings of the National Academy of Sciences USA*, **96**, 11904–9.

Gould, S. J. and Vrba, E. S. (1982). Exaptation – a missing term in the science of form. *Paleobiology*, **8**, 4–15.

Grant V. (1949). Pollination systems as isolating mechanisms in angiosperms. *Evolution*, **3**, 82–97.

Harder, L. D. and Wilson, W. G. (1998). A clarification of pollen discounting and its joint effects with inbreeding depression on mating system evolution. *American Naturalist*, **152**, 684–95.

Heilbuth, J. C. (2000). Lower species richness in dioecious clades. *American Naturalist*, **156**, 221–41.

Herrera, C.M. (1992). Historical effects and sorting processes as explanations for contemporary ecological patterns: Character syndromes in Mediterranean woody plants. *American Naturalist*, **140**, 421–46.

Hodges, S.A. (1997a). Floral nectar spurs and diversification. *International Journal of Plant Sciences*, **158**, S81–S88.

Hodges, S.A. (1997b). A rapid adaptive radiation via a key innovation in *Aquilegia*. In *Molecular Evolution and Adaptive Radiations*, ed. T. Givinish and K. Sytsma. Cambridge, MA: Cambridge University Press, pp. 391–405.

Holsinger, K. E. and Thomson, J. D. (1994). Pollen discounting in *Erythronium grandiflorum*: Mass-action estimates from pollen transfer dynamics. *American Naturalist*, **144**, 799–812.

Igic, B., Bohs, L. and Kohn, J.R. (2006). Ancient polymorphism reveals unidirectional breeding system shifts. *Proceedings of the National Academy of Sciences USA*, **103**, 1359–63.

Janzen, D. (1985). On ecological fitting. *Oikos*, **45**, 308–10.

Johnson, S.D. (2006). Pollinator-driven speciation in plants. In *Ecology and Evolution of Flowers*, ed. L. D. Harder and S. C. H. Barrett. Oxford, UK: Oxford University Press, pp. 295–310.

Jones, K. N. (2001). Pollinator-mediated assortative mating: Causes and consequences. In *Cognitive Ecology of Pollination: Animal Behavior and Floral Evolution*, ed. L. Chittka and J.D. Thomson. Cambridge, UK: Cambridge University Press, pp. 259–73.

Kalisz, S., Horth, L. and McPeek, M. A. (1997). Fragmentation and the role of seed banks in promoting persistence in isolated populations of *Collinsia verna*. In *Conservation in Highly Fragmented Landscapes*, ed. M. W. Schwartz. New York, NY: Chapman & Hall, pp. 286–312.

Kay, K. M., Voelckel, C., Yang, J. Y., Hufford, K. M., Kaska, D. D. and Hodges, S. A. (2006). Floral characters and species diversification. In *Ecology and Evolution of Flowers*, ed. L. D. Harder and S. C. H. Barrett. Oxford, UK: Oxford University Press, pp. 311–25.

Kelley, S. T. and Farrell, B. D. (1998). Is specialization a dead end? The phylogeny of host use in *Dendroctonus* bark beetles (Scolytidae). *Evolution*, **52**, 1731–43.

Kimball, S., Wilson, P. and Crowther, J. (2004). Local ecology and geographic ranges of plants in the Bishop Creek watershed of the eastern Sierra Nevada, California, USA. *Journal of Biogeogrpahy*, **31**, 1637–57.

Kondrashov, A. S. and Shpak, M. (1998). On the origin of species by means of assortative mating. *Proceedings of the Royal Society London B*, **265**, 2273–78.

Lande, R. and Schemske, D. W. (1985). The evolution of self-fertilization and inbreeding depression in plants. I. Genetic models. *Evolution*, **39**, 24–40.

Lavergne S., Thompson, J. D., Garnier, E. and Debussche, M. (2004). The biology and ecology of narrow endemic and widespread plants: A comparative study of trait variation in 20 congeneric pairs. *Oikos*, **107**, 505–18.

Lengyel, S., Gove, A. D., Latimer, A. M., Majer, J. D. and Dunn, R. R. (2009). Ants sow the seeds of global diversification in flowering plants. *PLoS ONE*, **4**, e5480.

Leppik, E. E. (1966). Floral evolution and pollination in the Leguminosae. *Annales Botanici Fennici*, **3**, 299–308.

Lester, S. E., Ruttenberg, B. I., Gaines, S. D. and Kinlan, B. P. (2007). The relationship between dispersal ability and geographic range size. *Ecology Letters*, **10**, 74558.

Levin, D. A. (2000). *The Origin, Expansion, and Demise of Plant Species*. Oxford, UK: Oxford University Press.

Lynch, M., Conery, J. and Reinhard, B. (1995). Mutational meltdowns in sexual populations. *Evolution*, **49**, 1067–1080.

McEwen, J. R. and Vamosi, J. C. (2010). Floral colour versus phylogeny in structuring subalpine flowering communities. *Proceedings of the Royal Society London B*, **277**, 2957–2965.

Mishler, B. D. (2010). Species are not unique biological entities. In *Contermporary Debates in Philosophy of Biology*, ed. F. J. Ayala and R. Arp. Oxford, UK: John Wiley & Sons, pp. 110–22.

Moore, B. R. and Donoghue, M. J. (2007). Correlates of diversification in the plant clade Dipsacales: Geographic movement and evolutionary innovations. *American Naturalist*, **170**, S28–S55.

Moran, R. C. (2004). *A Natural History of Ferns*. Portland, OR: Timber Press.

Mulcahy, D. L. (1979). The rise of the angiosperms: A genecological factor. *Science*, **206**, 20–3.

Nunney, L. (1989). The maintenance of sex by group selection. *Evolution*, **43**, 245–57.

Okasha, S. (2003). Does the concept of "clade selection" make sense? *Philosophy of Science*, **70**, 739–51.

Ollerton, J., Masinde, S., Meve, U., Picker, M. and Whittington, A. (2009). Fly pollination in *Ceropegia* (Apocynaceae: Asclepiadoideae): Biogeographic and phylogenetic perspectives. *Annals of Botany*, **103**, 1501–14.

Payne, J. L. and Finnegan, S. (2007). The effect of geographic range on extinction risk during background and mass extinctions. *Proceedings of the National Academy of Sciences USA*, **104**, 10506–11.

Prinzing, A., Durka, W., Klotz, S. and Bradl, R. (2001). The niche of higher plants: Evidence for phylogenetic conservatism. *Proceedings of the Royal Society London B*, **268**, 2383-89.

Rausher, M. D. (2006). The evolution of flavonoids and their genes. In *The Science of Flavonoids*, ed. E. Grotewold. New York, NY: Springer, pp. 175-211.

Rausher, M. D. (2008). Evolutionary transitions in floral color. *International Journal of Plant Sciences*, **169**, 7-21.

Ricklefs, R. E. and Renner, S. S. (2000). Evolutionary flexibility and flowering plant familial diversity: A comment on Dodd, Silvertown, and Chase. *Evolution*, **54**, 1061-1065.

Riedl R. (1975 translated 1978). *Order in Living Organisms: A Systems Analysis of Evolution.* New York, NY: Wiley.

Riedl, R. (1977). A systems–analytical approach to macroevolutionary phenomena. *Quarterly Review of Biology*, **52**, 351-70.

Roy, K., Hunt, G. and Jablonski, D. (2009). Phylogenetic conservatism of extinctions in marine bivalves. *Science*, **325**, 733-7.

Sargent, R. D. (2004). Floral symmetry affects speciation rates in angiosperms. *Proceedings of the Royal Society London B*, **271**, 603-8.

Schaefer, H. M., Schaefer, V. and Levey, D. J. (2004). How plant-animal interactions signal new insights in communication. *Trends in Ecology and Evolution*, **19**, 577-84.

Schiestl, F.P. and Schlüter, P.M. (2009). Floral isolation, specialized pollination, and pollinator behavior in orchids. *Annual Review of Entomology*, **54**, 425-46.

Slowinski, J. B. and Guyer. C. (1993). Testing whether certain traits have caused amplified diversification: An improved method based on a model of random speciation and extinction. *American Naturalist*, **142**, 1019-24.

Stebbins, G. L. (1952). Aridity as a stimulus to plant evolution. *American Naturalist*, **86**, 33-4.

Stebbins, G. L. (1957). Self-fertilization and population variability in the higher plants. *American Naturalist*, **91**, 337-54.

Stöcklin, J. and Fischer, M. (1999). Plants with longer-lived seeds have lower local extinction rates in grassland remnants 1950-1985. *Oecologia*, **120**, 539-43.

Takebayashi, N. and Morrell, P. L. (2001). Is self-fertilization an evolutionary dead end? Revisiting an old hypothesis with genetic theories and a macroevolutionary approach. *American Journal of Botany*, **88**, 1143-50.

Tiffney, B. H. and Mazer, S. J. (1995). Angiosperm growth habit, dispersal and diversification. *Evolutionary Ecology*, **9**, 93-117.

Vamosi, J. C. and Vamosi, S. M. (2005). Present day risk of extinction may exacerbate the lower species richness of dioecious clades. *Diversity and Distributions*, **11**, 25-32.

Valdespino, I. A. (1993). *Selaginellaceae.* Vol 2 of *Flora of North America North of Mexico*. New York, NY: Oxford University Press.

Vrba, E. S. and Gould, S. J. (1986). The hierarchical expansion of sorting and selection: Sorting and selection cannot be equated. *Paleobiology*, **12**, 217-28.

Wagner, G. P. and Laubichler, M. D. (2004). Rupert Riedl and the re-synthesis of evolutionary and developmental biology: Body plans and evolvability. *Journal of Experimental Zoology (Mol Dev Evol)*, **302B**, 92-102.

Waser, N. M. (2001). Pollinator behavior and plant speciation: Looking beyond the "ethological isolation" paradigm. In *Cognitive Ecology of Pollination: Animal Behavior and Floral Evolution*, ed. L. Chittka and J. D. Thomson. Cambridge, UK: Cambridge University Press, pp. 318–35.

Waser, N. M. and Price, M. V. (1981). Pollinator choice and stabilizing selection for flower color in *Delphinium nelsonii*. *Evolution*, **35**, 376–90.

Waser, N. M., Chittka, L., Price, M. V., Williams, N. M. and Ollerton, J. (1996). Generalization in pollination systems, and why it matters. *Ecology*, **77**, 1043–60.

Whitney, K. D. (2009). Comparative evolution of flower and fruit morphology. *Proceedings of the Royal Society London B*, **276**, 2941–47.

Williams, G. C. (1992). *Natural Selection: Domains, Levels, and Challenges*. Oxford, UK: Oxford University Press.

Wilson, P. and Stine, M. (1996). Floral constancy in bumblebees: Handling efficiency or perceptual conditioning? *Oecologia*, **106**, 493–99.

Wilson, P. and Thomson, J. D. (1996). How do flowers diverge? In *Floral Biology*, ed. D. Lloyd and S. C. H. Barrett. New York, NY: Chapman & Hall, pp. 88–111.

Wilson, P., Castellanos M. C., Wolfe, A. and Thomson, J. D. (2006). Shifts between bee- and bird-pollination among penstemons. In *Plant–Pollinator Interactions from Specialization to Generalization*, ed. N. Waser and J. Ollerton. Chicago, IL: University of Chicago Press, pp. 47–68.

Wilson, P., Wolfe, A. D., Armbruster, W. S. and Thomson, J. D. (2007). Constrained lability in floral evolution: Counting convergent origins of hummingbird pollination in *Penstemon* and *Keckiella*. *New Phytologist*, **176**, 883–90.

2

Pollination crisis, plant sex systems, and predicting evolutionary trends in attractiveness

TOM J. DE JONG

2.1 Introduction

Since publication of *The Forgotten Pollinators* by Buchmann and Nabhan (1997), the term pollination crisis has gained widespread currency. Catchy phrases like "silent springs" and "fruitless falls" have been adopted in both the scientific literature and newspapers. Sub-optimal pollination of crops incurs an economic cost; less pollination may also lead to profound changes in the species composition of ecosystems all over the world. However, Aizen et al. (2008) have recently challenged the related idea that colonies of honeybees are generally on the decline (Jacobsen 2008). Analyzing data obtained from the FAO, they noted a downward trend in the number of bee colonies in Europe and North America, but an upward trend in non-industrialized countries that more than compensated for the decline. While this is good news, it is not the whole story. Aizen et al. (2008) also noted a trend in the crops that are being grown. Traditionally, wind-pollinated grains (rice, maize, wheat, rye) make up most of the world's food supply. Now, insect-pollinated crops are on the rise – crops like Brazil nut, cocoa bean and oil palm. This creates a need for more honeybee colonies or other alternative pollinators, which is a challenge for the future. In this context it is useful to reflect on the likely effects of reduced

Evolution of Plant–Pollinator Relationships, ed S. Patiny. Published by Cambridge University Press. © The Systematics Association 2012.

pollination levels on natural ecosystems. Here I shall focus on plant sex systems and plant attractiveness in the context of reduced pollinator visitation, approaching the problem in the context of what is known about the evolutionary ecology of plants.

2.2 Expected effects of reduced pollination: dioecy and gynodioecy

The great majority of angiosperm species have perfect flowers. These flowers have both male parts that bear pollen, and female parts that receive pollen and later produce fruits and seeds. Combining the two sexes into a single flower is economic, sharing the costs of pollinator attraction and reward over the two sex functions. The proximity of the male and female organs has dual consequences, however. In plants that are self-compatible (SC), when pollinators are in short supply, selfing provides reproductive assurance, and this can be positive. Proximity may also be negative, though, when self-pollen on the stigma prevents outcrossing and reduces seed set, even in self-incompatible (SI) species (Webb and Lloyd 1986; Bertin 1993). These negative effects are known as pollen–stigma interference or pollen–pistil interference.

In dioecy, the sexes are completely separated over male and female individuals. Which factors favor the evolution of this sex system? Darwin (1878) argued that separation of the sexes effectively bars self-fertilization, and following Darwin's suggestion many models have been developed for the transition from hermaphroditism to dioecy (reviewed in Barrett 2010). If the driving force behind dioecy is indeed to bar self-pollination, one would expect that dioecy could only evolve from (i) an SC ancestor, (ii) with a high selfing rate, and that (iii) seeds resulting from self-pollination should suffer from high levels of inbreeding depression. For all these points there is now some support (reviewed in de Jong and Klinkhamer 2005). Charlesworth (1985) found that 22 dioecious species all had SC ancestors. Detailed studies on *Sagittaria latifolia* by Delesalle and Muenchow (1992), Dorken et al. (2002) and Barrett (2003) compared populations in which the species was dioecious and populations in which it was monoecious. They showed that self-pollination by pollen transfer to neighboring flowers (geitonogamy) occurs frequently and that selfed seeds suffer from considerable inbreeding depression, especially in the dioecious populations. The outcrossing hypothesis can further be tested for more species and can, in principle, be rejected when this new species does not meet the three criteria cited above. It seems that Darwin could be entirely correct in his suggestion that avoidance of self-pollination is the main selective force behind the evolution of dioecy.

Another related issue is the route from hermaphroditism to dioecy. Models developed in the last 40 years assume that a female is first to establish in the population.

This sex system with males and hermaphrodites is called gynodioecy. Females are 100 % outcrossed, which gives them an outcrossing advantage (Lloyd 1975) over partially selfing hermaphrodites. The great majority of gynodioecious species are SC (de Jong and Klinkhamer 2005), which is consistent with this idea. The outcrossing advantage helps the spread of a male-sterility mutation, regardless of whether this mutation resides in nuclear or in cytoplasmic DNA. These ideas are illustrated in a recent study by Kobuta and Ohara (2009), who recorded the frequency of female plants in SC and SI populations of *Trillium camschatcense*. The outcrossing advantage of females only holds in the SC populations and indeed the frequency of females was much higher (0–42 %) in SC than in SI populations (0–2 %).

Examples in the European flora of plant species that combine dioecy and insect-pollination include most *Salix* species, *Asparagus officinale, Bryonia dioica, Valerianella dioica, Silene dioica* and *S. alba*, and fruit crops like papaya and kiwi. Typically in such species, fruit set or seed set declines the farther female plants are removed from nearest males in the population (de Jong et al. 2005, Table 1). This declining seed set could have two causes. First, females receive fewer visits when farther from a male. Second, females receive visits of a lower quality when farther away from a male. When pollinators move between a male and a female of species *A*, they may visit flowers of species *B* on the way, all the more so as the distance between *A* individual's increases.

This can be illustrated by a few examples from our research on dioecious plants in the Dutch coastal sand dunes of Meijendel (cf. de Jong et al. 2005). When *Salix repens* flowers in April, it is the first major source of nectar and pollen for queen bumblebees. Bees go almost exclusively for *Salix* and have close to 100 % *Salix* pollen on their bodies. Visits are all of high quality and in *Salix* the observed decline in seed set with distance to the nearest male (Table 2.1) must therefore be due to fewer visits to isolated females.

In 2006, we studied pollination of the dioecious *Asparagus officinalis* in the coastal sand dunes. In that year, very few pollinators were observed. Fruits per flower showed a clear decline with distance from the nearest male: when the male was 12.9 m from the female, fruit set was only half the value of that when the male was adjacent to the female plant (Fig 2.1a). Fruit set was consistently below the maximum of six seeds per fruit in *Asparagus*, but showed a less steep decline with distance (Fig 2.1b). The data suggest that visited flowers set fruit, while those that are not visited do not. There are few indications of an additional role of pollination quality. Pollinators were scarce in 2006, the year that this research was carried out. The situation may well be different in other years when bees visiting female *Asparagus* carry relatively fewer conspecific pollen grains (Table 2.2). Although it does not seem to play a role in these two cases that we detailed, pollination quality and how this depends on various ecological factors is an important topic for further study (Mitchell et al. 2009; Vaughton and Ramsey 2010).

Table 2.1 Seed set declines with distance to the nearest male in various dioecious plant species.

Species#	50 % seed set distance (m)*
Valeriana dioica	2.3
Salix repens	5.3
Bryonia dioica	19.6
Asparagus officinale	16.1
Various tropical trees	13.3–125

\# For data and references see de Jong et al. (2005)
* Distance at which seed set is 50 % as compared to when a male was directly adjacent to the female plant, calculated using linear regression

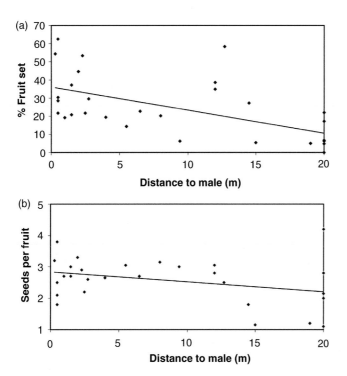

Fig 2.1 (*a*) Percentage fruit set of female *Asparagus officinale* plants declined with distance to the nearest male: $y = 0.414 - 0.016x$, Spearman $\tau = -0.66$, $P < 0.001$, $n = 30$. (*b*) Seeds per fruit showed a less steep decline: $y = 2.834 - 0.030x$, Spearman $\tau = -0.617$, $P < 0.001$.

Table 2.2 Percentage conspecific pollen[1] on bumblebees captured on female flowers.

Plant species	Year	% Conspecific pollen (range)	No. of bees captured
Asparagus officinale	2001	35.1 (31–43)	5
Asparagus officinale	2002	29.7 (14–52)	12
Asparagus officinale	2006	82.0 (38–100)	9
Bryonia dioica	2001	31.2 (21–96)	5
Bryonia dioica	2005	7.9 (0.1–50.8)	33

[1] Used as a proxy for "pollination quality"

Note that in *Bryonia dioica* visiting bees typically have few conspecific pollen grains on their body (Table 2.2), so this species would be a good candidate for further exploring the relation between pollination quality and seed set.

Because male and female individuals offer different pollinator rewards (nectar, pollen or both), there is also scope for flower visitors to specialize on one of the sexes. For instance, pollen-collecting beetles frequently visit male plants of *Salix repens* but avoid female plants. When queen bumblebees visit *S. repens* early in spring they need either nectar for energy or pollen for provisioning the first generation of larvae. The larvae need the pollen as a source of protein. It is therefore no surprise to see queen bumblebees in spring showing a clear preference for either male or female *Salix* plants. For wild strawberries, Ashman (2000) documented different flower visitors on male and female plants and suggested that this sets a limit on gender dimorphism. If male and female plants were to become too divergent, then only a few flower visitors would move between male and female plants, as required for pollination.

Females in dioecious and gynodioecious species thus appear to have an uneasy relationship with animal pollinators. A pollen donor is required for seed set, but when this donor is too distant, pollen becomes a limiting factor for seed production. Reduced pollination by animals could favor wind pollination. Reduced pollination could also favor hermaphroditism or monoecy, because these sex systems allow some reproductive assurance through selfing.

2.3 Expected effects of reduced pollination: monoecy

In monoecious species, separate male and female flowers are formed, which achieves spatial separation between the sexes. In Darwin's view, monoecy would be an adaptation that reduces selfing and promotes outcrossing. If so, one would expect an association between monoecy and SC. One would expect SI species, which have already "solved" their problem and are 100 % outcrossing, to be rarely monoecious. Contrary to this thinking, monoecy is equally common between SC and SI species (Bertin 1993). Bertin therefore suggested that monoecy reduces pollen–stigma interference. Even in an SI species self-pollen landing on the stigma may obstruct outcross pollen. Self-pollen may also germinate, fertilize ovules and produce embryos that are later aborted. This wastes ovules that are then no longer available for pollen from other plants. Monoecy leads to separation of the male and female function in space and reduces interference.

While many models have addressed the evolution of dioecy, hardly any work has been done on the transition from hermaphroditism to monoecy (but see Spalik 1991; de Jong et al. 2008). This is rather surprising since monoecy is also a complex change of the sex system in which genes are switched on or off, resulting in a developmental cascade leading to female flowers or to a different cascade leading to male flowers. Once the flower is functionally unisexual, subsequent mutations can change the characters of the male and female flowers and can reduce redundant or excessive structures. Darwin (1877) pointed out that in certain species, flowers contain the rudiment of the other sex (type I flowers), whereas in certain others such rudiments appear to be absent (type II flowers). This distinction does not mean these changes occurred in a single evolutionary event. Mitchell and Diggle (2005) showed that the loss of the opposite rudimentary sex organ evolved on at least four independent occasions in the angiosperms.

Monoecy may be beneficial in reducing pollen–stigma interference and may therefore increase seed production (Kawagoe and Suzuki 2005). There is also a cost, however, because hermaphroditic flowers are economic, sharing the cost of attraction over the male and female functions. Such a shared cost would be lowest in species with small flowers, with small petals and low nectar production rates. Indeed, monoecious species typically have much smaller flowers than hermaphrodites (on average six times smaller diameter; de Jong et al. 2008). De Jong et al. (2008) modelled the transition from hermaphroditic flowers to separate male and female flowers on the same individual. In that model, plants could optimize sex allocation, i.e., how much energy and resources to allocate to the male versus female function. In the simplest case (with linear fitness gain curves), outcrossing plants are selected to allocate as much to male as to female function (Fisher 1930). Monoecious plants can adjust sex allocation exactly to Fisher's 50:50 (or any other value) by adjusting the ratio of male to female flowers. When a female flower, including seed and fruit production, is far more costly than a male flower, the plant is selected to make many more male than female flowers. This phenomenon is widely observed among monoecious plant species (Ganeshaiah and Shaanker 1991). For hermaphroditic plants the sex allocation problem is far more complex, however. It is generally the case that, on a per-flower basis, female costs (seeds and fruit) far exceed male costs. There may be many reasons for this, but the simplest reason may be that, apart from adjusting sex allocation, plants also need to make units that "work." A fruit must have a certain minimum size to be picked up by a frugivore like a bird or mammal, while a seed must be of a certain size in order to survive the difficult seedling stage. Furthermore, a flower that matches the size of a bee will be the most efficient at exporting pollen. The result of meeting all these demands may well be that at the flower level there exists a strong female bias in sex allocation. When this bias becomes too strong, it becomes profitable for the plant not to fill each flower with a fruit. Instead the plant can abort some fruits. The "empty" flowers still contain pollen and count towards male reproductive success. Producing a certain number of empty flowers balances sex allocation at the individual plant level, although not to 50:50 (de Jong et al. 2008). Production of empty flowers is an inefficient strategy because female parts are produced and aborted seeds and fruits will incur some cost to the plant. When the abortion rate is always high, the plant can economize by making male flowers in which female parts never develop. Such a sex system with hermaphroditic and male flowers on the same plant is called andromonoecy. When female flowers are better at producing seeds than hermaphroditic flowers, andromonoecy could evolve to monoecy (de Jong et al. 2008). The transition is facilitated when there is pollen–stigma interference, when flowers are cheap to produce, and when fruits with seeds are much more costly than anthers with pollen. One study that deserves mention in this context is that of Miller and Diggle (2007). Following up on earlier studies on *Solanum* they quantified the fraction of male flowers in relation to fruit size. Sex allocation theory predicts that with a large costly fruit the plant is selected to produce many extra flowers. This is indeed the case (Fig 2.2); species with larger fruits produce relatively more male flowers.

With the model of de Jong et al. (2008) in mind, we can also now pose for monoecious plants the central question of this article, "What happens when pollination levels decline?" It turns out that in the model, high pollination levels facilitate the first step, the production of male flowers, and the second step, the forming of female instead of hermaphroditic flowers. When all hermaphroditic flowers make costly fruits and seeds, there is a strong female bias in allocation and an incentive for the production of more male flowers. When pollination levels are low, many flowers will be "empty," i.e. with pollen but without seeds; sex allocation at the level of the individual plant is not strongly biased and it may not pay to produce male flowers. Note also that for the fitness of a hermaphroditic plant with large fruits that optimizes sex allocation by, for instance, aborting 75 % of the developing embryos, it makes no difference in terms of female function whether fertilization is 100 or 25 %. After all, the hermaphroditic plant is flexible and can develop seeds in any flower that is fertilized. Only if less than 25 % of the ovules are fertilized does the plant become limited in its options. A monoecious plant immediately suffers from

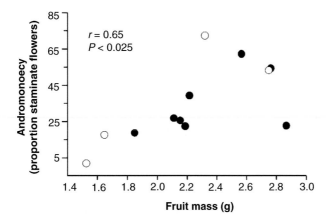

Fig 2.2 *Solanum* species with large fruits make more male flowers, thereby adjusting sex allocation. Reproduced with permission from Miller and Diggle (2007). Open circles: species from section *Acanthophora*; closed circles: section *Lasiocarpa*.

less than 100 % fertilization because it produces female flowers that are costly to the plant when they remain unfertilized.

2.4 Is seed set pollen-limited?

Do fewer pollinators result in lower seed set? That's the underlying assumption of the prophets of the pollination crisis. However, Bateman's principle states that female reproductive success is limited by how many offspring they can produce, while the number of matings limits male success. In such a situation, it does not matter whether some males are removed from the population, as long as there are others who can take over their role and mate with the females. Population growth depends on the females and their capacity to nurture as many young as they can, and is not limited by matings. Does Bateman's principle hold for plants? In a seminal article, "*The function of flowers,*" Bell (1985) addressed this problem with respect to plants with hermaphroditic flowers. There are two aspects to this problem, which in the original article by Bell (1985) are intertwined but which for the sake of clarity I shall keep separate.

2.4.1 Bateman's principle

First, plants produce many more pollen grains than there are ovules, in the same way that animals produce far more sperm than eggs. The problem, however, is that plants are not mobile and very few pollen grains will ever reach a stigma of a flower on a conspecific. It is therefore uncertain whether other pollen can take over the role of fertilizing an ovule when pollinator density declines. If seed set were to decline with distance to the nearest neighbor, this would be a clear indication that pollen is not super-abundant and isolated plants are pollen-limited in their seed set; in other words, for these plants Bateman's principle would not hold. The data in Table 2.1 indicate that this is typically the case for the insect-pollinated dioecious species listed, when male neighbors are more than a few meters away. It would be interesting to extend these results to females in gynodioecious species and SI species. If extra pollinator visits result in increased seed set, this is, by definition, pollen limitation and a deviation from Bateman's principle. However, in a breakthrough paper, Aizen and Harder (2007) emphasized that this logical experiment, allowing extra pollinator visits, is not what people typically do. Experimentalists typically test for pollen limitation by collecting an overdose of outcross pollen, in most cases from a single plant or a few plants nearby, and adding it to the stigma of a flower, preferably just after it has opened. This will result in fertilization of all ovules and maximum seed set. However, when a flower receives visits of low quality these visits do not only apply outcross pollen to the stigma but also self-pollen and pollen of different species. These pollen grains may interfere with newly arriving legitimate pollen on the stigma. They could also germinate and fertilize ovules and when the developing embryos are subsequently aborted, the ovules are no longer available. Aizen and Harder (2007) argued that, as a result of low-quality pollinator visits, the number of available ovules decreases over time. When no available ovules remain, attracting more pollinators or late experimental pollination with outcross pollen will have zero effect on seed set. Pollen is not limiting seed set. However, in the same situation the application of pure outcross pollen in a newly opened flower may well result in more seeds per flower. Aizen and Harder (2007) illustrated their idea with data on the hummingbird-pollinated mistletoe *Tristerix corymbosus* that showed c. 60 % seed set per flower over a very wide range of pollination intensities. This constant seed set suggests quality-limitation of seed set and shows that seed set is not pollen-limited, as extra pollinator visits did not result in higher seed set.

Several authors have argued against uncritically accepting Bateman's principle for plants (Wilson et al. 1994; Burd 1994). Haig and Westoby (1988) reasoned that if the seed production of a species is consistently provisioning-limited, individuals that allocate less to attraction are favored. Similarly, if seed set is consistently pollen-limited, natural selection favors individuals that allocate more resources to pollinator attraction. In their original paper, Haig and Westoby (1988) recognized that their idea does not work when competition to donate pollen sets pollen supply at much higher levels than is required for seed set. In an influential review, Burd (1994) reported that many species show increased seed set after pollen addition in some years and no effect in other years. Such data seems to support the Haig and Westoby (1988) view that many species are on the edge of where pollen is or is not limiting seed set. However, the criticism of Aizen and Harder (2007) applies to almost all data sets that Burd (1994) reviewed. With the correct experiment, the species that are now not pollen-limited remain in that category, while other species may move from the pollen-limited to the non-pollen-limited category. How often this will occur depends on pollination quality and how often it occurs that the "wrong" pollen at the stigma interferes with the "right" outcross pollen. Although there are exceptions, it is my opinion that in most cases pollen is not limiting seed set in SC plants and that, until the correct experiments prove me wrong, we should retain Bateman's principle.

2.4.2 Bell's principle

A second claim that Bell (1985) made was that drawing more pollinators to a flower benefits the male function of that flower more than the female function. For the female function, a single visit suffices, whereas pollen is not fully removed until after a flower has been visited many times (Fig 2.3). Note that this claim is similar to but also different from Bateman's principle. Whether seed set in the population is pollen-limited will depend on distances between plants, pollinator behavior and many other factors. It is likely to depend on the weather. Seed production of the whole plant is less likely to be pollen-limited when pollen removal rates are high and a single visit suffices for fertilization. However, this is not a one-to-one relationship and it is therefore wise to distinguish this idea, Bell's principle, from Bateman's principle because Bateman's principle could be taken to mean that seed set at the plant level is not limited by pollen availability. Bell's (1985) principle is not necessarily true. When a pollen-collecting bee visits a flower and strips it in a single visit of over 90 % of its pollen, yet misses the stigma, then surely the female function would benefit more from an extra visit than the male function.

At the extreme, Bell's principle states that just a few visits suffice for fertilization but many more visits are required for complete pollen removal from a flower. With low costs of attractiveness, plants may be selected to increase attractiveness to receive V visits per flower (Fig 2.3).

If we go from a situation with V visits per flower to fewer visits per flower, for instance as a result of a "pollination crisis," this would reduce male fitness but would have hardly any effect on female fitness (seed production). Fewer pollinators need not lead to less seed production. At the time of writing this chapter, Bell's (1985) paper had nearly 400 citations, demonstrating that it is a classic in this field. This is not the place to discuss in detail how many of the 400 papers that cite Bell (1985) agree or disagree with his idea. However, many studies support it. For instance, Stanton et al. (1986) showed that yellow-flowered plants of *Raphanus raphanistrum* receive more visits than white-flowered plants. These extra visits resulted in yellow plants being more often the father of seeds on their neighbors but had no positive effect on seed set. Bell and Cresswell (1998) found for *Brassica napus* that 13 hours after opening of the flowers, 50 % of the ovules were fertilized whereas only 10 % of the pollen was removed. When *B. napus* flowers open, pollen and stigma are simultaneously ripe, so that this result illustrates Bell's principle. On the other hand, De Jong and Klinkhamer (2005) found for *Echium vulgare*, little

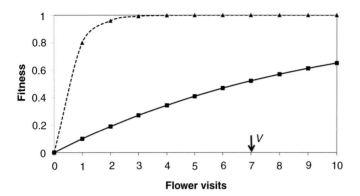

Fig 2.3 Bell's principle. More visits per flower have less effect on ovules fertilized than on pollen removal. When plant attractiveness is such that each flower receives V visits, then fewer visits will reduce male fitness (pollen removal, solid line) but not female fitness (seed production, broken line). Visits per flower is the product of attractiveness, a, and pollinators attracted per unit of attractiveness.

difference in how pollen removal and seed set depended on bumblebee visitation (see also Ashman and Schoen (1994)).

A corollary of Bell's principle is that plants should invest in attractiveness where it is most effective, i.e. in the male function. Bell (1985) had already indicated for 16 dioecious species that male flowers are in all cases larger than female flowers, with an average seven-fold difference in petal mass. Similarly, Bell reported that in gynodioecious species, flowers on female individuals are typically smaller than those on hermaphroditic individuals. For dichogamous flowers, Bell's principle means that flowers are open longer during the male phase and produce more nectar per unit of time during that period. Many insect-pollinated plants have protandrous flowers, which combine high nectar production in the male phase and a lower level of nectar production in the subsequent female phase (Cohen and Shmida 1993; Carlson and Harms 2006; Carlson 2007).

Again, pollen-limitation of seed set should be critically tested in appropriate fashion. If Bateman's and Bell's principles hold, which I would expect in the majority of cases, a moderate reduction in pollination services will have little effect on seed production of SC or SI plants.

2.5 Evolutionary effects of reduced pollination on attractiveness

Under Bell's principle there is an effect of reduced pollination levels on male fitness, i.e. successful pollen export, but not so edgy an effect on seed set. Plant densities might not be affected by reduced visitation. But what would be the evolutionary consequences of reduced pollination levels over many generations? Schoen and Ashman (1995) investigated an Evolutionarily Stable Strategy (ESS) model in which a plant could either allocate to increase longevity of the flower or make a new flower. They considered maintenance of the flower as a costly process. With frequent pollinator visits, pollen removal and fertilization were rapidly achieved, and this selected for flowers with a short life. With fewer pollinator visits, it took longer for pollen removal and fertilization to be completed and this caused selection for flowers with a long life. With long-lived flowers, more flowers are open simultaneously, so this character also leads to a larger floral display. Importantly, the Ashman and Schoen (1994) model predicted patterns in nature correctly: plant species in which daily male and female fitness accrual rates were low were the ones with the greatest flower longevity.

Several authors have developed models in which plants can allocate to male function (anthers with pollen), to female function (ovules, seeds and fruits), or to attractiveness (for a full explanation, see Chapter 14 of de Jong and Klinkhamer 2005). Charnov and Bull (1986) estimated the chance of removal of a pollen grain as $h = a^{\gamma}$, in which a is the fractional allocation to attractiveness. The value of γ is typically between 0 and 1. The model assumes that fertilization is assured, so that visitation only affects male fitness. The elegant result is that the EES for allocation to attraction is: $a^* = \gamma/(\alpha + \beta + \gamma)$ or $a^* = 1/(1+(\alpha + \beta)/\gamma)$, in which α and β are the exponents of the male and female gain curve, respectively. (The full model and ESS calculation are outlined on p. 241–244 of de Jong and Klinkhamer 2005.) When pollen removal decelerates strongly with attraction (γ low), a^* is low. This is a rather abstract result. The equation $h = a^{\gamma}$ is an oversimplification: pollen grains are removed by flower visitors like bees and the chain of events is attractiveness, pollinators, pollen removal. In the model, seed production is ensured and attractiveness only affects pollen removal, as in Bell's principle. Suppose that the common phenotype in the population has attractiveness a, with a reflecting the amount of nectar produced by a flower. Each common plant receives V visits per flower. If bees distribute themselves according to an ideal free distribution, then a mutant with attractiveness a_m should obtain $\frac{a_m}{a} V$ visits. In this way a flower with double nectar production receives twice as many visits and the reward for each bee is the same in all flowers. Now, for the chance that a pollen grain is dispersed we can take $h = V^{\gamma}$ for the common phenotype and $h = \left(\frac{a_m V}{a}\right)^{\gamma}$ for the rare mutant. Next suppose that, because of a pollination crisis, bee numbers are halved. This means that V would become half its original value. We can use the Charnov and Bull (1986) equation and substitute the new equation for h and then calculate the ESS value a^* in the original population with V visits and the new situation with $0.5V$ visits. In the fitness equation the visitation rate immediately cancels, so that fewer pollinator visits will have no effect on optimal allocation. This is perhaps a surprising result because several authors (e.g. Fishman and Willis 2008) have

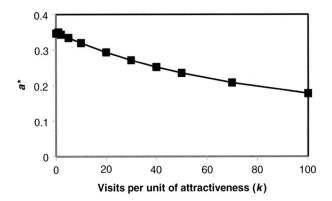

Fig 2.4 The ESS for allocation to attractiveness (a^*) decreases when pollination levels (per unit of attractiveness) increase; when pollinators become scarce one expects selection for increased attractiveness. Calculation based on the Charnov and Bull (1986) model with a mutant with attractiveness a_m in a population of a plants. The chance, h, that a pollen grain is removed from the anther is $h = 1 - p^V$. Per visit a fraction $1 - p$ of the pollen is removed from the anthers, with p pollen remaining in the anthers. V is pollinator visits per flower: $V = ka$. Parameter values: $p = 0.9$, $\alpha = \beta = 1$, $r = 0.3$. For further explanation of the equations used see de Jong and Klinkhamer (2005), p. 239–244.

asserted that there is strong theoretical support for attractiveness to increase when pollinator visits decline. Might the Charnov–Bull model be too simple to capture the essentials of pollen removal? Schoen and Ashman (1995) noted that when one pollinator visit removes a fraction $h = 1 - p$ pollen, leaving a fraction p pollen behind for the next visitor, then after V flower visits $h = 1 - p^V$ pollen grains have been exported. The assumption that each pollinator visit removes a fraction of the pollen might be more realistic, and it is also correct that between 0 to 100 % of the pollen in the anther can be exported. With this new formulation an analytical solution for a^* is no longer feasible. Nevertheless it is easy to calculate the ESS value of a^* numerically, making certain assumptions regarding the other parameters. Figure 2.4 shows that attractiveness does increase when a pollination crisis reduces V, the average visitation rate in the population. In other words, by adding realism to the model, it is plausible that scarcity of pollinators selects for greater allocation to attractiveness in plants.

Fishman and Willis (2008) grew plants of *Mimulus guttatus* under natural pollination conditions and in cages excluding all pollinators. Under pollinator exclusion, plants could only set seed through selfing. The authors noted that under cage conditions, without pollinators, certain floral phenotypes were selected, specifically those with narrow corollas and low stigma–anther distances that facilitated selfing. There was no selection for increased attractiveness. This result is unsurprising, since in this set-up there was no possibility for the attractive plants to receive more pollinator visits, which was pre-empted by the cage. An alternative experimental set-up would allow bumblebees limited access to the cages, thus reducing V. Such an experiment would be entirely feasible because bumblebees learn within a few hours to discriminate between plants with different nectar production rate and then remember locations (Klinkhamer et al. 2001). With extinction of the pollinator population, selection for increased selfing and reproductive assurance is of course an alternative route for the plant. In a commentary on the Fishman and Willis (2008) article, Mitchell and Ashman (2008) suggested that, "...in a world of declining populations the flowers themselves may begin to evolve to be less attractive and less reliant on pollinators, which might then reinforce pollinator declines." Contrast that to Buchmann and Nabhan's (1997) statement, "The brilliance and the showiness of the flower is but a visual reminder of the fact that pollinators are so often in short supply," which is a perfect summary of the model results sketched above. I would like to take an optimistic view that reduced pollination levels are not going to lead to a silent spring, that they do not necessarily reduce seed production or crop yield and may even be selecting for large, nectar-rich and colorful flowers. The study of these phenomena is important and future researchers should take pollination quality and sex system into account. In contrast to SC species with hermaphrodite flowers, SI (Burd 1994), gynodioecious and dioecious species are more likely to be pollen-limited in their seed set, especially at low densities, and therefore vulnerable to reduced visitation levels. These are likely to be the first species affected in a pollination crisis and are in need of most attention for conservation.

Acknowledgements

I thank Peter Klinkhamer and Avi Shmida for discussing the article, Kim ten Hoor and Pien van der Mark for data collection on *Asparagus* and Juliet Osborne for advice on the bumblebees. I am grateful to Nigel Harle for his revision of the English and thank Paul Wilson for his constructive comments.

References

Aizen, M. A. and Harder, L. D. (2007). Expanding the limits of the pollen-limitation concept: effects of pollen quantity and quality. *Ecology*, **88**, 271–81.

Aizen, M. A., Garibaldi, L. A., Cunningham, S. A. and Klein, A. M. (2008). Long-term global trends in crop yield and production reveal no current pollination shortage but increasing pollinator dependency. *Current Biology*, **18**, 1572–75.

Ashman, T. L. (2000). Pollinator selectivity and its implications for the evolution of dioecy and sexual dimorphism. *Ecology*, **81**, 2577–91.

Ashman, T. L. and Schoen, D. J. (1994). How long should flowers live? *Nature*, **371**, 788–91.

Barrett, S. C. H. (2003). Mating strategies in flowering plants: the outcrossing-selfing paradigm and beyond. *Philosophical Transactions of the Royal Society B, Biological Sciences*, **358**, 991–1004.

Barrett, S. C. H. (2010). Darwin's legacy: the forms, function and sexual diversity of flowers. *Philosophical Transactions of the Royal Society B, Biological Sciences*, **365**, 351–68.

Bell, G. (1985). On the function of flowers. *Proceedings of the Royal Society London B*, **224**, 223–65.

Bell, S. A. and Creswell, J. E. (1998). The phenology of gender in homogamous flowers: temporal change in the residual sex function of oilseed rape. *Oikos*, **98**, 375–84.

Bertin, R. I. (1993). Incidence of monoecy and dichogamy in relation to self-fertilization in Angiosperms. *American Journal of Botany*, **80**, 557–60.

Buchmann, S. L. and Nabhan, G. P. (1997). *The Forgotten Pollinators*. Washington, DC: Island Press.

Burd, M. (1994). Bateman's principle and plant reproduction: the role of pollen limitation in fruit and seed set. *Botanical Review*, **60**, 83–139.

Cohen, D. and Shmida, A. (1993). The evolution of flower display and reward. *Evolutionary Biology*, **27**, 197–243.

Carlson, J. E. (2007). Male-biased nectar production in a protandrous herb matches predictions of sexual selection theory. *American Journal of Botany*, **94**, 674–82.

Carlson, J. E. and Harms, K. E. (2006). The evolution of gender-biased production in hemaphrodite plants. *Botanical Review*, **72**, 179–205.

Charlesworth, D. (1985). The distribution of diocey and self-incompatibility in angiosperms. In *Evolution: Essays in Honour of John Maynard Smith*, ed. P. J. Greenwood, P. H. Harvey and M. Slatkin. Cambridge, MA: Cambridge University Press, pp. 237–268.

Charnov, E. L. and Bull, J. J. (1986). Sex allocation, pollinator attraction and fruit dispersal in cosexual plants. *Journal of Theoretical Biology*, **118**, 321–25.

Darwin, C. (1877). *The Different Forms of Flowers on Plants of the Same Species*. London, UK: Murray.

Darwin, C. (1878). *The Effect of Cross- and Self-Fertilization in the Vegetable Kingdom*. London, UK: Pickering.

Delesalle, V. A. and Muenchow, G. E. (1992). Opportunities for selfing and inbreeding depression in *Sagittaria* congeners (Alismataceae) with contrasting sexual systems. *Evolutionary Trends in Plants*, **6**, 81–91.

Dorken, M. E., Friedman, J. and Barrett, S. C. H. (2002). The evolution and maintenance of monoecy and dioecy in *Sagittaria latifolia* (Alismataceae). *Ecology*, **56**, 31–41.

Fisher, R. A. (1930). *The Genetical Theory of Natural Selection*. Oxford, UK: Clarendon Press.

Fishman, L. and Willis, J. H. (2008). Pollen limitation and natural selection on floral characters in the yellow monkeyflower, *Mimulus guttatus*. *New Phytologist*, **177**, 802–10.

Ganeshaiah, K. N. and Shaanker, R. U. (1991). Floral sex-ratios in monoecious species: why are trees more male-biased than herbs? *Current Science*, **60**, 319–21.

Haig, D. and Westoby, M. (1988). On limits to seed production. *American Naturalist*, **131**, 757–9.

Jacobsen, R. (2008). *Fruitless Fall: the Collapse of the Honeybee and the*

Coming Agricultural Crisis. London, UK: Bloomsbury Publishing.

de Jong, T. J. and Klinkhamer, P. G. L. (2005). *Evolutionary Ecology of Plant Reproductive Strategies*. Cambridge, UK: Cambridge University Press.

de Jong, T. J., Batenburg, J. C. and Klinkhamer, P. G. L. (2005). Distance-dependent pollen limitation of seed set in some insect-pollinated dioecious plants. *Acta oecologia*, **28**, 331–5.

de Jong, T. J., Shmida, A. and Thuijsman, F. (2008). Sex allocation in plants and the evolution of monoecy. *Evolutionary Ecology Research*, **10**, 1087–109.

Kawagoe, T. and Suzuki, N. (2005). Self-pollen on a stigma interferes with outcrossed seed production in a self-incompatible monoecious plant, *Akebia quinata* (Lardizabalaceae). *Functional Ecology*, **19**, 49–54.

Klinkhamer, P. G. L., de Jong, T. J. and Linnenbank, L. A. (2001). Small-scale spatial patterns determine ecological relationships: an experimental example using nectar production rates. *Ecology Letters*, **4**, 559–67.

Kobuta, S. and Ohara, M. (2009). Discovery of male sterile plants and their contrasting occurrence between self-compatible and self-incompatible populations of the hermaphroditic perennial *Trillium camschatcense*. *Plant Species Biology*, **24**, 169–78.

Lloyd, D. G. (1975). The maintenance of gynodioecy and androdioecy in angiosperms. *Genetica*, **45**, 325–39.

Miller, J. S. and Diggle, P. K. (2007). Correlated evolution of fruit size and sexual expression in andromonoecious *Solanum* sections *Acanthophora* and *Lasiocarpa* (Solanaceae). *American Journal of Botany*, **94**, 1706–15.

Mitchell, C. H. and Diggle, P. K. (2005). Evolution of unisexual flowers: morphological and functional convergence results from diverse developmental transitions. *American Journal of Botany*, **92**, 1068–76.

Mitchell, R. J. and Ashman, T. L. (2008). Predicting evolutionary consequences of pollinator declines: the long and short of floral evolution. *New Phytologist*, **177**, 576–9.

Mitchell, R. J., Flanagan, R. J., Brown, B. J., Waser, N. M. and Karron J. D. (2009). New frontiers in competition for pollination. *Annals of Botany*, **103**, 1403–13.

Schoen, D. J. and Ashman, T. L. (1995). The evolution of floral longevity: resource allocation to maintenance versus construction of repeated parts in modular organisms. *Evolution*, **49**, 131–9.

Spalik, K. (1991). On evolution of andromonoecy and "overproduction" of flowers: a resource allocation model. *Biological Journal of the Linnean Society*, **42**, 325–36.

Stanton, M. L., Snow, A. A. and Handel, S. N. (1986). Floral evolution: attractiveness to pollinators increases male fitness. *Science*, **232**, 1625–27.

Vaughton, G. and Ramsey, M. (2010). Floral emasculation reveals pollen quality limitation of seed output in *Bulbine bulbosa*. *American Journal of Botany*, **97**, 174–8.

Webb, C. J. and Lloyd, D. G. (1986). The avoidance of interference between the presentation of pollen and stigma in Angiosperms. *New Zealand Journal of Botany*, **24**, 163–78.

Wilson, P., Thomson, J. C., Stanton, M. L. and Rigney, L. P. (1994). Beyond floral Batemania: gender biases in selection for pollination success. *American Naturalist*, **143**, 283–96.

3

Evolution and ecological implications of "specialized" pollinator rewards

W. Scott Armbruster

3.1 Introduction

The transfer of pollen between flowers by animals or abiotic agents is a critical event in the reproduction of most flowering plant species, affecting both the number and quality of offspring (seeds) produced. Most plants attract animal pollinators to their flowers by offering nectar, pollen, or edible floral parts to these mutualists. A small but significant number of angiosperm species offer other kinds of rewards, which I will call "specialized" rewards. Yet other species offer no rewards at all, instead deceiving their pollinators, eliciting visitation without any compensation whatsoever.

While the list of plants known to offer specialized pollinator rewards has been growing, we still know relatively little about the evolution of these pollinator-attraction systems or their effects on community ecology. In particular, the ecological implications of such reward systems have not been factored into thinking about pollination networks or the relationship between specialization and vulnerability to anthropogenic disturbance. This chapter reviews these issues using published data and unpublished observations to assess the evolutionary dynamics and ecological importance of pollination relationships based on specialized rewards.

Evolution of Plant–Pollinator Relationships, ed S. Patiny. Published by Cambridge University Press. © The Systematics Association 2012.

3.2 How flowers attract pollinators

There are two basic systems for attracting pollinators to flowers. These are rewards (or "primary attractants") and advertisements ("secondary attractants"; Faegri and van der Pijl 1971; Fenster et al. 2004). Pollinator rewards are distinguished from advertisements in that the former constitute the primary or economic motivation for animals to visit flowers. In contrast, advertisements attract the attention of pollinators and promote associative learning. Common ("usual") floral rewards include nectar, pollen, and food bodies. Advertisements include bright floral colors, distinctive flower shapes, and strong, characteristic fragrances.

3.2.1 Nectar

Nectar, the most common reward attracting pollinators, is a nutritional liquid comprising a dilute to fairly concentrated aqueous solution of sugar and often amino acids. The main sugars produced are the hexoses, fructose and glucose, and the disaccharide, sucrose (Baker and Baker 1982). Sucrose-dominated nectars are typically associated with hummingbird, sphingid-moth, and large-bee pollination. Hexose-dominated nectars are associated with bat and perching-bird pollination and pollination by insects other than those listed above (Baker and Baker 1982). Nectar attracts a wide variety of animal acting as pollinators. These include bats, marsupials, rodents, birds, a few lizards, flies, bees, wasps, lepidopterans (butterflies and moths), and other insects.

3.2.2 Pollen

Pollen is the next most common reward attracting pollinators. Pollen is expensive for plants to lose to pollinators, not so much because it is highly nutritious, containing amino acids, starch and/or oils, but because pollen grains contain the male gametes. Thus pollen eaten or collected by pollinators represents gene copies not passed into the next generation. Pollen attracts a variety of vertebrates and invertebrates, although insects, particularly beetles, flies, and bees, are by far the most common.

Some pollen has been shown to be toxic or at least greatly retard the growth of bee larvae (Praz et al. 2008). The pollens of many Ranunculaceae and Asteraceae, for example, appear to be toxic and avoided by most bees. Those bees that do collect this pollen, such as bees that are oligolectic on Asteraceae (Müller and Kuhlmann 2008), are thus highly specialized in their pollen foraging. If pollen is the only reward, and oligolectic (pollen specialist) bees are good pollinators, then pollen can be regarded as a specialized reward (see below). Possible examples of this relationship have been reported in a couple Brazilian Leguminoseae (Cintra et al. 2003; de Carvalho and Message 2004), but these may be exceptions rather than the rule. Most toxic pollen is probably produced by flowers with nectar as the main reward (e.g. Asteraceae, Ranunculaceae). Thus, pollen toxicity is likely a defense mechanism in many, if not most, cases (see Hargreaves et al. 2009). More work is needed on the distribution of toxic pollens. How often are toxic pollens found in pollen-reward versus nectar-reward flowers? Only in the former can we begin to suspect that selection for pollinator specialization has led to the evolution of toxic pollen. Theoretical expectations are that toxic or repellent pollens should evolve more often when primary pollinators do not consume pollen (Hargreaves et al. 2009). Indeed, inspection of the literature suggests that most toxic pollen is found in nectar-reward or wind-pollinated flowers, and hence toxicity is probably more commonly defensive. I will not therefore consider "specialized" pollen rewards further, but more research is, however, needed into the evolutionary dynamics of this relationship.

3.2.3 Food bodies

Specialized food bodies, ranging from bracts and perianth parts to edible trichomes, are rich in starch and/or proteins and easily eaten or collected. This reward class has been reported across a scattering of plant families, e.g. Nymphaceae, Calycanthaceae (Faegri and van der Pijl 1971; Thien et al. 2009), Orchidaceae (Simpson and Neff 1981), and Pandanaceae (Cox 1982). The main groups reported to be attracted to floral food bodies are beetles, but pollinating bats and/or birds have also been reported to be attracted by, and to feed on fleshy bracts in *Freycinetia* (Pandanaceae; Cox 1982) and fleshy pedicel/peduncle tissues in male *Hura* (Euphorbiaceae) inflorescences (Steiner 1982). Because these rewards are, as far as we know, broadly nutritious, I do not treat them as specialized rewards in the next section.

3.3 Pollinator attraction by production of specialized rewards

3.3.1 Specialization in flowers and pollinators

Before beginning a review of specialized rewards, it is necessary to discuss and define specialization in the context of floral evolution and pollination. This has been a somewhat controversial topic in recent decades (cf. Waser et al. 1996; Ollerton 1996; Johnson and Steiner 2000), although recent reviews suggest the common ground is actually quite substantial once semantic confusion is resolved (Armbruster et al. 2000; Fenster et al. 2004).

Specialized pollination can be defined as an ecological phenomenon in which a flower species is pollinated by one or a few species or functional groups of animals ("ecological specialization" of flower pollination). Ecological specialization of the pollinator similarly refers to it visiting only one or a few flowers for food or resources. A good example of this is oligolectic bees, which visit only one or a few species of related plants for pollen. Floral specialization refers to flowers and flower features that limit the diversity of animals that can visit and pollinate the flowers. The "specialized" rewards described below are good examples. Finally, evolutionary specialization of plant–pollinator relationships refers to the evolutionary processes that increase ecological specialization, generally (but not necessarily) in response to specializing selection. This process presumably generated nearly all extant specialized plant–pollinator relationships.

The first specialized reward system to be described was brood-sites (Riley 1872; Darwin 1874; Kerner 1898), followed by resin-rewards (Müller 1879; Cammerloher 1931; Skutch 1971), and much later, fragrance-reward (Vogel 1966), and floral-oil mutualisms (Vogel 1969, 1974).

3.3.2 Brood site rewards

Brood-site rewards involve relationships wherein the pollinator lays eggs on the flowers and larvae eat a portion of the developing seeds or sometimes other tissues. These relationships are similar to those involving the consumption of food bodies or other floral parts (above), but differ in a couple of important ways. First, gametes are usually consumed, as in pollen rewards, and this creates evolutionary dynamics that are very different from food bodies, e.g. selective abortion to punish excessively "greedy" mothers (Pellmyr and Huth 1994; Goto et al. 2010). Second, brood-site reward relationships are usually highly specialized.

Plants involved in brood-site pollination relationships may experience either passive or active ("intentional") pollination. The latter involves pollen collection and active placement on the stigmas. In either case, pollination by the female insect results in production of seeds, often some portion of which are fed upon by its offspring. In some brood-site mutualism, other nearby floral or vegetative tissues are fed upon by the larvae (see review in Dufay and Anstett 2003). Classic brood-site relationships include figs (*Ficus*, Moraceae) and fig wasps (Hymenoptera: Agaonidae; Wiebes 1979; Rønsted et al. 2005), yuccas (*Yucca*, Agavaceae/ Liliaceae) and yucca moths (Lepidoptera: Prodoxidae: *Tegiticula*; Pellmyr et al. 1996), which all involve active pollination; and *Trollius* (Ranunculaceae) and *Chiastochaeta* flies (Diptera: Anthomyidae), which involves passive pollination (Pellmyr 1989, 1992). More recently described relationships include *Shorea* trees (Dipterocarpaceae) and thrips (Thysanoptera; Appanah 1990), *Eupomatia laurina* trees (Eupomatiaceae) and *Elleschodes* weevils (Coleoptera: Curculionidae; Irvine and Armstrong 1990), *Siparuna* trees (Siparunaceae) and *Asynapta* flies (Diptera: Cecidomyiidae; Feil 1992), *Silene* herbs (Caryophyllaceae) and *Hadena* moths (Lepidoptera: Noctuidae; Petersson 1991), *Lithophragma* herbs (Saxifragaceae) and *Greyia* moths (Lepidoptera: Prodoxidae; Thompson and Pellmyr 1992); senita cacti (*Lophocereus schottii*, Cactaceae) and *Upiga* moths (Lepidoptera: Pyralidae; Holland and Fleming 1999); Phyllantheae trees and shrubs (*Breynia, Glochidion, Phyllanthus*; Phyllanthaceae) and *Epicephala* moths (Lepidoptera: Gracillariidae; Kato et al. 2003; Svensson et al. 2010; Kawakito 2010); *Macaranga* trees (Euphorbiaceae) and *Neoheegeria* thrips (Thysanoptera: Phlaeothripidae;

Moog et al. 2002) or anthocorid and mirid bugs (Heteroptera: Anthocoridae, Miridae; Ishida et al. 2009); and *Chamaerops* palms (Arecaceae) and *Derelomus* wevils (Coleoptera: Curculionidae; Dufay 2010).

In those plant species that experience active pollination (minimally figs, yuccas, senita cacti) and those that lack copollinators (minimally figs, yuccas, *Chaemerops*, *Trollius*) we see the most specialized plant–pollinator relationships ever described (see review in Dufay and Anstett 2003). One or a few pollinator species provide pollination services to a species of plant and each pollinating insect species has only one or a few host-plant species. Unlike most plant–pollinator relationships, specialized brood-site mutualisms commonly reflect a high degree of symmetry, mutual dependence, non-nested structure of interaction webs, and ecosystem sensitivity to species loss (see discussion below).

3.3.3 Oil

Flowers producing oil rewards for pollinators ("oil flowers") were discovered only relatively recently, with initial elucidation by Vogel (1974). Subsequent studies have documented 11 plant families (Buchmann 1987; Vogel 2009; Renner and Schaefer 2010) and at least 28 evolutionary origins of this mode of attracting and rewarding pollinators (Renner and Schaefer 2010). Plant families with large proportions of species involved in this relationship include Malpighiaceae (primarily the New World species; 36 genera), Krameriaceae (*Krameria;* Fig 3.1), and Calceolariaceae (*Calceolaria*). Other families with numerous species involved are the Scrophulariaceae (*sensu lato*), especially in the neotropics (four genera) and southern Africa (five to six genera), Iridaceae (four genera, including South American species of *Sisyrinchium*; see Cocucci and Vogel 2001), and Orchidaceae (ca. ten genera). Floral oil secretions produced by these diverse plants are, surprisingly, all energy-rich, long-chain (14–18 carbons), free acyloxy-fatty acids (Vogel 2009).

Bees collect these oils for larval provisioning, and in some cases for lining the nests, and the relationship appears to be highly specialized and obligatory rather than facultative. Bees collecting floral oils are found in two families, Apidae (modern sense; Michener 2007) and Melittidae. Systematic relationships within the huge and diverse family Apidae are complex, but there are clearly several distinct tribes involved in oil collection, representing three to four or more origins of oil collection (Schaeffer and Renner 2008; Cardinal et al. 2010). These include members of tribes Centridini (*Centris*, *Epicharis*), Tapinotaspidini (*Monoeca*, *Trigonopedia*, *Tapinotaspis*, *Tapinotaspoides*, *Paratetrapedia*, and probably *Arhysoceble*, *Chalepogenus* and *Caenonomada*), and Tetrapediini (*Tetrapedia*) (Neff and Simpson 1981; Buchmann 1987; Michener 2007). Finally, *Ctenoplectra* species (Apidae: Ctenoplectrini; Michener 2007) collect oils from some cucurbits (*Momordica*, *Thladiantha*) in tropical Africa and Asia. The other important family is Melittidae, with the southern African genus *Rediviva* being an especially

Fig 3.1 Some representative flowers offering specialized rewards. A. *Yucca* cf. *schedigera*, brood-site reward flower pollinated by *Tegiticula* moths. B. *Krameria* sp. (Krameriaceae) offering oil to attract pollinating centridine bees (Hymenoptera: Apidae). C. *Dalechampia spathulata* being polinated by a fragrance-collecting male *Eulaema* sp. (Apidae: Euglossini) in Costa Rica (photo by W. Hallwachs, reproduced with permission). D. *Euglossa viridissima* (Euglossini) collecting eugenol from a blotter-paper bait, Quintana Roo, Mexico. E. *Dalechampia tiliifolia*, which offers a bright orange resin as the pollinator reward. F. *Trigona* (sensu lato) sp. (Apidae: Meliponini) collecting resin from a staminate *Clusia* sp. (Clusiaceae) in Suriname. The resin is secreted from staminodia in the center of the flower. Note balls of resin already placed in the bee's corbiculae (pollen baskets). See plate section for color version.

important mutualist with several Scrophulariaceae (*s.l.*) and Orchidaceae genera (Steiner and Whitehead 1990, 1991a; Pauw 2006). *Macropis* bees (also Melittidae) collect oils from, and pollinate, *Lysimachia* (Myrsinaceae/Primulaceae) in Eurasia and North America (Michez and Patiny 2005; Michez et al. 2008, 2009).

3.3.4 Fragrance

Perhaps the most extensively studied, but still incompletely understood, special-ized-reward relationship is between male euglossine bees (Apidae: Euglossini) and the flowers from which they collect terpenoid and aromatic fragrances. It is clear that these bees collect fragrances from flowers (as well as non-floral sources, such as fungi and vegetative parts of plants) and pollinate flowers in the process. It is not as clear just how these fragrances are used by the bees. The fragrances are apparently used to attract and/or impress females for mating. It is not known, however, to what extent they are modified into pheromones, or, as now seems more likely, simply expressed as admixtures of the original floral compounds (Eltz et al. 1999). The collection of a diversity of compounds is probably favored through sex-ual selection: females probably choose males on the basis of the composition and complexity of the fragrance bouquets they have assembled. If heritable, the bou-quet would indicate the male's genetic quality for traits influencing foraging abil-ity, which is an important capability for both the sons and daughters of the female choosing a mate. There is evidence that females mate only once (Zimmermann et al. 2009a), so this choice is extremely important. Species recognition and repro-ductive isolation are either by-products or possibly contributing selective drivers of this system (Zimmermann et al. 2009b).

Flowers involved in mutualisms with fragrance-collecting male euglossine bees include a numbers of orchid genera, notably nearly all species in Catesetinae (e.g. *Catesetum*) and Stanhopinae (e.g. *Stanhopea*), plus another 50+ genera for a total of more than 650 species (Dressler 1972). Other monocots whose flow-ers are exploited by male euglossines for fragrances include *Spathiphyllum* and *Anthurium* (Araceae), *Xyphidium* (Haemodoraceae; Dressler 1972), and pos-sibly one or more palms, including *Geonoma* (Arecaceae; Listabarth 1993). Dicot participants include *Drymonia*, *Gloxinia* (Gesneriaceae), *Bignonia*, *Saritaea*, (Bignoniaceae), *Cyphomandra* (Solanaceae), *Mandevilla* (Apocynaceae), *Tovomita* (Clusiaceae), and *Dalechampia* (Euphorbiaceae; Dressler 1972; Armbruster et al. 1992; Armbruster 1993; Nogueira et al. 1998; Ramirez et al. 2002). Male eugloss-ines are also reported visiting for fragrance and probably pollinating flowers in the Iridaceae, Liliaceae/Amaryllidaceae, and Theaceae (Ramirez et al. 2002). These obviously represent many independent origins of the mutualism on the plant side.

In contrast, it appears that the fragrance-collection behavior of euglossines orig-inated only once and no reversals have occurred. This is based on two observations:

1) the tribe Euglossini (Hymenoptera: Apidae) is monophyletic (Ramirez et al. 2010); and 2) all members of the tribe collect fragrances from flowers and/or non-floral sources (Ramirez et al. 2002). These include not only genera with "normal," independent life histories (*Euglossa* [ca. 100 spp.], *Eufriesea* [ca. 60 spp.], *Eulaema* [15 spp.]) but also both genera with nest-parasitic life histories (*Exaerete* [5 spp.], *Aglae* [1 sp.]). Males of all genera except *Aglae* have been observed visiting flowers for fragrance and pollinating them, whereas males of the infrequently observed *Aglae* have, to date, only been observed collecting fragrances at artificial baits (Ramirez et al. 2002).

3.3.5 Resin

The use of resin in nest construction and defense by hymenopterans is widely known (Schwarz 1948; Dodson 1966; Krombein 1967; Stephen et al. 1969; Armbruster 1984). Resin sources include many groups of plants that produce resin from wounds or broken stems, or as coatings of vegetative structures (e.g. Burseraceae, Clusiaceae, Leguminoseae, Pinaceae, Zygophyllaceae; Langenheim, 2003). Construction use includes employment of resin as constituents of outer walls, combs, and cells, binder of other materials, and sealant of gaps. Protective use includes applying resin to enemies (e.g. on ants by stingless bees, *Trigona*; Skutch 1971) and as defense against microbes attacking stored food and developing larvae (Messer 1985; Lokvam and Braddock 1999). What has been less appreciated, however, is the use of flowers or inflorescences as a source of resin.

The earliest reports of resin being secreted by flowers and attracting bees were made by Müller (1879) and Cammerloher (1931) who reported stingless bees (Apidae: Meliponini) collecting resin from *Dalechampia* (Euphorbiaceae) blossoms for nest construction in Brazil and Java, respectively. Neither report documented pollination, however. Indeed, stingless bees are often too small to effect pollination in many flowers, acting instead as thieves. Skutch (1971) was apparently the first to report pollination by stingless bees collecting resin from *Clusia* flowers in Costa Rica (although he called it secretion gum, it is now known to be a polyisoprenylated benzophenone resin; Porto et al. 2000). A later review of this reward system clarified the role of resin in attracting pollinating insects that need resin in nest construction (Armbruster 1984), although this does not exclude the alternative hypothesis that resin sometimes glues pollen to pollinators for secure transport (Bittrich and Amaral 1997; see discussion in Armbruster 1984).

To date only three families and five genera of flowering plants are known to produce terpenoid or isoprenylated floral resins for attracting pollinators: *Dalechampia* (Euphorbiaceae; ca 100 species have resin rewards); *Clusia, Tovomitopsis* and *Clusiella* (Clusiaceae; ca 200 species); and *Maxillaria* (Orchidaceae; a few species; Armbruster 1984, 1993; Gustafsson and Bittrich 2002; Flach et al. 2004; Singer and Koehler 2004; Whitten et al. 2007). While the three resin-reward Clusiaceae and

orchids are restricted to the Neotropics, resin-reward *Dalechampia* occur in tropical Asia, Africa and Madagascar, as well as the neotropics. Given the importance
of resin in the life history of many bees, other taxa with resin-reward flowers are
likely to be found.

Bees reported to collect resin from flowers occur in two families: Apidae and
Megachilidae. In the Apidae, the most frequently observed pollinators of resin flow-
ers are the non-parasitic female euglossine bees, *Euglossa*, *Eulaema*, and *Eufriesea*
(Euglossini; Armbruster 1984, 1993). These are almost exclusively neotropical in
distribution (marginally into the New World subtropics). Stingless bees (Apidae:
Meliponini) are also common resin-collecting floral visitors and sometimes pollinators, especially *Trigona* (*sensu lato*) (*Dalechampia, Clusia*) and less often *Melipona*
(*Clusia*; Armbruster unpublished data). Honeybees (Apidae: *Apis mellifera*) have
also been reported collecting floral resin (*Clusia* flowers) in Hawaii, where both
have been introduced (H. G. Baker, personal communication, 1992). Among the
megachilids, *Hypanthidium* (Megachilidae: Anthidiini) is the most common pollinator of small-blossomed *Dalechampia* in the Neotropics (Armbruster 1993). The
primary pollinators of *Dalechampia* in Africa are *Pachyanthidium* (Megachilidae:
Anthidiini) and *Heriades* (Megachildae: Megachilini; Armbruster and Mziray 1987;
Steiner and Whitehead 1991b, Armbruster and Steiner 1992), and at least the latter
is likely to be important in tropical Asia as well. Fieldwork in southwest China has
shown recently that *Megachile* (*Callomegachile*) (=*Chalicodoma*) spp. also collect
floral resins and pollinate *Dalechampia* (Armbruster et al. 2011).

There is good evidence that female euglossines, which are medium-sized to
large bees, visit flowers offering large amounts of resin (i.e. *Dalechampia* and
Clusia with large resin glands; Armbruster 1984). In contrast, most of the megachilids and *Trigona* are smaller bees and visit *Dalechampia* with smaller glands
(as well as species with larger glands, although here larger gland–stigma distances
may preclude regular pollination; Armbruster 1988). Beyond the size relationship,
however, it is unclear if the association between certain resin-collecting bees and
certain flower species is a biological signal or the result of limited field observation (i.e. inadequate sampling). For example, to date, no megachilids have been
observed visiting *Clusia* flowers for resin, and *Melipona* has not been observed
visiting *Dalechampia* (W. S. Armbruster, unpublished data).

Some degree of specificity would not be surprising because not only do floral
resins from different plants differ chemically, but different bees use floral resins in
different ways. For example, the megachilid bees, so far as we know, use resin primarily to cement pebbles together to make the nest walls and otherwise seal the
nest. *Eufriesea* use resin to cement together strips of bark to build the nest walls
and also resin in construction of the cells. *Eulaema* uses resin in cell construction and in sealing the nest, isolating it from the surrounding matrix. *Euglossa* that
nest in cavities use resin much as do *Eulaema*, but *Euglossa* that build nests in the

open use resin for the outer wall, the cells, and also for nocturnal closure of the entrance (Fig 3.2; W. S. Armbruster, unpublished data). Meliponines mix resins with secreted wax and use it to reinforce comb structures and seal the hive. *Apis* use resin or a mixture of resin and secreted wax to seal the hive. Several megachilid and apid bees collect resin and deploy it in the nest apparently to inhibit the growth of bacteria and fungi on food stores and growing larvae (Messer 1985). This last use is difficult to assess and may be much more common than we realize. *Trigona* (*sensu lato*) are reported to repel attacking ants by gluing balls of sticky resin onto them (Skutch 1971; Lehmberg et al. 2008).

We can thus infer patterns of resin use that range from potentially generalized to highly specialized. Resin collected by bees for sealing alone may be very different from resin used by other species in different ways. Similarly, different resin sources may be used by a single bee to fulfil different applications. For example, resin that is used by megachilids to cement pebbles together solidifies and hardens quickly. The same is probably true of resin used to seal nests and to form outside walls. In contrast, resin used for cells must remain reasonably soft and workable for an extended period, so that the new adult can emerge and that materials can be recycled in some cases. The same is true for the front "door" of a *Euglossa* nest (Fig 3.2). Resins used for protection against bacteria and fungi must obviously have antibiotic properties, probably through emission of volatile components. Thus bee

Fig 3.2 *Euglossa* nest constructed entirely of plant resins, with "front door" open, hanging from branch of cf. *Lycopodium*, Parque Nacional Soberanía, Panama. The resin used in the hard outer wall is most likely to be a different chemical composition to that of the internal resins and the "door" resin, which need to stay liquid and malleable. Note female *Euglossa* sp. in the entrance. This opening is closed with a resin "door" by night. See plate section for color version.

species may vary dramatically in the sources of resins used, and many bees collect several types of resin from multiple sources.

Another "service" provided to plants by resin-collecting bees is dispersal of seeds. Several cases have been reported where bees that use resin in nest construction visit resiniferous fruits for resin and disperse seeds in the process of transporting the resin back to the nest (e.g. *Corymbia/Eucalyptus* [Myrtaceae]; Wallace and Trueman 1995; Wallace et al. 2008; Wallace and Lee 2010; *Vismia* [Clusiaceae], Roubik 1989). Whether this reflects plant adaptation (or exaptation/preadaptation) for seed dispersal is unclear, though there is some evidence from *Corymbia* that suggests this possibility. It seems more likely for *Vismia*, if not both genera, that this is largely an incidental by-product of bees foraging for resins that protect seeds, rather than a dispersal adaptation.

3.3.6 Waxes

There have been a few reports of the production of waxy materials by the labellum of flowers of the *Ornithidium* orchids (Porsch 1905) and *Maxillaria* orchids (Singer and Koehler 2004; Whitten et al. 2007). These waxes are presumably collected by pollinating bees, although detailed observations are largely lacking or unpublished. Meliponines are known to be wax collectors and these are likely to be the pollinators.

Insight into this system is yielded by consideration of wax-collecting meliponine bees visiting *Coussapoa* (Cecropiacae) fruits in Brazil. In a system apparently parallel to the resin-fruits described above, bees disperse seeds incidentally in association with collecting wax and transporting it to their nests (Garcia et al. 1992; Nunez et al. 2008). This supports the supposition that there is a potential pollination niche, in which wax-collecting stingless bees provide pollination services.

3.4 Evolutionary patterns

3.4.1 Evolutionary lability and homoplasy

Perusal of published macroevolutionary studies of pollination reveals a general pattern in the evolution of specialized rewards. There is a large amount of evolutionary lability (frequent transitions) and abundant homoplasy (parallelisms and reversals) in the evolution of specialized rewards. This includes evolutionary transitions from "normal" rewards (pollen or nectar) to specialized rewards, evolutionary shifts between different specialized rewards, and reversals from specialized rewards back to "normal" rewards.

In the evolution of yucca moths and their hosts, for example, there is good phylogenetic evidence that brood-site mutualisms originated multiple times on both sides of the partnership. That is, multiple lineages of the moth independently evolved active pollination and oviposition relationships with particular plants groups (Pellmyr 2003), and multiple lineages of agavaceous plants entered into brood-site pollination mutualisms independently (Bogler et al 1995; Smith et al. 2008). This is at first surprising given the tight relationship and mutual dependence. However, the likelihood of parallelism was probably increased by preaptations being in place long before the mutualism evolved (Pellmyr et al 1996; Pellmyr 1997). Additional homoplasy is exhibited by the repeated loss of mutualistic behaviors by the moths (i.e. reversals; Pellmyr et al. 1996, Pellmyr 2003).

Similarly high levels of homoplasy are seen in the evolution of oil rewards. A recent macroevolutionary survey by Renner and Schaefer (2010) indicated that, across all angiosperms, oil-reward systems show a striking level of homoplasy, with at least 28 parallel origins and 36–40 losses (generally reversals back to pollen or nectar). These authors date the origin of the oil–flower and oil–bee relationship back to at least 56 millions of years before present.

Oil foraging in bees also exhibited rampant parallelism, with at least six origins (Renner and Schaefer 2010). There is strong evidence of homoplasy in the origins of oil collection within the Apidae; there were at least four origins of oil collection (Schaeffer and Renner 2008; Cardinal et al. 2010). In the case of the Melittidae, it has been hypothesized that these relationships with oil flowers evolved twice independently (Michez et al. 2009; but see Steiner and Cruz 2006). There is clear evidence of direct or diffuse coevolution between oil flowers and oil bees. Nearly all oil bees have specialized setae (e.g. Melittidae: *Rediviva, Macropis*) or scrapers (e.g. Apidae: *Centris, Epicharis*) on fore and mid-legs, often corresponding to the number and position of specialized oil glands in their host flowers (e.g. fore-leg and mid-leg scapers in *Centris* and four sepalar oil glands in Malpighiaceae, fore-leg setae in *Rediviva* and oil-glands in a paired spurs in *Diascia*). In turn, there is very strong evidence that *Rediviva* and *Diascia* have undergone coevolution at both the within- and among-species levels, such that leg-length in the female bees corresponds closely with the spur length in the host flowers. Curiously, an apparent genetic correlation between leg lengths in male and female bees means that male bee legs also covary with spur length, although the legs are not as long as in females and the male bees have no interactions with the flowers (Steiner and Whitehead 1990, 1991a).

Fragrance-reward systems also show multiple origins (parallelism) in the Orchidaceae (e.g. Chase and Hills 1992) and Euphorbiaceae (Armbruster 1993), and probably in other plant families. Even within one genus, *Dalechampia*, with fewer than a dozen fragrance-reward species (out of ca. 130 spp. total), there is good evidence of three to four independent origins of fragrance rewards and male-euglossine pollination (Fig 3.3). In two to three cases this involves fragrance secretion by the stigmatic surface of the pistils, and in one case it involves a modified "resin gland" secreting monoterpene fragrances instead (Armbruster et al. 1992; Armbruster 1993).

Fig 3.3 Bayesian estimate of *Dalechampia* (Euphorbiaceae) phylogeny, showing evolution of pollinator rewards. The branch lengths are proportional to divergence in the ITS sequence. See Armbruster et al. (2009) for estimation details. Pollinator rewards are mapped onto the tree using parsimony, trace-across-all-trees, and MPRs modes

Resin-reward mutualisms show four or more origins in three families: one origin and at least four losses in *Dalechampia* (Fig 3.3; Armbruster 1993), two to four origins and at least three losses in Clusiaceae (Gustafsson and Bittrich 2002), and one to several origins in *Maxillaria* and close relatives (Orchidaceae; Whitten et al. 2007). This is a surprising amount of homoplasy, especially in the Clusiaceae. As mentioned previously, the high frequency of parallelism may reflect preaptations being in place, a topic I address next.

3.4.2 Exaptation

Another recurrent pattern evident from the phylogenetic analysis of pollinator interactions with plants offering specialized rewards is the importance of exaptation (≈ preadaptation; see Gould and Vrba 1982). In the evolution of the brood-site mutualism with yuccas, the moth partners actually evolved most of the necessary traits (e.g. local host specificity, laying eggs in flowers, and limited destruction of seeds by larvae) *before* the mutualistic relationship was established. Hence preaptations being in place made the final transition to a full-fledged brood-site mutualism only a small step – the evolution of active pollination (Pellmyr et al. 1996; Pellmyr 1997).

The evolution of fragrance rewards attracting male euglossines also probably reflects exaptation in most cases. Most flowers produce fragrances as advertisements and some of these may attract euglossines incidentally. Once visiting, these bees may sometimes exert strong enough selective pressure to shift system to fragrance reward (with loss of the original reward). Euglossines hence started out as fragrance thieves and possibly incidental pollinators, becoming copollinators, and eventually exclusive pollinators. This process almost certainly explains the three to four shifts to male-euglossine pollination in *Dalechampia* (Fig 3.3). We know that stigmas secrete advertisement fragrances in most species, including sisters to the two of the male-euglossine clades. Similarly, there is a biochemical link between monoterpene fragrance synthesis and triterpene resin synthesis, in that both are derived from isoprenoid precursors. Thus, resin synthesis may be a preaptation for monoterpene fragrance synthesis (Armbruster 1993) and vice versa in other cases.

The origin of resin rewards in both Clusiaceae and Euphorbiaceae was probably also by exaptation. Resin in most plants plays a defensive role and probably was the original function of resins and/or latex produced by flowers in both groups of plants (Armbruster 1984; Armbruster et al. 2009). Interesting, the same compounds that have antibacterial activity in floral resins of *Clusia* are found also in the stem and leaf latex (Lokvam and Braddock 1999; Porto et al. 2000).

Fig 3.3 (*cont.*)

in Mesquite (Maddison and Maddison, 2009). The width of each color on each branch indicates the proportion reconstructions with that character state, across 35 trees sampled from the posterior distribution of 5×10^6 retained trees. The proportion of trees lacking that branch and node is indicated in red. See plate section for color version.

3.4.3 Specialized rewards and species diversity

There has been a general expectation since Darwin that floral specialization promotes diversification and hence clade species richness. Some evidence supports this, e.g. the observation that lineages of plants with long nectar spurs have more specialized pollination and are more diverse than sister lineages without (Hodges and Arnold 1995). Similarly, bilaterally symmetrical flowers (more specialized) are more species-rich than corresponding sister lineages with radially symmetrical flowers (less specialized; Sargent 2004; Kay and Sargent 2009; Vamosi and Vamosi 2010). We might similarly expect specialized rewards to promote diversification, and indeed this argument has been made, although apparently never tested, for euglossine-pollinated orchids. This hypothesis has been tested explicitly in yuccas (brood-place reward). A very thorough study by Smith et al. (2008) indicates that the specialized brood-site pollination system involving yucca moths has *not* increased diversification rates over those of sister lineages in the Agavaceae. A similar pattern seems to hold in Euphorbiaceae, wherein probable sister genera *Tragia* (generalist pollen reward) and *Dalechampia* (specialized resin reward) comprise similar numbers of species.

3.5 Network characteristics: structure, redundancy, and resilience

Specialized pollinator rewards have important effects on the structure of interaction webs. As alluded to above, brood-site mutualisms that lack copollinators create tight, even one-to-one, relationships between plant and pollinator. This generates clear mutual dependencies and sensitivity to local extinction of one partner, e.g. secondary extinction. Other specialized rewards and even other brood-site rewards do not necessarily create such strong specialization and mutual dependency. Instead these relationships may resemble, at least superficially, more generalized pollination systems that abound in both temperate (Waser et al. 1996) and many tropical regions (cf. Johnson and Steiner 2000; Armbruster 2006).

As an example, I will consider in detail euglossine bees as a focal interactor. These bees are among the most important, long-distance pollinators of trees, orchids, and other plants in neotropical forests (Dressler 1972; Roubik 1989). A single species will visit several to many species of flowers for nectar (males and females), some additional species for pollen (females only), other species for fragrances (and/or non-flower sources; males only), and yet other flowers for resin (and/or non-flower sources; females only). This leads to a network structure that looks highly connected and generalized (Fig 3.4).

However, the various network connections are not equivalent. Circles connected to the same box (Fig 3.4) may be redundant as is normally expected with multiple

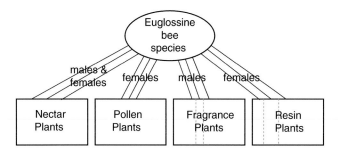

Fig 3.4 Simplified network diagram showing links between euglossine bees and the four major floral resource groups they must access.

connectors. Circle connections to different boxes are not redundant or parallel, but instead can be viewed as a set of serial connectors. The bee species must connect to each of the four boxes at least once. Conceivably, one plant species could be in more than one box although, except for nectar and pollen sources, this is probably very rare (but see Cappellari et al. 2009). This is because all bees must supply their energy needs with nectar, all males must collect floral (and other) fragrances in order to engage females for mating, all females must collect resin in order to build nests, and all females must provide pollen for their larvae. In reality, the boxes may need to be further subdivided (dotted lines, Fig 3.4); females may need to collect several types of resin; male euglossines need to collect a variety of compounds, not all of which come from a single source. Thus, there may be many more obligatory links in this web, the absence of which would cause extinction (or at least evolution) of the bee.

More generally, animal reliance on specialized rewards reduces the apparent redundancy of plant–pollinator links, relative to the impression we would get from a simple network analysis. This pattern extends beyond euglossines. Megachilid bees that use resin depend on nectar, pollen, and resin sources for survival and reproduction. Similar multiple dependencies and specialization can be seen in oil-collecting apid bees: they need to visit nectar plants for their own food (females and males), pollen plants for larval food provisions (females), and oil plants also for larval food provisions (females).

The dependency of animals on one or several special floral rewards is further accentuated by the fact that special rewards also increase "within-box" specialization. That is, for a given pollinator, there are usually many potential nectar and pollen hosts (except for oligolectic bees; see Michener 2007), but usually only one or a few species of plants provide special rewards locally. For oil bees, there are commonly only a few oil flowers in the local habitat (but see Bezerra et al. 2009). For fragrance-collecting bees, there may be few or many fragrance sources, depending on the specific fragrance needed, which varies with species and with individual history. Resin-collecting bees often have only one or two floral resin sources in the local habitat, although non-floral sources are important for some bees.

This increase in specialization and dependence on additional types of resources is expected to increase the sensitivity of the system to disturbance, such as loss of certain members of the network (Memmott et al. 2004; Pemberton 2010). Indeed, a recent review indicated that specialists are generally more at risk of extinction than generalists, at least as based on ecological and paleoecological studies (Colles et al. 2009). Regarding pollinators, Biesmijer et al. (2006) found that more specialized bees in the UK and Netherlands were suffering greater declines in abundance than were more generalized bees. Curiously, however, one study of an oil-flower/oil-bee subweb suggested the opposite; oil bees and flowers were actually more resilient than the average full pollination web (Bezerra et al. 2009). These conflicting results on resilience underscore the critical need for additional research on the effect of specialization on sensitivity to disturbance.

The cryptic specialization associated with special rewards and the insects dependent on them means that apparently redundant links in interaction webs are requisite, not redundant, as noted above. This observation underscores the importance of estimating interaction webs in ways that capture qualitative structure as well as quantitative linkages. Earlier breakthroughs in estimating interaction and food webs incorporated calculation and depiction of abundance of participating species and the numerical strengths of those interactions (Memmott et al. 1994; Memmott and Waser 2002; Memmott et al. 2004; Bascompte and Jordano 2007). However, often missing from these analyses are the natural history details of each link. We see from an examination of specialized rewards in pollination that these details can sometimes be very important, making substantial differences in our interpretation of specialization, redundancy, and ecosystem resilience.

Acknowledgements

This research was supported by grant no. DEB-9020265, DEB-9318640, DEB-9596019, DEB-0444745 from the US National Science Foundation, a grant from the National Geographic Society, and several grants from the Norwegian Research council. The ideas incorporated in this manuscript benefited from contributions from, and discussions with, numerous postgraduates and postdoctorates. I also thank an anonymous reviewer for comments on the manuscript.

References

Appanah S. (1990). Plant–pollinator interactions in Malaysian rain forests. In *Reproductive Ecology of Tropical Forest Plants,* ed. K. S. Bawa and M. Hadley. Paris, France: Unesco, pp. 85–102.

Armbruster, W. S. (1984). The role of resin in angiosperm pollination: ecological and chemical considerations. *American Journal of Botany*, **71**, 1149–60.

Armbruster, W. S. (1988). Multilevel comparative analysis of morphology, function, and evolution of *Dalechampia* blossoms. *Ecology*, **69**, 1746–61.

Armbruster, W. S. (1993). Evolution of plant pollination systems: hypotheses and tests with the neotropical vine *Dalechampia*. *Evolution*, **47**, 1480–1505.

Armbruster, W. S. (2006). *Evolutionary and ecological perspectives on specialization: from the arctic to the tropics*. In *Plant–Pollinator Interactions: From Specialization to Generalization*, ed. N. Waser and J. Ollerton. Chicago, MI: University of Chicago Press, pp. 260–82.

Armbruster, W. S. and Mziray, W. R. (1987). Pollination and herbivore ecology of an African *Dalechampia* (Euphorbiaceae): comparisons with New World species. *Biotropica*, **19**, 64–73.

Armbruster, W. S. and Steiner, K. E. (1992). Pollination ecology of four *Dalechampia* species (Euphorbiaceae) in northern Natal, South Africa. *American Journal of Botany*, **79**, 306–13.

Armbruster, W. S., Fenster, C. B. and Dudash, M. R. (2000). Pollination "principles" revisited: Specialization, pollination syndromes, and the evolution of flowers. *Det Norske Videnskapsacademi I. Matematisk Naturvidenskapelige Klasse Skrifter, Ny Serie*, **39**, 139–48.

Armbruster, W. S., Gong, Y. B. and Huang, S. Q. (2011). Are pollination "syndromes" predictive? Asian *Dalechampia* fit neotropical models. *American Naturalist*, **178**, 135–43.

Armbruster, W. S., Herzig, A. L. and Clausen, T. P. (1992). Pollination of two sympatric species of *Dalechampia* (Euphorbiaceae) in Suriname by male euglossine bees. *American Journal of Botany*, **79**, 1374–81.

Armbruster, W. S., Lee, J. and Baldwin, B. G. (2009). Macroevolutionary patterns of defense and pollination in *Dalechampia* vines: adaptation, exaptation, and evolutionary novelty. *Proceedings of the National Academy of Sciences USA*, **106**, 18085–90.

Baker, H. G. and Baker, I. (1982). *Chemical constituents of nectar in relation to pollination mechanisms and phylogeny*. In *Biochemical Aspects of Evolutionary Biology*, ed. M. H. Nitecki. Chicago, MI: University of Chicago Press, pp. 131–71.

Bascompte, J. and Jordano, P. (2007). Plant–animal mutualistic networks: The architecture of biodiversity. *Annual Review of Ecology Evolution and Systematics*, **38**, 567–93.

Bezerra, E. L. S., Machado, I. C. and Mello, M. A. R. (2009). Pollination networks of oil-flowers: a tiny world within the smallest of all worlds. *Journal of Animal Ecology*, **78**, 1096–101.

Biesmeijer, J.-C., Roberts, S. P. M., Reemer, M., Ohlemüller, R., Edwards, M., Peeters, T., Schaffers, A. P., Potts, S. G., Kleukers, R., Thomas, C. D., Settele, J. and Kunin, W. E. (2006). Parallel declines in pollinators and insect-pollinated plants in Britain and the Netherlands. *Science*, **313**, 351–54.

Bittrich V. and Amaral M. C. E. (1997). Floral biology of some Clusia species from Central Amazonia. *Kew Bulletin*, 52, 617–35.

Bogler, D. J., Neff, J. L. and Simpson, B. B. (1995). Multiple origins of the yucca-yucca moth association. *Proceedings of the National Academy of Sciences USA*, **92**, 6864–67.

Buchmann, S. L. (1987). The ecology of oil flowers and their bees. *Annual Review of Ecology and Systematics*, **18**, 343–69.

Cardinal, S., Straka, J. and Danforth, B. N. (2010). Comprehensive phylogeny of apid bees reveals the evolutionary origins and antiquity of cleptoparasitism. *Proceedings of the National Academy of Sciences USA*, **107**: 16207–211.

Cammerloher, H. (1931). *Blütenbiologie* I. Berlin: Gerbriider Borntraeger.

Cappellari, S. C., Harter-Marques, B., Aumeier, P. and Engels, W. (2009). *Mecardonia tenella* (Plantaginaceae) attracts oil-, perfume-, and pollen-gathering bees in southern Brazil. *Biotropica*, **41**, 721–29.

Chase, M. W. and Hills, H. G. (1992). Orchid phylogeny, flower sexuality, and fragrance-seeking. *Bioscience* **42**, 43–49.

Cintra, P., Malaspina, O. and Bueno, O. C. (2003). Toxicity of barbatimão to *Apis mellifera* and *Scaptotrigona postica*, under laboratory conditions. *Journal of Apicultural Research*, **42**, 9–12.

Cocucci, A. A. and Vogel, S. (2001). Oil-producing flowers of *Sisyrinchium* species (Iridaceae) and their pollinators in southern South America. *Flora*, **196**, 26–46.

Colles, A., Liow, L. H. and Prinzing, A. (2009). Are specialists at risk under environmental change? Neoecological, paleoecological and phylogenetic approaches. *Ecology Letters*, **8**, 849–63.

Cox, P. A. (1982). Vertebrate pollination and the maintenance of dioecism in *Freycinetia*. *American Naturalist*, **120**, 65–80.

Darwin, C. (1874). Letter to J. D. Hooker, April 7, 1874. In *A calendar of the correspondence of Charles Darwin, 1821-1882*, ed. F. Burkhardt and S. Smith. Cambridge, UK: The Press Syndicate of the University of Cambridge.

de Carvalho, A. C. P. and Message, D. (2004). A scientific note on the toxic pollen of *Stryphnodendron polyphyllum* (Fabaceae, Mimosoideae), which causes sacbrood-like symtoms. *Apidologie*, **35**, 89–90.

Dodson, C. H. (1966). Ethology of some bees of the tribe Euglossini. *Journal of the Kansas Entomological Society*, **39**, 607–29.

Dressler, R. L. (1972). Biology of the orchid bees (Euglossini). *Annual Review of Ecology and Systematics*, **13**, 373–94.

Dufay, M. (2010). Impact of plant flowering phenology on the cost/benefit balance in a nursery pollination mutualism, with honest males and cheating females. *Journal of Evolutionary Biology*, **23**, 977–86.

Dufay, M. and Anstett, M.-C. (2003). Conflicts between plants and pollinators that reproduce within inflorescences: evolutionary variations on a theme. *Oikos*, **100**, 3–14.

Eltz, T., Whitten, W. M., Roubik, D. W. and Linsenmair, K. E. (1999). Fragrance collection, storage, and accumulation by individual male orchid bees. *Journal of Chemical Ecology*, **25**, 157–76.

Faegri, K. and van der Pijl, L. (1971). *The Principles of Pollination Ecology*, 2nd edn. Oxford, UK: Pergamon.

Feil, J. P. (1992). Reproductive ecology of dioecious *Siparuna* (Monimiaceae) in Ecuador – a case of gall midge pollination. *Botanical Journal of the Linnean Society*, **110**, 171–203.

Fenster, C. B., Armbruster, W. S., Wilson, P., Dudash, M. R. and Thomson, J. T. (2004). Pollination syndromes and floral specialization. *Annual Review of Ecology and Systematics*, **35**, 375–403.

Flach, A., Dondon, R. C., Singer, R. B., Koehler, S., Amaral, M. D. E. and Marsaioli, A. J. (2004). The chemistry of pollination in selected Brazilian maxillariinae orchids: Floral rewards and fragrance. *Journal of Chemical Ecology*, **30**, 1045–56.

Garcia, M. V. B., deOliveira, M. L. and Campos, L. A. D. (1992). Use of seeds of *Coussapoa asperifolia magnifolia* (Cecropiaceae) by stingless bees in the central Amazonian forest (Hymenoptera, Apidae, Meliponinae). *Entomologia Generalis*, **17**, 255–8.

Goto, R., Okamoto, T., Kiers, E. T., Kawakita, A. and Kato, M. (2010). Selective flower abortion maintains moth cooperation in a newly discovered pollination mutualism. *Ecology Letters*, **13**, 321–9.

Gould, S. J. and Vrba, E. S. (1982). Exaptation – a missing term in the science of form. *Paleobiology*, **8**, 4–15.

Gustafsson, M. H. G. and Bittrich, V. (2002). Evolution of morphological diversity and resin secretion in flowers of *Clusia* (Clusiaceae): insights from ITS sequence variation. *Nordic Journal of Botany*, **22**, 183–203.

Hargreaves, A. L., Harder, L. D. and Johnson, S. D. (2009). Consumptive emasculation: the ecological and evolutionary consequences of pollen theft. *Biological Reviews*, **84**, 259–76.

Hodges, S. A. and Arnold, M. L. (1995). Spurring plant diversification: are floral nectar spurs a key innovation? *Proceedings of the Royal Society. B*, **262**, 343–8.

Holland, J. N. and Fleming, T. H. (1999). Mutualistic interactions between *Upiga virescens* (Pyralidae), a pollinating seed-consumer, and *Lophocereus schottii* (Cactaceae). *Ecology*, **80**, 2074–84.

Ishida, C., Kono, M. and Sakai, S. (2009). A new pollination system: brood-site pollination by flower bugs in *Macaranga* (Euphorbiaceae). *Annals of Botany*, **103**, 39–44.

Irvine, A. K. and Armstrong, J. E. (1990). *Beetle pollination in tropical forests of Australia*. In *Reproductive ecology of tropical forest plants*, ed. K. S. Bawa and M. Hadley. Paris, France: Unesco, pp. 135–50.

Johnson, S. D. and Steiner, K. E. (2000) Generalization versus specialization in plant pollination systems. *Trends in Ecology & Evolution*, **15**, 140–3.

Kato, M., Takimura, A. and Kawakita, A. (2003). An obligate pollination mutualism and reciprocal diversification in the tree genus *Glochidion* (Euphorbiaceae). *Proceedings of the National Academy of Sciences USA*, **100**, 5264–67.

Kawakita, A. (2010). Evolution of obligate pollination mutualism in the tribe Phyllantheae (Phyllanthaceae). *Plant Species Biology*, **25**, 3–19.

Kay, K. M. and Sargent, R. D. (2009). The role of animal pollination in plant speciation: Integrating ecology, geography, and genetics. *Annual Review of Ecology Evolution and Systematics*, **40**, 637–56.

Kerner, A. (1898). *Pflanzenleben II*, 2nd edn. Leipzig, Wien: Bibliographisches Institut.

Krombein, V. (1967). *Trap-Nesting Wasps and Bees*. Washington, DC: Smithsonian Press.

Langenheim, J. H. (2003). *PlantRresins. Chemistry, Evolution,Eecology,*

Ethnobotany. Portland, OR: Timber Press.

Lehmberg, L., Dworschak, K. and Bluthgen, N. (2008). Defensive behavior and chemical deterrence against ants in the stingless bee genus *Trigona* (Apidae, Meliponini). *Journal of Apicultural Research*, **47**, 17–21.

Listabarth, C. (1993). Pollination in *Geonoma macrostachys* and 3 congeners, *G. acaulis*, *G. gracilis*, and *G. interrupta*. *Botanica Acta*, **106**, 496–506.

Lokvam, J. and Braddock, J. F. (1999). Anti-bacterial function in the sexually dimorphic pollinator rewards of *Clusia grandiflora* (Clusiaceae). *Oecologia*, **119**, 534–40.

Maddison, W. P. and Maddison, D. R. (2009). *Mesquite. A modular system for evolutionary analysis*. http://mequiteproject.org

Memmott, J. and Waser, N. M. (2002). Integration of alien plants into a native flower–pollinator visitation web. *Proceedings of the Royal Society, B*, **269**, 2395–99.

Memmott, J., Godfray, H. C. J. and Gauld, I. D. (1994). The structure of a tropical host parasitoid community. *Journal of Animal Ecology*, **63**, 521–40.

Memmott, J., Waser, N. M. and Price, M. V. (2004). Tolerance of pollination networks to species extinctions. *Proceedings of the Royal Society, B*, **271**, 2605–11.

Messer, A. C. (1985). Fresh dipterocarp resins gathered by megachilid bees inhibit growth of pollen-associated fungi. *Biotropica*, **17**, 175–6.

Michener, C. D. (2007). *The Bees of the World*, 2nd edn. Baltimore, MD: Johns Hopkins University Press.

Michez, D. and Patiny, S. (2005). World revision of the oil-collecting bee genus *Macropis* Panzer 1809 (Hymenoptera, Apoidea, Melittidae) with a description of a new species from Laos. *Annales de la Société entomologique de France (n. s.)*, **41**, 15–28.

Michez, D., Patiny, S. and Danforth, B. N. (2009). Phylogeny of the bee family Melittidae (Hymenoptera: Anthophila) based on combined molecular and morphological data. *Systematic Entomology*, **34**, 574–97.

Michez, D., Patiny, S., Rasmont, P., Timmermann, K. and Vereecken, N. J. (2008). Phylogeny and host-plant evolution in Melittidae s.l. (Hymenoptera: Apoidea). *Apidologie*, **39**, 146–62.

Moog U., Fiala B., Federle W. and Maschwitz U. (2002). Thrips pollination of the dioecious ant plant *Macaranga hullettii* (Euphorbiaceae) in Southeast Asia. *American Journal of Botany*, **89**, 50–9.

Müller, A. and Kuhlmann, M. (2008). Pollen hosts of western palaearctic bees of the genus *Colletes* (Hymenoptera: Colletidae): the Asteraceae paradox. *Biological Journal of the Linnean Society*, **95**, 719–33.

Müller, H. (1879). Die Wechselbeziehungen zwischen den Blumen und den ihre Kreuzung vermittelnden Insekten. In *Handbuch der Botanik I*, ed. A. Schenck. Breslau: Eduard Trewent, pp. 1–112.

Neff, J. L. and Simpson, B. S. (1981). Oil-collecting structures in the Anthophoridae (Hymenoptera): Morphology, function, and use in systematics. *Journal of the Kansas Entomological Society*, **54**, 95–123.

Nogueira, P. C. D., Marsaioli, A. J., Amaral, M. D. E. and Bittrich, V. (1998). The fragrant floral oils of *Tovomita* species. *Phytochemistry*, **49**, 1009–12.

Nunez, C. V., de Oliveira, M. L., Lima, R.D., Diaz, I. E. C., Sargentini, E., Pereira, O. L.

and Araujo, L. M. (2008). Chemical analyses confirm a rare case of seed dispersal by bees. *Apidologie*, **39**, 618–26.

Ollerton, J. (1996). Reconciling ecological processes with phylogenetic patterns: The apparent paradox of plant-pollinator systems. *Journal of Ecology*, **84**, 767–69.

Pauw A. (2006). Floral syndromes accurately predict pollination by a specialized oil-collecting bee (*Rediviva peringueyi*, Melittidae) in a guild of South African orchids (Coryciinae). *American Journal of Botany*, **93**, 917–26.

Pellmyr, O. (1989). The cost of mutualism: interactions between *Trollius europaeus* and its pollinating parasites. *Oecologia*, **78**, 53–9.

Pellmyr, O. (1992). The phylogeny of a mutualism: evolution and coadaptation between *Trollius* and its seed-parasitic pollinators. *Biological Journal of the Linnean Society*, **47**, 337–65.

Pellmyr, O. (1997). Pollinating seed eaters: Why is active pollination so rare? *Ecology*, **78**, 1655–60.

Pellmyr, O. (2003). Yuccas, yucca moths, and coevolution: A review. *Annals of the Missouri Botanical Garden*, **90**, 35–55.

Pellmyr, O. and Huth, C. J. (1994). Evolutionary stability of mutualism between yuccas and yucca moths. *Nature*, **372**, 257–60.

Pellmyr, O., Thompson, J. N., Brown, J. and Harrison, R. G. (1996). Evolution of pollination and mutualism in the yucca moth lineage. *American Naturalist*, **148**, 827–47.

Pemberton, R.W. (2010). Biotic resource needs of specialist orchid pollinators. *Botanical Review* **76**, 275–92.

Pettersson, M. W. (1991). Pollination by a guild of fluctuating moth populations:

option for unspecialization in *Silene vulgaris*. *Journal of Ecology*, **79**, 591–604.

Porsch, O. (1905). Beitrage zur "histologischen Bliutenbiologie". I. Über zwei neue Insektenanlockungsmittel der Orchideen-blüte. *Österreichische Botanische Zeitschrift*, **55**, 165–73, 227–35, 253–60.

Porto, A. L. M., Machado, S. M. F., de Oliveira, C. M. A., Bittrich, V., Amaral, M. C. E, and Marsaioli, A.J. (2000). Polyisoprenylated benzophenones from *Clusia* floral resins. *Phytochemistry*, **55**, 755–68.

Praz, C. J., Müller, A. and Dorn, S. (2008). Specialized bees fail to develop on non-host pollen: do plants chemically protect their pollen? *Ecology* **89**, 795–804.

Ramírez, S., Dressler, R. L. and Ospina, M. (2002). Abejas euglosinas (Hymenoptera: Apidae) de la Región Neotropical: Listado de especies con notas sobre su biología. *Biota Colombiana*, **3**, 7–118.

Ramírez, S., Roubik, D. W., Skov, C. and Pierce, N. E. (2010). Phylogeny, diversification patterns and historical biogeography of euglossine orchid bees (Hymenoptera: Apidae). *Biological Journal of the Linnean Society*, **100**, 552–72.

Renner, S. S. and Schaefer, H. (2010). The evolution and loss of oil-offering flowers: new insights from dated phylogenies for angiosperms and bees. *Philosophical Transactions of the Royal Society B-Biological Sciences*, **365**, 423–35.

Riley, C. V. (1872). The fertilization of the yucca plant by *Pronuba yuccasella*. *Canadian Entomologist*, **4**, 182.

Rønsted, N., Weiblen, G. D., Cook, J. M., Salamin, N., Machado, C. A. and

Savolainen, V. (2005). 60 million years of codivergence in the fig-wasp symbiosis. *Proceedings of the Royal Society London B*, **272**, 2593–99.

Roubik, D. W. (1989). *Ecology and Natural History of Tropical Bees*. Cambridge, UK: Cambridge University Press.

Sargent, R. D. (2004). Floral symmetry affects speciation rates in angiosperms. *Proceedings of the Royal Society London B*, **271**, 603–8.

Schaefer, H. and Renner, S. (2008). A phylogeny of the oil bee tribe Ctenoplectrini (Hymenoptera: Anthophila) based on mitochondrial and nuclear data: Evidence for Early Eocene divergence and repeated out-of-Africa dispersal. *Molecular Phylogenetics and Evolution*, **47**, 799–811.

Schwarz, F. (1948). Stingless bees (Meliponidae) of the Western Hemisphere. *Bulletin of the American Museum of Natural History*, **90**, 1–546.

Simpson, B. and Neff, J. L. (1981). Floral rewards: alternatives to pollen and nectar. *Annals of the Missouri Botanical Garden*, **68**, 301–22.

Singer, R. B. and Koehler, S. (2004). Pollinarium morphology and floral rewards in Brazilian Maxillariinae (Orchidaceae). *Annals of Botany*, **93**, 39–51.

Skutch, A. F. (1971). *A Naturalist in Costa Rica*. Gainesville, FL: University of Florida Press.

Smith, C. I., Pellmyr, O., Althoff, D. M., Balcazar-Lara, M., Leebens-Mack, J. and Segraves, K. A. (2008). Pattern and timing of diversification in *Yucca* (Agavaceae): specialized pollination does not escalate rates of diversification. *Proceedings of the Royal Society B*, **275**, 249–58.

Steiner K. E. (1982). *Mammal Pollination of Neotropical Euphorbiaceae*. PhD dissertation, University of California, Davis.

Steiner, K. E. and Cruz, C. B. (2006). The evolution of oil collection and oil-collecting structures in the Melittidae. *Abstracts International Union for the Study of Social Insects, 31 July 2006, Washington, DC*. http://www.ulb. ac.be/ sciences/ecoevol/docs/Michez_ IUSSI_06.pdf

Steiner, K. E. and Whitehead, V. B. (1990). Pollinator adaptation to oil-secreting flowers – *Rediviva* and *Diascia*. *Evolution*, **44**, 1701–7.

Steiner, K. E. and Whitehead, V. B. (1991a). Oil flowers and oil bees – further evidence for pollinator adaptation. *Evolution*, **45**, 1493–1501.

Steiner, K. E. and Whitehead, V. B. (1991b). Resin collection and pollination of *Dalechampia capensis* (Euphorbiaceae) by *Pachyanthidium cordatum* (Hymenoptera: Megachilidae) in South Africa. *Journal of the Entomological Society of South Africa*, **54**, 67–72.

Stephen, W. P., Bohart, G. E. and Torchio, P. F. (1969). *The Biology and External Morphology of Bees*. Corvallis Oregon, Agricultural Experimental Station, Oregon State University.

Svensson, G. P., Okamoto, T., Kawakita, A., Goto, R. and Kato, M. (2010). Chemical ecology of obligate pollination mutualisms: testing the "private channel" hypothesis in the *Breynia-Epicephala* association. *New Phytologist*, **186**, 995–1004.

Thien, L. B., Bernhardt, P., Devall, M. S., Chen, Z.-D., Luo, Y.-B., Fan, J.-H., Yuan, L.-C. and Williams, J. H. (2009). Pollination biology of basal angiosperms (ANITA grade). *American Journal of Botany*, **96**, 166–82.

Thompson, J. N. and Pellmyr, O. (1992). Mutualism with pollinating seed parasites amid copollinators – constraints on specialization. *Ecology*, **73**, 1780–1791.

Vamosi, J. C. and Vamosi, S. M. (2010). Key innovations within a geographical context in flowering plants: towards resolving Darwin's abominable mystery. *Ecology Letters*, **13**, 1270–79.

Vogel, S. (1966). Parfümsammelnde Bienen als Bestauber von Orchideen und Gloxinia. *Österreichesches botanisches Zietschrift*, **113**, 302–361.

Vogel, S. (1969). Flowers offering fatty oil instead of nectar. *Abstracts of the XI International Botanical Congress*, p. 229.

Vogel, S. (1974). Ölblumen und ölsammelnde Bienen. *Tropische und Subtropische Planzenwelt*, **7**, 1–267.

Vogel, S. (1988). The oil-bee oil-flower relationships – parallelism and other aspects of their evolution in space and time. *Zeitschrift fur Zoologische Systematik und Evolutionsforschung*, **26**, 341–62.

Vogel, S. (2009). The non-African oil-flowers and their bees: A brief survey. *South African Journal of Botany*, **75**, 389–90.

Wallace, H. M. and Lee, D. J. (2010). Resin-foraging by colonies of *Trigona sapiens* and *T. hockingsi* (Hymenoptera: Apidae, Meliponini) and consequent seed dispersal of *Corymbia torelliana* (Myrtaceae). *Apidologie*, **41**, 428–35.

Wallace, H. M. and Trueman, S. J. (1995). Dispersal of *Eucalyptus torelliana*

seeds by the resin-collecting stingless bee, *Trigona carbonaria*. *Oecologia*, **104**, 12–6.

Wallace, H. M., Howell, M. G. and Lee, D. J. (2008). Standard yet unusual mechanisms of long-distance dispersal: seed dispersal of *Corymbia torelliana* by bees. *Diversity and Distributions*, **14**, 87–94.

Waser, N. M., Chittka, L., Price, M. V., Williams, N. M. and Ollerton, J. (1996). Generalization in pollination systems, and why it matters. *Ecology*, **77**, 1043–60.

Wiebes, J. T. (1979). Coevolution of figs and their insect pollinators. *Annual Review of Ecology and Systematics*, **10**, 1–10.

Whitten, W. M., Blanco, M. A., Williams, N. H., Koehler, S., Carnevali, G., Singer, R. B., Endara, L. and Neubig, K. M. (2007). Molecular phylogenetics of *Maxillaria* and related genera (Orchidaceae : Cymbidieae) based on combined molecular data sets. *American Journal of Botany*, **94**, 1860–89.

Zimmermann, Y., Ramirez, S. R. and Eltz, T. (2009a). Chemical niche differentiation among sympatric species of orchid bees. *Ecology*, **90**, 2994–3008.

Zimmermann, Y., Roubik, D. W., Quezada-Euan, J. J. G., Paxton, R. J. and Eltz, T. (2009b). Single mating in orchid bees (*Euglossa*, Apinae): implications for mate choice and social evolution. *Insectes Sociaux*, **56**, 241–249.

4

Fig–fig wasp mutualism: the fall of the strict cospeciation paradigm?

Astrid Cruaud, James Cook, Yang Da-Rong,
Gwenaëlle Genson, Roula Jabbour-Zahab, Finn Kjellberg,
Rodrigo Augusto Santinelo Pereira, Nina Rønsted,
Otilene Santos-Mattos, Vincent Savolainen,
Rosichon Ubaidillah, Simon van Noort, Peng Yan-Qiong
and Jean-Yves Rasplus

4.1 Introduction

At least three classic systems of nursery pollination mutualism are known: the fig (*Ficus*, Moraceae) – agaonid (Hymenoptera, Chalcidoidea) association (Cook and Rasplus 2003), the yucca (Yucca, Hesperoyucca; Agavaceae) – yucca moths (Tegeticula, Parategeticula; Lepidoptera, Prodoxidae) association (Pellmyr 2003) and the Glodichion (Phyllanthaceae) – Epicephala moths (Lepidoptera, Gracillariidae) association (Kato et al. 2003). All these mutualisms are obligate, which means that each partner depends on the other for its own reproductive success. The insect pollinates the flowers and oviposits in the plant ovaries where the insect larvae subsequently feed on a subset of the developing seeds. A shift from mutualism to parasitism by the pollinating insect would lead to reproduction failure of the plant and, without host shift, to the extinction of both lineages. Therefore, the speciation of mutualistic pollinators is generally believed to be driven by the speciation of their host-plants. In this hypothesis, when an ancestral plant species splits into two daughter species, its mutualistic pollinator also splits. This scenario

Evolution of Plant–Pollinator Relationships, ed S. Patiny. Published by Cambridge University Press. © The Systematics Association 2012.

should result in perfect congruence of hosts and pollinator phylogenies (Farenholz's rule) (Farenholz 1913). However, this seems increasingly unlikely. Indeed, more and more studies on different coevolved associations show that a strict Farenholz's rule is not respected, even when a high level of host specificity exists (e.g. Paterson and Banks 2001; Desdevises et al. 2002; Charleston and Perkins 2006).

Topological incongruence between host and associate phylogenetic trees can result from processes like host switching, sorting events (extinction and lineage sorting), duplication events (speciation of the parasite independent of the host), and failure of the associate to diverge when the host diverges ("missing the boat") (Page 1991; Page 1994; Page and Charleston 1998; Legendre et al. 2002; Charleston and Perkins 2006).

To be validated, a strict cospeciation hypothesis requires that (i) the tree topologies are congruent, and (ii) the timing of speciation in both lineages, inferred from these trees, is synchronous (a correlation may only imply phylogenetic tracking) (Lopez-Vaamonde et al. 2001; Percy et al. 2004; Light and Hafner 2008; Jousselin et al. 2009). Consequently, rigorous cophylogenetic analyses require: (i) exhaustive sampling of hosts and associates (extant and when possible extinct), (ii) reliable and fully resolved phylogenetic hypotheses, and iii) accurate cospeciation and dating analyses. Assessing cospeciation between lineages is consequently a difficult task and results must always be taken with caution.

In the case of nursery pollination mutualisms, few studies have investigated the level of cocladogenesis between coarse phylogenies of the mutualistic partners. While no cocladogenesis study has been published on the *Yucca* mutualism (Smith et al. 2008), the only study so far of the *Glodichion* and *Epicephala* (Kawakita et al. 2004) suggest that both cospeciation and host shift have played an important role.

4.2 Study system: the figs and the fig wasps

Agaonid fig wasps (Hymenoptera, Chalcidoidea) are associated strictly with *Ficus* (Moraceae) (see Weiblen 2002 and Cook and Rasplus 2003 for a review on the biology of the mutualism and Table 4.1 for details concerning *Ficus* classification and worldwide distribution). With the exception of two known cases of "cuckoo" agaonid wasps (Compton et al. 1991; Peng et al. 2008), all species are effective pollinators of figs. Most agaonid fig wasps are only associated with one fig species (the "one-to-one rule") and there are only a few exceptions (Rasplus 1996; Cook and Segar 2010). When the same agaonid species pollinates more than one fig species, these *Ficus* are closely related and belong to the same *Ficus* section. In contrast, *Ficus* species are pollinated by one to four wasp species (Michaloud et al. 1985; Kerdelhué et al. 1997; Lopez-Vaamonde et al. 2002; Molbo et al. 2003; Haine et al. 2006; Su et al. 2008). The number of cases where one fig species has multiple

Table 4.1 Classification and worldwide distribution of *Ficus* genus following (Berg and Corner, 2005). The genus is subdivided in six subgenera. Sections grouping monoecious (resp. dioecious) figs are tagged with an "M" (resp. "D"). Agaonid genera that pollinate each subgenus are indicated in the last column. 1274 species of birds and mammals belonging to 92 families eat figs making *Ficus* a keystone resource for tropical ecosystems (Shanahan et al., 2001).

Fig classification of Berg (2005)

Subgenera	Sections / Subsections	Distribution	Associated pollinators (genera)
Pharmacosycea (82 sp)	*Pharmacosycea / Bergianae* **M** *Petenses* **M**	Neotropics	*Tetrapus*
	Oreosycea / Glandulosae **M** *Pedunculatae* **M**	Afrotropics, Oriental, Australasia	*Dolichoris*
Ficus (61 sp)	*Ficus* **D**	Afrotropics, Oriental, Australasia	*Blastophaga*
	Eriosycea **D**		*Valisia*
Synoecia (74 sp)	*Kissosycea* **D**	Oriental, Australasia	*Wiebesia*
	Rhizocladus **D**		
Sycidium (109 sp)	*Sycidium* **D**	Afrotropics, Oriental, Australasia	*Kradibia*
	Palaeomorphe **D**		
Sycomorus (142 sp)	*Sycomorus / Sycomorus* **M** *Neomorphe* **D**	Afrotropics, Oriental, Australasia	*Ceratosolen*
	Sycocarpus / Sycocarpus **D** *Macrostyla* **D**		
	Adenosperma **D**		
	Hemicardia **D**		
	Papuasyce **M** + **D**		
	Bosscheria **D**		
	Dammaropsis **D**		
Urostigma (288 sp)	*Americana* **M**	Neotropics	*Pegoscapus*
	Galoglychia **M**	Afrotropics	*Agaon, Alfonsiella, Allotriozoon, Courtella, Elisabethiella, Nigeriella, Paragaon*
	Stilpnophyllum / Stilpnophyllum **M** *Malvanthera* **M**	Australasia	*Pleistodontes*
	Urostigma / Urostigma **M** *Conosycea* **M**	Afrotropics, Oriental, Australasia	*Urostigma : Platyscapa Conosycea: Eupristina, Deilagaon, Waterstoniella*

Agaonidae pollinators is increasing with study effort (Cook and Segar 2010). To our knowledge at least 50 species of fig are pollinated by more than one pollinator and ca. 340 species of fig are known to be pollinated by one exclusive pollinator (Cruaud and Rasplus unpublished data).

This long-lived, one-to-one rule (where most sections are pollinated by different pollinator genera that are attracted by specific volatile compounds and manage to enter the figs thanks to numerous specific morphological adaptations (Hossaert-McKey et al. 1994; Borges et al. 2008), and the observation that closely related sections were supposed to be pollinated by closely related agaonids (Ramirez 1974; Wiebes 1982; Ramirez 1991) has led to a paradigm of strict cospeciation between *Ficus* and their pollinators at a coarse systematic scale (sections of figs and genera of wasps). However, cophylogenetic studies between figs and their pollinators are still scant and conclusions have varied depending on the taxonomic level analyzed (Cook and Segar 2010).

At a coarse taxonomic scale, recent fig phylogenies based on molecular data (fig 4.1 and Table 4.2) are roughly paralleled by most fig wasp phylogenies (fig 4.2 and Table 4.3), therefore, strict cospeciation between both partners during the last 60 million years (Rønsted et al., 2005) is generally accepted by researchers. However, at a coarse taxonomic scale, the cocladogenesis hypothesis has been formally tested only once on a limited dataset (12 species of figs and their agaonid pollinators) using tree reconciliation analyses with subsequent randomization tests (Jackson 2004). The results support roughly the current consensus that pollinators show significant cospeciation with their hosts, but switches and losses are still required to reconcile fig and pollinator phylogenetic trees. Moreover, dating analyses are lacking and cospeciation cannot be demonstrated.

At a fine taxonomic scale, more studies have explored coevolution/cospeciation between figs and their pollinating wasps (Weiblen and Bush 2002; Silvieus et al. 2007; Jackson et al. 2008; Jousselin et al. 2008). There is evidence of cospeciation between *Ceratosolen* and Papuan figs of the subgenus *Sycomorus* (Weiblen and Bush 2002; Silvieus et al. 2007). However, Afrotropical fig wasps (several genera of wasps, Jousselin et al. 2008) and *Pegoscapus* (Machado et al. 2005; Marussich and Machado 2007; Jackson et al. 2008) often do not cospeciate with their monoecious figs (section *Galoglychia* and *Americana*, respectively). Indeed, Jousselin et al. (2008), using tree-reconciliation methods, show that host switching and duplication followed by asymmetrical lineage extinction may have occurred during Afrotropical pollinator diversification. Therefore, fine-scale studies from figs and wasps in different parts of the world cast considerable doubt on the generality of strict cospeciation and strongly suggest significant roles for other evolutionary processes (Cook and Segar 2010).

Finally, a clear weakness of all these cospeciation analyses is that they used a relatively low number of species that represent lineages of figs or fig wasps

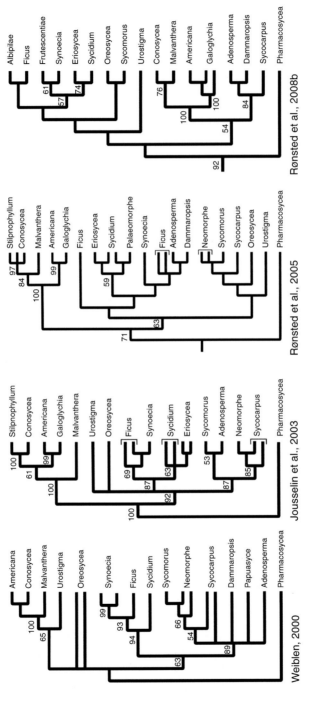

Fig 4.1 Most recent phylogenetic hypotheses of *Ficus*. See Table 4.2 for corresponding materials and methods (BP at nodes).

Table 4.2 Review of materials and methods used in the previous studies that attempted to reconstruct fig phylogeny.

Publications	Species	Subgenera	Sections	Nucleotides	Markers	Phylogenetic methods used
(Yokohama, 1994)	7	4	6	–	RFLP, chloroplasts	–
(Herre et al. 1996)	15	6	10	1800	rbcL, tRNA Leu intron, tRNA spacer	Parsimony
(Weiblen, 2000)	46	6	14	643	ITS	Parsimony
(Jousselin et al. 2003)	41	6	17	1147	ITS, ETS	Parsimony
(Rønsted et al. 2005)	146	6	19	1230	ITS, ETS	Parsimony/Bayesian
(Rønsted et al. 2008b)	102	6	18	2097	ITS, ETS, *G3pdh*	Parsimony

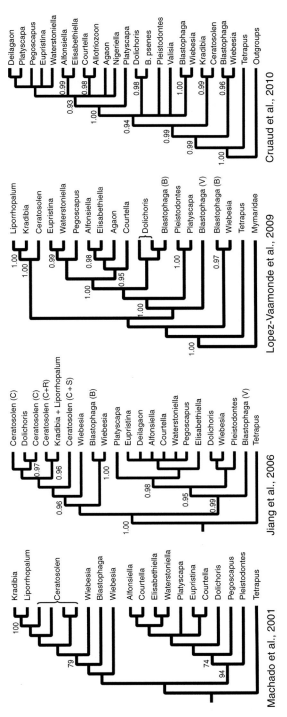

Fig 4.2 Most recent phylogenetic hypotheses of fig wasps. See Table 4.3 for corresponding materials and methods (BP/PP at nodes).

Table 4.3 Review of materials and methods used in the previous studies that attempted to reconstruct agonid phylogeny.

Publications	Individuals/species/ genera	Outgroups (genera)	Markers and nucleotides	Morphological study?	Phylogenetic methods used
(Herre et al. 1996)	12/12/6	2 : *Idarnes, Critogaster*	12S, 350 bp COII, 276 bp	no	Parsimony
(Weiblen, 2001)	43/43/10	1 : *Apocryptophagus*	COI-tRNA(Leu)-COII, 1724 bp	yes	Maximum Likelihood/ Parsimony
(Machado et al. 2001)	32/32/15	4 : *Philocaenus, Seres, Critogaster* (2)	COI, 816 bp	no	Maximum Likelihood
(Jiang et al. 2006)	101/31/15	0	COI, 816 bp	yes	Bayesian
(Lopez-Vaamonde et al. 2009)	64/64/15	1 : *Anaphes*	COI, 824 bp 28S, 1313 bp	no	Bayesian
(Cruaud et al. 2010)	101/101/19	4 : *Sycophaga, Ficomila, Megastigmus, Trichograma*	COI, 696 bp Cytb, 739 bp Wg, 481 bp 18S rRNA, 786 bp 28S rRNA, 1480 bp	no	Parsimony/Bayesian

separated by several million years of independent evolution. Furthermore, fig and fig wasp phylogenies are mostly based on one or two genes and a low number of individuals per species and/or lineages (rarely more than one) so that analyses are biased towards overestimating cospeciation. It therefore appears that a strict cospeciation hypothesis between *Ficus* and their pollinators has yet to be tested statistically.

4.3 How to test for cospeciation

Host and associate phylogenies can be assessed for similarity using event-based and global-fit methods. Event-based methods attempt to use five coevolutionary scenarios: codivergence/cospeciation, duplication, host-shifts, lineage sorting, and "failure to diverge" (Page and Charleston 1998; Charleston and Perkins 2006) to map the associate phylogeny to the host one. Any resulting incongruence in the mapping is reconciled by attribution to coevolutionary events previously mentioned. Global-fit methods use statistical methods to assess the level of congruence between host and associate phylogenies and to identify specific associations that contribute to cophylogeny.

The most common event-based method is known as tree-reconciliation and is implemented in TreeMap 1.0b (Page 1994). This method aims at finding the least costly reconstruction of host-associated relationships by maximizing the number of putative cospeciation events. The major weakness of TreeMap 1.0b is that it adds host switches a posteriori. This weakness has been adjusted in TreeMap 2.02β (Charleston and Page 2002). Moreover, TreeMap 2.02β allows users to assign different costs to the diversification events, and implements the Jungle algorithm that allows exploring all possible mappings of one tree to another (Charleston 1998). However, the complexity of the Jungle algorithm causes calculation limitations because of exponential running time. Therefore, TreeMap 2.02β is primarily useful for relatively small trees. Indeed, even on pruned topologies that included two representatives of each host section and respective pollinators, TreeMap 2.02β reached calculation limitation and crashed. Therefore, because of the large size of the host and associate phylogenies (89 taxa each), event-based analyses were performed with TreeMap 1.0b. We also used the recent software Jane (Conow et al. 2010). Jane's algorithm is fast enough to find solutions to relatively large problems in a few days. Jane's runtime grows linearly (not exponentially) with the number of solves (invocations of the dynamic programming solver). Moreover, Jane provides functionalities not found in other programs. These include the ability to (i) set time zone ranges on both the host and parasite trees, (ii) limit host switch distance, and (iii) define regions in the host tree and specify different host switch costs between each pair of regions. However, Jane does not implement randomization test.

4.4 Aims of the study

In this chapter we use an already-published large sets of sequences from figs (Rønsted et al. 2008b) and an unpublished dataset of pollinating fig wasps that was extracted from our large worldwide database of agaonids (about 350 species sequenced) to (i) reconstruct new molecular phylogenies for figs and their pollinating wasps (Agaonidae) at a coarse systematic level, (ii) assess the extent of cospeciation in this association, and (iii) propose an interpretation of the observed pattern.

4.5 New phylogenetic hypotheses for figs and fig wasps

The sampling analyzed represents all *Ficus* sections recognized by Berg and Corner (2005) and most of the genera of Agaonidae. When possible, agaonids and figs are true associates, i.e. although they may not have been collected together, the agaonid species is the pollinating wasp associated with this fig species. In a few cases, the corresponding agaonid was not available in our collection and instead we used the pollinator of a closely related species (Fig 4.4). A comprehensive list of all sampled species is given in Table 4.4. True associates (66 % of the sampling) are shaded in grey. Our final fig matrix contained 88 taxa and 2141 bp (ITS = 864 bp, ETS = 515 bp, *G3pdh* = 762 bp). Of these, 854 bp were variable and 503 bp (23 %) were parsimony informative. The final alignment of agaonid dataset contained 89 taxa and 4763 bp (COI + Cytb = 2250 bp, EF = 516 bp, Wg = 403 bp, 28S core and stems = 839 bp, 28S loops and clustal-aligned parts = 755 bp). Of these, 2627 bp were variable and 1995 bp (42 %) were parsimony informative. Alignment of exons revealed no indels. For all partitions the GTR + I + G was determined as the best-fitting model by MrAIC (see 4.8 Materials and methods).

Whatever the analytical method used, maximum likelihood (ML) or Bayseian inferences (BI), the recovered topologies are similar (Fig 4.3).

Black lines between taxa indicate *Ficus*-agaonid associations. To preserve clarity only some of these associations are given (see Table 4.4 for details). Node supports are mentioned above branches (ML bootstrap support > 70 % / Bayesian posterior probabilities > 0.95).

Our *Ficus* phylogeny (Fig 4.3) appears congruent with the main results of Rønsted et al. (2008b) (Fig 4.1, Table 4.2), and clearly rejects Berg's classification

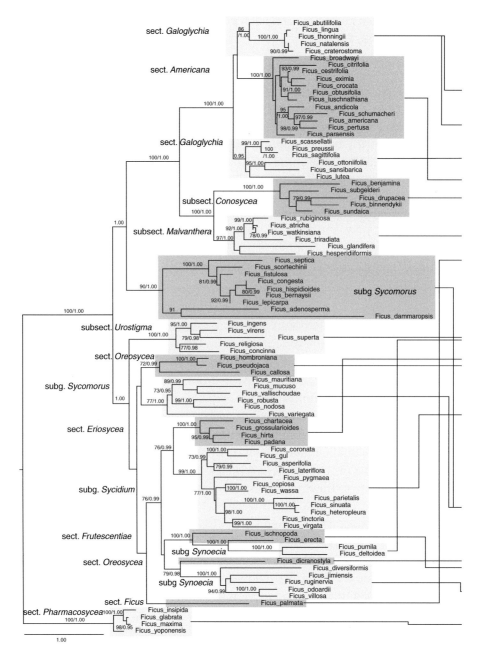

Fig 4.3 Comparison of fig and fig wasp phylogenies.

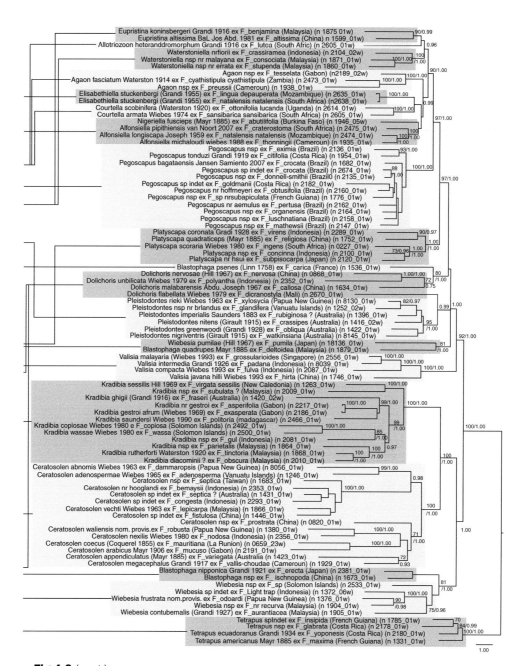

Fig 4.3 (*cont.*)

Table 4.4 Taxonomic sampling used in the present study. True *Ficus*/agaonid associates are shaded in gray.

Ficus phylogeny (Rønsted et al. 2008a)	Agaonid phylogeny (this study)					
	Voucher	Genera	Species	Host	Country	Province
F. ingens	0227_01w	*Platyscapa*	*soraria* (Wiebes, 1980)	*F. ingens*	South Africa	
F. mauritiana	0659_23w	*Ceratosolen*	*coecus* (Coquerel, 1855)	*F. mauritiana*	La Réunion	
F. scortechinii	0820_01w	*Ceratosolen*	undescribed species	*F. prostrata*	China	
F. pseudojaca	0868_01w	*Dolichoris*	*nervosae* (Hill, 1967)	*F. nervosa*	China	
F. adenosperma	1246_01w	*Ceratosolen*	*adenospermae* (Wiebes, 1965)	*F. adenosperma*	Vanuatu Islands	
F. glandifera	1252_02w	*Pleistodontes*	undescribed species nr *blandus*	*F. glandifera*	Vanuatu Islands	
F. virgata	1263_01w	*Kradibia*	*sessilis* Hill, (1969)	*F. virgata sessilis*	New Caledonia	
F. maxima	1331_01w	*Tetrapus*	*americanus* (Mayr, 1885)	*F. maxima*	French Guiana	
F. villosa	1372_06w	*Wiebesia*	undescribed species	Light trap	Indonesia	Sulawesi
F. odoardii	1376_01w	*Wiebesia*	*frustrata nom. provis.* (Weiblen)	*F. odoardi*	Papua New Guinea	
F. robusta	1380_01w	*Ceratosolen*	*waliensis nom. provis.* (Weiblen)	*F. robusta*	Papua New Guinea	
F. rubiginosa	1396_01w	*Pleistodontes*	*imperialis* (Saunders, 1883)	*F. rubiginosa*	Australia	
F. triradiata	1416_02w	*Pleistodontes*	*nitens* (Girault, 1915)	*F. crassipes*	Australia	

Fig species	Code	Wasp genus	Wasp species	Host fig	Country
F. coronata	1420_02w	Kradibia	ghigii (Grandi, 1916)	F. fraseri	Australia
F. atricha	1422_01w	Pleistodontes	greenwoodi (Grandi, 1928)	F. obliqua	Australia
F. variegata	1423_01w	Ceratosolen	appendiculatus (Mayr, 1885)	F. variegata	Australia
F. congesta	1431_01w	Ceratosolen	sp nr notus	F. septica ?	Australia
F. fistulosa	1446_01w	Ceratosolen	constrictus hewitti (Waterston, 1921)	F. fistulosa	China
F. palmata	1536_01w	Blastophaga	psenes (Linné, 1758)	F. carica	France
F. drupacea	1599_01w	Eupristina	altissima (Bal., Jos and Abd., 1981)	F. altissima	China
F. callosa	1634_01w	Dolichoris	malabarensis (Abdu. and Joseph, 1967)	F. callosa	China
F. ischnopoda	1673_01w	Blastophaga	undescribed species	F. ischnopoda	China
F. andicola	1682_01w	Pegoscapus	bagataensis Jansen and Sarmiento, 2007	F. andicola	Colombia
F. septica	1683_01w	Ceratosolen	undescribed species	F. septica	Taiwan
F. hirta	1745_01w	Valisia	javana hilli (Wiebes, 1993)	F. hirta	China
F. religiosa	1752_01w	Platyscapa	quadraticeps (Mayr, 1885)	F. religiosa	China

Table 4.4 (cont.)

Ficus phylogeny (Rønsted et al. 2008a)	Agaonid phylogeny (this study)					
	Voucher	Genera	Species	Host	Country	Province
F. abutilifolia	1765_05w	Nigeriella	fusciceps (Mayr, 1885)	F. abutilifolia	Burkina Faso	
F. americana	1776_01w	Pegoscapus	undescribed species	sp nr F. subapiculata	French Guiana	
F. insipida	1785_01w	Tetrapus	costaricanus (Grandi, 1925)	F. insipida	French Guiana	
F. pumila	1836_01w	Wiebesia	pumilae (Hill, 1967)	F. pumila	Japan	
F. sundaica	1860_01w	Waterstoniella	undescribed species nr errata	F. stupenda	Malaysia	Sarawak
F. parietalis	1864_01w	Kradibia	undescribed species	F. parietalis	Malaysia	Sarawak
F. lepicarpa	1866_01w	Ceratosolen	vechti (Wiebes, 1963)	F. lepicarpa	Malaysia	Sarawak
F. tinctoria	1868_01w	Kradibia	rutherforti (Waterston, 1920)	F. tinctoria	Malaysia	Sarawak
F. subgelderi	1871_01w	Waterstoniella	undescribed species nr malayana	F. consociata	Malaysia	Sarawak
F. benjamina	1875_01w	Eupristina	koninsbergeri (Grandi, 1916)	F. benjamina	Malaysia	Sarawak
F. deltoidea	1879_01w	Blastophaga	quadrupes (Mayr, 1885)	F. deltoidea	Malaysia	Sarawak
F. diversiformis	1904_01w	Wiebesia	undescribed species	F. nr recurva	Malaysia	Sarawak
F. ruginervia	1905_01w	Wiebesia	contubernalis (Grandi, 1927)	F. aurantiacea	Malaysia	Sarawak

Fig species	Code	Genus	Pollinator species	Associated Ficus	Country	Region
F. vallischoudae	1929_01w	Ceratosolen	megacephalus (Grandi, 1917)	F. vallis-choudae	Cameroun	
F. thonningii	1935_01w	Alfonsiella	michaloudi (Wiebes, 1988)	F. thonningii	Cameroun	
F. preussii	1938_01w	Agaon	undescribed species	F. preussii	Cameroun	
F. citrifolia	1954_01w	Pegoscapus	tonduzi (Grandi, 1919)	F. citrifolia	Costa Rica	
F. sinuata	2009_01w	Kradibia	undescribed species	F. rubromitotis	Malaysia	Sarawak
F. heteropleura	2010_01w	Kradibia	giacominii ?	F. obscura	Malaysia	Sarawak
F. gul	2081_01w	Kradibia	undescribed species	F. gul	Indonesia	Sulawesi
F. chartacea	2087_01w	Blastophaga	compacta (Wiebes, 1993)	F. fulva	Indonesia	Sulawesi
F. concinna	2100_01w	Platyscapa	undescribed species	F. concinna	Indonesia	Sulawesi
F. binnendykii	2104_02w	Waterstoniella	nr fiorii	F. crassiramea	Indonesia	Sulawesi
F. superba	2120_01w	Platyscapa	nr hsui	F. subpisocarpa	Japan	RyuKyu Islands
F. schumacheri	2135_01w	Pegoscapus	undescribed species	F. donnell-smithii	Brazil	
F. eximia	2136_01w	Pegoscapus	undescribed species	F. eximia	Brazil	
F. broadwayi	2147_01w	Pegoscapus	undescribed species	F. mathewsii	Brazil	
F. luschnathiana	2158_01w	Pegoscapus	undescribed species	F. luschnatiana	Brazil	
F. obtusifolia	2160_01w	Pegoscapus	nr hoffmeyeri	F. obtusifolia	Brazil	
F. pertusa	2162_01w	Pegoscapus	nr aemulus	F. pertusa	Brazil	

Table 4.4 (cont.)

Ficus phylogeny (Rønsted et al. 2008a)	Agaonid phylogeny (this study)					
	Voucher	Genera	Species	Host	Country	Province
F. cestrifolia	2164_01w	Pegoscapus	undescribed species	F. organensis	Brazil	
F. glabrata	2178_01w	Tetrapus	undescribed species	F. glabrata	Costa Rica	
F. yoponensis	2180_01w	Tetrapus	ecuadoranus (Grandi, 1934)	F. yoponensis	Costa Rica	
F. paraensis	2182_01w	Pegoscapus	sp	F. goldmanii	Costa Rica	
F. pygmaea	2186_01w	Kradibia	gestroi afrum (Wiebes, 1969)	F. exasperata	Gabon	
F. sagittifolia	2189_02w	Agaon	undescribed species	F. tesselata	Gabon	
F. mucuso	2191_01w	Ceratosolen	arabicus (Mayr, 1906)	F. mucuso	Gabon	
F. asperifolia	2217_01w	Kradibia	nr gestroi	F. asperifolia	Gabon	
F. virens	2289_01w	Platyscapa	coronata (Grandi, 1928)	F. virens	Indonesia	Java
F. hispidioides	2293_01w	Ceratosolen	sp	F. congesta	Indonesia	Papua Barat
F. hombroniana	2352_01w	Dolichoris	umbilicata (Wiebes, 1979)	F. polyantha	Indonesia	Papua Barat
F. bernaysii	2353_01w	Ceratosolen	nr hooglandi	F. bernaysii	Indonesia	Papua Barat
F. nodosa	2356_01w	Ceratosolen	nexilis (Wiebes, 1980)	F. nodosa	Indonesia	Papua Barat

Fig species	Code	Genus	Wasp species	Host fig	Locality
F. erecta	2381_01w	*Blastophaga*	*nipponica* Grandi, (1921)	*F. erecta*	Japan
F. lateriflora	2466_01w	*Kradibia*	*saundersi* (Wiebes, 1990)	*F. politoria*	Madagascar
F. scassellatii	2473_01w	*Agaon*	*fasciatum* (Waterston, 1914)	*F. cyathistipula cyathistipula*	Zambia
F. natalensis	2474_01w	*Alfonsiella*	*longiscapa* (Joseph, 1959)	*F. natalensis natalensis*	Mozambique
F. craterostoma	2475_01w	*Alfonsiella*	*pipithiensis* (van Noort, 2007)	*F. craterostoma*	South Africa
F. copiosa	2492_01w	*Kradibia*	*copiosae* (Wiebes, 1980)	*F. copiosa*	Solomon Islands Guadalcanal
F. wassa	2500_01w	*Kradibia*	*wassae* (Wiebes, 1980)	*F. wassa*	Solomon Islands Guadalcanal
F. jimiensis	2533_01w	*Wiebesia*	undescribed species	*F. sp*	Solomon Islands Gatokae
F. grossularioides	2556_01w	*Valisia*	*malayana* (Wiebes, 1993)	*F. grossularioides*	Singapore
F. lutea	2602_01w	*Allotriozoon*	*heterandromorphum* (Grandi, 1916)	*F. lutea*	South Africa
F. sansibarica	2605_01w	*Courtella*	*armata* (Wiebes, 1974)	*F. sansibarica sansibarica*	South Africa
F. ottoniifolia	2614_01w	*Courtella*	*scobinifera* (Waterston, 1920)	*F. ottonifolia lucanda*	Uganda

Table 4.4 (cont.)

Ficus phylogeny	Agaonid phylogeny (this study)					
(Rønsted et al. 2008a)	Voucher	Genera	Species	Host	Country	Province
F. lingua	2635_01w	Elisabethiella	stuckenbergi (Grandi, 1955)	F. lingua depauperata	Mozambique	
F. natalensis	2638_01w	Elisabethiella	stuckenbergi (Grandi, 1955)	F. natalensis natalensis	South Africa	
F. dicranostyla	2670_01w	Dolichoris	flabellata (Wiebes, 1979)	F. dicranostyla	Mali	
F. crocata	2674_01w	Pegoscapus	Sp	F. crocata	Brazil	
F. padana	8039_01w	Valisia	intermedia (Grandi, 1926)	F. padana	Indonesia	Java
F. dammaropsis	8056_01w	Ceratosolen	abnormis (Wiebes, 1963)	F. dammaropsis	Papua New Guinea	
F. hesperidiiformis	8130_01w	Pleistodontes	rieki (Wiebes, 1963)	F. xylosycia	Papua New Guinea	East New Britain
F. watkinsiana	8145_01w	Pleistodontes	nigriventris (Girault, (1915)	F. watkinsiana	Australia	

of *Ficus* (Berg and Corner 2005). As already observed in previous studies, the tree is not resolved in the deeper nodes and our analyses failed to resolve the relationships within the dioecious figs.

The main difference between the present analyses and the Rønsted et al. (2008b) study, is the presence of a strongly supported clade (BP 76, PP 0.99) including subgenus *Sycidium*, section *Eriosycea*, subgenus *Synoecia* and subsection *Frustescentiae* (including *F. pumila* whose current classification within *Rhizocladus* section is doubtful). In other words, the subgenera *Synoecia* (canopy lianas occurring in Malaysian Dipterocarpaceae forests) and *Sycidium* (small standing trees, rarely hemiepiphytes, that mostly occur in Asia) render the subgenus *Ficus* paraphyletic (forest standing trees or bushes occurring in Asia, often in secondary/disturbed vegetation). Hence, without considering the unresolved position of *F. palmata*, the present subgenera *Synoecia*, *Sycidium* and *Ficus* could be part of an extended subgenus *Ficus*, which includes most of the dioecious figs to the exception of *Sycomorus*. Subgenus *Sycidium*, which is recovered monophyletic, and section *Eriosycea* appear as sister groups (BP 76, PP 0.97), confirming results by Jousselin et al. (2003) and Rønsted et al. (2005 and 2008a). Morphologically, the relationships between these fig groups are difficult to assess. Subgenus *Sycidium* shares few characters with *Ficus* and *Synoecia*: presence of *Terminalia*-habit branching, sapling leaves larger, and more toothed and lobed than adult ones (Berg and Corner 2005). However, a sister-group relationship between *Sycidium* and section *Eriosycea* is supported by the evolutionary pathway that gynodioecy was acquired in both groups and may be difficult to reject (Kjellberg et al. 2005).

In our analysis, the subgenus *Sycomorus* (standing trees growing along streams and rivers, in Asia and Africa) is recovered as polyphyletic, a result also found by Rønsted et al. (2005), but not by Jousselin et al. (2003) and Weiblen (2000) (Fig 4.1). Sections *Sycomorus* and *Neomorphe* group together (PP 1.00) in a first clade basal to the dioecious figs, whereas sections *Adenosperma*, *Dammaropsis* and *Sycocarpus* cluster in a second clade (PP 1.00) basal to most *Urostigma* species. Subgenus *Sycomorus* is well-defined by several morphological apomorphies: (i) with the exception of few fig species with small syconia, the staminate flowers are enveloped by two large bracteoles, (ii) the perianth is tubular, and (iii) the perianth lobe is hooded and consequently entirely enclose the stamens. Consequently as suggested by Rønsted et al. (2005), the polyphyly of subgenus *Sycomorus* is difficult to explain and could be artifactual due to lack of resolution.

The subgenus *Pharmacosycea* appeared polyphyletic, a result already observed in all previous studies. Sections *Pharmacosycea* and Oreosycea never clustered together. Furthermore, in all molecular studies where representatives of both subsections *Glandulosae* and *Pedunculatae* are present, section *Oreosycea* (small to large trees occurring at low density mostly in the Oriental and Pacific regions, highly diversified in New Caledonia) is always recovered as paraphyletic or polyphyletic (Weiblen 2000; Cruaud et al. present study; Jousselin et al. 2003; Rønsted et al. 2008b). This result is not surprising given the high morphological diversity of the group that makes subsection delimitation difficult. In our analyses, representatives of subsection *Pedunculatae* (*F. dicranostyla* and *F. callosa*) cluster with subgenus *Synoecia* (BP 79, PP 0.98) and subsection *Glandulosae* (BP 72, PP 0.99), respectively.

The diverse subgenus *Urostigma* is not recovered as monophyletic due to the unresolved position of subsection *Urostigma*, which does not group with other *Urostigma* sections and subsections. This result was also observed in most previous phylogenetic studies (Fig 4.1) and is corroborated morphologically by the presence of ostiolar staminate flowers in the subsection *Urostigma*. The large monoecious clade comprising all other *Urostigma* sections is subdivided into two strongly supported clades (BP > 95, PP 1.00): [*Conosycea*, (large stranglers of Malaysian forests, some of them like *F. benjamina* may grow in your living room) + *Malvanthera*, (Australian hemi-epiphytic figs)] and [*Americana*, (Neotropical hemi-epiphytes or stranglers) + *Galoglychia* (Afrotropical hemi-epiphytic figs)]. Although these relationships were not recovered by Jousselin et al. (2003) (*Malvanthera* appeared basal to other sections, Fig 4.1), this branching pattern was already observed by Rønsted et al. (2005). Section *Americana* renders section *Galoglychia* paraphyletic to section *Americana*, as was also found by Rønsted et al. (2007), while in earlier analyses based on ITS and ETS alone, *Americana* was recovered as sister to *Galoglychia* (Jousselin et al. 2003; Rønsted et al. 2005). Morphologically, these groups share few common characters (e.g. presence of only two basal bracts (Corner 1958)) and probably belong to closely related but different entities (Renoult et al. 2009). This result may be due to lack of resolution, potentially caused by limited sampling and/or hybridization between sections and we suspect that future work may confirm the monophyly of both sections as predicted by Renoult et al. (2009).

The coarse topology of the agaonid phylogeny (Fig 4.3) is congruent with the previous observations of Lopez-Vaamonde et al. (2009) and Cruaud et al. (2010) (fig 4.2). Apart from *Wiebesia*, *Blastophaga* and *Dolichoris*, all agaonid genera represented by at least two species are recovered as monophyletic (BP > 90, PP > 0.95). However, concerning the genus *Platyscapa*, results must be considered with caution because they rely on non-exhaustive sampling. Indeed, Cruaud et al. (2010) have shown that including more *Platyscapa* species, especially species from continental Asia associated with relict figs (*F. orthoneura, F. hookeriana*), or species from Madagascar (associated with *F. menabeensis*), renders the genus polyphyletic. Moreover, two groups of *Platyscapa* can be recognized on the basis of morphology and host taxonomy. Concerning the polyphyly of *Wiebesia*, *Blastophaga* and *Dolichoris*, our observations are congruent with morphology (Cruaud et al. 2010) and previous molecular studies (Fig 4.2). They strongly suggest that nomenclatural rearrangements are required.

As for previous studies, ML-analysis does not resolve the basal nodes of the tree, which appear as a trichotomy including: (i) subfamily Kradibiinae (*Ceratosolen* + *Kradibia sensu* Cruaud et al. 2010), (ii) a clade that includes most *Wiebesia* species and *Blastophaga* pollinating subsection *Frustescentiae*, and (iii) a well-supported (BP 92, PP 1.00) monophyletic clade grouping all other genera. In Bayesian analysis, Kradibiinae and the clade grouping *Wiebesia* and *Blastophaga* species are sister taxa (PP 1.00) whereas in ML-phylogeny this relationship is poorly supported (BP 64).

Within the third clade, relationships are poorly resolved, a result that confirms difficulties encountered by previous studies. However, the present hypothesis is the best resolved. The basal relationships within this group appear as a polytomy including genera *Valisia*, *Pleistodontes*, *Dolichoris* + *B. psenes* (BP 80, PP 1.00), *D. flabellatus*, a clade clustering pollinators of *F. pumila* and *F. deltoidea* (BP 81, PP 1.00), and a clade grouping all other genera (BP 97, PP 1.00). Moreover, subgenus *Valisia* could be sister to all other genera (BP 68, PP 1.00), as suggested by Jiang et al. (2006). However this result needs to be confirmed as it is not supported by morphology. *Dolichoris* species (*D. flabellatus* excepted) and *Blastophaga psenes* (type-species of the genus *Blastophaga*) are recovered as sister taxa with strong support (BP 80, PP 1.00) corroborating both molecular (Lopez-Vaamonde et al. 2009; Cruaud et al. 2010) and morphological (Cruaud et al. 2010) hypotheses.

Pollinators of subgenus *Urostigma* (*Pleistodontes* excepted) cluster together in a well-supported clade (BP 97, PP 1.00). Relationships within this clade are better resolved and supported than in any previous study (Fig 4.2). *Platyscapa* is sister to all other genera (BP 97, PP 1.00), and *Pegoscapus* is sister (BP 97, PP 1.00) to a highly supported clade (BP 90, PP 1.00) that clusters pollinators of *Conosycea* and *Galoglychia* figs. Pollinators associated with subsection *Conosycea* and section *Galoglychia* do not form monophyletic groups, and their respective species are intermixed, a result already observed by Lopez-Vaamonde et al. (2009). In Bayesian analysis, all Afrotropical genera (*Allotriozoon* excepted) cluster in a strongly supported clade (PP 0.99) whereas in ML-analysis these genera cluster poorly (BP 63). This result can be compared to the results of Erasmus et al. (2007), who found that the genus *Allotriozoon* was the sister taxon to all other pollinators of *Galoglychia* figs.

A clade intermixing pollinators of *Conosycea* and *Galoglychia* figs is not supported by morphology. Indeed, *Eupristina* share more characters with *Pegoscapus* than with any other genus (Cruaud et al. 2010). *Eupristina* species does not exhibit (i) the same structure of the first funicular segment as *Galoglychia* pollinators, and (ii) a longitudinally divided pronotum, a character shared by most Afrotropical genera. Finally, *Eupristina* or *Waterstoniella* males are strongly different from males of *Galoglychia* pollinators. Consequently, further studies are required to better infer phylogenetic relationships between pollinators of *Urostigma* figs.

4.6 Facts about cospeciation

Global tests using Parafit resulted in rejection of random association between host and pollinator taxa (P = 0.001). Sixty-three of the 89 tests of individual host-associate pairs (shallower nodes) resulted in significant associations between figs and their agaonid pollinators (P = 0.001). Reconciliation analyses using TreeMap 1.0b suggested 50 cospeciation events (on 89 nodes) and randomization indicated significant cospeciation between *Ficus* and their pollinating wasps (P = 0.001). Most of the putative cospeciation events appear to occur on shallower nodes (data available upon request). Reconciliation analyses using Jane suggested only 33 putative cospeciation events (Fig 4.4).

In our analyses, the deeper nodes of *Ficus* and agaonid phylogenies are mostly unresolved, especially for fig topology. Due to computational limitation, a cross-testing of alternative topologies was impractical, so the true level of congruence could not be estimated accurately. Global tests using Parafit and Treemap are significant. Jane infers less cospeciation events than TreeMap. We can therefore imagine a mixed structure with parts of the two trees coevolving whereas other parts are not. Strict cospeciation (simultaneous radiation of both lineages) probably did occur, but other processes have probably played an important role during the mutualism diversification.

Almost all authors have previously agreed in assessing long-time strict cospeciation between *Ficus* and their pollinators (but see recent studies by Kjellberg et al. 2005; Machado et al. 2005; Jackson et al. 2008; Cruaud et al. 2010). Formal tests by Jackson (2004) revealed cospeciation within a "*Ficus* microcosm;" however taxon coverage was poor. Fig and pollinator phylogenies were based on one species (two for *Sycomorus*) per fig section or subgenus so that distances between lineages may have overestimated cospeciation. In the present study, although deeper nodes of *Ficus* and agaonid phylogenies are only partly resolved, a visual examination of both topologies reveals strong discrepancies between diversification patterns of *Ficus* and their agaonid pollinators.

Ceratosolen and *Kradibia* are undoubtedly sister taxa, whereas their host *Ficus* subgenera (*Sycomorus* and *Sycidium* respectively) are never recovered as sister taxa (Fig 4.3). Instead, *Sycidium* is always nested, mostly with strong support, within a clade of dioecious figs belonging to the subgenera *Ficus* and *Synoecia*. Morphologically, subgenus *Sycidium* appears to share fewer characters with *Ficus* and *Synoecia* than with *Sycomorus*. *Sycidium* and *Sycomorus* species share (i) the presence of bracts on the syconium, (ii) the structure of the ostiole, (iii) the presence of ostiolar stamens, (iv) the figs are often cauliflorous, and (v) the leaves are frequently asymmetric (especially in *Sycocarpus*), but this last character could be a convergent adaptation (Kjellberg et al. 2005). Further, peduncular bracts (two to four) are present in a number of *Sycidium* species as well as in *Adenosperma* species

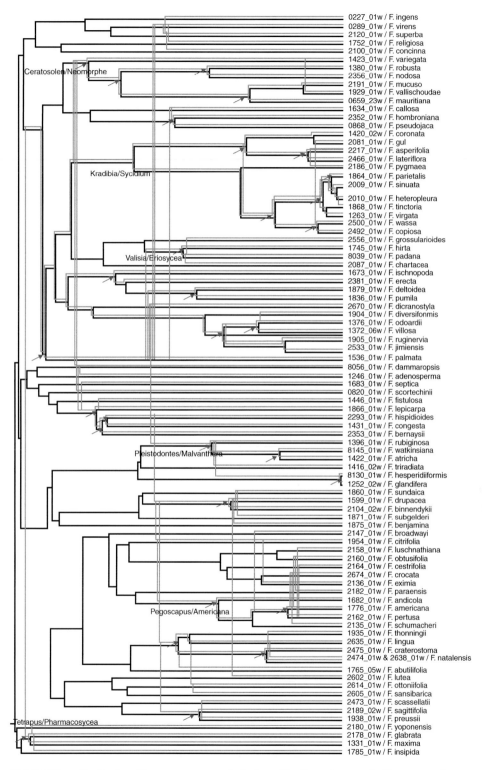

Fig 4.4 Hypothetical codivergence scenario inferred by Jane. Arrows indicate putative cospeciation event. See plate section for color version.

(occasionally in some *Oreosycea*). Finally, subgenera *Sycidium* and *Sycomorus* differ from all other subgenera by the absence of subulate stigma. However, a close relationship between *Sycidium* and *Eriosycea* is recovered in all studies and is possibly corroborated by a common origin of dioecy (Kjellberg et al. 2005), so it is difficult to rule out this relationship (Kjellberg et al. 2005; Machado et al. 2005). Consequently, the incongruent phylogenetic pattern between pollinators and their host figs could suggest an ancestral shift of pollinators. Host-switching of pollinators may have been facilitated by the common ancestral areas (probably New Guinea) and the similar habitats (forest along streams) where *Sycomorus* and *Sycidium* occurred.

One strongly supported clade clusters *Galoglychia*, *Americana*, *Malvanthera* and *Conosycea* figs, but not *Urostigma*. All these groups are recovered as monophyletic, mostly with a high support value. *Malvanthera* and *Conosycea* are recovered as sister taxa and *Americana* renders *Galoglychia* paraphyletic. In contrast, this part of the tree is poorly resolved in all agaonid phylogenies published to date (Fig 4.2). In most analyses (our study included) *Pleistodontes* and pollinators of other *Ficus* subgenera (i.e. *Dolichoris*, *Wiebesia*) form a polyphyletic group basal to the all other *Urostigma* pollinators. *Platyscapa* is sister to all other genera and *Pegoscapus* is sister to a group clustering the pollinators of *Conosycea* and *Galoglychia*. Furthermore, the pollinators of *Conosycea* and *Galoglychia* do not form monophyletic groups. However, these results need to be confirmed using a broader taxonomic sampling for both groups.

Within *Galoglychia*, cospeciation is clearly not the rule (Erasmus et al. 2007; Rønsted et al. 2007; Jousselin et al. 2008; Rønsted et al. 2008b; Renoult et al. 2009). Some subsections are pollinated by a combination of unrelated genera and the discrepancies between wasps and *Ficus* classification cannot be completely explained by the difficulties in classifying the figs (Kjellberg et al. 2005). Instead, host switches and duplications followed by asymmetrical lineage extinction may have occurred during agaonid and fig diversification (Jousselin et al. 2008). It consequently appears that a simple strict cospeciation model does not adequately describe the codivergence of the fig/agaonid mutualism.

Previous studies provided evidence of cospeciation between *Ceratosolen* and Papuan figs of subgenus *Sycomorus* (Weiblen and Bush 2002; Jackson 2004). However, this result may be due to undersampling of both taxa. Indeed, phylogenies of Afrotropical fig wasps (several genera of wasps) (Jousselin et al. 2008) and *Pegoscapus* (Machado et al. 2005; Jackson et al. 2008) are not congruent with the topology of their hosts (section *Galoglychia* and *Americana*, respectively). Another example is the case of the *Pleistodontes/Malvanthera* association. To date, two phylogenies of *Pleistodontes* (Lopez-Vaamonde et al. 2001; Cook et al. 2004) and one phylogeny of *Malvanthera* have been published (Rønsted et al. 2008a). A visual examination of phylogenies reveals incongruences between resolved nodes meaning that *Pleistodontes/Malvanthera* codivergence cannot be explained by strict cospeciation, but a formal statistical test is still missing. In the present study, the best taxonomic samplings concern section *Americana* and subgenus *Sycidium* (12 species each with 58 % of true associates). For both groups, Treemap indicates only a few strict cospeciation events between figs and their pollinators. Consequently, at a fine systematic scale, there is little evidence to support the strict cospeciation hypothesis.

Discordance between phylogenies could be explained as follows. While high host specificity and evidence of phenotypic coevolution of both partners, especially on characters involved in the pollination syndromes (active versus passive) have been mentioned (Jousselin and Kjellberg 2001; Jousselin et al. 2003; Weiblen 2004; Kjellberg et al. 2005), several cases of *Ficus* species sharing the same pollinators in sympatry have been reported (*F. auriculata* and *F. oligodon*, *F. natalensis* and *F. thonningii*, *F. exasperata* and *F. asperifolia*, *F. popenoei* and *F. bullenei*). In such cases, lineage sorting of wasps has not yet been achieved. Furthermore, we know about 50 species of figs are pollinated by more than one pollinator (Cruaud and Rasplus unpublished data). These cases reveal either speciation of fig wasps by duplication on the same host (in case of sister taxa relationships) or host shift from another fig species (Cruaud and Rasplus unpublished data), two cases that are clearly not cospeciation events. In most cases of agaonid sister taxa, both species exhibit coevolved traits with the figs, e.g. *Ceratosolen flabellatus* and *C. silvestrianus* (Kerdelhué et al. 1997); *Pleistodontes* species (Haine et al. 2006), *Pegoscapus hoffmeyeri* A and B (Molbo et al. 2003), whereas some sister taxa exhibit different traits (small versus large pollen pockets and presence versus absence of coxal combs (Peng et al. 2008)) that suggest breakdown of the mutualism and evolution of cheating habits.

4.7 Conclusion and prospects

It clearly appears from our study that a simple strict cospeciation model does not adequately describe the diversification of *Ficus*/agaonid mutualism at either fine or coarse systematic scales. Indeed, a visual examination of phylogenies reveals incongruence between resolved nodes; therefore, host switching and duplication, followed by asymmetrical lineage extinction, must have occurred during agaonid diversification. The sampling of the present study covers only 1/8 of the species diversity of worldwide *Ficus*/agaonid associations. However, due to computational limitations of available software, this number is already too high to allow us to statistically test the strict cospeciation hypothesis. Moreover, both fig and pollinator relationships remain partly unresolved creating another hindrance for assessing the degree of cophylogeny between the two groups. Increasing the taxon sampling and the number of informative characters for both groups should facilitate the analyses. We are working on both objectives and aim to propose robust phylogenetic hypotheses for both *Ficus* and agaonids in the near future. This will allow for an accurate test of the degree of cospeciation between these two large datasets, and the strict cospeciation paradigm between *Ficus* and their pollinators will finally be comprehensively assessed.

4.8 Materials and methods

4.8.1 Laboratory protocols (agaonids)

Extraction protocol follows Cruaud et al. (2010). All voucher specimens are deposited at the Center for Biology and Management of Populations (CBGP), INRA, Montferrier-sur-Lez, France. In the present study, we combined two nuclear protein-coding genes, wingless (Wg, 403 bp), F2 copy of elongation factor-1α (EF-1α, 516 bp), two mitochondrial protein-coding genes Cytochrome oxydase I (COI, 1503 bp), Cytochrome b (Cytb, 747 bp) and 28S rRNA (D2-D3 and D4-D5 expansion regions, 1594 bp). Choice of the markers (slow or fast evolving) was guided by the wish to resolve both intra and inter-generic level of phylogenetic relationships (Wiens et al. 2005). Primers and PCR conditions are described in Cruaud et al. (2011). PCR products were purified using ExonucleaseI and Phosphatase and sequenced directly using the BigDyeTerminator V3.1 kit (Applied biosystem) and an ABI3730XL sequencer, whereas some other were cloned prior to sequencing (especially for Wg). Both strands for each overlapping fragment were assembled using the sequence editing software Geneious v3.7 (Drummond et al. 2007). All the sequences have been deposited in GenBank.

4.8.2 Sequence alignment

All details for *Ficus* sequence alignment are provided in Rønsted et al. (2008b). Concerning agaonids, protein-coding genes (COI, Cytb, EF, Wg) were aligned using ClustalW 1.81 (Thompson et al. 1994) with default gap opening, extension and substitution costs. For confirmation, alignments were translated to amino acids using MEGA version 4 (Tamura et al. 2007). Alignment of sequences encoding rRNA was based on secondary structure models (Gillespie et al. 2005; Gillespie et al. 2006) using the terminology developed by Kjer (1995) and Gillespie et al. (2004). Hypervariable regions have been aligned using ClustalW with default settings and included in further analyses. Structural model of rRNA gene fragments and alignment details are available in Cruaud et al. (2010).

4.8.3 Phylogenetic analyses

Phylogenetic trees were estimated using maximum likelihood and Bayesian methods. All the analyses have been conducted on a Linux Cluster (quadral-processor Xeon X5472, 3GHz computers) at CBGP, Montpellier, France.

In order to allow cophylogenetic analyses with tree-reconciliation methods, fig and agaonid trees were rooted on subgenus *Pharmacosycea* and genus *Tetrapus*, respectively. Indeed these taxa were recovered basal to all other fig/agaonid species in numerous studies.

We undertook partitioned analyses implementing separate nucleotide substitution models for subsets of the data more likely to have experienced similar evolutionary processes (mitochondrial genes, each nuclear genes and rRNA stems and loops). The best fitting model for each partition was identified using the Akaike information criterion (Akaike 1973) as implemented in MrAIC 1.4.3 (Nylander 2004).

We performed maximum likelihood analyses (ML) and associated non-parametric bootstrap analyses using the MPI-parallelized version of RA×ML 7.0.4 (Stamatakis, 2006b). Given the computational difficulties associated to ML bootstrapping, the GTRCAT approximation of models has been used in the second part of RAxML analysis (Stamatakis 2006a). Robustness of topologies was assessed by bootstrap procedures by using 1000 replicates. Following Felsenstein and Kishino 1993, a bootstrap percent (BP) > 95 % is considered a strong and a value <70 % indicated a weak support for the clade.

Bayesian inferences (BI) were conducted using a parallel version of MrBayes v.3.1.1 (Huelsenbeck and Ronquist 2001). Following Nylander et al. (2004), Marshall et al. (2006) and McGuire et al. (2007), we assumed a cross-partition heterogeneity (APRV = Among Partition Rate Variation) in model parameters by considering the parameter m (rate multiplier, reflecting the overall rate of substitutions of the given partitions). However, we assumed proportional branch lengths among the different partitions. Parameter values for the model were initiated with default uniform priors, and branch lengths were estimated using default exponential priors (no molecular clock). We used four Metropolis-coupled MCMC, one cold and three heated, with incremental heating to improve mixing of the cold chain and to avoid it becoming stocked on local optima (Huelsenbeck and Ronquist 2001). To address the question of a good approximation of the target distribution, we carried out two independent runs starting from different and randomly chosen topologies. Each run was allowed to work for ten million generations. All values were sampled every 1000 generations. For the initial determination of burn-in, we examined the plot of overall model likelihood against generation number to find the point where the likelihood started to fluctuate around a stable value (Huelsenbeck and Ronquist 2001). We then discarded the points sampled prior to convergence of the chains. We used a range of MCMC convergence and good mixing diagnostics following Huelsenbeck and Ronquist 2001; Huelsenbeck et al. 2002; Rambaut and Drummond 2003; Nylander et al. 2004; Nylander et al. 2008; Cruaud et al. 2010). Finally, the results were based on the pooled samples from the stationary phases of the two independent runs. Given that PP may overestimate clade support for reasons discussed elsewhere (Erixon et al. 2003; Simmons et al. 2004), only clades with PP > 0.95 were considered strongly supported.

4.8.4 Cophylogenetic analyses

For the analyses with TreeMap 1.0b, we used heuristic searches to find optimal solutions. Randomization of the agaonid tree has been conducted using the proportional to distinguishable model with 10 000 replicates. The observed number of putative codivergence events was compared to the null distribution of codivergence events derived from this randomization procedure to determine whether the number of codivergence events recovered from the reconciliation analysis was significant.

For the analyses with Jane, we used default cost settings (0 for codivergence events, 1 for host switching and duplications and 2 for loss) and 900 solves (30 30).

We used the global-fit method, ParaFit, developed by Legendre et al. (2002) implemented in the program CopyCat (Meier-Kolthoff et al. 2007). ParaFit evaluates the global hypothesis of host-associate cospeciation with a matrix permutation test of codivergence. This test crosses three types of information: the associate phylogeny and the host phylogeny, both described by their respective matrices of patristic distances, and the observed host-associate associations. Each matrix representing associates and hosts are transformed into a matrix of principal coordinates. The association is then described by a matrix that crosses both matrices of principal coordinates and the matrix of association. Patristic distances were computed from fig and agaonid ML-phylogenies. Tests of random association (null hypothesis) were performed using 999 permutations globally across both phylogenies.

Acknowledgements

A. C. and J. Y. R. thank Denis Bourguet (CBGP Montpellier, France), Stefan Ungricht (ETH Zurich, Switzerland), Jeremy Bouyer (Burkina Faso), Paul Hanson (San Jose University, Costa Rica), Philippe Birnbaum (AMAP Montpellier, France), Jenny Underhill (Kirstenbosch RC Cape Town, South Africa), Zhi-hui Su (Osaka University, Japan), and George D. Weiblen (University of Minnesota, USA) for contributing samples and help in the field. A. C. and J. Y. R. also thank all their guides in Borneo, Sulawesi, Papua Barat, Gabon, and Costa Rica, especially Jaman, Lary and Mado. R. A. S. P. thanks Fernando Farache, Ludmila Teixeira, Luis Coelho, Michele Medeiros, and Monise Cerezini for their help in the field in Brazil. A. C. expresses her sincere gratitude to Sylvain Piry, Alexandre Dehne Garcia, and Arnaud Estoup (CBGP Montpellier, France) for their help with cluster computers.

Financial support was provided by grants from the ANR (National Research Agency) that supports the "NiceFigs" and "Biofigs" projects, led by M. Hossaert-McKey

(CNRS Montpellier, France), the French sequencing center Genoscope, an NRF grant GUN 61497 to S.v.N, a Marie Curie Fellowship to N. R. and a Biota/Fapesp grant (04/10299–4) to R. A. S. P.

References

Akaike, H. (1973). Information Theory and an Extension of the Maximum Likelihood Principle. In *Second International Symposium on Information Theory*, ed. P. N. Petrov and F. Csaki. Budapest, Hungary: Akad. Kiado, pp. 267–81.

Berg, C. C. and Corner, E. J. H. (2005). *Moraceae – Ficus*. Fora Malesiana, Series I, 17/2. Leiden.

Borges, R. M., Bessiere, J. M. and Hossaert-McKey, M. (2008). The chemical ecology of seed dispersal in monoecious and dioecious figs. *Functional Ecology*, **22**, 484–93.

Charleston, M. A. (1998). Jungles: a new solution to the host/parasite phylogeny reconciliation problem. *Mathematical Biosciences*, **149**, 191–223.

Charleston, M. A. and Page, R. D. M. (2002). TREEMAP (v2.0). Applications for Apple Macintosh. University of Oxford, Oxford, UK.

Charleston, M. A. and Perkins, S. L. (2006). Traversing the tangle: algorithms and applications for cophylogenetic studies. *Journal of Biomedical Informatics*, **39**, 62–71.

Compton, S. G., Holton, K. C., Rashbrook, V. K., Noort, S. v., Vincent, S. L. and Ware, A.B. (1991). Studies of *Ceratosolen galili* a non-pollinating agaonid fig wasp. *Biotropica*, **23**(2), 188–94.

Conow, C., Fielder, D., Ovadia, Y. and Libeskind-Hadas, R. (2010). Jane: a new tool for the cophylogeny reconstruction problem. *Algorithms for Molecular Biology*, **5**, 16.

Cook, J. M. and Rasplus, J.-Y. (2003). Mutualists with attitude: coevolving fig wasps and figs. *Trends in Ecology and Evolution*, **18**, 241–248.

Cook, J. M. and Segar, S. T. (2010). Speciation in fig wasps. *Ecological Entomology*, **35**, 54–66.

Cook, J. M., Bean, D., Power, S. A. and Dixon, D. J. (2004). Evolution of a complex coevolved trait: active pollination in a genus of fig wasps. *Journal of Evolutionary Biology*, **17**, 238–46.

Corner, E. J. H. (1958). An introduction to the distribution of Ficus. *Reinwardtia*, **4**, 325–55.

Cruaud, A., Jabbour-Zahab, R., Genson, G., Cruaud, C., Couloux, A., Kjellberg, F., van Noort, S. and Rasplus, J.-Y. (2010). Laying the foundations for a new classification of Agaonidae (Hymenoptera: Chalcidoidea), a multilocus phylogenetic approach. *Cladistics*, **26**, 359–87.

Cruaud, A., Jabbour-Zahab, R., Genson, G., Couloux, A., Yan-Qiong, P., Da Rong, Y., Ubaidillah, R., Pereira, R. A. S., Kjellberg, F., Van Noort, S., Kerdelhué, C. and Rasplus, J.-Y. (2011). Out-of-Australia and back again: the worldwide historical biogeography of non-pollinating fig wasps (Hymenoptera: Sycophaginae). *Journal of Biogeography, Special Paper*, doi:10.1111/j.1365-2699.2010.02429.x

Desdevises, Y., Morand, S., Jousson, O. and Legendre, P. (2002). Coevolution between lamellodiscus (monogenea:diplectanidae) and sapridae (Telestostei): the study of a complex host–parasite system. *Evolution*, **56**, 2459–71.

Drummond, A. J., Ashton, B., Cheung, M., Heled, J., Kearse, M., Moir, R., Stones-Havas, S., Thierer, T. and Wilson, A. (2007). Geneious v3.8.5, Available from http://www.geneious.com/

Erasmus, C., van Noort, S., Jousselin, E. and Greef, J. (2007). Molecular phylogeny of fig wasps pollinators (Agaonidae, Hymenoptera) of Ficus section Galoglychia. *Zoologica Scripta*, **36**, 61–78.

Erixon, P., Svennblad, B., Britton, T. and Oxelman, B. (2003). Reliability of bayesian posterior probabilities and bootstrap frequencies in phylogenetics. *Systematic Biology*, **52**, 665–73.

Farenholz, H. (1913). Ectoparasiten und Abstammungslehre. *Zoologische Anzeiger*, **41**, 371–4.

Felsenstein, J. and Kishino, H. (1993). Is there something wrong with the bootstrap on phylogenies? A reply to Hillis and Bull. *Systematic Biology*, **42**, 193–200.

Gillespie, J. J., Cannone, J. J., Gutell, R. R. and Cognato, A. (2004). A secondary structural model of the 28S rRNA expansion segments D2 and D3 from rootworms and related leaf beetles (Coleoptera : Chrysomelidae; Galerucinae). *Insect Molecular Biology*, **13**, 495–518.

Gillespie, J. J., Johnston, J. S., Cannone, J. J. and Gutell, R. R. (2006). Characteristics of the nuclear (18S, 5.8S, 28S and 5S) and mitochondrial (12S and 16S) rRNA genes of *Apis mellifera* (Insecta: Hymenoptera): structure, organization and retrotransposable elements. *Insect Molecular Biology*, **15**, 657–86.

Gillespie, J. J., Munro, J. B., Heraty, J. M., Yoder, M. J., Owen, A. K. and Carmichael, A.E. (2005). A secondary structural model of the 28S rRNA expansion segments D2 and D3 for chalcidoid wasps (Hymenoptera : Chalcidoidea). *Molecular Biology and Evolution*, **22**, 1593–608.

Haine, E. R., Martin, J. and Cook, J. M. (2006). Deep mtDNA divergences indicate cryptic species in a fig-pollinating wasp. *BMC Evolutionary Biology*, **6, 83.**

Herre, E. A., Machado, C. A., Bermingham, E., Nason, J. D., Windsor, D. M., McCafferty, S., Van Houten, W. and Bachmann, K. (1996). Molecular phylogenies of figs and their pollinator wasps. *Journal of Biogeography*, **23**, 521–30.

Hossaert-McKey, M., Giberneau, M. and Frey, J. E. (1994). Chemosensory attraction of fig wasps to substances produced by receptive figs. *Entomologia Experientia et Applicata*, **70**, 185–91.

Huelsenbeck, J. P. and Ronquist, F. (2001). MrBayes: Bayesian inferences of phylogeny. *Bioinformatics*, **17**, 754–5.

Huelsenbeck, J. P., Larget, B., Miller, R. E. and Ronquist, F. (2002). Potential applications and pitfalls of Bayesian inference in phylogeny. *Systematic Biology*, **51**, 673–88.

Jackson, A. P. (2004). Cophylogeny of the Ficus microcosm. *Biological Reviews*, **79**, 751–68.

Jackson, A. P., Machado, C. A., Robbins, N. and Herre, E. A. (2008). Multi-locus phylogenetic analysis of neotropical figs does not support cospeciation with the pollinators: The importance of systematic scale in fig.wasp cophylogenetic studies. *Symbiosis*, **45**, 57–72.

Jiang, Z. F., Huang, D. W., Zhu, C. D. and Zhen, W. Q. (2006). New insights into the phylogeny of fig pollinators using Bayesian analyses. *Molecular Phylogenetics and Evolution*, **38**, 306–15.

Jousselin, E. and Kjellberg, F. (2001). The functional implications of active and passive pollination in dioecious figs. *Ecology Letters*, **4**, 151–8.

Jousselin, E., Rasplus, J. Y. and Kjellberg, F. (2003). Convergence and coevolution in a mutualism: evidence from a molecular phylogeny of *Ficus*. *Evolution*, **57**, 1255–72.

Jousselin, E., Desdevises, Y. and Coeur d' Acier, A. (2009). Fine-scale cospeciation between Brachycaudus and Buchnera aphidicola: bacterial genome helps define species and evolutionary relationships in aphids. *Proceedings of the Royal Society B*, **276**, 187–96.

Jousselin, E., van Noort, S., Rasplus, J. Y., Rønsted, N., Erasmus, C. and Greeff, J. (2008). One tree to bind them all: host conservatism in a fig wasp community unraveled by cospeciation analyses among pollinating and nonpollinating fig wasps. *Evolution*, **62**, 1777–97.

Kato, M., Takimura, A. and Kawakita, A. (2003). An obligate pollination mutualism and reciprocal diversification in the tree genus Glochidion (Euphorbiaceae). *Proceedings of the National Acacemy of Sciences USA*, **100**, 5264–7.

Kawakita, A., Takimura, A., Terachi, T., Sota, T. and Kato, M. (2004). Cospeciation analysis of an obligate pollination mutualism: Have Glochidion trees (Euphorbiaceae) and pollinating Epicephala moths (Gracillariidae) diversified in parallel? *Evolution*, **58**, 2201–14.

Kerdelhué, C., Hochberg, M. and Rasplus, J. Y. (1997). Active pollination of Ficus sur by two sympatric fig wasp species in West Africa. *Biotropica*, **29**, 69–75.

Kjellberg, F., Jousselin, E., Hossaert-McKey, M. and Rasplus, J. Y. (2005). Biology, ecology and evolution of fig-pollinating wasps (Chalcidoidea, Agaonidae). In *Biology, Ecology, and Evolution of Gall-Inducing Arthropods*, ed. A. Raman, C. W. Schaefer and T. M. Withers. Enfield, NH: Sciences Publishers, Inc., pp. 539–72.

Kjer, K. M. (1995). Use of rRNA secondary structure in phylogenetic studies to identify homologous positions: an example of alignment and data presentation from the frogs. *Molecular Phylogenetics and Evolution*, **4**, 314–30.

Legendre, P., Desdevises, Y. and Bazin, E. (2002). A statistical test for host-parasite coevolution. *Systematic Biology*, **51**, 217–34.

Light, J. E. and Hafner, M. S. (2008). Codivergence in heteromyid rodents (Rodentia:Heteromyidae) and their sucking lice of the genus Fahrenholzia (Phthiraptera:Anoplura). *Systematic Biology*, **57**, 449–65.

Lopez-Vaamonde, C., Dixon, D. J., Cook, J. and Rasplus, J. Y. (2002). Revision of the Australian species of Pleistodontes (Hymenoptera:Agaonidae) fig-pollinating wasps and their host plant associations. *Zoological Journal of the Linnean Society*, **136**, 637–83.

Lopez-Vaamonde, C., Rasplus, J. Y., Weiblen, G. D. and Cook, J. M. (2001). Molecular phylogenies of fig wasps: partial cocladogenesis of pollinators and parasites. *Molecular Phylogenetics and Evolution*, **21**, 55–71.

Lopez-Vaamonde, C., Cook, J. M., Rasplus, J.-Y., Machado, C. A. and Weiblen, G. (2009). Molecular dating and biogeography of fig-pollinating wasps. *Molecular Phylogenetics and Evolution*, **52**, 715–26.

Machado, C. A., Robbins, N., Gilbert, M. T. P. and Herre, E. A. (2005). Critical review of host specificity and its coevolutionary implications in the fig/ fig-wasp mutualism. *Proceedings of the National Academy of Sciences USA*, **102**, 6558–65.

Machado, C. A., Jousselin, E., Kjellberg, F., Compton, S. and Herre, E. A. (2001). Phylogenetic relationships, historical biogeography and character evolution of fig-pollinating wasps. *Proceedings of the Royal Society of London B*, **268**, 685–94.

Marshall, D. C., Simon, C. and Buckley, T. R. (2006). Accurate branch length estimation in partitioned Bayesian analyses requires accommodation of among-partition rate variation and attention to branch length priors. *Systematic Biology*, **55**, 993–1003.

Marussich, W. A. and Machado, C. A. (2007). Host-specificity and coevolution among pollinating and non-pollinating New World fig wasps. *Molecular Ecology*, **16**, 1925–46.

McGuire, J. A., Witt, C. C., Altshuler, D. L. and Remsen, J. V. J. (2007). Phylogenetic systematics and biogeography of hummingbirds: bayesian and maximum likelihood analyses of partitioned data and selection of an appropriate partitioning strategy. *Systematic Biology*, **56**, 837–56.

Meier-Kolthoff, J. P., Auch, A. F., Huson, D. H. and Goker, M. (2007). CopyCat: cophylogenetic analysis tool. *Bioinformatics*, **23**, 898–900.

Michaloud, G., Michaloud-Pelletier, S., Wiebes, J. T. and Berg, C. C. (1985). The co-occurrence of two pollinating species of fig wasp on one species of fig. *Proceedings of the Koninklijke Nederlandse Akademie van Wetenschappen, (C)*, **88**, 93–119.

Molbo, D., Machado, C. A., Sevenster, J. G., Keller, L. and Herre, E. A. (2003). Cryptic species of fig-pollinating wasps: implications for the evolution of the fig–wasp mutualism, sex allocation, and precision of adaptation. *Proceedings of the National Academy of Sciences USA*, **100**, 5867–5872.

Nylander, J. A. A. (2004). MrAIC.pl. Program distributed by the author. Evolutionary Biology Centre, Uppsala University.

Nylander, J. A. A., Ronquist, F., Huelsenbeck, J. P. and Nieves-Aldrey, J. L. (2004). Bayesian phylogenetic analysis of combined data. *Systematic Biology*, **53**, 47–67.

Nylander, J. A. A., Wilgenbusch, J. C., Warren, D. L. and Swofford, D. L. (2008). AWTY (are we there yet?): a system for grafical exploration of MCMC convergence in Bayesian phylogenetics. *Bioinformatics*, **24**, 581–4.

Page, R. D. M. (1991). Clocks, clades, and cospeciation: comparing rates of evolution and timing of cospeciationn events in host-parasite assemblages. *Systematic Zoology*, **40**, 188–98.

Page, R. D. M. (1994). Parallel phylogenies – reconstructing the history of host-parasite assemblages. *Cladistics*, **10**, 155–73.

Page, R. D. M. and Charleston, M. A. (1998). Trees within trees: phylogeny and historical associations. *Trends in Ecology and Evolution*, **13**, 356–9.

Paterson, A. M. and Banks, J. B. (2001). Analytical approaches to measuring cospeciation of host and parasites: through a glass, darkly. *International Journal of Parasitology*, **31**, 1012–22.

Pellmyr, O. (2003). Yuccas, yucca moths, and coevolution: a review. *Annals of the Missouri Botanical Garden*, **90**, 35–55.

Peng, Y. Q., Duan, Z.-B., Yang, D. R. and Rasplus, J. Y. (2008). Co-occurrence of two Eupristina species on Ficus altissima in Xishuangbanna, SW China. *Symbiosis*, **45**, 9–14.

Percy, D. M., Page, R. D. M. and Cronk, Q. C. B. (2004). Plant–insect interactions: double-dating associated insect and plant lineages reveals asynchronous radiations. *Systematic Biology*, **53**, 120–7.

Rambaut, A. and Drummond, A. J. (2003). Tracer v1.5. Available online at http://tree.bio.ed.ac.uk/software/tracer/

Ramirez, W.B. (1974). Coevolution of *Ficus* and Agaonidae. *Annals of the Missouri Botanical Garden*, **61**, 770–780.

Ramirez, W. B. (1991). Evolution of the mandibular appendage in fig wasps (Hymenoptera: Agaonidae). *Revista de Biologia Tropical*, **39**(1), 87–95.

Rasplus, J. Y. (1996). *The one-to-one species-specificity of the Ficus-Agaoninae mutualism : how casual?* In *The Biodiversity of African Plants*, ed. L. J. G. Van Der Maesen, X. M. Van Der Burgt and J. M. Van Medenbach De Rooy. Wageningen, The Netherlands: Kluwer Academic Publishers, pp. 639–649.

Renoult, J. P., Kjellberg, F., Grout, C., Santoni, S. and Khadari, B. (2009). Cyto-nuclear discordance in the phylogeny of *Ficus* section *Galoglychia* and host shifts in plant–pollinator associations. *BMC Evolutionary Biology*, **9**, 248.

Rønsted, N., Salvo, G. and Savolainen, V. (2007). Biogeographical and phylogenetic origins of African fig species (Ficus section Galoglychia). *Molecular Phylogenetics and Evolution*, **43**, 190–201.

Rønsted, N., Weiblen, G., Savolainen, V. and Cook, J. (2008a). Phylogeny, biogeography, and ecology of Ficus section Malvanthera (Moraceae). *Molecular Phylogenetics and Evolution*, **48**, 12–22.

Rønsted, N., Weiblen, G., Clement, W. L., Zerega, N. J. C. and Savolainen, V. (2008b). Reconstructing the phylogeny of figs (Ficus, Moraceae) to reveal the history of the fig pollination mutualism. *Symbiosis*, **45**, 45–55.

Rønsted, N., Weiblen, G. D., Cook, J. M., Salamin, N., Machado, C. A. and Savolainen, P. (2005). 60 million years of codivergence in the fig-wasp symbiosis. *Proceedings of the Royal Society B: Biological Sciences*, **272**, 2593–99.

Shanahan, M., So, S., Compton, S. and Corlett, R. (2001). Fig-eating by vertebrate frugivores: a global review. *Biological Reviews of the Cambridge Philosophical Society*, **76**, 529–70.

Silvieus, S. I., Clement, W. L. and Weiblen, G. D. (2007). Cophylogeny of figs, pollinators, gallers and parasitoids. In *Specialization, Speciation, and Radiation: the Evolutionary Biology of Herbivorous Insects*, ed. K. J. Tilmon. Berkeley, CA: University of California Press, pp. 225–39.

Simmons, M. P., Pickett, K. M. and Miya, M. (2004). How meaningful are Bayesian support values? *Molecular Biology and Evolution*, **21**, 188–99.

Smith, C. I., Pellmyr, O., Althoff, D. M., Balcazar-Lara, M., Leebens-Mack, J. and Segraves, K. A. (2008). Pattern and timing of diversification in Yucca (Agavaceae): specialized pollination does not escalate rates of diversification. *Proceedings of the Royal Society* B, **275**, 249–58.

Stamatakis, A. (2006a). Phylogenetic models of rate heterogeneity: a high performance computing perspective. International Parallel and Distributed Processing Symposium (IPDPS 2006) Rhodes Island, Greece, 8 pp.

Stamatakis, A. (2006b). RAxML-VI-HPC: maximum likelihood-based phylogenetic analyses with thousands of taxa and mixed models. *Bioinformatics*, **22**, 2688–90.

Su, Z. H., Iino, H., Nakamura, K., Serrato, A. and Oyama, K. (2008). Breakdown of the one-to-one rule in Mexican fig-wasp associations inferred by molecular phylogenetic analysis. *Symbiosis*, **45**, 73–81.

Tamura, K., Dudley, J., Nei, M. and Kumar, S. (2007). MEGA4: Molecular Evolutionary Genetics Analysis (MEGA) software version 4.0. *Molecular Biology and Evolution*, **24**, 1596–99.

Thompson, J. D., Higgins, D. G. and Gibson, J. T. (1994). CLUSTAL W: improving the sensitivity of progressive multiple sequence alignment through sequence weighting, position specific gap penalties and weight matrix choice. *Nucleic Acids Research*, **22**, 4673–80.

Weiblen, G. D. (2000). Phylogenetic relationships of functionally dioecious Ficus (Moraceae) based on ribosomal DNA sequences and morphology. *American Journal of Botany*, **87**, 1342–57.

Weiblen, G. D. (2001). Phylogenetic relationships of fig wasps pollinating functionally dioecious Ficus based on mitochondrial DNA sequences and morphology. *Systematic Biology*, **50**, 243–67.

Weiblen, G. D. (2002). How to be a fig wasp? *Annual Review of Entomology*, **47**, 299–330.

Weiblen, G. D. (2004). Correlated evolution in fig pollination. *Systematic Biology*, **53**, 128–39.

Weiblen, G. D. and Bush, G. L. (2002). Speciation in fig pollinators and parasites. *Molecular Ecology*, **11**, 1573–78.

Wiebes, J. T. (1982). The phylogeny of the Agaonidae (Hymenoptera, Chalcidoidea). *Netherlands Journal of Zoology*, **32**, 395–411.

Wiens, J. J., Fetzner, J. W., Parkinson, C. L. and Reeder, T. W. (2005). Hylid frog phylogeny and sampling strategies for speciose clades. *Systematic Biology*, **54**, 719–48.

Yokohama, J. (1994). Molecular phylogeny and coevolution. *Plant Species Biology*, **9**, 163–7.

5

Fossil bees and their plant associates

DENIS MICHEZ, MARYSE VANDERPLANCK
AND MICHAEL S. ENGEL

5.1 Introduction

The bees comprise a derived monophyletic group (Anthophila) of pollen-consuming (secondarily phytophagous) wasps of the superfamily Apoidea, and that diverged from a grade of predatory apoid wasps (formerly "Sphecidae") sometime in the mid Cretaceous (~120–125 megaannum) (Michener 1944, 1979, 2007; Brothers 1975, 1998; Alexander 1992; Ronquist 1999; Engel 2001a, 2011; Danforth et al. 2006). Seven contemporary families are usually acknowledged: Andrenidae, Apidae, Colletidae, Halictidae, Melittidae, Megachilidae and Stenotritidae, including ~1200 genera and ~20 000 species (Michener 2007; Engel 2005, 2011). Two fossil families are also described: Paleomelittidae from middle Eocene Baltic amber, and a stem-group, Melittosphecidae from Cretaceous Burmese amber which, as discussed below, may or may not be a bee (Engel 2001a; Poinar and Danforth 2006; Ohl and Engel 2007). Bees likely arose concomitantly with the diversification of flowering plants (angiosperms) (Michener 1979; Grimaldi 1999; Engel 1996, 2001a; Crepet et al. 2004; Grimaldi and Engel 2005). Represented by more than 250 000 described species, angiosperms are the most diversified group of vascular plants, covering nearly all terrestrial and many aquatic habitats (Soltis and Soltis 2004). The congruent rise of flowering plants and numerous phytophagous insect lineages, such as bees, ditrysian Lepidoptera, and various flowering-visiting beetles and flies, has fuelled the

Evolution of Plant–Pollinator Relationships, ed S. Patiny. Published by Cambridge University Press. © The Systematics Association 2012.

notion of coradiation between these lineages. Such a conclusion is supported by the observation of flowers with specific combinations of traits that are correlated with particular pollinators (Bronstein et al. 2006). Selection for insect-pollinated clades is also supported by the fact that deliverance by pollinators of unconsumed pollen to the host plant's female reproductive organs is clearly less stochastic and more efficient than alternative ancestral wind, water or gravity dispersive methods (Labandeira 1998). Lastly, association with pollinators increases opportunities for the evolution of specialization and subsequent diversification (Vamosi and Vamosi 2010).

To test hypotheses regarding the macroevolutionary dynamics of plants and their insect associates, the examination of fossil plant–insect interactions is essential. This chapter is focused on the peculiar mutualistic interactions between angiosperms and their major pollinators, the bees, as well as a consideration of those traces of their past interactions preserved in the geological records. We first describe modern bee–plant interactions and their syndromes. From there we synthesize some methodologies for studying past ecological associations. The bulk of this contribution is an elaboration of the main fossil records for bees in the context of their contemporaneous environmental factors, such as climate, habitat, and likely host plants. A general catalogue of bee fossils is presented in Table 5.1 and constitutes the current state of affairs for paleomelittology. Naturally, much revisionary work remains to be undertaken for all of these deposits, particularly the historical accounts of paleofaunas such as Florissant, and a re-evaluation of these may result in considerable changes of taxonomic affinity. Such changes are beyond the scope of the present work and require careful revisions of historical type material. Relationships between bee fossils and their likely host plants are discussed in the context of higher relationships as proposed in Grimaldi (1999) and Danforth et al. (2006). Refer to Engel (2004b), Grimaldi and Engel (2005), and Ohl and Engel (2007) for an alternative set of phylogenetic relationships in association with the fossil records.

5.2 Modern evidence of bee–plant interactions

Pollinators develop adaptive morphological features to forage on plant rewards while plants develop traits to announce such rewards. These adaptive character syndromes are described as "pollination syndromes." They are morphologically convergent adaptive trends exhibited by both the floral features of pollinated plants and the mouthpart structures as well as other flower-interactive features of their respective pollinators (Proctor et al. 1996; Bronstein et al. 2006). Bees and bee-pollinated angiosperms show obvious pollination syndromes.

Table 5.1 Bee body fossils and traces are listed by geological eras and then in alphabetical order. All species described from inclusions in amber as well as compressions are extinct. All species described from copal are still extant with the exceptions of *Liotrogona vetula* and *Euglossa cotylisca*. All extinct supraspecific taxa are noted by an asterix (*). The classification follows that of Michener (2007). A = Amber. C = Compression. Co = Copal. IN = ichnospecies, nest trace (we do not consider nest traces included in the ichnogenera *Brownichnus*, *Celliforma* and *Palmiraichnus* because of their simple structure, poorly relatable to bees). ILD = ichnospecies, leaf damage. F = Female. FW = Female worker. M = Male. Upp = Upper. Low = Lower. Mid = Middle. Num = Numerous. Note that the trace fossil *"Atta praecursor"* from the Late Cretaceous of Bohemia (Frič and Bayer 1901) was considered a *Megachile* trace by Baroni Urbani (1980) but the specimen does not appear to be a bee trace fossil (personal observation).

Family	Subfamily	Tribe	Species (A, C) or Ichnospecies (IN, ILD)	Age	Deposit locality	N	References
Cretaceous							
Apid.	Apinae	Meliponini	*Cretotrigona* * *prisca* (Michener and Grimaldi 1988) (A)	~70	Kinkora (USA)	1FW	Michener and Grimaldi (1988a, 1988b); Engel (2000b)
Hal.	Incerta Sedis	Incerta Sedis	*Cellicalichnus chubutensis* Genise 2000 (IN)	~70	Chubut (Argentina)	Num	Genise (2000); Engel and Peñalver (2006)
Hal.	Incerta Sedis	Incerta Sedis	*C. dakotensis* (Elliot and Nations 1998) (IN)	~94	Arizona (USA)	15	Elliott and Nations (1998); Genise (2000); Engel and Peñalver (2006)
Hal.	Incerta Sedis	Incerta Sedis	*Corimbatichnus fernandezi* Genise and Verde 2000 (IN)	~70	Nueva Palmira (Urug.)	Num	Genise (2000); Genise and Verde (2000); Engel and Peñalver (2006)
Hal.	Incerta Sedis	Incerta Sedis	*Ellipsoideichnus meyeri* Roselli 1987 (IN)	~70	Nueva Palmira (Urug.)	1	Genise (2000); Engel and Peñalver (2006)

Table 5.1 (cont.)

Family	Subfamily	Tribe	Species (A, C) or Ichnospecies (IN, ILD)	Age	Deposit locality	N	References
Hal.	Incerta Sedis	Incerta Sedis	Uruguay auroranormae Roselli 1938 (IN)	–70	Nueva Palmira (Urug.)	Num	Genise (2000); Engel and Peñalver (2006)
Hal.	Incerta Sedis	Incerta Sedis	Uruguay rivasi (Roselli 1987) (IN)	–70	Nueva Palmira (Urug.)	Num	Genise (2000); Engel and Peñalver (2006)
Melitt.*	Melittosphe.	Melittosphecini	Melittosphex burmensis Poinar and Danforth 2006 (A)	–100	Hukawng valley (Bur.)	1M	Poinar and Danforth (2006); Ohl and Engel (2007)
Paleocene							
Apid.	Apinae	Anthophorini	Paleohabropoda* oudardi Michez and Rasmont 2009 (C)	–60	Menat (France)	1F	Michez et al. (2009a)
Meg.	Megachilinae	Incerta Sedis	Probombus* hirsutus Piton 1940 (C)	–60	Menat (France)	1F	Nel and Petrulevicius (2003)
Meg.	Megachilinae	Megachilini	Phagophytichnus spp. (ILD)	–60	Menat (France)	1	Wedmann et al. (2009)
Eocene – Baltic amber							
Apid.	Apinae	Electrapini*	Electrapis krishnorum Engel 2001 (A)	–48	Baltic basin	4FW	Engel (2001a)
Apid.	Apinae	Electrapini*	E. martialis (Cockerell 1908) (A)	–48	Baltic basin	2FW	Cockerell (1908c); Engel (2001a)
Apid.	Apinae	Electrapini*	E. meliponoides (Buttel-Reepen 1906) (A)	–48	Baltic basin	1FW	Engel (2001a)
Apid.	Apinae	Electrapini*	E. tornquisti Cockerell 1908 (A)	–48	Baltic basin	3FW	Cockerell (1908c); Engel (2001a)

Apid.	Apinae	Electrapini*	Protobombus basilaris Engel 2001 (A)	-48	Baltic basin	1FW	Engel (2001a)
Apid.	Apinae	Electrapini*	P. fatalis (Cockerell 1908) (A)	-48	Baltic basin	2FW	Cockerell (1908c); Engel (2001a)
Apid.	Apinae	Electrapini*	P. hirsutus (Cockerell 1908) (A)	-48	Baltic basin	4FW	Cockerell (1908c); Engel (2001a)
Apid.	Apinae	Electrapini*	P. indecisus (Cockerell 1908) (A)	-48	Baltic basin	4FW	Cockerell (1908c); Engel (2001a)
Apid.	Apinae	Electrapini*	P. tristellus Cockerell 1909 (A)	-48	Baltic basin	1FW	Engel (2001a)
Apid.	Apinae	Electrapini*	Thaumastobombus andreniformis Engel 2001 (A)	-48	Baltic basin	2FW	Engel (2001a); Patiny et al. (2007)
Apid.	Apinae	Electrobombini*	Electrobombus samlandensis Engel 2001 (A)	-48	Baltic basin	2FW	Engel (2001a)
Apid.	Apinae	Melikertini*	Melikertes clypeatus Engel 2001 (A)	-48	Baltic basin	3FW	Engel (2001a, 2004b, unpublished data)
Apid.	Apinae	Melikertini*	M. proavus (Menge 1856) (A)	-48	Baltic basin	2FW	Engel (2001a)
Apid.	Apinae	Melikertini*	M. stilbonotus (Engel 1998) (A)	-48	Baltic basin	9FW	Engel (1998b, 2001a, 2004b)
Apid.	Apinae	Melikertini*	Melissites trigona Engel 2001 (A)	-48	Baltic basin	1FW	Engel (2001a)

Table 5.1 (cont.)

Family	Subfamily	Tribe	Species (A, C) or Ichnospecies (IN, ILD)	Age	Deposit locality	N	References
Apid.	Apinae	Melikertini*	Roussyana palmnickenensis (Roussy 1937) (A)	–48	Baltic basin	3FW	Engel (2001a)
Apid.	Apinae	Melikertini*	Succinapis goeleti Engel 2001 (A)	–48	Baltic basin	4FW	Engel (2001a, 2004b)
Apid.	Apinae	Melikertini*	S. micheneri Engel 2001 (A)	–48	Baltic basin	30FW	Engel (2001a, 2004b)
Apid.	Apinae	Melikertini*	S. proboscidea Engel 2001 (A)	–48	Baltic basin	3FW	Engel (2001a)
Apid.	Apinae	Meliponini	Liotrigonopsis rozeni Engel 2001 (A)	–48	Baltic basin	1FW	Engel (2001a)
Apid.	Apinae	Meliponini	Kelneriapis eocenica Kelner-Pillault 1969 (A)	–48	Baltic basin	1FW	Engel (2001a)
Apid.	Xylocopinae	Boreallodapini*	Boreallodape baltica Engel 2001 (A)	–48	Baltic basin	9F	Engel (2001a, 2004b)
Apid.	Xylocopinae	Boreallodapini*	B. mollyae Engel 2001 (A)	–48	Baltic basin	6F	Engel (2001a, 2004b)
Apid.	Xylocopinae	Boreallodapini*	B. striebichi Engel 2001 (A)	–48	Baltic basin	14F	Engel (2001a)
Hal.	Halictinae	Halictini	Electrolictus* antiquus Engel 2001 (A)	–48	Baltic basin	1F	Engel (2001a)
Mel.	Melittinae	Macropidini	Eomacropis* glaesaria Engel 2001 (A)	–48	Baltic basin	1F	Engel (2001a); Michez et al. (2009b)

	Subfamily	Tribe	Species	Age	Locality	Specimens	Reference
Meg.	Megachilinae	Protolithurgini*	*Protolithurgus ditomeus* Engel 2001 (A)	~48	Baltic basin	1F	Engel (2001a, 2005); Engel and Perkovsky (2006)
Meg.	Megachilinae	Ctenoplectrellini*	*Ctenoplectrella cockerelli* Engel 2001 (A)	~48	Baltic basin	3F	Engel (2001a); Engel and Perkovsky (2006)
Meg.	Megachilinae	Ctenoplectrellini*	*C. gorskii* Engel 2008 (A)	~48	Baltic basin	2F	Engel (2008)
Meg.	Megachilinae	Ctenoplectrellini*	*C. grimaldii* Engel 2001 (A)	~48	Baltic basin	1F	Engel (2001a); Engel and Perkovsky (2006)
Meg.	Megachilinae	Ctenoplectrellini*	*C. viridiceps* Cockerell 1909 (A)	~48	Baltic basin	12F/2M	Cockerell (1909c); Engel (2001a, 2004b); Engel and Perkovsky (2006)
Meg.	Megachilinae	Ctenoplectrellini*	*Glaesosmia* genalis* Engel 2001 (A)	~48	Baltic basin	1F	Engel (2001a); Engel and Perkovsky (2006)
Meg.	Megachilinae	Glyptapini*	*Glyptapis densopunctata* Engel 2001 (A)	~48	Baltic basin	4F	Engel (2001a); Engel and Perkovsky (2006)
Meg.	Megachilinae	Glyptapini*	*G. disareolata* Engel 2001 (A)	~48	Baltic basin	2F	Engel (2001a); Engel and Perkovsky (2006)
Meg.	Megachilinae	Glyptapini*	*G. fuscula* Cockerell 1909 (A)	~48	Baltic basin	9F	Engel (2001a); Engel and Perkovsky (2006)
Meg.	Megachilinae	Glyptapini*	*G. mirabilis* Cockerell 1909 (A)	~48	Baltic basin	5F	Cockerell (1909c); Engel (2001a, 2004b); Engel and Perkovsky (2006)
Pal.*	Paleomettinae	Paleomelittini	*Paleomelitta nigripennis* Engel 2001 (A)	~48	Baltic basin	2F	Engel (2001a)

Table 5.1 (cont.)

Eocene – Other than Baltic deposit

Family	Subfamily	Tribe	Species (A, C) or Ichnospecies (IN, ILD)	Age	Deposit locality	N	References
Apid.	Apinae	Electrapini*	*Electrapis electrapoides* (Lutz 1993) (C)		Messel (Germany)	3FW	Lutz (1993); Wappler and Engel (2003)
Apid.	Apinae	Electrapini*	*E. micheneri* Wappler and Engel 2003 (C)	−44	Eckfeld (Germany)	1FW	Wappler and Engel (2003)
Apid.	Apinae	Electrapini*	*E. prolata* Wappler and Engel 2003 (C)	−44	Eckfeld (Germany)	1FW	Wappler and Engel (2003)
Apid.	Apinae	Electrapini*	*Protobombus messelensis* Wappler and Engel 2003 (C)	−49	Messel (Germany)	1FW	Wappler and Engel (2003)
Apid.	Apinae	Electrapini*	*P. pristinus* Wappler and Engel 2003 (C)	−44	Eckfeld (Germany)	1FW	Wappler and Engel (2003)
Apid.	Apinae	Electrapini*	*P.* spp. (A)	−52	Cambay basin (India)	1FW	Rust et al. (2010); Engel (unpublished data)
Apid	Apinae	Melikertini*	*Melikertes* spp. (A)	−52	Cambay basin (India)	3FW	Rust et al. (2010); Engel (unpublished data)
Apid.	Apinae	Incerta Sedis	*Pygomelissa* * *lutetia* Engel and Wappler 2003 (C)	−49	Messel (Germany)	1F	Wappler and Engel (2003)
Hal.	Halictinae	Halictini	*Halictus?savenyei* Engel and Archibald 2003 (C)	−53	Quilchena (Canada)	1F	Engel and Archibald (2003); Engel and Peñalver (2006)

Mel.	Melittinae	Macropidini	*Paleomacropis* eocenicus Michez and Nel 2007 (A)	−53	Oise (France)	1F	Michez et al. (2007)
Meg.	Megachilinae	Ctenoplectrellini*	*Friccomelissa schopowi* Wedmann et al. 2009 (C)	−49	Messel (Germany)	1F	Wedmann et al. (2009)
Meg.	Megachilinae	Ctenoplectrellini*	*Ctenoplectrella zherkhini* Engel and Perkovsky 2006 (A)	−48	Rovno (Ukraine)	1F	Engel and Perkovsky (2006)
Meg.	Megachilinae	Megachilini	*Phagophytichnus pseudocircus* Sarzetti et al. 2008 (ILD)	−44	Eckfeld (Germany)	3	Wappler and Engel (2003); Wedmann et al. (2009)
Meg.	Megachilinae	Megachilini	*P. pseudocircus* Sarzetti et al. 2008 (ILD)	−49	Messel (Germany)	1	Wedmann et al. (2009)
Meg.	Megachilinae	Megachilini	*P. pseudocircus* Sarzetti et al. 2008 (ILD)	Mid	Puryear (USA)	1	Brooks (1955); Labandeira (2002a); Wedmann et al. (2009)
Meg.	Megachilinae	Megachilini	*P. pseudocircus* Sarzetti et al. 2008 (ILD)	−48	Rio Pichi-Leufú (Arg.)	1	Sarzetti et al. (2008)
Meg.	Megachilinae	Megachilini	*Phagophytichnus* spp. (ILD)	−44	Eckfeld (Germany)	4	Wappler and Engel (2003); Wedmann et al. (2009)
Meg.	Megachilinae	Megachilini	*Phagophytichnus* spp. (ILD)	−49	Messel (Germany)	4	Wedmann et al. (2009)
Meg.	Megachilinae	Megachilini	*Phagophytichnus* sp. (ILD)	−49	Republic (USA)	2	Labandeira (2002a)

Table 5.1 (cont.)

Family	Subfamily	Tribe	Species (A, C) or Ichnospecies (IN, ILD)	Age	Deposit locality	N	References
Meg.	Megachilinae	Megachilini	Phagophytichnus spp. (ILD)	−49	MacAbee (Canada)	1	Labandeira (2002a)
Meg.	Megachilinae	Megachilini	Phagophytichnus spp. (ILD)	Mid	Viola (USA)	1	Wedmann et al. (2009)
Oligocene – Florissant shale							
And.	Andreninae	Andrenini	Andrena? clavula Cockerell 1906 (C)	−32	Florissant (USA)	1F	Cockerell (1906)
And.	Andreninae	Andrenini	A. grandipes Cockerell 1911 (C)	−32	Florissant (USA)	1F	Cockerell (1911b)
And.	Andreninae	Andrenini	A. hypolitha Cockerell 1908 (C)	−32	Florissant (USA)	1F	Cockerell (1908c)
And.	Andreninae	Andrenini	A. percontusa Cockerell 1914 (C)	−32	Florissant (USA)	1F	Cockerell (1914)
And.	Andreninae	Andrenini	A. septula Cockerell 1906 (C)	−32	Florissant (USA)	1F	Cockerell (1906)
And.	Andreninae	?	Lithandrena* saxorum Cockerell 1906 (C)	−32	Florissant (USA)	1F	Cockerell (1906)
And.	Andreninae	?	Pelandrena* reducta Cockerell 1909 (C)	−32	Florissant (USA)	1F	Cockerell (1909b)
And.	Panurginae	?	Libellulapis* antiquorum Cockerell 1906 (C)	−32	Florissant (USA)	2F	Cockerell (1906, 1913b)

And.	Panurginae	?	L.* wilmattae Cockerell 1913 (C)	-32	Florissant (USA) 1F	Cockerell (1913b)
Apid.	Apinae	Anthophorini	Anthophora melfordi Cockerell 1908 (C)	-32	Florissant (USA) 1?	Cockerell (1908c); Michez et al. (2009a)
Apid.	Bombinae	Bombini	Calyptapis* florissantensis Cockerell 1906 (C)	-32	Florissant (USA) 2?	Cockerell (1906, 1908c)
Apid.	Xylocopinae	Ceratinini	Ceratina disrupta Cockerell 1906 (C)	-32	Florissant (USA) 1?	Cockerell (1906); Engel (2001b)
Apid.	Xylocopinae	Xylocopini	Xylocopa gabrielae Engel 2001 (C)	-32	Florissant (USA) 1F	Engel (2001b); Nel and Petruvecius (2003)
Apid.	Apinae	Melectini	Protomelecta* brevipennis Cockerell 1908 (C)	-32	Florissant (USA) 1?	Cockerell (1908a); Engel (2004a)
Hal.	Halictinae	Halictini	Cyrtapis* anomalus (Cockerell 1908) (C)	-32	Florissant (USA) 1?	Cockerell (1908a); Engel (2002a); Engel and Peñalver (2006)
Hal.	Halictinae	Halictini	Kronolictus* scudderiellus Cockerell 1906 (C)	-32	Florissant (USA) 1F	Cockerell (1906); Engel (2002a); Engel and Peñalver (2006)
Hal.	Halictinae	Halictini	K.* vulcanus Engel 2002 (C)	-32	Florissant (USA) 1F	Engel (2002a); Engel and Peñalver (2006)
Hal.	Halictinae	Halictini	Ocymoromelitta* florissantella Cockerell 1906 (C)	-32	Florissant (USA) 1F	Cockerell (1906); Engel (2002a); Engel and Peñalver (2006)

Table 5.1 (cont.)

Family	Subfamily	Tribe	Species (A, C) or Ichnospecies (IN, ILD)	Age	Deposit locality	N	References
Hal.	Halictinae	Halictini	O.* miocenica (Cockerell 1909) (C)	−32	Florissant (USA)	1F	Cockerell (1909b); Engel (2002a); Engel and Peñalver (2006)
Hal.	Halictinae	Halictini	O.* sorella Engel 2002 (C)	−32	Florissant (USA)	1F	Engel (2002a); Engel and Peñalver (2006)
Mel.	Melittinae	Melittini	Melitta willardi Cockerell 1909 (C)	−32	Florissant (USA)	1F	Cockerell (1909a); Michez et al. (2007)
Meg.	Megachilinae	Anthidiini	Anthidium exhumatum Cockerell 1906 (C)	−32	Florissant (USA)	1M	Cockerell (1906); Engel (1999c); Engel and Perkovsky (2006)
Meg.	Megachilinae	Anthidiini	A. scuderri Cockerell 1906 (C)	−32	Florissant (USA)	1?	Engel (1999c); Engel and Perkovsky (2006)
Meg.	Megachilinae	Anthidiini	Dianthidium? tertiarium Cockerell 1906 (C)	−32	Florissant (USA)	1M	Cockerell (1906); Engel (1999c); Engel and Perkovsky (2006)
Meg.	Megachilinae	Anthidiini	Lithanthidium pertriste Cockerell 1911 (C)	−32	Florissant (USA)	1?	Cockerell (1911a); Engel (1999c); Engel and Perkovsky (2006)
Meg.	Megachilinae	Megachilini	Megachile praedicta Cockerell 1908 (C)	−32	Florissant (USA)	1F	Cockerell (1908b); Engel (1999c); Engel and Perkovsky (2006)
Meg.	Megachilinae	Megachilini	Phagophytichnus pseudocircus Sarzetti et al. 2008 (ILD)	−32	Florissant (USA)	1	Cockerell (1910); Sarzetti et al. (2008); Wedmann et al. (2009)

Meg.	Megachilinae	Osmiini	*Heriades bowditchi* Cockerell 1906 (C)	–32	Florissant (USA)	2F	Cockerell (1906); Engel (1999c); Engel and Perkovsky (2006)
Meg.	Megachilinae	Osmiini	*H. halictinus* Cockerell 1906 (C)	–32	Florissant (USA)	1F	Cockerell (1906); Engel (1999c); Engel and Perkovsky (2006)
Meg.	Megachilinae	Osmiini	*H. laminarum* Cockerell 1906 (C)	–32	Florissant (USA)	1?	Cockerell (1906); Engel (1999c); Engel and Perkovsky (2006)
Meg.	Megachilinae	Osmiini	*H. mersatus* Cockerell 1923 (C)	–32	Florissant (USA)	1?	Cockerell (1923); Engel (1999c); Engel and Perkovsky (2006)
Meg.	Megachilinae	Osmiini	*H. mildredae* Cockerell 1925 (C)	–32	Florissant (USA)	1?	Engel (1999c); Engel and Perkovsky (2006)
Meg.	Megachilinae	Osmiini	*H. priscus* Cockerell 1917 (C)	–32	Florissant (USA)	1F	Cockerell (1917); Engel (1999c); Engel and Perkovsky (2006)
Meg.	Megachilinae	Osmiini	*H. saxosus* Cockerell 1913 (C)	–32	Florissant (USA)	1M	Cockerell (1913a); Engel (1999c); Engel and Perkovsky (2006)

Oligocene – other than Florissant shale

Apid.	Apinae	Anthophorini?	*Anthophora effosa* Heyden 1862 (C)	Upp	Rott (Germany)	1?	Cockerell (1908c); Michez et al. (2009a)

Table **5.1** (cont.)

Family	Subfamily	Tribe	Species (A, C) or Ichnospecies (IN, ILD)	Age	Deposit locality	N	References
Apid.	Apinae	Apini	Apis (Synapis*) henshawi Cockerell 1907 (C)	Upp	Rott (Germany)	1FW/5?	Cockerell (1907); Meunier (1920); Arillo et al. (1996); Engel (1998a, 1999b, 2006); Kotthoff et al. (2011)
Apid.	Apinae	Apini	A. (S.*) henshawi Cockerell 1907 (C)	Upp	Marseille (France)	2?	Nel et al. (1999a)
Apid.	Apinae	Apini	A. (S.*) henshawi Cockerell 1907 (C)	Upp	Aix-en-Provence (Fr.)	3FW/2?	Nel et al. (1999a); Engel (1998a, 1999b, 2006)
Apid.	Apinae	Apini	A. (S.*) cuenoti Théobald 1937 (C)	Upp	Céreste (France)	2FW/1?	Engel (1998a); Nel et al. (1999a); Kotthoff et al. (2011)
Apid.	Apinae	Apini	A. (S.*) vetusta Engel 1998 (C)	Upp	Rott (Germany)	1FW	Engel (1998a, 2006)
Apid.	Apinae	Apini	Apis spp. C (C)	Upp	Aix-en-Provence (Fr.)	1?	Nel et al. (1999a); Kotthoff et al. (2011)
Apid.	Apinae	Eucerini	Tetralonia berlandi Théobald 1937 (C)	Upp	Céreste (France)	1?	Zeuner & Manning (1976)
Apid.	Xylocopinae	Xylocopini	Xylocopa celinae Nel and Petrulevicius 2003 (C)	Upp	Camoins-ls-bains (Fr.)	1?	Nel and Petrulevicius (2003)
Apid.	Xylocopinae	Xylocopini	X. friesei Statz 1936 (C)	Upp	Rott (Germany)	1?	Nel and Petrulevicius (2003)

				Age	Locality	Count	References
Hal.	Halictinae	Halictini	*Halictus? ruissatelensis* Timon-David 1944 (C)	Upp	Marseilles (France)	1?	Timon-David (1944); Engel (2002a)
Hal.	Halictinae	Halictini	*Lasioglossum celinae* Nel and Petrulevicius 2003 (C)	Mid	Bois d'Asson (France)	1F	Nel and Petrulevicius (2003); Engel and Peñalver (2006)
Hal.	Incerta Sedis	Incerta Sedis	*Cellicalichnus ficoides* (Retallack 1984) (IN)	Upp	Badlans (USA)	Num	Retallack (1984); Engel (1996); Genise (2000); Engel and Peñalver (2006)
Meg.	Megachilinae	Anthidiini	*Anthidium mortuum* (Meunier 1920) (C)	Upp	Rott (Germany)	1F	Meunier (1920); Engel and Perkovsky (2006)
Meg.	Megachilinae	Osmiini?	*Osmia? carbonum* Heyden 1862 (C)	Upp	Rott (Germany)	1?	Engel and Perkovsky (2006)
Miocene – Dominican amber							
And.	Panurginae	Protandrenini	*Heterosarus (Het.) eickworti* Rozen 1996 (A)	–19	Dominican Republic	1M	Rozen (1996)
Apid.	Apinae	Euglossini	*Euglossa moronei* Engel 1999 (A)	–19	Dominican Republic	1F	Engel (1999d)
Apid.	Apinae	Euglossini	*Eufriesea melissiflora* (Poinar 1998) (A)	–19	Dominican Republic	2F	Poinar (1998); Engel (2000)
Apid.	Apinae	Euglossini	*Paleoeuglossa* * undescribed species (A)	–19	Dominican Republic	1F?	Poinar (2010)
Apid.	Apinae	Meliponini	*Proplebeia* * *dominicana* (Wille and Chandler 1964) (A)	–19	Dominican Republic	>1000	Engel (2009)

Table 5.1 (cont.)

Family	Subfamily	Tribe	Species (A, C) or Ichnospecies (IN, ILD)	Age	Deposit locality	N	References
Apid.	Apinae	Meliponini	P.* tantilla Camargo et al. 2000 (A)	~19	Dominican Republic	2M/2F	Camargo et al. (2000)
Apid.	Apinae	Meliponini	P.* vetusta Camargo et al. 2000 (A)	~19	Dominican Republic	3F	Camargo et al. (2000)
Col.	Xeromelissinae	Xeromelissini	Chilicola (Hyl.) electrodominica Engel 1999 (A)	~19	Dominican Republic	1F	Engel (1999a)
Col.	Xeromelissinae	Xeromelissini	C. (H.) gracilis Michener and Poinar 1996 (A)	~19	Dominican Republic	1M	Michener and Poinar (1996)
Hal.	Halictinae	Augochlorini	Augochlora (Electr.) leptoloba Engel 2000 (A)	~19	Dominican Republic	1F	Engel (2000a, 2002a); Engel and Peñalver (2006)
Hal.	Halictinae	Augochlorini	Augochloropsis sp. (A)	~19	Dominican Republic	?	Poinar (2004); Engel (2002a); Engel and Peñalver (2006)
Hal.	Halictinae	Augochlorini	Neocorynura electra Engel 1995 (A)	~19	Dominican Republic	1F	Engel (1995, 2002a); Engel and Peñalver (2006)
Hal.	Halictinae	Augochlorini	Oligochlora* (Oligochlora) eickworti Engel 1996 (A)	~19	Dominican Republic	1F	Engel (1996, 2002a); Engel and Peñalver (2006)
Hal.	Halictinae	Augochlorini	O.* (O.) grimaldii Engel 1997 (A)	~19	Dominican Republic	2F	Engel (1997, 2002a); Engel and Peñalver (2006)

Hal.	Halictinae	Augochlorini	O.* (O.) micheneri Engel 1996 (A)	–19	Dominican Republic	1F	Engel (1996, 2002a); Engel and Peñalver (2006)
Hal.	Halictinae	Augochlorini	O.* (O.) semirugosa Engel 2009 (A)	–19	Dominican Republic	1F	Engel (2009)
Hal.	Halictinae	Augochlorini	O.* (Soliapis) marquetorum Engel & Rightmyer 2000 (A)	–19	Dominican Republic	2F	Engel and Rightmyer (2000); Engel (2002a); Engel and Peñalver (2006)
Hal.	Halictinae	Augochlorini	O.* (S.) rozeni Engel 2000 (A)	–19	Dominican Republic	1F	Engel (2000a, 2002a); Engel and Peñalver (2006)
Hal.	Halictinae	Caenohalictini	Eickwortapis* dominicana Michener and Poinar 1996 (A)	–19	Dominican Republic	1M/2F	Michener and Poinar (1996); Engel (2002a); Engel and Peñalver (2006)
Hal.	Halictinae	Caenohalictini	Nesagapostemon* moronei Engel 2009 (A)	–19	Dominican Republic	1F	Engel (2009)
Meg.	Megachilinae	Megachilini	Megachile (Chalicod.) glaesaria Engel 1999 (A)	–19	Dominican Republic	1F	Engel (1999c)

Miocene – Other than Dominican amber

And.	Andreninae	Andrenini	Andrena? primaeva Cockerell 1909 (C)	Upp	Oeningen (Germany)	1?	Cockerell (1909c)
Apid.	Apinae	Anthophorini?	Anthophorites* longaeva Heer 1867 (C)	Low	Radoboj (Croatia)	5?	Cockerell (1909c); Zeuner and Manning (1976)

Table 5.1 (*cont.*)

Family	Subfamily	Tribe	Species (A, C) or Ichnospecies (IN, ILD)	Age	Deposit locality	N	References
Apid.	Apinae	Anthophorini?	*A. mellona* Heer 1849 (C)	Upp	Oeningen (Germany)	3?	Heer (1849); Engel and Perkovsky (2006)
Apid.	Apinae	Anthophorini?	*A. thoracica* Heer 1867 (C)	Low	Radoboj (Croatia)	1?	Zeuner and Manning (1976)
Apid.	Apinae	Anthophorini?	*A. titiana* Heer 1849 (C)	Upp	Oeningen (Germany)	2?	Heer (1849); Cockerell (1909c)
Apid.	Apinae	Anthophorini?	*A. tonsa* Heer 1849 (C)	Upp	Oeningen (Germany)	1?	Heer (1849)
Apid.	Apinae	Anthophorini?	*A. veterana* Heer 1849 (C)	Upp	Oeningen (Germany)	2?	Heer (1849)
Apid.	Apinae	Apini	*Apis (Cascapis*) armbrusteri* Zeuner 1931 (C)	Upp	Böttingen (Germany)	>4FW	Engel (1998a, 1999b, 2006); Kotthoff et al. (2011)
Apid.	Apinae	Apini	*A. (C.*) armbrusteri* Zeuner 1931 (C)	Low	Rubielos (Spain)	1FW	Nel et al. (1999a); Kotthoff et al. (2011)
Apid.	Apinae	Apini	*A. (C.*) armbrusteri* Zeuner 1931 (C)	Upp	Parschlug (Austria)	2?	Nel et al. (1999a); Kotthoff et al. (2011)
Apid.	Apinae	Apini	*A. (C.*) armbrusteri* Zeuner 1931 (C)	Upp	Lleida (Spain)	9?	Nel et al. (1999a); Kotthoff et al. (2011)
Apid.	Apinae	Apini	*A. (C.*) armbrusteri* Zeuner 1931 (C)	Upp	Sainte-Reine (France)	2?	Nel et al. (1999a); Kotthoff et al. (2011)

Apid.	Apinae	Apini	A. (C.*) armbrusteri Zeuner 1931 (C)	Low	Bilina Mine (Cz. Rep.)	6?	Prokop and Nel (2003); Engel et al. (2009); Kotthoff et al. (2011)
Apid.	Apinae	Apini	A. (C.*) nearctica Engel et al. 2009 (C)	–14	Stewart valley (USA)	1FW	Engel et al. (2009); Kotthoff et al. (2011)
Apid.	Apinae	Apini	A. (Megapis) lithothermaea Engel 2006 (C)	Mid	Iki Island (Japan)	1FW	Nel et al. (1999a); Engel (2006); Kotthoff et al. (2011)
Apid.	Apinae	Apini	A. (Synapis*) longtibia Zhang 1990 (C)	Upp	Shandong (China)	1FW	Engel (1998a, 2006)
Apid.	Apinae	Apini	A. (S.*) miocenica Hong 1983 (C)	Upp	Shandong (China)	4FW	Nel et al. (1999a), Engel (1998a, 2006)
Apid.	Apinae	Apini	A. (S.*) petrefacta (Říha 1973) (C)	Low	Kundratice (Cz. Rep.)	1FW	Nel et al. (1999a); Engel (1998a, 2006)
Apid.	Apinae	Apini	A. (S.*) henshawi Cockerell 1907 (C)	Low	Izarra (Spain)	1FW	Arillo et al. (1996); Engel (1998a, 1999b, 2006); Kotthoff et al. (2011)
Apid.	Apinae	Apini	Apis "Miocene 1" (C)	Upp	Sainte-Reine (France)	3?	Nel et al. (1999a); Kotthoff et al. (2011)
Apid.	Apinae	Apini	Apis "Miocene 1" (C)	Upp	Andance (France)	8?	Nel et al. (1999a); Kotthoff et al. (2011)
Apid.	Apinae	Apini	Apis "Miocene 1" (C)	Upp	Andance (France)	2?	Nel et al. (1999a); Kotthoff et al. (2011)
Apid.	Apinae	Bombini	Bombus spp. (C)	Low	Bilina Mine (Cz. Rep.)	1?	Prokop and Nel (2003)

Table 5.1 (*cont.*)

Family	Subfamily	Tribe	Species (A, C) or Ichnospecies (IN, ILD)	Age	Deposit locality	N	References
Apid.	Apinae	Bombini	*B. anacolus* Zhang et al. 1994 (C)	Upp	Shandong (China)	1F	Zhang et al. (1994)
Apid.	Apinae	Bombini	*B. dilectus* Zhang et al. 1994 (C)	Upp	Shandong (China)	1F	Prokop and Nel (2003)
Apid.	Apinae	Bombini	*B. luianus* Zhang 1990 (C)	Upp	Shandong (China)	1F	Rasnitsyn and Michener (1991)
Apid.	Apinae	Bombini	*B. proavus* Cockerell 1931 (C)	Upp	Latah (USA)	1F	Rasnitsyn and Michener (1991)
Apid.	Apinae	Bombini	*B. vetustus* Rasnitsyn and Michener 1991 (C)	Upp	Botchi river (Russia)	1M	Rasnitsyn and Michener (1991)
Apid.	Apinae	Bombini	*B.? pristinus* Unger 1867 (C)	Low	Euboea (Greece)	1?	Rasnitsyn and Michener (1991)
Apid.	Apinae	Bombini	*B.? crassipes* Novak 1877 (C)	Low	Krottensee (Cz. Rep.)	1?	Zeuner and Manning (1976)
Apid.	Apinae	Meliponini	*Meliponorytes sicula* Tosi 1896 (A)	~20	Sicily (Italy)	1FW	Tosi (1896); Zeuner and Manning (1976); Engel (2001a)
Apid.	Apinae	Meliponini	*M. succini* Tosi 1896 (A)	~20	Sicily (Italy)	>8	Tosi (1896); Zeuner and Manning (1976); Engel (2001a)
Apid.	Apinae	Meliponini	*Nogueirapis silacea* Wille 1959 (A)	~20	Chiapas (Mexico)	Num	Wille (1959)

Apid.	Apinae	Incerta Sedis	Paraelectrobombus* patriciae Nel and Petrulevicius 2003 (C)	−22.5	Bes-Konak Basin (Tu.)	1F	Nel and Petrulevicius (2003); likely a synonym of Bombus
Apid.	Apinae	Incerta Sedis	Oligoapis* beskonakensis Nel and Petrulevicius 2003 (C)	−22.5	Bes-Konak Basin (Tu.)	1FW	Nel and Petrulevicius (2003); likely a synonym of Bombus
Apid.	Xylocopinae	Xylocopini	Xylocopa abavus (Heer 1849) (C)	Upp	Oeningen (Germany)	4F	Heer (1849); Cockerell (1909c)
Apid.	Xylocopinae	Xylocopini	X. diatoma Zhang 1990 (C)	Upp	Shandong (China)	1F	Zhang (1990); Engel (2001b)
Apid.	Xylocopinae	Xylocopini	X. hydrobiae Zeuner 1938 (C)	Upp	Biebrich (Germany)	2F	Engel (2001b); Prokop and Nel (2003)
Apid.	Xylocopinae	Xylocopini	X. jurinei (Heer 1865) (C)	Upp	Oeningen (Germany)	2F	Cockerell (1909c); Engel (2001b)
Apid.	Xylocopinae	Xylocopini	X. obata Zhang et al. 1994 (C)	Upp	Shandong (China)	1F	Engel (2001b)
Apid.	Xylocopinae	Xylocopini	X. senilis Heer 1849 (C)	Upp	Oeningen (Germany)	8?	Heer (1849); Zeuner and Manning (1976); Engel (2001b)
Apid.	Xylocopinae	Xylocopini	X. veta Zhang et al. 1994 (C)	Upp	Shandong (China)	1F	Engel (2001b)
Hal.	Halictinae	Halictini	Halictus petrefactus Engel and Peñalver 2006 (C)	Low	Rubielos (Spain)	1F	Engel and Peñalver (2006)

Table 5.1 (cont.)

Family	Subfamily	Tribe	Species (A, C) or Ichnospecies (IN, ILD)	Age	Deposit locality	N	References
Hal.	Halictinae	Halictini	*H. schemppi* (Armbuster 1938) (C)	Upp	Randeck (Germany)	4?	Engel (2002a)
Hal.	Halictinae	Halictini	*Halictus* spp. (C)	Low	Euboea (Greece)	1F	Bachmayer et al. (1971); Engel and Peñalver (2006)
Hal.	Halictinae	Incerta Sedis	Halictinae spp. (C)	Low	Izzara (Spain)	1?	Arillo et al. (1996)
Hal.	Incerta Sedis	Incerta Sedis	*Rosellichnus patagonicus* Genise and Bown 1996 (IN)	Upp	Puesto L.S. (Argen.)	2	Genise and Bown (1996); Genise (2000); Engel and Peñalver (2006)
Hal.	Incerta Sedis	Incerta Sedis	*Cellicalichnus habari* (Thackray 1994) (IN)	Upp	Rusinga (Kenya)	?	Genise (2000)
Meg.	Lithurginae	Lithurgini	*Lithurgus? adamiticus* (Heer 1865) (C)	Upp	Oeningen (Germany)	3?	Cockerell (1909c); Engel (1999c); Engel and Perkovsky (2006)
Meg.	Megachilinae	Anthidiini	*Anthidium basalticum* Zhang 1989 (C)	Upp	Shandong (China)	1F	Zhang (1989); Engel (1999c); Engel and Perkovsky (2006)
Meg.	Megachilinae	Megachilini	*Megachile amaguensis* Cockerell 1925 (C)	Low	Kudia river (Russia)	1?	Cockerell (1925); Engel (1999c); Engel and Perkovsky (2006)
Meg.	Megachilinae	Megachilini	*M. shanwangae* Zhang 1989 (C)	Upp	Shandong (China)	1F	Zhang (1989); Engel (1999c); Engel and Perkovsky (2006)

Meg.	Megachilinae	Osmiini	*Osmia? antiqua* Heer 1849 (C)	Upp	Oeningen (Germany)	1?	Heer (1849); Cockerell (1909C); Engel (1999c); Engel and Perkovsky (2006)
Meg.	Megachilinae	Osmiini	*O. nigra* Zeuner and Manning 1976 (C)	Upp	Oeningen (Germany)	2?	Zeuner and Manning (1976); Engel (1999c); Engel and Perkovsky (2006)
Meg.	Megachilinae	incertae sedis	Megachilinae spp. Indet (C)	Upp	Latah (USA)	1?	Engel (2004a); Engel and Perkovsky (2006)
Mel.	Melittinae	Macropidini	*Macropis basaltica* (Zhang 1989) (C)	Upp	Shandong (China)	1F	Zhang (1989); Michez et al. (2007)

Pliocene/Pleistocene

Apid.	Apinae	Apini	*Apis mellifera* L. 1758 (Co)	Pleis	East Africa	2FW	Cockerell (1909c); Zeuner and Manning. (1976); Engel (1998a, 2001a)
Apid.	Apinae	Apini	Comb of *Apis cerana* L. (I, comb)	Pleis	Batu (Malaysia)	1	Engel (1996)
Apid.	Apinae	Meliponini	*Hypotrigona gribodoi* (Magretti 1884) (Co)	Pleis	East Africa	19?	Zeuner and Manning (1976); Engel (2001a)
Apid.	Apinae	Meliponini	*Liotrigona vetula* Moure and Camargo 1978 (Co)	Pleis	East Africa	1FW	Moure & Camargo (1978); Engel (2001a)

Table **5.1** (cont.)

Family	Subfamily	Tribe	Species (A, C) or Ichnospecies (IN, ILD)	Age	Deposit locality	N	References
Apid.	Apinae	Meliponini	*Meliponula erythra* (Schletterer 1891) (Co)	Pleis	East Africa	1FW	Zeuner and Manning (1976); Engel (2001a)
Apid.	Apinae	Meliponini	*Trigona iridipennis* Smith 1854 (Co)	Pleis	Hukong (Myanmar)	2FW	Zeuner and Manning (1976); Engel (2001a)
Apid.	Apinae	Meliponini	*T. lurida* Smith 1854 (Co)	–2,5	Santander (Colombia)	2FW	Engel (2001a)
Apid.	Apinae	Meliponini	*Trigonisca* spp. (Co)	–2,5	Santander (Colombia)	>5FW	Engel (2001a)
Apid.	Apinae	Euglossini	*Euglossa cotylisca* Hinojosa-Diaz and Engel 2007 (Co)	–2,5	Santander (Colombia)	1M	Hinojosa-Diaz and Engel (2007)
Hal.	Halictinae	Halictini	*Dialictus coeruleus* (Robertson 1893) (C)	Pleis	Lockport (USA)	Num	Miller and Morgan (1982); Engel (2002a); Nel and Petrulevicius (2003)
Hal.	Incerta Sedis	Incerta Sedis	*Rosellichnus arabicus* Genise and Bown 1996 (IN)	0.28	Abu Dhabi (U.A.E.)	5	Genise and Bown (1996); Genise (2000); Engel and Peñalver (2006)
Meg.	Megachilinae	Osmiini	*Heriades* spp. (Co)	Pleis	East Africa	1?	Engel (2001a)

5.2.1 Bee adaptations

Bees forage on plants to collect various rewards: pollen, nectar, oil, perfumes, resin, and some material for nesting like pieces of leaves and petals (Fig 5.1; Wcislo and Cane 1996; Labandeira 2000; Pouvreau 2004; Michener 2007). Foraging on plant rewards is a mechanically complex activity that requires certain handling skills, which differ from species to species (Westerkamp and Claßen-Bockhoff 2007). Various foraging strategies have been described among bees mainly based on pollen foraging behavior. Some taxa display floral specificity, restricting their flower visits to closely related plant taxa (pollen specialists) while other bee species are more opportunistic, exploiting a wide range of different flowers (pollen generalists) (Roberston 1925; Westrich 1989; Müller 1996a; Cane and Sipes 2006; Müller and Kuhlmann 2008; Dötterl and Vereecken 2010). To characterize the degree of bee-host plant specialization, different terms were progressively introduced by several authors to better reflect the reality of a continuum in bee-host breadth, from extreme specialization to extreme generalization (Roberston 1925; Rasmont 1988; Cane and Sipes 2006; Müller and Kuhlmann 2008). According to Müller and Kuhlmann (2008), the three main categories are:

(1) monolecty
(2) oligolecty, which is differentiated into three subcategories – narrow oligolecty, broad oligolecty, and eclectic olygolecty
(3) polylecty *sensu lato* which is also differentiated into three subcategories – polylecty with strong preference, mesolecty and polylecty *sensu stricto*.

Females have a wide range of morphological and behavioral features for collecting and transporting pollen. Although some females carry pollen internally in the crop, most exhibit external modifications and behaviors for capturing pollen, and for grooming it from their bodies and loading into scopae (*sensu* Engel 2001a, *contra* Michener 1999) for transport. Several studies have demonstrated that oligolectic species often have specific structures for gathering pollen (Thorp 1979, 2000). These highly modified morphological structures are used by bees to gather pollen that is otherwise difficult to access (Müller 2006) or of large size (Pasteels and Pasteels 1979; Thorp 1979). For example, some species possess hooked hairs on their mouthparts or forelegs to collect pollen from hidden anthers (Shinn 1967; Thorp 1979; Parker and Tepedino 1982; Houston 1990, 1991; Harder and Barrett 1993; Müller 1995; Thorp 2000). Others have additional clusters of hairs, usually on the head, to accumulate pollen from nototribic flowers (Thorp 2000). Some specialized behaviors, such as vibration of flowers (buzz pollination), are also displayed by females to enhance the uptake of pollen, especially in plants with poricidal anthers like Ericaceae or Solanaceae (Michener 1962; Macior 1986, 1995; Buchmann 1983; Houston and Thorp 1984; Gottsberger and Silberbauer-Gottsberger 1988; Neff and

Fig 5.1 Modern bee–plant interactions. A. *Andrena hattorfiana* female foraging on pollen of *Scabiosa* sp. (Schrophulariaceae) (picture Edith Tempez). B. *Macropis europaea* female foraging on oil and pollen of *Lysimachia vulgaris* (Myrsinaceae) (picture Yvan Barbier). C. *Anthophora plumipes* male foraging on nectar of Lamiaceae (picture Jean-Marc Michalowski). D. *Megachile circumcincta* with peace of leaf for cell linning (picture Nicolas J. Vereecken). E. Male of orchid bee collecting fragrances (picture Günter Gerlach). F. Worker of stingless bee *Melipona* cf. *rufiventris* (Meliponini) carrying resin (picture Claus Rasmussen). See plate section for color version.

Simpson 1988; Proença 1992; Müller et al. 1997). These specialized morphological stuctures and behaviors have evolved several times independently during the evolution of bees and in widely divergent taxa of both oligolectic and polylectic forms (Thorp 2000). However, many oligolectic bees do not show any evident morphological adaptations. Oligolecty is more often based on a combination of restricted phenology and behavior rather than any particular morphology attribute (Michez et al. 2008).

Pollen is not the only source of rewards for insects. Vascular plants produce fluid rewards such as nectar and oils. Their extraction and gathering can also require morphological adaptations. The morphological feature used for nectar collection is the labiomaxillary complex that may be differentially shortened or elongated to reach nectar reserves in shallow flowers or concealed in tubular corollas (Wcislo and Cane 1996). The bee mouthparts consist of a glossa and associated clasping structures from the labium and maxillae, and represent one of the most complicated apparati for fluid feeding among insects (Labandeira 2000). Moreover a rich diversity of prominent morphological specializations has originated for gathering floral oils, such as conspicuous setal brushes and combs on the bee's tarsi and sometimes, abdominal sterna (Vogel 1981, 1986). One of the more amazing innovations is the elongate forelegs of some *Rediviva* species, which are used to collect oils from the twinned elongate spurs of *Diascia* flowers (Vogel and Michener 1985; Steiner and Whitehead 1990; Wcislo and Cane 1996). Morphological specializations for the processes of floral fluids are also well known in bees, particularly the collection of floral components and their transfer to male metatibial organs in orchid bees (Sakagami 1965; Vogel 1966; Whitten et al. 1989; Kimsey 1984; Cruz-Landim and Franco 2001). Such structures are not also without their independent origin elsewhere among bees, such as the metafemoral organs of Arabian *Eoanthidium*, which have a remarkably similar morphological structure and may also be used for processing of plant fragrances (Engel 2004c).

Another resource for bees is the plant itself, especially its tissues. Some bee species are closely associated with plants as a source of materials for nest construction e.g. corbiculate bees and Megachilinae (Wedmann et al. 2009). They use resins, masticated leaves, cut petals, trichomes, or other plant materials sometimes along with mud to construct nests in cavities or in the soil (Müller 2011). Females can also use plant fragments like circular excisions of leaves and petals to line their brood cells e.g. some *Megachile* s.l. species, and some Osmiini, or masticated leaves to hide the nest e.g. some *Osmia* species (Rozen et al. 2010).

5.2.2 Plant adaptations

As pollen plays a predominant role in plant reproductive processes, repeated returns to the same plant species not only provide advantages for the forager but are also an inevitable prerequisite for reliable cross-pollination (Westerkamp and Claßen-Bockhoff 2007). However, the pollen grains are removed in great quantity by bees to ensure their own reproduction (as nest cell provisions for their developing brood). Moreover, the pollen-gathering efficiency can be such that all pollen is entirely removed from a flower, leaving next to nothing for pollination (Westrich 1989; Müller 1996a; Müller et al. 2006; Westerkamp and Claßen-Bockhoff 2007). For example, Schlindwein et al. (2005) reported that 95.5 % of the pollen produced by flowers of *Campanula rapunculus* were collected by its oligolectic pollinators, while only 3.7 % contributed to pollination. Another study showed that among 41 bee species, 85 % required the whole pollen content of more than 30 flowers to rear a single larva. The pollen of more than 1000 flowers is needed for some species (Müller et al. 2006). After each flower visitation, the female bee carefully grooms her body and transfers the pollen grains into the scopae, making them inaccessible for pollination (Westerkamp 1996). This huge quantity of pollen withdrawn from flowers for bee reproduction conflicts with the successful pollination of the host plant, resulting in a strong rivalry. Accordingly, the ecological relationship between bees and flowers may not be merely mutualistic (Inouye 1980; Westerkamp 1996, 1997; Thorp 2000; Irwin et al. 2001) but may be better viewed as a "balanced mutual exploitation" (Westerkamp 1996) wherein flowers must continually balance the need to attract bees for pollination on the one hand, and to restrict pollen losses on the other (Praz et al. 2008). Plants must therefore minimize pollen loss by narrowing the spectrum of their pollen-feeding visitors.

Plant and flower size, color and constriction of the corolla, presence of a landing platform, quantity and quality of nectar, scent, timing of flowering as well as morphology of the reproductive system are the consequences of selective pressure imposed by particular pollinators attracted by floral rewards. Several morphological traits of flowers are currently considered adaptations to prevent excessive pollen harvesting:

(1) heteranthery (Vogel 1993)

(2) anther dissimulation in nototribic flowers (Müller 1996a; Houston 2000; Thorp 2000), in narrow floral tubes (Thorp 1979, 2000; Parker and Tepedino 1982; Müller 1995; Müller and Kuhlmann 2003; Neff 2004; Müller 2006) or in keel flowers (Westerkamp 1997)

(3) concealment of pollen in poricidal anthers (Buchmann 1983; Harder and Barclay 1994)

(4) progressive pollen release (Erbar and Leins 1995; Schlindwein et al. 2005)

(5) zygomorphy (Vamosi and Vamosi 2010).

These adaptations can also maximize the contact between the stigma and the visitors e.g. nototribic flowers. These highly specialized bee flowers are mechanically complex, and gathering pollen from such flowers requires some force and a coordinated movement of many external bee structures. Their access is thus limited to a guild of specialized and faithful pollinators (Westerkamp 1997).

5.3 Palaeoecology of bees and plants

Ancient associations can be explored in two principle ways: phylogeny linked to ecology and palaeobiology linked to modern biology. Under the first, a cladogram derived from diverse data sources can be used to reconstruct ecological or behavioral attributes such as relationships with host plants, insect herbivores, degree of sociality, nest architecture, etc. (Engel 2001c; Michez et al. 2007, 2008; Sedivy et al. 2008). However phylogenetic data are not always required to shed light on past plant–insect associations. Ecological interactions can be inferred through morphological analogies with extant organisms and systems. For example, the presence of palm bugs (Heteroptera: Thaumastocoridae) and fig wasps (Hymenoptera: Agaonidae) indicates the presence of palm and fig trees, respectively (Grimaldi 1996; Peñalver et al. 2006). Similarly, the presence of orchid bee fossils in a fauna can indicate the presence of Orchidaceae even in the absence of definitive orchid fossils (Engel 1999d). This secondary approach relies on both a detailed knowledge of modern biology coupled with paleobiology, and is centered around comparisons between fossil and extant material. Evidence of past plant–insect associations consists of three distinct but linked fossil records: insect body-fossils, plant body-fossils, and trace fossils of their associations. This latter archive of associations is based on six principal types of evidence:

(1) plant reproductive biology indicating insect association
(2) insect-mediated plant damage
(3) dispersed insect coprolites
(4) insect gut contents
(5) plant-related structure of insect mouthparts and ovipositors
(6) taxonomic assignment to a modern descendant for which reliable ecological data exist (Labandeira 2000).

Unlike the trilobite that has left a prodigious fossil record, insects are more limited to numerous Lagerstätte and form a highly discontinuous record despite the tremendous number that could have been preserved. In many cases, the fossils are fragmentary showing few critical characters, so that studying them is often frustrating. Nonetheless, the fossil record of insects is growing and sheds considerable insight into the various phases of hexapod evolution (Grimaldi and Engel 2005). The reason for the relative scarcity of insect remains is their more infrequent preservation in sedimentary matrix owing to various taphonomic factors, the degradation of the exoskeleton in some oxygenated environments, their diminutive proportions relative to the sediment grain size, and the generally lower number of freshwater (e.g. lacustrine) relative to marine deposits. The most notable exception is preservation in amber, which constitutes the most valuable record for insect fossils owing to its unique fidelity of preservation, with even the finest (even life-like) details faithfully conserved (Grimaldi and Engel 2005). Given that many insects were too small to escape entrapment when coming into contact with sticky plant exudates, their frequent preservation as biological inclusions is easily understood. Although amber is known from as far back as the Carboniferous, the earliest fossiliferous resins with insect inclusions date to the Early Cretaceous and about 125 megaannum. The amber record represents the last 30 % of terrestrial plant and arthropod history, and is useful for the earlier evolution of otherwise extant clades. To investigate earlier episodes of insect history, deposits with greater geochronological persistence are necessary, such as lacustrine sediments, extending into the Late Palaeozoic (Labandeira 2002b; Grimaldi and Engel 2005). The earliest evidence of pollination is provided by compression-type material but this mutualism remains difficult to demonstrate because of the very indirect nature of the evidence for plant entomophily and insect pollen transfer (Labandeira 2002b). During the Late Jurassic to Early Cretaceous, the first pollinators of early angiosperms were probably generalized insects without adaptations for flower feeding, such as wasps, moths, thrips, beetles, and flies, from other seed plants possessing flower-like structures (Gnetales, Bennettitales and cycads). More plant-dependent insects followed in later stages such as bees in the Cretaceous and butterflies in the Tertiary. Unfortunately, fossils of bees are exceptionally rare, particularly in Cretaceous deposits (Table 5.1 and following sections). The scarcity of bees can be at least partially explained by their habitat preferences (Bennett and Engel 2006). Most species live in xeric areas outside of those forests that typically produced amber, or oustside of anoxic areas that produce most Lagerstätte (Michener 1979, 2007; Engel 2001a, 2004b; Grimaldi and Engel 2005). Fossil records of Apoidea are therefore too patchy to give great precision into the origin of each clade of bees. However, they are very useful for documenting minimal ages for particular clades and for studying their morphological and ecological evolution. The major deposits with bee fossils are known from the Cenozoic:

Dominican amber from the Early Miocene (~19 megaannum)

Florissant shale from the Oligocene (~34 megaannum)

Baltic amber from the middle Eocene (~45 megaannum)

These three deposits have produced the largest bee paleofaunas (Zeuner and Manning 1976; Engel 2001a, 2004b). Excluding these, only six older body fossils have been discovered from isolated sites scattered around the world, and only two of which are from the Mesozoic era. Accordingly, plant body fossil morphology is critical for assessing the possibility of insect-mediated pollination, especially the structure of reproductive units. Many attributes in the plant fossil record have been inferred to indicate the presence of biotic pollination: accessibility and modifications of flower reproductive structures to attract insects, presence of rewards such as food, nesting material or others to lure potential pollinators, features that promote transfer of pollen or enhance certain pollination types, and the size and surface properties of pollen provide circumstantial evidence for insect pollination (Labandeira 2000, but see previous chapter).

Direct reliable trace fossils of bee–plant associations are quite rare. In many herbivorous clades, the insect-mediated plant damages are the most useful and common records of past relationships, e.g. galling, mining. But bee damages are very uncommon except for damages for nest construction. Given their interesting relationship with plants as nesting resources, the diversity of megachiline bees (*Megachile* and related genera and tribes) in past epochs can be ascertained from not only the remains of actual bee specimens but also from the record of their activities on the surrounding flora (Sarzetti et al. 2008; Wedmann et al. 2009). Such evidence may further help to expand our current understanding of the diversity of these tribes in the past, despite the usual paucity of bee specimens in the fossil records (Wedmann et al. 2009). Preservation of nesting activities is also observed for some lineages like Halictidae digging nests in the soil (described in the ichnofamily Celliformidae, for a review see Genise 2000). However, such paleoichnological data (leaf damage, fossil nests) must be carefully considered before any definitive conclusions from misidentifications become common (Engel 2001a, 2004a).

Records of pollen grains on fossil insects and in coprolites provide additional circumstantial evidence for ancient bee–flower interactions. But the presence of pollen on the fossilized body does not exclude the possibility of flower visitation without pollination. Because evidence of plant–pollinator interactions is exceedingly rare in the fossil record, our current knowledge of ancient pollination is mainly indirectly inferred from specialized morphological features of fossilized insects (Grimaldi 1999; Ramirez et al. 2007; Michez et al. 2007) and flowers (Crepet 1979; Crepet et al. 1991; Gandolfo et al. 2004).

5.4 The "proto-bee" and the Cretaceous record of bees

5.4.1 The "proto-bee"

Hypotheses about the origin of the first bee are based on (i) the oldest bee fossil records, (ii) the origin and fossil record of their closest relatives (Crabronidae, spheciform Apoidea), and (iii) the origin of their likely host plants (Angiosperm). While the oldest spheciforms are those species of *Angarosphex* from the Barremian of Brazil's Crato Formation (~125–130 megaannum) and other Early Cretaceous deposits (Grimaldi and Engel 2005), the putative sister group of bees, Crabronidae, are not known until the Early mid Cretaceous (Antropov 2000; Bennett and Engel 2006). The diversification of spheciform Apoidea occurred during the Early Cretaceous (Grimaldi and Engel 2005; Bennett and Engel 2006) (Fig 5.2). Based on the record of fossil Crabronidae (Antropov 2000; Bennett and Engel 2006) and these other factors, a rational timing supposes the origin of bees around the Early mid Cretaceous, or about 125–120 megaannum (Engel 2001a, 2004b, 2011; Grimaldi and Engel 2005; Ohl and Engel 2007). The oldest bee trace is from the Cenomanian of Arizona (94 megaannum), although *Cellicalichnus dakotensis* is quite contemporary in its form (Table 5.1, Elliott and Nations 1998). The descriptions of fossil bee nests from the Triassic were incorrect (Lucas et al. 2010). Angiosperms are hypothesized as having first originated in the xeric interior of Gondwanaland and during the earliest Cretaceous (Raven and Axelrod 1974; Taylor and Hickey 1992), and this is likely also where bees first diverged from their common ancestor with Crabronidae (Engel 2001a, 2004b). In summary, bees likely diverged from among the apoid wasps sometime in the late Early Cretaceous and in the Southern Hemisphere (Engel 2001a, 2004b; Grimaldi and Engel 2005). Molecular phylogenies of Apidae and Halictidae associated with estimates of divergence times support this conclusion and also that bee diversification took place during the Early mid Cretaceous (Danforth et al. 2004; Cardinal et al. 2010; Ware et al. 2010).

5.4.2 From carnivorous predator to phytophagous pollen forager

Pollen consumption has generally been the evolutionary precursor to pollination (Labandeira 1998). Pollen contains vitamins, starch, lipids, proteins, and amino acids, which provide nutritional requirements for most animal species (Roulston and Cane 2000). Apoid wasps have a predatory diet high in protein. The protein value of pollen is high enough for apoid wasp nutritional requirements, ranging from 12–60 % (Roulston et al. 2000). They also assimilate cholesterol from their prey, and some pollen contains equivalent sterols (Dötterl and Vereecken 2010). Moreover, several bodies of evidence suggest that apoid wasps already displayed attributes suitable to becoming pollen consumers and foragers, as well as pollinators of angiosperms.

Apoid wasps existed and diversified when the first angiosperms appeared. The niche of pollen food was probably not yet overexploited when some spheciforms initially diverged to become bees.

Apoid wasps had mandibulate mouthparts more suitable for chewing pollen than piercing-sucking mouthparts (Crepet 1979).

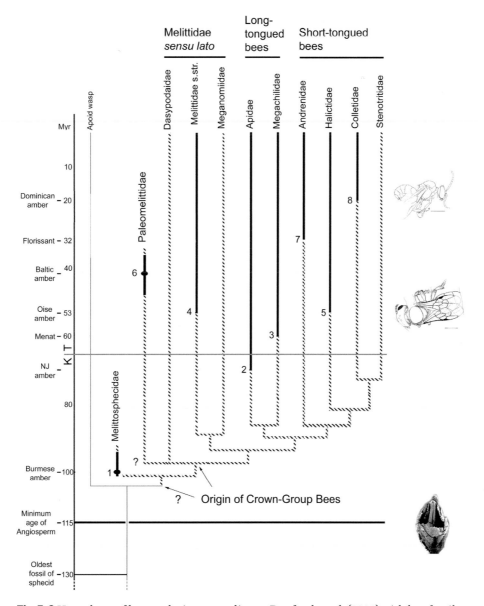

Fig 5.2 Hypothese of bee evolution according to Danforth et al. (2006) with bee fossils mapped on. For an alternative interpretation, refer to Engel (2004b) and Ohl and Engel (2007). 1. *Melittosphex burmensis*. 2. *Cretotrigona prisca*. 3. *Probombus hirsutus*. 4. *Paleomacropis eocenicus*. 5. *Halictus?savenyei*. 6. *Paleomelitta nigripennis*. 7. *Andrena? clavula*. 8. *Chilicola electrodominica*. Drawings from the top to the bottom: *Proplebeia dominicana* (from Camargo et al. 2000; scale = 1mm); *Paleomacropis eocenicus* (from Michez et al. 2007; scale = 1mm); *Divisestylus brevistamineus* (from Crepet et al. 2004).

Apoid wasps flew very well and are good at hovering, allowing them to forage rapidly on many flowers.

Apoid wasps constructed nests and transported food (arthropod prey) to feed their offspring. They were then able to substitute prey transport for pollen transport.

Plumose setae, often integral in pollen-collecting structures, may have been present in spheciforms before subsequent cooption of their original function. For example, plumose setae were likely originally used for thermoregulation (Heinrich 1996).

Plesiomorphically apoid wasps, like most aculeate Hymenoptera, are cold-blooded and live mainly in hot xeric climates. Plumose setae are useful for capturing heat and permitting the body to warm up more quickly after the torpor of cold desert nights. This character would have appeared randomly and could be considered an exaptation. Much like the origin of feathers among therapod dinosaurs for thermoregulation prior to their eventual use in flight by their descendents – the birds, so too, does it appear that branched hairs may have predated pollen collection and even bees, representing an earlier evolutionary solution to a completely different problem.

Ancestral specialist behavior could have been a key feature allowing the proto-bee to promote fixation of its new foraging behavior i.e. pollen foraging. Indeed, a bee's foraging behavior exhibits particular constraints:

pollen-collecting behavior is very complex (Wcislo and Cane 1996) and proto-bees could have been cognitively limited to use a large range of hosts

bees invest strongly in their offspring. Females of bees lay only a few eggs and generalist risk-takers could have been selected against. Many pollen contents could have been unsuitable for adult and larval feeding (Praz et al. 2008).

host perception seems more complex than in other phytophagous insects. Bees detect color, shape, size and scent of flowers (Raine et al. 2006).

specialization can increase the selection of pollen-foraging efficiency (Strickler 1979).

All these characteristics have probably forced the specialization and its inheritance.

Although there does exist some clear examples of transitions from polylecty to oligolecty, growing evidence suggests that oligolecty might be the ancestral state in bees (Danforth et al. 2006). Firstly, many generalist bee species have evolved from oligolectic ancestors. For example, in several anthidiine bees as well as in pollen-collecting masarine wasps, polylecty appears to be a derived trait (Müller 1996b; Mauss et al. 2006). In the genus *Andrena*, oligolecty is also assumed to be the plesiomorphic condition and polylecty has evolved independently several times (Larkin et al. 2008). Shifts from oligolecty to polylecty are more frequent than the reverse (Müller 1996b; Michez et al. 2008). Secondly, basal clades of most bee lineages such as Dasypodainae, Melittinae, Fideliinae, Rophitinae etc. include a high proportion of oligoleges (McGinley and Rozen 1987; Westricht 1989; Wcliso and Cane 1996; Engel 2004d; Danforth et al. 2006; Patiny et al. 2008; Michez et al. 2008). These facts could be a hint that, in general, polylecty is a derived foraging strategy that has evolved multiple times among bees. Plant associations mapped across bee phylogenies have only recently begun to become more and more prominent in studies (Müller 1996b; Engel 2002b, 2004d; Michez et al. 2008; Sedivy et al. 2008) so the body of evidence for this is continuing to grow but it does appear that the Danforth et al. (2006) hypothesis of polarity is widely supported.

Once foraging behaviors took place and larvae were able to digest the pollen as protein and lipid resources, it was perhaps not long before pollen foraging became a significant advantage for survival. Searching for regularly and conspicuously displayed pollen rewards would have been easier than seeking and subduing mobile prey, which generally tried to conceal itself (Engel 2001a, 2004b). Under this scenario, strong selective pressures would quickly stabilize any lineage toward becoming "bees".

5.4.3 To bee or not to bee? Burmese amber and the conceptual challenge of stem groups

The report of the Burmese amber apoid, *Melittosphex*, is an immensely interesting and important contribution to the fossil history of the superfamily, alongside the detailed monographic treatment of Antropov (2000). This particular fossil was introduced with much fanfare and bravado as the "oldest fossil bee." However, several conceptual challenges make it difficult to determine whether this species truly provisioned its nests with pollen and nectar, and that its larva was an obligate consumer of such resources. As noted many times before, bees are those ecologically dominant, apoid wasps that, as adults, collect pollen and nectar for exclusive consumption by their larvae. Accordingly, any lineage after the evolutionary novelty of obligate pollen-feeding larvae is a bee, while any lineage diverging before this is not. Thus, is this fossil species actually a bee, or rather a predatory apoid wasp sister to bees? This is not a specific criticism of *Melittosphex* but instead a philosophical question highlighting the difficulty of identifying any true, stem-group bee. At what point must we accept ignorance as our answer rather than a definitive attribution to the ecological clade we recognize as bees? As noted by Ohl and Engel (2007), it would appear that in order to make *Melittosphex* a bee, the definition of "bee" has to be set on a restricted set of features, expanding "bees" to include organisms that may or may not actually perform the ecological role that has made the group so very famous – namely, their mutualistic association with flowering plants. Rather than the traditional concept of bees as essentially vegetarian derivatives of the apoid wasps, that is, adults collecting pollen and nectar which are then consumed by the larvae (features entirely unknown for the fossil), the attribution as a "bee" is based solely on branched hairs and a slightly enlarged hind tarsal article, rather than positive evidence of pollen feeding. Thus, *Melittosphex*, which may have been predatory like other wasps, is accorded bee status simply by its close relation to actual bees rather than for an exhibition of true bee-like habits or ecology. The assertion that branched-hairs automatically indicates pollen collection is erroneous as some predatory wasps, such as sapygids, have identical hairs while masarine wasps and a crabronid wasp (*Krombeinictus*) lack them, yet successfully collect and provision pollen. Moreover, some bees, like hylaeines and euryglossines, effectively lack such plumose hairs and transport pollen in the crop (Michener 1965, 2007; Torchio 1984). The only available specimen of *Melittosphex* is a poorly preserved male, the sex that does not collect pollen and provision if it was a bee, and so any pollen (if actually present) on its body must be incidental, just as occurs on many specimens of male and female apoid wasps. Male and female apoid wasps are often captured with pollen on their bodies since they visit flowers to consume nectar, but they do not store pollen or feed the pollen to their larvae (otherwise they would be no different from bees). There is no evidence that *Melittosphex* provisioned its brood cells with pollen. Naturally, this is an issue with all stem-group fossils and begs the question as to at what point we arbitrarily decide what is and what is not a "bee" (Ohl and Engel 2007). Interestingly, this trap with regard to Cretaceous, stem-group Anthophila was predicted by Engel (2001a, p. 155–8), and this represents a critical conceptual and practical challenge to the designation of any stem-group Anthophila as being "bees" rather than predatory wasps. In our estimation, the sole feature that distinguishes a bee from the wasps they evolved from is the specialization of the larva for consumption of a mixture of pollen, nectar, and/or floral oils; and the subsequent morphological adaptation of the female to provision the larva, otherwise they are merely anatomically peculiar wasps. Thus, in the absence of data on larval feeding (including features of the female conclusively associated with provisioning the larva with such resources), any fossil with a placement outside of the crown-group clade of true bees must be considered for what it truly is ... merely an apoid wasp that resembles and is allied to bees, and which may or may not be a "bee." From a conservative epistemological position, we cannot deem any stem group as a "bee" without such positive evidence and must remain agnostic as to their biological/ecological identity.

Regardless of these challenges, Burmese amber is dated as to near the Albian–Cenomanian boundary (~100 megaannum) (Grimaldi et al. 2002; Cruickshank and Ko 2003; Ross et al. 2010). Grimaldi et al. (2002) consider that the environment was tropical within an average temperature range of 32–55 °C. Ross et al. (2010) listed arthropod records of 36 orders, 216 families, and 228 species, mainly Diptera, Coleoptera and Hymenoptera. *Melittosphex burmensis* from these deposits has been placed in the monospecific family Melittosphecidae, apparently sharing only some synapomorphies with contemporary bees (Poinar and Danforth 2006). Branched hairs are putatively present on the body which, as mentioned, the authors use as a proxy for indicating pollen foraging behavior, but the only known specimen is a male and so this is a speculative conclusion. Nonetheless, the small size of *M. burmensis*, at around 3 mm, is consistent with the small size of contemporary angiosperms (Poinar and Danforth 2006; Crepet et al. 2004). Further exploration of mid Cretaceous deposits in Myanmar and elsewhere will hopefully bring evidence to more positively resolve the identity of this and any other stem-group Anthophila.

5.4.4 Raritan amber: the first definitive bee remains

Raritan amber occurs throughout Cretaceous outcrops of the Atlantic Coastal Plain of eastern North America and is Turonian (~90 megaannum) in age (Grimaldi et al. 2000; Grimaldi and Nascimbene 2010). Origin of the amber is now understood to be from the Cupressaceae. The paleohabitat was an inter-distributary system of shallow, brackish water channels that formed a delta in the mid to Late Cretaceous (Grimaldi and Nascimbene 2010). The proximity of freshwater is indicated by the diversity of adult insects that breed in freshwater such as Trichoptera. All indications are that New Jersey was at this time tropical or subtropical in climate and that angiosperms comprized a substantial part of the flora. Based on macrofossils and pollen analyses, there were more than 130 angiosperm taxa, including some entomophilous lineages like Clusiaceae, Ericales, Lauraceae, Palmae and Protaceae (Michener and Grimaldi 1988a). There is a total of 104 described species and 59 families of arthropods (Grimaldi and Nascimbene 2010). Only one bee fossil is described from these deposits, *Cretotrigona prisca* (Michener and Grimaldi 1988a, 1988b; Engel 2000b; but see Rasnitsyn and Michener 1991). This species is included in a derived taxon of the corbiculate Apinae – the Meliponini. This tribe includes extant eusocial species showing polylectic and resin-foraging behaviors. The presence of a corbicula indicates that *C. prisca* likely shared the same foraging and carrying behavior as its modern meliponine counterparts. Specific relationships to any plant family described from the deposit is unlikely as *C. prisca* would have been a generalist pollinator of many entomophilous taxa, like its modern relatives. Nonetheless, *C. prisca* demonstrates that bees with highly modified and specialized structures were already well-established by the Late Cretaceous.

5.5 Paleocene and Eocene bee fossils

5.5.1 Paleocene and Eocene characteristics

The Cenozoic began with the Paleocene epoch (65–55 megaannum) and was one of the warmest periods in Earth's history. At this time, "tropical" lineages were nearly ubiquitous in geographic distribution (Grimaldi and Engel 2005). Europe was composed of multiple land masses and archipelagos separated from Asia by an epicontinental seaway (Turgai Strait). Southern Europe bordered the northwestern bays of the Tethys Ocean, which later formed the Mediterranean Sea to its west. North America was also crossed by a deep sea and India was still isolated from the other continents. Climatically, the Eocene (55–38 megaannum) was the most dramatic episode in the Tertiary. During the Early Eocene, no large, standing regions of ice were present, even at the poles, but by the end of this epoch, the glaciation of Antarctica had begun. This global cooling had a critical impact on the global distributions of many plants and animals, including bees. There are presently thirteen documented deposits with bee fossils from the Paleocene and Eocene: Menat (-60 megaannum, France), Oise (-53 megaannum, France), Quilchena (-53 megaannum, Canada), Cambay (-52 megaannum, India), Messel/Eckfel (-49/-44 megaannum, Germany), MacBee/Republic (-49 megaannum, Canada/USA), Baltic region (-48/-45 megaannum, northern Europe), Rovno (-48/-45 megaannum, Ukraine), Rio Pichi-Leufú (-48 megaannum, Argentina), Puryear/Viola (middle Eocene, USA). Four contemporary bee families (Apidae, Halictidae, Melittidae and Megachilidae) and one extinct family (Paleomelittidae) have been described from these deposits based on 51 species and 178 body fossils specimens. It appears that the K/T transition had little effect on bee diversity at a higher level, although certainly those bees in the zones of the various extraterrestrial impacts would have been considerably effected, leading to perhaps localized extinctions of particular faunal elements but without global alterations of the families or subfamilies of Apoidea. However, the global cooling at the end of the Eocene, a noted period of mass extinction (Eocene–Oligocene transition), probably caused the extinction of many corbiculate lineages principally known from Baltic amber (Engel 2001c), as well as from Cambay amber (Rust et al. 2010; Engel, unpublished data).

5.5.2 Menat, France: an ancient "bee community"

The Menat deposit is of primary importance because it is the oldest deposit with more than one isolated bee fossil and it is the only one presently with such material from the Paleocene. Two fossils of long-tongued bees, *Probombus hirsitus* (Megachilidae) and *Paleohabropoda oudardi* (Apidae) and one ichnospecies have been described from this deposit (Table 5.1). These taxa probably lived in a wet and very warm climate. The area of Menat (~60 megaannum) was likely characterized by a forest of oak and willow trees distributed around a crater lake (Piton 1940). The fauna comprized crocodiles, numerous large Mantodea (Chaeteessidae), Blattodea, Coleoptera (Buprestidae and Cerambycidae), Odonata (Megapodagrionidae) and very diverse Hemiptera (Fulgoroidea); all indicative of a warm palaeoclimate and a forest palaeoenvironment (Piton 1940; Nel and Roy 1996; Nel et al. 1997). *Paleohabropoda oudardi* was included in the Anthophorini, an old lineage where polylectic species are dominant (Iuga 1958; Michez et al. 2009a; Cardinal et al. 2010). Thus, association with a particular pollen host is unlikely. The presence of Megachilidae is more informative about potential host-plant associations. The description of one typical *Megachile* excision on a willow leaf could indicate a potential association between Megachilidae and Salicaceae at this time (Wedmann et al. 2009). Salicaceae could have been as an important a pollen resource for these fossil species as willow pollen is today for contemporary spring bees in Europe (Westrich 1989).

5.5.3 Cambay Basin, India: Early Eocene bees from an "isolated" subcontinent

Recently abundant Ypresian-aged amber has been discovered and reported on from the Cambay Basin in western India (Rust et al. 2010). This amber is rich in biological inclusions and comes from a time almost immediately before the complete connection of the Indian subcontinent with Asia when the subcontinent was still separated from Asia by oceanic waters and connecting archipelagos. This amber is of dipterocarp origin and is quite chemically distinctive (Rust et al. 2010). Interestingly, while work has only just begun on these deposits, four bee specimens are already known from such a limited sampling of inclusions (Rust et al. 2010; Engel, unpublished data). What is more interesting, is that the available material is representative of taxa otherwise known in Baltic amber, including a large fragmentary electrapine (perhaps of the genus *Protobombus*) and three melikertines of perhaps two distinct species. These are all eusocial lineages of corbiculate Apinae (Engel 2001a), and as such were likely polylectic and resin-collecting. Indeed, given the abundance of dipterocarps in this forest, it is highly probable that these species visited Dipterocarpaceae among other plant families for numerous resources including resins for nest construction. Given the immensity of the Cambay amber deposits (Rust et al. 2010), the potential for critical insights into Paleogene bees from a biogeographically and floristically distinct region is considerable.

5.5.4 Oise, France: early oil-collecting bees

Until relatively recently, fossil resins were poorly known from France (Nel and Brasero 2010). The situation changed dramatically after the discovery of an important 53 megaannum amber deposit in the department of Oise (Nel et al. 1999b). The dominance of an arborescent amber-producing species and the presence of freshwater in the French Eocene environs suggest a semi-deciduous forest with a mosaic of gallery-forest mixed with dryer plant communities, in a deltaic paratropical region (De Franceschi and De Ploëg 2003). The amber-producing tree was deduced as *Aulacoxylon sparnacense* (Combretaceae or Fabaceae-Caesalpinaceae). Brasero et al. (2009) provided an inventory of insects already described from the Oise deposit, with 59 families and 78 species, among them one female bee of *Paleomacropis eocenicus* (Melittidae, Macropidini) (Michez et al. 2007). *Paleomacropis eocenicus* was characterized by dense plumose setae on the inner and outer surfaces of the mesobasitarsus and long, erect setae on the metasoma. These kinds of setae could be linked to the collection of oil and pollen, as in the contemporary oil-collecting bee genus *Macropis*. Indeed, the oil of *Lysimachia* flowers (Primulaceae) is harvested by *Macropis* females using specialized setae on the inner surface of their pro- and mesobasitarsi (Vogel 1976; Cane et al. 1983; Michez and Patiny 2005). Moreover, dry pollen is initially held at the same time by the simple, long, erect setae of the metasomal sterna (Cane et al. 1983). Based on molecular clock analyses, the stem of *Lysimachia* was dated to around 41 megaannum (28–52 megaannum) (Renner and Schaeffer 2010). The plausible temporal coincidence between *Paleomacropis* and proto-*Lysimachia* could support the notion that *Macropis* and *Lysimachia* coevolved from the onset, even if the oldest fossil record of *Lysimachia* consists of fossil seeds from late mid Miocene of Jutland, Denmark (Hao et al. 2004). Other plant families producing oil, like Malpighiaceae (Davis et al. 2002), are known from the Eocene and could have been alternative host plants for *Paleomacropis eocenicus*.

5.5.5 Baltic amber: a diverse and abundant paleofauna for bees

Baltic amber represents the most diverse paleofauna and paleoflora described in the world (Weitschat and Wichard 2010). The dating of this deposit was debated for a long time, but today most evidence support Baltic amber as either middle Eocene (~45 megaannum) or late mid Eocene to Early Eocene (48–50 megaannum). The source plant of the amber has also been debated, with evidence supporting either a Pinaceae producer (based on inclusions such as wood remains, male cones and needles) or Sciadopityaceae (based on FTIR chemotaxonomic inferences) (Engel 2001a; Wolfe et al. 2009). The palaeohabitat was probably very similar to contemporary pine forests of the South Atlantic Coastal Plain of North America (from North Carolina to Florida). At lower elevations the amber forest was adapted to paratropical climates (savannah woods) with a population of conifer and palm trees. Moreover, the forest must have had lightly wooded areas with various different habitat types (Weitschatt and Wichard 2010). Among plant inclusions, branched epidermal trichomes of oak tree leaves and flower buds are very common. With more than 98 % of all embedded animal inclusions, the Arthropoda are most frequently represented in Baltic amber. Weitschat and Wichard (2010) listed 539 families including 1535 genera and 3068 species. Engel (2001a) listed

36 species and 18 genera of bees, with one species of *Ctenoplectrella* subsequently added (Engel 2008). These descriptions were based on 160 specimens (Table 5.1), although even more are known in collections today, which represent the second most important and abundant deposit for bee body fossils after Dominican amber. Only four specimens, representing three species, are short-tongued bees; perhaps not surprising as most are not resin collectors, typically nest in the soil, and are often more diverse in xeric habitats. Most are long-tongued taxa of the Apidae or Megachilidae, and all are known from females except *Ctenoplectrella viridiceps*, where both sexes are known. All species from Baltic amber exhibit morphological structures for pollen collection:

a corbicula for those Electrapini, Electrobombini, Melikertini and Meliponini

a metasomal scopa for those Ctenoplectrellini, Glyptapini and Prolithurgini

a hind leg scopa for those Boreallodapini, Halictini, Macropidini and Paleomelittidae.

Based on comparisons with extant taxa, the eusocial species (Electrapini, Electrobombini, Melikertini), the solitary Xylocopinae (Boreallodapini) and Halictini from Baltic amber were probably polylectic while *Eomacropis* could have been oligolectic as modern Melittidae (Michez et al. 2008). Engel (2001a) describes *Eomacropis* without any particular morphological adaptation to collect oil as the actual *Macropis* do, but the unusual long glossa of the female could have been a particular adaptation to a special host plant. Just as extant polylectic taxa are dominant in tropical and subtropical regions (Michener 1979), so too were these groups apparently dominant in number and diversity during the middle Eocene of Europe (Engel 2001a, 2004b; Wappler and Engel 2003).

5.6 Oligocene bee fossils

5.6.1 Oligocene characteristics

The Eocene–Oligocene transition is a well-documented episode of climate change and extinction. Average global temperature dropped from near 22 °C to 12 °C. Many lineages of corbiculate bees disappeared during this global cooling event (Engel 2001c), and it appears that the bee fauna underwent a considerable shift around this time, from principally ancient lineages to largely modern forms (Engel 2004b). Eight deposits with bee remains or traces have been documented: Florissant (-34 megaannum, USA), Badlands (Upper Oligocene, USA), the French deposits of Aix-en-Provence, Bois d'Asson, Camoins-les-Bains, Céreste and Marseille (Middle to Upper Oligocene, France), and Rott (Upper Oligocene, Germany). Total material represents 62 body fossil specimens, 47 species, and two ichnospecies. All extant families are present in these faunas with the exception of Stenotritidae, for which there is no fossil record, and all are preserved as compressions with little to no relief, particularly those from the diverse and prolific Florissant deposits of Colorado.

5.6.2 Florissant, Colorado: glimpses into the first "modern" bee fauna

Fossils from Florissant, Colorado, USA are compressions with low to no relief, formed in fine volcanic ash (Engel 2001a). This deposit is dated about 34 megaannum (Epis and Chapin 1974). The extant bee families represented in the Baltic amber (Apidae, Halictidae, Melittidae and Megachilidae) have also been described from this deposit with the addition of Andrenidae. Engel (2002a) revised the Florissant fauna of Halictidae, but the primary information for this fauna comes from the publications of T. D. A. Cockerell and is quite dated (see list in Table 5.1). Even if numerous and diverse taxa are present in the deposit – at least 36 body fossils, 34 species, 19 genera (Table 5.1) – body fossils are typically preserved only by wing venation or some structures of the legs and thorax, so accurate comparisons with living species as well as other fossil deposits are not entirely possible. Some attributions of species to extant genera are quite speculative and some, such as *Ceratina disrupta*, may be assignable only to Apoidea or even Aculeata incertae sedis. Furthermore, morphological structures associated with pollen collection are not discernible in the available material except under uncommon circumstances. Some species considered as cleptoparasites, based on the apparent absence of setae patches recognizable as scopae, need confirmation (e.g. *Protomelecta brevipennis*) and may represent poor preservation rather than definitive absence.

5.7 Neogene and quaternary bee fossils

5.7.1 Neogene and quaternary characteristics

The Neogene began 23.8 megaannum ago with the Miocene epoch. Climatic conditions were similar to the end of the Oligocene. As far as known, bee body fossils or traces have been described from 27 Miocene deposits: Bes-Konak Basin (22.5 megaannum, Turkey), Dominican amber (20 megaannum, Dominican Republic), Mexican amber (20 megaannum, Mexico), Sicilian amber (20 megaannum, Italy), Stewart valley (-14 megaannum, USA), Bilina Mine (Lower Miocene (LM), Czech Republic), Euboea (LM, Greece), Izarra (LM, Spain), Krottensee (LM, Czech Republic), Kudia River (LM, Russia), Kundratice (LM, Czech Republic), Radoboj (LM, Croatia), Rubielos (LM, Spain), Iki Island (Middle Miocene, Japan), Andance (Upper Miocene (UM), France), Biebrich (UM, Germany), Botchi river (UM, Russia), Böttingen (UM, Germany), Latah (UM, USA), Lleida (UM, Spain), Oeningen (UM, Germany), Parschlug (UM, Austria), Puesto Los Sauces (UM, Argentina), Randeck

(UM, Germany), Rusinga (UM, Kenya), Sainte-Reine (UM, France), Shandong (UM, China); and six Pliocene/Pleistocene deposits: Santander (2.5 megaannum, Colombia), Abu Dhabi (0.28 megaannum, UAE), Batu (Pleistocene, Malaysia), Eastern African copal (Pleistocene), Hukong (Pleistocene, Myanmar), Lockport (Pleistocene, USA). Based on the description of 84 species and 5 ichnospecies, all extant families are present except Stenotritidae with no described fossil at all (Table 5.1). The bee fauna is clearly modern in the Miocene deposits. As most of described species from Copal still exist, evolutionary significance of the Pliocene and Pleistocene concerns the origins of modern species.

5.7.2 Dominican and Mexican amber: the most recent paleofauna of bees

Dominican amber has been dated as Burdigalian in age (between 20.43 ± 0.05 megaannum and 15.97 ± 0.05 megaannum), the first and longest warming period of the Miocene (Poinar 2010). Dominican amber preserves the most diverse described bee fauna among Miocene deposits. All bee families have been recorded with the exception of Melittidae *s.l.* and Stenotritidae (Engel 2009). Among the twelve-recorded genera, only three genera (*Augochlora*, *Euglossa* and *Megachile*) are still present in the Greater Antilles, but only four are now extinct at a global scale (*Eickwortapis*, *Nesagapostemon*, *Oligochlora* and *Proplebeia*), although with close relatives among Central and South American taxa. The absence of Melittidae and Stenotrotidae in Dominican amber is expected as they are also absent today in Central and South America (Michener 2007; Michez et al. 2009b; Almeida and Danforth 2009). The bee fauna preserved in Dominican amber is largely equivalent to the modern Neotropical fauna, and quite distinct from those of Asia, Europe, and Africa. Mexican amber is roughly contemporaneous age and similarly harbors an abundant bee fauna, albeit mostly representing a single species, *Nogueirapis silacea* (Solórzano-Kraemer 2007).

Apidae are the most numerous in number of specimens (among the thousands) but only five species have been described (Table 5.1; Poinar 1998; Camargo et al. 2000; Engel 2009). All described apids that were probably resin collectors, but the high number of apine specimens is truly due to only one eusocial species, *Proplebeia dominicana*. Presence of foraged resin on the corbiculae of some specimens is indirect evidence that *Proplebeia* species routinely and actively gathered resin from extinct *Hymenaea* (Fabaceae, resin source of Dominican amber) (Camargo et al. 2000). Moreover, one specimen was described with an attached pollinium of *Meliorchis caribea* (Orchidaceae) (Ramirez et al. 2007). This fossil constitutes a rare, direct observation of plant–pollinator interaction. Moreover, as the staminal filaments are fused to the style in the Orchidaceae, the anatomical match required for a pollinator to remove the pollinium is nearly identical to that necessary for its subsequent delivery (Ramirez et al. 2007). *Proplebeia dominicana* was therefore probably a pollinator of *Meliorchis caribea*. But pollinia do not constitute an alimentary resource for sting-less bee workers. Visitors of orchids forage on nectar or fragrances. But orchids can also deceive their visitors by not producing rewards and only mimicking alternative alimentary rewards of neighboring host plants (Vereecken and McNeil 2010). As *M. caribea* is included in the subtribe Goodyerinae, the fossil species probably offered nectar similar to some modern species of this tribe (Singer and Sazima 2001). Two other fossils probably foraged on Orchidaceae during the Miocene, *Euglossa moronei* and *Eufriesea melissiflora* (Engel 1999d). These bees are orchid bees (Euglossini) where the males of contemporary species mainly forage on orchids to collect fragrances (Dressler 1982; Michener 2007; Ramirez 2009). However, *M. caribea* is not a good candidate as a fragrance host plant for euglossines since Goodyerinae do not produce suitable scents. Euglossini instead are mainly associated with the orchid taxa Gongoreae, Catasetinae, Zygopetallinae, Lycastinae, Bifrenariinae and Oncidiinae (Dressler 1982).

In their degree of pollen specialization, all apid lineages described from Dominican amber are polylectic clades (Dressler 1982; Michener 2007). Among short-tongued bees, Halictidae are the most diverse with 11 species but known from only 13 specimens. The rarity of halictids in the Dominican amber record may be partly due to the fact that most species nest in the ground and do not collect resin as other bees, thus making contact and preservation unlikely. This is also true for the sole panurgine (Andrenidae) in Dominican amber (Rozen 1996). Although Panurginae are greatly oligolectic, the Anthemurgini show little diversification in such behavior.

5.8 General conclusion

5.8.1 Bee fossil diversity

Knowledge about bee fossils has improved dramatically during the last 15 years. About one-third of the 184 described fossil species have been documented within this time period, and historical species discovered prior to this are gradually undergoing revision and reassessment (Engel 2000b, 2002a; Nel and Petruvelicius 2003; Michez et al. 2009a; Kotthoff et al. 2011). As paleontology continues to experience a current Renaissance among the entomological community, the value of such data will only become more significant and refined. Even for lineages with relatively sparse records, such as bees (in comparison to the more prolific records of flies or beetles), profound improvements are undoubtedly in store as more and more deposits are discovered and more attention is focused on these resources. Already the scant data is overturning some elements of current dogma, such as the discovery of native fossil honeybees in North America (Engel et al. 2009). The revelation that *Apis* was once native to the New World during the Neogene with subsequent post-Miocene extinction, revises traditional concepts of apine biogeography, with honeybees mimicking the pattern observed in the more completely understood records of horses (*Equus caballus*), gingkos, and the Chinese Tree of Heaven (*Ailanthus*), among many others. The prospect for future revelations of similar nature, ones perhaps entirely unexpected, is considerable. Accordingly, any study ignoring the fossil record, regardless of how meager, does so to its own jeopardy. If concerted efforts during such a brief span of time as 15 years can increase the available record by 33 %, then it is staggering to conceive how much more fully complete this record may be 15–25 years from today. Truly we are only at the earliest dawn of paleomelittology.

While 59 deposits include at least one bee fossil and/or one likely bee trace, only ten deposits have revealed more than three bee fossils, highlighting the scarcity of bees in fossil deposits (Table 5.1). Moreover, there are presently three principle biases in this record: geographical, biological, and habitat based.

Firstly, the geological history of bees is "northern biased" (Fig 5.3; Engel 2004b; Engel and Peñalver 2006). There are merely five deposits in the Southern Hemisphere (Table 5.1), with all other localities distributed in the Northern Hemisphere. A growing number of suitable deposits are continuously being discovered in the Southern Hemisphere or from regions that were once south of the Equator but are no longer. Aside from suitable compression sites already known from South American and southern Africa (e.g. Late Cretaceous of Botswana), perhaps the most exciting are those southern amber locales such as Peru, Ecuador, Ethiopia (which during the Cenomanian was obviously more southerly), and the rich outcrops of Australia. Intense exploration of these and others is only just beginning. As already noted, our record of fossil bees will change profoundly in the years to come, hopefully eliminating at least this first bias.

Secondly, resin-foraging bees are likely to be over-represented in amber and copal deposits, which include both 41 % of the described bee fossils (see examples in Fig 5.4). Excluding the hyperabundant stingless bee, *Proplebeia dominicana*, apids still represent 61 % of species and 71 % of specimens in amber and copal. In the modern fauna, apids represent 29 % of species globally and can represent 35 % or more of the species in some tropical habitats (Gonzalez and Engel 2004). Resin collectors more frequently come into contact with such substances and, although they are more adept at handling this resource, they are still significantly more likely to become entrapped, particularly when considering eusocial species where the increased numbers of individuals make the potential for "accidents" greater, pushing the numbers of such bees in amber higher. As such, resin foraging behavior could explain a large portion of this bias, although some component certainly does reside in the third, and last, obvious bias.

Unfortunately, the last bias may represent a hurdle more difficult to clear than the others. Large components of bee diversity are found in xeric habitats,

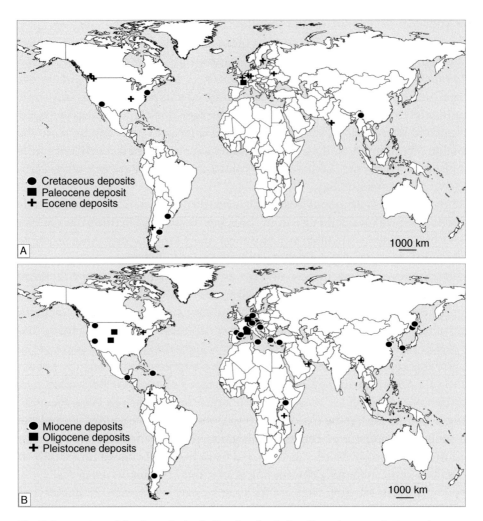

Fig 5.3 Mapping of the deposits including bee body fossils and traces. A. Deposits from Cretaceous, Paleocene and Eocene. B. Deposits from Oligocene, Miocene and Pleistocene.

regions not frequently represented in suitable fossil deposits and, when they are, they frequently lack the fine-scale sedimentary deposition necessary to preserve remains with sufficient detail to permit confident identification and comparison with modern counterparts. The small size of many bees, at least relative to many sedimentary sites that faithfully preserve larger animals such as vertebrates, and particularly the diminutive proportions predicted for the earliest of bees, may mean that the elimination of this habitat bias will be a long time in the works. For the foreseeable future, our record may be largely confined to more tropical, even wet tropical, habitats rather than the deserts that harbor our beloved objects of investigation.

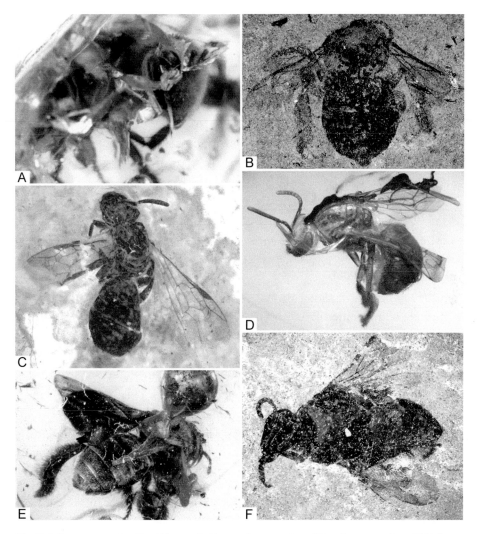

Fig 5.4 Representative fossil bees. A. *Cretotrigona prisca* (New Jersey amber, USA, late Cretaceous; Engel 2000b). B. *Paleohabropoda oudardi* (Menat, France, Paleocene; Michez et al. 2009). C. *Halictus petrefactus* (Rubielos de Mora basin, Spain, Early Miocene; Engel & Peñalver 2006). D. *Oligochlora eickworti* (Dominican amber, Early Miocene; Engel 1996). E. *Thaumastobombus andreniformis* (Baltic amber, middle Eocene; Patiny et al. 2007). F. *Protobombus messelensis* (Messel, Germany, middle Miocene; Wappler & Engel 2003).

5.8.2 Evidence of bee–plant interactions

Among the six principal types of evidence of past association between insects and plants (see previous chapter), two are quite common in bee fossil records: bee-mediated plant damage (Megachile damage for nest construction) and plant-related structure of bee-body fossil (morphological structure associated to pollen, nectar and oil foraging). We can synthesize the succession of plant-associated features:

(1) first evidence of plumose hairs in *Melittosphex burmensis* (-100 megaannum). This feature is present in all following bees (Michener 2007).

(2) small size likely adapted to small flower in *Melittosphex burmensis* (-100 megaannum). The size increased maybe linked to the evolution of polylecty (Thorp 1979).

(3) long tongue in *Cretotrigona prisca* (-70 megaannum) to collect nectar in deep flower. This feature is present in all extant Megachilidae and Apidae and some "short-tongued bees" (Roig-Alsina and Michener 1993; Alexander and Michener 1995).

(4) first corbicula in *Cretotrigona prisca* (-70 megaannum). This feature likely derived from hind-leg scopa and is present in the clade of corbiculate bees (Kawakita et al. 2008).

(5) earliest evidence of scopa in *Paleohabropoda oudardi* (-60 megaannum). Hind-leg scopa is present in most non-cleptoparasite taxa (Michener 2007).

(6) first evidence of oil-collecting setae in *Paleomacropis eocenicus* (-53 megaannum). Features associated with oil foraging are present in some unrelated clades of modern bees like Melittidae and Apidae (Renner and Schaefer 2010).

(7) metasomal scopae in Baltic Megachilid bees

(8) earliest evidence of modified male hind tibia in *Euglossa cotylisca* (2.5 megaannum). This feature is present in all males of extant orchid bees (Ramirez et al. 2010).

We also characterize past association thanks to taxonomic assignment to a modern descendant for which reliable ecological data exists, but this evidence is more speculative. The other archives of associations, plant reproductive biology indicating narrow bee-association, dispersed coprolites, and gut contents are no longer available in the present records.

5.8.3 Importance of paleobiological studies

Fossils provide a rare opportunity to study not only the origin and (co)-evolution of clades, but also their ecology, offering a unique window on past interactions. The functioning of ancient ecosystems has a direct bearing on the evolution of entire guilds, the diversification of specific lineages, and the ability of communities to respond to extrinsic changes such as climatic shifts. While much can be deduced from extending knowledge of current ecosystem operations and plant-host-herbivore associations into past epochs, at least for those in which the individual operators are presumed to have functioned similar to their modern counterparts, the power of this exercise pales in comparison to any direct views into ancient communities themselves. This is all the more critical with progressively more antique ecologies in which there may be guilds and lineages represented that left no survivors or ecologically-analogous taxa. The paleontological forefront is as important today as is the application of tools such as molecular and developmental biology, comparative anatomy and physiology, and the biochemistry and energy flow of entire ecosystems. Ignoring fossils compromises understanding of evolution.

Acknowledgement

Thanks to Bernard Vaissière and Claus Rasmussen for sending references. Edith Tempez, Yvan Barbier, Jean-Marc Michalowski, Nicolas Vereecken, Günter Gerlach and Claus Rasmussen generously provided pictures of modern bees. Thibaut De Meulemeester kindly proofread an early version of the manuscript.

References

Alexander, B. A. (1992). An exploratory analysis of cladistic relationships within the superfamily Apoidea, with special reference to sphecid wasps (Hymenoptera). *Journal of Hymenoptera Research*, **1**, 25–61.

Alexander, B. A. and Michener, C. D. (1995). Phylogenetic studies of the families of short-tongued bees (Hymenoptera:Apoidea). *The University of Kansas Science Bulletin*, **55**, 377–424.

Almeida, E. A. B. and Danforth, B. N. (2009). Phylogeny of colletid bees (Hymenoptera: Colletidae) inferred from four nuclear genes. *Molecular Phylogenetics and Evolution*, **50**, 290–309.

Antropov, A. V. (2000). Digger wasps (Hymenoptera, Sphecidae) in Burmese amber. *Bulletin of the Natural History Museum, Geology Series*, **56**, 59–77.

Arillo, A., Nel, A. and Ortuño, V. (1996). Two fossil bees from the Oligocene of Izarra (Alava, Spain) (Hymenoptera, Apoidea). *Bulletin de la Société Entomologique de France*, **101**, 59–64.

Bachmayer, F., Symeonidis, N. and Theodoropoulos, D. (1971). Einige Insektenreste aus den Jungtertiären Süsswasserablagerungen von Kumi (Insel Euboea, Griechenland). *Annales Géologiques des Pays Helléniques*, **23**, 165–174.

Baroni Urbani, C. (1980). First description of fossil gardening ants (amber collection Stuttgart and Natural History Museum Basel; Hymenoptera: Formicidae. I: Attini). *Stuttgarter Beiträge zur Naturkunde, Serie B (Geologie und Paläontologie)*, **54**, 1–13.

Bennett, D. J. and Engel, M. S. (2006). A new moustache wasp in Dominican Amber, with an account of apoid wasp evolution emphasizing Crabroninae (Hymenoptera: Crabonidae). *American Museum Novitates*, **3529**, 1–10.

Brasero, N., Nel, A. and Michez, D. (2009). Insects from the Early Eocene amber of Oise (France): diversity and palaeontological significance. *Denisia*, **26**, 41–52.

Bronstein, J. L., Alarcòn, R. and Geber, M. (2006). Transley review: the evolution of plant–insect mutualisms. *New Phytologist*, **172**, 412–28.

Brooks, H. K. (1955). Healed wounds and galls on fossil leaves from the Wilcox deposits (Eocene) of Western Tennessee. *Psyche*, **62**, 1–9.

Brothers, D. J. (1975). Phylogeny and classification of the aculeate Hymenoptera, with special reference to Mutillidae. *University of Kansas Science Bulletin*, **50**, 483–648.

Brothers, D. J. (1998). Phylogeny and evolution of wasps, ants and bees (Hymenoptera, Chrysidoidea, Vespoidea and Apoidea). *Zoologica Scripta*, **28**, 233–49.

Buchmann, S. L. (1983). Buzz pollination in angiosperms. In *Handbook of Experimental Pollination Biology*, ed. C. E. Jones. New York, NY: Van Nostrand Reinhold, pp. 73–113.

Camargo, J. M. F. d., Grimaldi, D. A. and Pedro, S. R. M. (2000). The extinct fauna of stingless bees (Hymenoptera, Apidae, Meliponini) in Dominican amber: two new species and redescription of the male of *Problebeia dominicana* (Wille and Chandler). *American Museum Novitates*, **3293**, 1–24.

Cane, J. H. and Sipes, S. D. (2006). Characterizing floral specialization by bees: analytical methods and a revised lexicon for oligolecty. In *Plant-Pollinator Interactions: Specialization and Generalization*, ed. N. M. Waser and J. Ollerton. Chicago, MI: University of Chicago Press, pp. 99–122.

Cane, J. H., Eickwort, G. C., Wesley, F. R. and Spielholz, J. (1983). Foraging, grooming and mate-seeking behaviors of *Macropis nuda* (Hymenoptera, Melittidae) and use of *Lysimachia ciliata* (Primulaceae) oils in larval provisions and cell linings. *American Midland Naturalist*, **110**, 257–64.

Cardinal, S., Straka, J. and Danforth, B. N. (2010). Comprehensive phylogeny of apid bees reveals the evolutionary origins and antiquity of cleptoparasitism. *Proceedings of the National Academy of Sciences USA*, **107**, 16207–211.

Cockerell, T. D. A. (1906). Fossil hymenoptera from Florissant, Colorado. *Bulletin of the Museum of Comparative Zoology*, **50**, 33–58.

Cockerell, T. D. A. (1907). A fossil honey-bee. *The Entomologist*, **40**, 227–9.

Cockerell, T. D. A. (1908a). Descriptions and records of bees. XIX. *Annals and Magazine of Natural History*, **8(1)**, 337–44.

Cockerell, T. D. A. (1908b). A fossil leaf-cutting bee. *Canadian Entomologist*, **40**, 31–2.

Cockerell, T. D. A. (1908c). Descriptions and records of bees. XX. *Annals and Magazine of Natural History*, **8(2)**, 323–34.

Cockerell, T. D. A. (1909a). New North American bees. *The Canadian Entomologist*, **41**, 393–95.

Cockerell, T. D. A. (1909b). Two fossil bees. *Entomological News*, **20**, 159–61.

Cockerell, T. D. A. (1909c). Some European fossil bees. *The Entomologist*, 313–17.

Cockerell, T. D. A. (1910). A Tertiary leaf-cutting bee. *Nature*, **82**, 429.

Cockerell, T. D. A. (1911a). Descriptions and records of bees. XXXIV. *Annals and Magazine of Natural History*, **8(7)**, 225–37.

Cockerell, T. D. A. (1911b). Fossil insects from Florissant, Colorado. *Bulletin of the American Museum of Natural History*, **30**, 71–82.

Cockerell, T. D. A. (1913a). Some fossils insects from Florissant, Colorado. *The Canadian Entomologist*, **45**, 229–33.

Cockerell, T. D. A. (1913b). Some fossils insects from Florissant, Colorado. *Proceedings of the United States National Museum*, **44**, 341–6.

Cockerell, T. D. A. (1914). Miocene fossil insects. *Proceedings of the Academy of Natural Sciences of Philadelphia*, **66**, 634–48.

Cockerell, T. D. A. (1917). New Tertiary insects. *Proceedings of the United States National Museum*, **52**, 373–84.

Cockerell, T. D. A. (1923). Two fossil Hymenoptera from Florissant (Vespidae, Megachilidae). *Entomological News*, **34**, 270–71.

Cockerell, T. D. A. (1925). Tertiary insects from Kudia River, Maritime Province, Siberia. *Proceedings of the United States National Museum*, **68**, 1–16.

Crepet, W. L. (1979). Insect pollination: a paleontological perspective. *BioScience*, **29**, 102–8.

Crepet, W. L., Friis, E. M. and Nixon, K. C. (1991). Fossil Evidence for the Evolution of Biotic Pollination. *Philosophical Transactions of the Royal Society of London Series B-Biological Sciences*, **333**, 187–95.

Crepet, W. L., Nixon, K. C. and Gandolfo, M. A. (2004). Fossil evidence and phylogeny: The age of major angiosperm clades based on mesofossil and macrofossil evidence from cretaceous deposits. *American Journal of Botany*, **91**, 1666–82.

Cruickshank, R. D. and Ko, K. (2003). Geology of an amber locality in the Hukawng Valley, northern Myanmar. *Journal of Asian Earth Sciences*, **21**, 441–55.

Cruz-Landim, C. and Franco, A. C. (2001). Light and electron microscopic aspects of glands and pseudoglandular structures in the legs of bees (Hymenoptera, Apinae, Euglossini). *Brazilian Journal of Morphological Sciences*, **18**, 81–90.

Danforth, B. N., Brady, S. G., Sipes, S. D. and Pearson, A. (2004). Single-copy nuclear genes recover Cretaceous-age divergences in bees. *Systematic Biology*, **53**, 309–26.

Danforth, B. N., Sipes, S. D., Fang, J. and Brady, S. G. (2006). The history of early bee diversification based on five genes plus morphology. *Proceedings of the National Academy of Sciences USA*, **103**, 15118–23.

Davis, C. C., Bell, C. D., Mathews, S. and Donoghue, M. J. (2002). Laurasian migration explains Gondwanan disjunction: Evidence from Malpighiaceae. *Proceedings of the National Academy of Sciences USA*, **99**, 6833–37.

De Franceschi, D. and De Ploëg, G. (2003). Origine de l'ambre des faciès sparnaciens (Eocène inférieur) du bassin de Paris: le bois de l'arbre producteur. *Geodiversitas*, **25**, 633–47.

Dötterl, S. and Vereecken, N. J. (2010). The chemical ecology and evolution of bee flower interactions: a review and perspectives. *Canadian Journal of Zoology*, **88**, 668–97.

Dressler, R. L. (1982). Biology of the orchid bees (Euglossini). *Annual Review of Ecology and Systematics*, **13**, 373–94.

Elliott, D. K. and Nations, J. D. (1998). Bee burrows in the Late Cretaceous (Late Cenomanian) Dakota Formation, northeastern Arizona. *Ichnos*, **5**, 243–53.

Engel, M. S. (1995). *Neocorynura electra*, a new fossil bee species from Dominican amber (Hymenoptera: Halictidae). *Journal of the New York Entomological Society*, **103**, 317–23.

Engel, M. S. (1996). New augochlorine bees (Hymenoptera: Halictidae) in Dominican amber, with a brief review of fossil Halictidae. *Journal of the Kansas Entomological Society*, **69**, 334–45.

Engel, M. S. (1997). A new fossil bee from the Oligo–Miocene Dominican amber (Hymenoptera: Halictidae). *Apidologie*, **28**, 97–102.

Engel, M. S. (1998a). Fossil honeybees and evolution in the genus *Apis* (Hymenoptera: Apidae). *Apidologie*, **29**, 265–81.

Engel, M. S. (1998b). A new species of the Baltic amber bee genus *Electrapis* (Hymenoptera: Apidae). *Journal of Hymenoptera Research*, **7**, 94–101.

Engel, M. S. (1999a). A new Xeromelissine bee in Tertiary amber of the Dominican Republic (Hymenoptera: Colletidae). *Entomologica Scandinavica*, **30**, 453–8.

Engel, M. S. (1999b). The taxonomy of recent and fossil honeybees (Hymenoptera: Apidae; *Apis*). *Journal of Hymenoptera Research*, **8**, 165–96.

Engel, M. S. (1999c). *Megachile glaesaria*, the first megachilid bee fossil from amber (Hymenoptera: Megachilidae). *American Museum Novitates*, **3276**, 1–13.

Engel, M. S. (1999d). The first fossil *Euglossa* and phylogeny of the orchid bees (Hymenoptera: Apidae; Euglossini). *American Museum Novitates*, **3272**, 1–14.

Engel, M. S. (2000a). Classification of the bee tribe Augochlorini (Hymenoptera: Halictidae). *Bulletin of the American Museum of Natural History*, **250**, 1–90.

Engel, M. S. (2000b). A new interpretation of the oldest fossil bee (Hymenoptera: Apidae). *American Museum Novitates*, **3296**, 1–11.

Engel, M. S. (2001a). A monograph of the Baltic Amber bees and evolution of the Apoidea (Hymenoptera). *Bulletin of the American Museum of Natural History*, **259**, 1–192.

Engel, M. S. (2001b). The first large carpenter bee from the Tertiary of North America, with a consideration of the geological history of Xylocopinae. *Transactions of the American Entomological Society*, **127**, 245–54.

Engel, M. S. (2001c). Monophyly and extensive extinction of advanced eusocial bees: insights from an unexpected Eocene diversity. *Proceedings of the National Academy of Sciences USA*, **98**, 1661–1664.

Engel, M. S. (2002a). Halictine bees from the Eocene-Oligocene boundary of Florissant, Colorado (Hymenoptera: Halictidae). *Neues Jahrbuch für Geologie und Paläontologie, Abhandlungen*, **225**, 251–73.

Engel, M. S. (2002b). Phylogeny of the bee tribe Fideliini (Hymenoptera: Megachilidae), with the description of a new genus from Southern Africa. *African Entomology*, **10**, 305–13.

Engel, M. S. (2004a). Notes on a megachiline bee (Hymenoptera: Megachilidae) from the Miocene of Idaho. *Transactions of the Kansas Academy of Sciences*, **107**, 97–100.

Engel, M. S. (2004b). Geological history of the bees (Hymenoptera: Apoidea). *Revista de Tecnologia e Ambiente*, **10**, 9–33.

Engel, M. S. (2004c). A new species of the bee genus *Eoanthidium* with extraordinary male femoral organs from the Arabian Peninsula

(Hymenoptera: Megachilidae). *Scientific Paper of the Natural History Museum University of Kansas*, **34**, 1-6.

Engel, M. S. (2004d). Fideliine phylogeny and classification revisited (Hymenoptera: Megachilidae). *Journal of the Kansas Entomogical Society*, **77**, 821-36.

Engel, M. S. (2005). Family-group names for bees (Hymenoptera: Apoidea). *American Museum Novitates*, **3476**, 1-33.

Engel, M. S. (2006). A giant honeybee from the middle Miocene of Japan (Hymenoptera: Apidae). *American Museum Novitates*, **3504**, 1-12.

Engel, M. S. (2008). A new species of *Ctenoplectrella* in Baltic amber (Hymenoptera: Megachilidae). *Acta Zoologica Academiae Scientiarum Hungaricae*, **54**, 319-24.

Engel, M. S. (2009). Two new Halictine bees in Miocene amber from the Dominican Republic (Hymenoptera, Halictidae). *Zookeys*, **29**, 1-12.

Engel, M. S. (2011). Systematic melittology: where to from here? *Systematic Entomology*, **36**, 2-15.

Engel, M. S. and Archibald, S. B. (2003). An Early Eocene bee (Hymenoptera: Halictidae) from Quilchena, British Columbia. *The Canadian Entomologist*, **135**, 63-9.

Engel, M. S. and Peñalver, E. (2006). A Miocene Halictine bee from Rubielos de Mora Basin, Spain (Hymenoptera: Halictidae). *American Museum Novitates*, **3503**, 1-12.

Engel, M. S. and Perkovsky, E. E. (2006). An Eocene bee in Rovno amber, Ukraine (Hymenopetra: Megachilidae). *American Museum Novitates*, **3506**, 1-12.

Engel, M. S. and Rightmyer, M. G. (2000). A new augochlorine bee species in Tertiary amber from the Dominican Republic (Hymenoptera:Halictidae). *Apidologie*, **31**, 431-6.

Engel, M. S., Hinojosa-Díaz, I. A. and Rasnitsyn, A. P. (2009). A honeybee from the Miocene of Nevada and the biogeography of *Apis* (Hymenoptera:Apidae:Apini). *Proceedings of the California Academy of Sciences*, **60**, 23-38.

Epis, R. C. and Chapin, C. E. (1974). Stratigraphic nomenclature of the Thirtynine Mile volcanic field, central Colorado, US. *Geological Survey Bulletin*, **1395-C**, 1-23.

Erbar, C. and Leins, P. (1995). Portioned pollen release and the syndromes of secondary pollen presentation in the Campanulales–Asterales complex. *Flora*, **190**, 323-38.

Frič, A. and Bayer, E. (1901). Studien im Gebiete der Böhmischen Kreideformation. Palaeontologische Untersuchungen der einzelnen Schichten. *Archiv der Naturwissenschaftlichen Landesdurchforschung von Böhmen*, **11**, 1-184.

Gandolfo, M. A., Nixon, K. C. and Crepet, W. L. (2004). Cretaceous flowers of Nymphaeaceae and implications for complex insect entrapment pollination mechanisms in early angiosperms. *Proceedings of the National Academy of Sciences USA*, **101**, 8056-60.

Genise, J. F. (2000). The ichnofamily Celliformidae for *Celliforma* and allied ichnogenera. *Ichnos*, **7**, 267-82.

Genise, J. F. and Bown, T. M. (1996). *Uruguay* Roselli 1938 and *Rosellichnus*, N. Ichnogenus: two ichnogenera for clusters of fossil bee cells. *Ichnos*, **4**, 199-217.

Genise, J. F. and Verde, M. (2000). *Corimbatichnus fernandezi*: a cluster of fossil bee cells from the late cretaceous-early tertiary of Uruguay. *Ichnos*, **7**, 115–25.

Gonzalez, V. H. and Engel, M. S. (2004). The tropical Andean bee fauna (Insecta:Hymenoptera:Apoidea), with examples from Colombia. *Entomologische Abhandlungen*, **62**, 65–75.

Gottsberger, G. and Silberbauer-Gottsberger, I. (1988). Evolution of flower structures and pollination in Neotropical Cassiinae (Caesalpiniaceae) species. *Phyton (Austria)*, **28**, 293–320.

Grimaldi, D. (1996). *Amber: Window to the Past*. New York, NY: American Museum of Natural History.

Grimaldi, D. (1999). The coradiations of pollinating insects and angiosperms in the Cretaceous. *Annals of the Missouri Botanical Garden*, **86**, 373–406.

Grimaldi, D. and Engel, M. S. (2005). *Evolution of the Insects*, Cambridge, UK: Cambridge University Press.

Grimaldi, D. and Nascimbene, P. C. (2010). Raritan (New Jersey) amber. In *Biodiversity of Fossils in Amber from the Major World Deposits*, ed. D Penney. Manchester, UK: Siri Scientific Press, pp. 167–91.

Grimaldi, D., Lillegraven, J. A., Wampler, T. P., Bookwalter, D. and Shedrinsky, A. (2000). Amber from Upper Cretaceous through Paleocene strata of the Hanna Basin, Wyoming, with evidence for source and taphonomy of fossil resins. *Rocky Mountain Geology*, **35**, 163–204.

Grimaldi, D., Engel, M. S. and Nascimbene, P. C. (2002). Fossiliferous Cretaceous amber from Myanmar (Burma): its rediscovery, biotic diversity, and paleontological significance. *American Museum Novitates*, **3361**, 1–71.

Hao, G., Yuan, Y.-M., Hu, C.-M., Ge, X.-J. and Zhao, N.-X. (2004). Molecular phylogeny of *Lysimachia* (Myrsinaceae) based on chloroplast *trn*L-F and nuclear ribosomal ITS sequences. *Molecular Phylogenetics and Evolution*, **31**, 323–39.

Harder, L. D. and Barclay, R. M. R. (1994). The functional-significance of poricidal anthers and buzz pollination: controlled pollen removal from dodecatheon. *Functional Ecology*, **8**, 509–17.

Harder, L. D. and Barrett, S. C. H. (1993). Pollen removal from Tristylous Pontederia Cordata: effects of anther position and pollinator specialization. *Ecology*, **74**, 1059–72.

Heer, O. (1849). *Die Insektenfauna der Tertiärgebilde von Oeningen und von Radoboj in Croatien*, Leipzig, Germany: W. Engelmann.

Heinrich, B. (1996). *The Thermal Warriors: Strategies of Insect Survival*. Cambridge, MA: Harvard University Press.

Hinojosa-Díaz, I. A. and Engel, M. S. (2007). A new fossil orchid bee in Colombian copal (Hymenoptera: Apidae). *American Museum Novitates*, **3589**, 1–7.

Houston, T. F. (1990). Descriptions of new paracolletine bees associated with flowers of Eremophila (Hymenoptera: Colletidae). *Records of the Western Australian Museum*, **14**, 583–621.

Houston, T. F. (1991). Two new and unusual species of the bee genus *Leiproctus* Smith (Hymenoptera: Colletidae), with notes on their behaviour. *Records of the Western Australian Museum*, **15**, 83–96.

Houston, T. F. (2000). *Native bees on wildflowers in western Australia. A synopsis of bee visitation of wildflowers based on the bee collection of the*

Western Australian Museum, Perth, Australia: Western Australian Insect Study Society.

Houston, T. F. and Thorp, R. W. (1984). Bionomics of the bee *Stenotritus greavesi* and ethological characteristics of the Stenotritidae (Hymenoptera). *Records of the Western Australian Museum*, **11**, 375–85.

Inouye, D. W. (1980). The terminology of floral larceny. *Ecology*, **61**, 1251–3.

Irwin, R. E., Brody, A. K. and Waser, N. M. (2001). The impact of floral larceny on individuals, populations, and communities. *Oecologia*, **129**, 161–8.

Iuga, V. G. (1958). *Insecta. Volumul IX. Fascicula 3. Hymenoptera Apoidea Fam. Apidae subfam. Anthophorinae*, Bucarest, Hungary: Academia Republicii Populare Romîne.

Kawakita, A., Ascher, J. S., Sota, T., Kato, M. and Roubik, D. W. (2008). Phylogenetic analysis of the corbiculate bee tribes based on 12 nuclear protein-coding genes (Hymenoptera:Apoidea:Apidae). *Apidologie*, **39**, 163–75.

Kimsey, L. S. (1984). The behavioural and structural aspects of grooming and related activities in euglossine bees (Hymenoptera:Apidae). *Journal of Zoology*, **204**, 541–50.

Kotthoff, U., Wappler, T. and Engel, M. S. (2011). Miocene honey bees from the Randeck Maar of southwestern Germany (Hymenoptera, Apidae). *ZooKeys*, 11–37.

Labandeira, C. C. (1998). How old is the flower and the fly. *Science*, **280**, 57–9.

Labandeira, C. C. (2000). The paleobiology of pollination and its precursors. *Paleontological Society Papers*, **6**, 233–69.

Labandeira, C. C. (2002a). Paleobiology of middle Eocene plant-insect associations from the Pacific

Northwest: a preliminary report. *Rocky Mountain Geology*, **37**, 31–59.

Labandeira, C. C. (2002b). The history of associations between plants and animals. In *Plant–Animal Interactions*, ed. M. Herrera and O. Pellmyr. Oxford, UK: Blackwell Publishing, pp. 26–74.

Larkin, L. L., Neff, J. L. and Simpson, B. B. (2008). The evolution of a pollen diet: Host choice and diet breadth of *Andrena* bees (Hymenoptera: Andrenidae). *Apidologie*, **39**, 133–45.

Lucas, S. G., Minter, N. J. and Hunt, A. P. (2010). Re-evaluation of alleged bees' nests from the Upper Triassic of Arizona. *Palaeogeography, Palaeoclimatology, Palaeoecology*, **286**, 194–201.

Lutz, H. (1993). *Eckfeldapis electrapoides* nov. gen. n. sp., eine "Honigbiene" aus dem Mittel-Eozän des "Eckfelder Maares" bei Manderscheid/Eifel, Deutschland (Hymenoptera:Apidae, Apinae). *Mainzer Naturwissenschaftliches Archiv*, **31**, 177–99.

Macior, L. W. (1986). Pollination ecology and endemism of *Pedicularis pulchella* Pennell (Scrophulariaceae). *Plant Species Biology*, **1**, 173–80.

Macior, L. W. (1995). *Pedicularis*, a valuable information resource for plant conservation. In *Environmental Factors and Pollination*, ed. R. C. Sihag. Hisar: Rajendra Science Publications, pp. 8–19.

Mauss, V., Müller, A. and Prosi, R. (2006). Mating, nesting and flower association of the east Mediterranean pollen wasp *Ceramius bureschi* in Greece (Hymenoptera:Vespidae:Masarinae). *Entomologia Generalis*, **29**, 1–26.

McGinley, R. J. and Rozen, J. G. (1987). Nesting biology, immature stages, and phylogenetic placement of the palaearcti bee *Pararhophites* (Hymenoptera:Apoidea). *American Museum Novitates*, **2903**, 1-21.

Meunier, F. (1920). Quelques insectes de l'Aquitanien de Rott, Sept-Monts (Prusse rhénane). *Verhandelingen der Koninklijke Akademie van Wetenschappen*, **22**, 727-37.

Michener, C. D. (1944). Comparative external morphology, phylogeny, and classification of the bees (Hymenoptera). *Bulletin of the American Museum of Natural History*, **82**, 1-326.

Michener, C. D. (1962). An interesting method of pollen collecting by bees from flowers with tubular anthers. *Revista de Biologia Tropical*, **10**, 167-75.

Michener, C. D. (1965). A classification of the bees of the Australian and South Pacific regions. *Bulletin of the American Museum of Natural History*, **130**, 1-789.

Michener, C. D. (1979). Biogeography of the bees. *Annals of the Missouri Botanical Garden*, **66**, 277-342.

Michener, C. D. (1999). The corbiculae of bees. *Apidologie*, **30**, 67-74.

Michener, C. D. (2007). *The Bees of the World, 2nd edn*. Baltimore, MD: The Johns Hopkins University Press.

Michener, C. D. and Grimaldi, D. (1988a). A *Trigona* from late Cretaceous amber of New Jersey (Hymenoptera:Apidae:Meliponinae). *American Museum Novitates*, **2917**, 1-12.

Michener, C. D. and Grimaldi, D. (1988b). The oldest fossil bee: apoid history, evolutionary stasis, and antiquity of social behavior. *Proceedings of the National Academy of Sciences USA*, **85**, 6424-26.

Michener, C. D. and Poinar, G., Jr. (1996). The known bee fauna of Dominican amber. *Journal of the Kansas Entomological Society*, **69**, 353-61.

Michez, D. and Patiny, S. (2005). World revision of the oil-collecting bee genus *Macropis* Panzer 1809 (Hymenoptera, Apoidea, Melittidae) with a description of a new species from Laos. *Annales de la Société entomologique de France (n. s.)*, **41**, 15-28.

Michez, D., Patiny, S. and Danforth, B. N. (2009b). Phylogeny of the bee family Melittidae (Hymenoptera: Anthophila) based on combined molecular and morphological data. *Systematic Entomology*, **34**, 574-97.

Michez, D., Nel, A., Menier, J.-J. and Rasmont, P. (2007). The oldest fossil of a melittid bee (Hymenoptera:Apiformes) from the Early Eocene of Oise (France). *Zoological Journal of the Linnean Society*, **150**, 701-9.

Michez, D., Patiny, S., Rasmont, P., Timmermann, K. and Vereecken, N. J. (2008). Phylogeny and host-plant evolution in Melittidae *s.l.* (Hymenoptera:Apoidea). *Apidologie*, **39**, 146-62.

Michez, D., De Meulemeester, T., Nel, A., Rasmont, P. and Patiny, S. (2009a). New fossil evidence of the early diversification of bees: *Paleohabropoda oudardi* from the French Paleocene (Hymenoptera, Apidae, Anthophorini). *Zoologica Scripta*, **38**, 171-81.

Miller, D. F. and Morgan, A. V. (1982). A postglacial coleopterous assemblage from Lockport gulf, New York. *Quaternary research*, **17**, 258-74.

Moure, J. S. and Camargo, J. M. F. (1978). A fossil stingless bee from Copal (Hymenoptera: Apidae). *Journal of* the Kansas Entomological Society, **51**, 560-66.

Müller, A. (1995). Morphological specializations in central European bees for the uptake of pollen from flowers with anthers hidden in narrow corolla tubes (Hymenoptera:Apoidea). *Entomologia Generalis*, **20**, 43-57.

Müller, A. (1996a). Convergent evolution of morphological specializations in Central European bee and honey wasp species as an adaptation to the uptake of pollen from nototribic flowers (Hymenoptera, Apoidea and Masaridae). *Biological Journal of the Linnean Society*, **57**, 235-52.

Müller, A. (1996b). Host-plant specialization in Western Palearctic anthidiine bees (Hymenoptera:A poidea:Megachilidae). *Ecological Monographs*, **66**, 235-57.

Müller, A. (2006). Unusual host-plant of *Hoplitis pici*, a bee with hooked bristles on its mouthparts (Hymenoptera:Mega chilidae:Osmiini). *European Journal of Entomology*, **103**, 497-500.

Müller, A. (2011). Palaearctic Osmiine Bees. ETH Zürich, http://blogs.ethz. ch/osmiini

Müller, A. and Kuhlmann, M. (2003). Narrow flower specialization in two European bee species of the genus Colletes (Hymenoptera:Apoidea:Collet idae). *European Journal of Entomology*, **100**, 631-35.

Müller, A. and Kuhlmann, M. (2008). Pollen hosts of western palaearctic bees of the genus *Colletes* (Hymenoptera:Colletidae): the Asteraceae paradox. *Biological Journal of the Linnean Society*, **95**, 719-33.

Müller, A., Krebs, A. and Amiet, F. (1997). *Bienen, Mitteleuropäische Gattungen, Lebensweise, Beobachtung*, Augsburg, Germany: Naturbuch Verlag.

Müller, A., Diener, S., Schnyder, S., Stutz, K., Sedivy, C. and Dorn, S. (2006). Quantitative pollen requirements of solitary bees: Implications for bee conservation and the evolution of bee-flower relationships. *Biological Conservation*, **130**, 604-15.

Neff, J. L. (2004). Hooked hairs and not so narrow tubes: two new species of *Colletes* Latreille from Texas (Hymen optera:Apoidea:Colletidae). *Journal of Hymenoptera Research*, **13**, 250-61.

Neff, J. L. and Simpson, B. B. (1988). Vibratile pollen-harvesting by *Megachile mendica* Cresson (Hymenoptera:Megachilidae). *Journal of Kansas Entomological Society*, **61**, 242-24.

Nel, A. and Brasero, N. (2010). Oise Amber. In *Biodiversity of Fossils in Amber from the Major World Deposits*, ed. D. Penney. Manchester, UK: Siri Scientific Press, pp. 137-48.

Nel, A. and Petrulevicius, J. F. (2003). New Palaeogene bees from Europe and Asia. *Alcheringa*, **27**, 227-93.

Nel, A. and Roy, R. (1996). Revision of the fossil "mantid" and "ephemerid" species described by Piton from the Palaeocene of Menat (France) (Mantodea:Chaeteessidae, Mantidae; Ensifera:Tettigonioidea). *European Journal of Entomology*, **93**, 223-34.

Nel, A., Martínez-Delclòs, X., Papier, F. and Oudard, J. (1997). New Tertiary fossil Odonata from France. (Sieblosiidae, Lestidae, Coenagrioniidae, Megapodagrionidae, Libellulidae). *Deutsche Entomologische Zeitschrift*, **44**, 231-58.

Nel, A., Martínez-Delclòs, X., Arillo, A. and Peñalver, E. (1999a). A review of the Eurasian fossil species of the bee *Apis*. *Paleontology*, **42**, 243-285.

Nel, A., De Ploëg, G., Dejax, J., Dutheil, D., de Franceschi, D., Gheerbrant, E.,

Godinot, M., Hervet, S., Menier, J.-J.,
Augé, M., Bignot, G., Cavagnetto, C.,
Duffaud, S., Gaudant, J., Hua, S.,
Jossang, A., de Lapparent de Broin, F.,
Pozzi, J.-P., Paicheler, J.-C., Bouchet, F.
and Rage, J.-C. (1999b). Un gisement
sparnacien exceptionnel à plantes,
arthropodes et vertébrés (Éocène
basal, MP7): Le Quesnoy (Oise,
France). *Comptes Rendus de l'Académie
des Sciences, Sciences de la terre et des
planètes*, **329**, 65–72.

Ohl, M. and Engel, M. S. (2007). Die
Fossilgeschichte der Bienen
und ihrer nächsten Verwandten
(Hymenoptera:Apoidea). *Denisia*, **20**,
687–700.

Parker, F. D. and Tepedino, V. J. (1982).
A nest and pollen collection records
of *Osmia sculleni* Sandhouse, a bee
with hooked hairs on mouthparts
(Hymenoptera:Megachilidae). *Journal
of the Kansas Entomological Society*,
51, 145–73.

Pasteels, J. J. and Pasteels, J. M. (1979).
Etude au microscope électronique à
balayage des scopas collectrices de
pollen chez les Andrenidae (Hymenop
tera:Apoidea:Andrenidae). *Archives de
Biologie*, **90**, 113–30.

Patiny, S., Michez, D. and Danforth, B. N.
(2008). Phylogenetic relationships and
host-plant evolution within the basal
clade of Halictidae (Hymenoptera,
Apoidea). *Cladistics*, **24**, 255–69.

Patiny, S., Engel, M. S., Vanmarsenille, P.
and Michez, D. (2007). A new
record of *Thaumastobombus
andreniformis* Engel in Eocene amber
(Hymenoptera:Apidae). *Annales de la
Société entomologique de France (n. s.)*,
43, 505–8.

Peñalver, E., Engel, M. S. and Grimaldi, D.
(2006). Fig wasps in Dominican amber

(Hymenoptera: Agaonidae). *American
Museum Novitates*, **3541**, 1–16.

Piton, L. (1940). Paléontologie du gisement
éocène de Menat (Puy-de-Dôme)
(Flore et faune). *Mémoire de la Société
d'Histoire Naturelle d'Auvergne*, **1**,
1–303.

Poinar, G. J. (1998). *Paleoeuglossa
melissiflora* gen. n., sp. n.
(Euglossinae:Apidae), fossil orchid
bees in Dominican amber. *Journal of
the Kansas Entomological Society*, **71**,
29–34.

Poinar, G. O. J. (2004). Evidence of
parasitism by Strepsiptera in
Dominican amber. *Biocontrol*, **49**,
239–44.

Poinar, G. J. (2010). Palaeoecological
perspectives in Dominican amber.
*Annales de la societe entomologique de
France (n.s.)*, **46**, 23–52.

Poinar, G. J. and Danforth, B. N. (2006).
A fossil bee from early Cretaceous
Burmese amber. *Science*, **314**, 614.

Pouvreau, A. (2004). *Les Insectes
Pollinisateurs*, Paris, France:
Delachaux et Niestlé.

Praz, C. J., Müller, A. and Dorn, S. (2008).
Specialized bees fail to develop on
non-host pollen: do plants chemically
protect their pollen? *Ecology*, **89**,
795–804.

Proctor, M., Yeo, P. and Lack, A. (1996). *The
Natural History of Pollination*, London,
UK: Harper Collins Publishers.

Proença, C. E. B. (1992). Buzz pollination:
older and more widespread than we
think? *Journal of Tropical Ecology*, **8**,
115–20.

Prokop, J. and Nel, A. (2003). New fossil
Aculeata from the Oligocene of the
Ceske Stredohori Mts. and the Lower
Miocene of the Most Basin in nothern
Czech Republic (Hymenoptera:

Apidae, Vespidae). *Acta Musei Nationalis Pragae, Series B, Natural History*, **59**, 163–71.

Raine, N. E., Ings, T. C., Dornhaus, A., Saleh, N. and Chittka, L. (2006). Adaptation, genetic drift, pleiotropy, and history in the evolution of bee foraging behavior. *Advances in the Study of Behavior*, **36**, 305–54.

Ramirez, S. R. (2009). Orchid bees. *Current Biology*, **19**, R1061–3.

Ramirez, S. R., Roubik, D. W., Skov, C. and Pierce, N. E. (2010). Phylogeny, diversification patterns and historical biogeography of euglossine orchid bees (Hymenoptera:Apidae). *Biological Journal of the Linnean Society*, **100**, 552–72.

Ramirez, S. R., Gravendeel, B., Singer, R. B., Marshall, C. R. and Pierce, N. E. (2007). Dating the origin of the Orchidaceae from a fossil orchid with its pollinator. *Nature*, **448**, 1042–5.

Rasmont, P. (1988). *Monographie écologique et biogéographique des bourdons de France et de Belgique (Hymenoptera, Apidae, Bombinae)*, Gembloux, Belgium: Faculté Universitaire des Sciences Agronomiques de Gembloux.

Rasnitsyn, A. P. and Michener, C. D. (1991). Miocene fossil bumble bee from the Soviet far east with comments on the chronology and distribution of fossil bees (Hymenoptera:Apidae). *Annals of Entomological Society of America*, **84**, 583–9.

Raven, P. H. and Axelrod, I. (1974). Angiosperm biogeography and past continental movements. *Annals of the Missouri Botanical Garden*, **61**, 539–673.

Renner, S. S. and Schaefer, H. (2010). The evolution and loss of oil-offering flowers: new insights from dated phylogenies for angiosperms and bees. *Philosophical Transactions of the Royal Society B, Biological Sciences*, **365**, 423–35.

Retallack, G. J. (1984). Trace fossils of burrowing beetles and bees in an Oligocene paleosol, Badlands National Park, South Dakota. *Journal of Paleontology*, **58**, 571–92.

Robertson, C. (1925). Heterotrophic bees. *Ecology*, **6**, 412–36.

Roig-Alsina, A. and Michener, C. D. (1993). Studies of the phylogeny and classification of long-tongued bees (Hymenoptera: Apoidea). *The University of Kansas Science Bulletin*, **55**, 123–73.

Ronquist, F. (1999). Phylogeny of the Hymenoptera (Insecta): the state of the art. *Zoologica Scripta*, **28**, 3–11.

Ross, A., Mellish, C., York, P. and Crighton, B. (2010). Burmese amber. In *Biodiversity of Fossils in Amber from the Major World Deposits*, ed. D. Penney. Manchester, UK: Siri Scientific Press, pp. 208–35.

Roulston, T. H. and Cane, J. H. (2000). Pollen nutritional content and digestibility for animals. *Plant Systematics and Evolution*, **222**, 187–209.

Roulston, T. H., Cane, J. H. and Buchmann, S. L. (2000). What governs protein content of pollen: pollinator preferences, pollen-pistil interactions, or phylogeny. *Ecological Monographs*, **70**, 617–43.

Rozen, J. G. (1996). A new species of the bee *Heterosarus* from Dominican amber (Hymenoptera:Andrenidae; Panurginae). *Journal of the Kansas Entomological Society*, **69**, 346–52.

Rozen, J. G., Ozbek, H., Ascher, J. S., Sedivy, C., Praz, C., Monfared, A. and Muller, A. (2010). Nests, petal usage, floral preferences, and immatures of *Osmia* (*Ozbekosmia*) *avosetta* (Megachilidae:Megachilinae:Osmiini), including biological comparisons with other Osmiine bees. *American Museum Novitates*, 1–22.

Rust, J., Singh, H., Rana, R. S., McCann, T., Singh, L., Anderson, K., Sarkar, N., Nascimbene, P. C., Stebner, F., Thomas, J. C., Kraemer, M. S., Williams, C. J., Engel, M. S., Sahni, A. and Grimaldi, D. (2010). Biogeographic and evolutionary implications of a diverse paleobiota in amber from the early Eocene of India. *Proceedings of the National Academy of Sciences USA*, **107**, 18360–65.

Sakagami, S. F. (1965). Über dem Bau der männlichen Hinterschiene von *Eulaema nigrita* Lepeletier (Hymenoptera, Apidae). *Zoologischer Anzeiger*, **175**, 347–54.

Sarzetti, L. C., Labandeira, C. C. and Genise, J. F. (2008). A leafcutter bee trace fossil from the middle Eocene of Patagonia, Argentina, and a review of Megachilid (Hymenoptera) ichnology. *Palaeontology*, **51**, 933–41.

Schlindwein, C., Wittmann, D., Martins, C. F., Hamm, A., Siqueira, J. A., Schiffer, D. and Machado, I. C. (2005). Pollination of *Campanula rapunculus* L. (Campanulaceae): how much pollen flows into pollination and intro reproduction of oligolectic pollinators? *Plant Systematics and Evolution*, **250**, 147–56.

Sedivy, C., Praz, C. J., Muller, A., Widmer, A. and Dorn, S. (2008). Patterns of host-plant choice in bees of the genus *Chelostoma*: the constraint hypothesis of host-range evolution in bees. *Evolution*, **62**, 2487–507.

Shinn, A. F. (1967). A revision of the bee genus *Calliopsis* and the biology and ecology of *C. andreniformis*

(Hymenoptera:Andrenidae). *University of Kansas Science Bulletin*, **46**, 753–936.

Singer, R. B. and Sazima, M. (2001). Flower morphology and pollination mechanism in three sympatric Goodyerinae orchids from southeastern Brazil. *Annals of Botany*, **88**, 989–97.

Solórzano-Kraemer, M. M. S. (2007). Systematic, palaeoecology, and palaeobiogeography of the insect fauna from Mexican amber. *Palaeontographica Abteilung a-Palaozoologie-Stratigraphie*, **282**, 1–133.

Soltis, P. S. and Soltis, D. E. (2004). The origin and diversification of Angiosperms. *American Journal of Botany*, **91**, 1614–26.

Steiner, K. E. and Whitehead, V. B. (1990). Pollinator adaption to oil-secreting flowers – *Rediviva* and *Diascia*. *Evolution*, **44**, 1701–07.

Strickler, K. (1979). Specialization and foraging efficiency of solitary bees. *Ecology*, **60**, 998–1009.

Taylor, D. W. and Hickey, L. J. (1992). Phylogenetic evidence for the herbaceous origin of angiosperms. *Plant Systematics and Evolution*, **180**, 137–56.

Thorp, R. W. (1979). Structural behavioral, and physiological adaptations of bees (Apoidea) for collecting pollen. *Annals of the Missouri Botanical Garden*, **66**, 788–812.

Thorp, R. W. (2000). The collection of pollen by bees. *Plant Systematics and Evolution*, **222**, 211–23.

Timon-David, J. (1944). Insectes fossiles de l'Oligocène inférieur Camoins (Bassin de Marseille) II. Hyménoptères. *Bulletin de la Société Entomologique de France*, **49**, 40–5.

Torchio, P. F. (1984). The nesting biology of *Hylaeus bisinuatus* Forster and development of its immature forms

(Hymenoptera: Colletidae). *Journal of the Kansas Entomological Society*, **57**, 276–97.

Tosi, A. (1896). Di un nuevo genere di Apiaria fossile nell' ambra di Sicilia (*Meliponorytes succini – M. sicula*). *Rivista Italiana di Paleontologia*, **2**, 352–6.

Vamosi, J. C. and Vamosi, S. M. (2010). Key innovations whitin a geographical context in flowering plants: towards resolving Darwin's abominable mystery. *Ecology Letters*, **13**, 1270–9.

Vereecken, N. J. and McNeil, J. N. (2010). Cheaters and liars: chemical mimicry at its finest. *Canadian Journal of Zoology*, **88**, 725–52.

Vogel, S. (1966). Parfümsammelnde Bienen als Bestäuber von Orchidaceen und *Gloxinia*. *Österreichischen Botanischen Zeitschrift*, **113**, 302–61.

Vogel, S. (1976). *Lysimachia*: Ölblumen der Holarktis. *Naturwissenschaften*, **63**, 44–5.

Vogel, S. (1981). Abdominal oil-mapping: a new type of foraging in bees. *Naturwissenschaften*, **68**, 627–8.

Vogel, S. (1986). Olblumen und ölsammelnde Bienen II. *Lysimachia* und *Macropis*. *Tropische und Subtropische Pfanzenwelt*, **54**, 147–312.

Vogel, S. (1993). Betrug bei Pflanzen: Die Täuschblumen. *Abhandlungen der Akademie der Wissenschaften und der Literatur Mainz*, **1**, 1–48.

Vogel, S. and Michener, D. C. (1985). Long bee legs and oil-producting floral spurs, and a new *Rediviva* (Hymenoptera, Melittidae;Scrophulariaceae). *Journal of the Kansas Entomological Society*, **58**, 359–64.

Wappler, T. and Engel, M. S. (2003). The middle Eocene bee faunas of Eckfeld and Messel, Germany (Hymenoptera:Apoidea). *Journal of Paleontology*, **77**, 908–21.

Ware, J. L., Grimaldi, D. A. and Engel, M. S. (2010). The effects of fossil placement and calibration on divergence times and rates: an example from the termites (Insecta: Isoptera). *Arthropod Structure & Development*, **39**, 204–19.

Wcislo, W. T. and Cane, J. H. (1996). Floral resource utilization by solitary bees (Hymenoptera:Apoidea) and exploitation of their stored foods by natural enemies. *Annual Review of Entomology*, **41**, 257–86.

Wedmann, S., Wappler, T. and Engel, M. S. (2009). Direct and indirect fossil records of megachilid bees from the Paleogene of Central Europe (Hymenoptera:Megachilidae). *Naturwissenschaften*, **96**, 703–12.

Weitschatt, W. and Wichard, W. (2010). Baltic amber. In *Biodiversity of Fossils in Amber from the Major World Deposits*, ed. D. Penney. Manchester, UK: Siri Scientific Press, pp. 80–115.

Westerkamp, C. (1996). Pollen in bee–flower relations: some considerations on melittophily. *Botanica Acta*, **109**, 325–32.

Westerkamp, C. (1997). Flowers and bees are competitors-not partners. Towards a new understanding of complexity in specialised bee flowers. *Acta Horticulturae*, **437**, 71–4.

Westerkamp, C. and Claßen-Bockhoff, R. (2007). Bilabiate flowers: the ultimate response to bees? *Annals of Botany*, **100**, 361–74.

Westrich, P. (1989). *Die Wildbienen Baden-Württembergs: Allgemeiner Teil, pp. 1–43; Spezieller Teil: Die Gattungen Und Arten, pp. 437–972*, Stuttgart, Germany: Eugene Ulmer.

Whitten, W. M., Young, A. M. and Williams, N. H. (1989). Function of glandular secretions in fragrance collection by male euglossine bees (Apidae:Euglossini). *Journal of Chemical Ecology*, **15**, 1285–95.

Wille, A. (1959). A new fossil stingless bee (Meliponini) from the amber of Chiapas, Mexico. *Journal of Paleontology*, **33**, 849–52.

Wolfe, A. P., Tappert, R., Muehlenbachs, K., Boudreau, M., McKellar, R. C., Basinger, J. F. and Garrett, A. (2009). A new proposal concerning the botanical origin of Baltic amber. *Proceedings of the Royal Society B, Biological Sciences*, **276**, 3403–12.

Zeuner, F. E. and Manning, F. J. (1976). A monograph on fossil bees (Hymenoptera: Apoidea). *Bulletin of the British Museum (Natural History), Geology*, **27**, 149–268.

Zhang, J.-F. (1989). *Fossil Insects from Shanwang, Shandong, China*. Jinan, China: Shandong Science and Technology Publishing House.

Zhang, J.-F. (1990). New fossil species of Apoidea (Insecta: Hymenoptera). *Acta Zootaxonomica Sinica*, **15**, 83–91. [In Chinese, with English summary].

Zhang, J.-F., Sun, B. and Zhang, X.-Y. (1994). *Miocene Insects and Spiders from Shanwang, Shandong*. Beijing, China: Science Press. [In Chinese, with English summary].

6

Pollen evidence for the pollination biology of early flowering plants

SHUSHENG HU, DAVID L. DILCHER
AND DAVID WINSHIP TAYLOR

6.1 Introduction

Angiosperms are the dominant and most diverse plant group living today. They are also found in the greatest number of terrestrial ecosystems on Earth of any group of plants (Judd et al. 2002; Soltis and Soltis 2004). They provide human beings and other terrestrial animals, directly or indirectly, with the majority of their nutrition (e.g. Theissen and Melzer 2007). Much of these foods, such as fruits, nuts, seeds, and grains, are the direct products of flowers, and pollination is an essential step in their formation. Pollination biology has long been an interest of biologists and agricultural scientists (e.g. Faegri and van der Pijl 1979; Proctor et al. 1996; Aizen et al. 2009; Lonsdorf et al. 2009; Mitchell et al. 2009). However, our understanding of the early phases of the evolution of angiosperm pollination is still limited and attempts to reconstruct the history of the interactions between angiosperms and pollinators are challenging (Hu et al. 2008; Taylor and Hu 2010). Evolutionary biologists have attempted to deduce the possible histories of pollination syndromes (summarized in Taylor and Hu, 2010) based upon usually incomplete and limited early angiosperm flower fossil records (e.g. Dilcher 1979; Retallack and Dilcher 1981; Crane et al. 1986; Herendeen et al. 1995; Crepet and Nixon 1996; Friis et al. 1999, 2000, 2006; Crepet 2008), limited insect fossil records

Evolution of Plant–Pollinator Relationships, ed S. Patiny. Published by Cambridge University Press. © The Systematics Association 2012.

(e.g. Grimaldi 1999; Labandeira 2000, 2002; Grimaldi and Engel 2005; Ren et al. 2009), parsimony analysis (e.g. Hu et al. 2008; Friedman and Barrett 2008; Taylor and Hu 2010), investigation on pollination biology of the most basal angiosperms (e.g.Thien et al. 2009), and angiosperm pollen fossil records (e.g. Hu et al. 2008; Taylor and Hu 2010).

Currently there are three hypotheses regarding early angiosperm pollination biology (Taylor and Hu, 2010):

(1) Ancestral angiosperms were insect pollinated (e.g. Crepet and Friis 1987; Wing and Boucher 1998; Friis et al. 1999; Feild and Arens 2005).

(2) During the mid Cretaceous there were increases in advanced pollination syndromes (e.g. Crepet et al. 1991; Crepet 2008; Hu et al. 2008, Taylor and Hu 2010).

(3) Wind pollination (anemophily) is derived (e.g. Culley et al. 2002; Hu et al. 2008, Taylor and Hu 2010).

There also are two recent tentative hypotheses concerning early angiosperm pollination. The first is that angiosperm pollen grains were initially dry with no sticky substances to cause pollen clumping (Hu et al. 2008, Taylor and Hu 2010). The second is that the earliest angiosperm flowers may have been capable of being pollinated by insects and wind, a type called ambophilous pollination (Taylor and Hu 2010).

Because of the strong association between dry pollen and wind pollination, and sticky pollen resulting in pollen clumps (Hesse 1979a, 1979b, 1981a, 1981b; Hu 2006; Hu et al. 2008; Taylor and Hu 2010), it has been assumed that early angiosperm pollen was sticky (Pacini 2000). Taylor and Hu (2010) showed that clumping is not common until the mid Cretaceous, based on a study of pollen from Eastern and mid Western USA localities in which pollen samples were gently processed to increase the likelihood that clumps would be preserved. In addition, research on living angiosperms from the ANITA (Amborellaceae, Nymphaeales, and Austrobaileyales) grade indicates they are apparently not sticky (Thien et al. 2009).

Research on modern pollination systems indicates that wind pollination can be an important factor for some insect-pollinated angiosperms when pollinators are scarce (Cox, 1991; Culley et al. 2002). Wind pollination is likely to evolve in angiosperms with small flowers and dry pollen. Considering the presence of early small fossil flowers (Friis et al. 2006; Crepet 2008) that may have had non-sticky pollen (Hu et al. 2008; Taylor and Hu 2010), some early angiosperm flowers may have been both insect and wind pollinated (ambophily). Especially in newly colonized habitats, species have a relative high probability of ambophily (Culley et al.

2002; Friedman and Barrett 2009). Proctor (1978) also indicated that ambophily is best suited to early successional plants. Considering that many early angiosperms were probably early successional plants (Retallack and Dilcher 1981) or lived in unstable environments (Taylor and Hickey 1996) and that the Early Cretaceous is a period of rapid radiation and diversification for angiosperms, some early angiosperms may have utilized ambophily in newly invaded habitats. Friedman and Barrett (2008, 2009) also mentioned that angiosperms with small non-showy flowers, which are usually pollinated by flies and small bees, may be more likely to evolve wind pollination. The majority of extant basal angiosperms have small flowers and are pollinated by flies, and some are also pollinated by bees (Thien et al. 2009). These data support the hypotheses that early angiosperm pollen did not clump and that some early angiosperms are ambophilous based on their small flowers with non-sticky pollen and the floral and pollen features of extant basal angiosperms.

Understanding the pollination biology of the earliest angiosperms is constrained by the limited data available. The earliest flowers (Sun et al. 2002, Friis et al. 2006, Dilcher et al. 2007; Taylor, 2010) have morphologies not comparable to living angiosperms. The fossil insect record shows the presence of pollinators but cannot conclusively show what plants were pollinated, or quantify the types of pollination. Pollen sculpturing has been shown to be a useful proxy for separating wind and animal pollination types (e.g. Whitehead 1969,1983; Doyle and Hickey 1976; Batten 1986; Crane 1986; Proctor et al. 1996; Ackerman 2000; Tanaka et al. 2004; Hu et al. 2008, Taylor and Hu 2010). Correlation of sculpturing types to types of animal pollination remains uncertain (e.g., Sannier et al. 2009 and references therein). Based on the current research, mostly directed at derived angiosperms, Table 6.1 summarizes the association between pollen morphology and basic pollination types (Taylor and Hu 2010).

To further elucidate our understanding of early pollination biology, we apply pollen–pollinator associations to produce data from two sources. Considering the available extensive pollen records, we use worldwide early angiosperm pollen records (from the late Valanginian to the Aptian) in order to infer the pollination biology of the angiosperms that produced the pollen. We compare pollen size, sculpturing, aperture type, and potential clumping of pollen from the Valanginian, Hauterivian, Barremian, and Aptian. In addition, to estimate the ancestral pollination biology of living angiosperms, we examine the phylogenetic distribution of pollen characteristics on a current molecular phylogenetic tree of basal angiosperms (APG III 2009; Stevens 2001). On this tree, we assess the distribution of pollen characters such as pollen size, sculpturing, and aperture type and any associations to ovary type, gynoecium type, and presence of nectaries to the known pollination syndromes for these families. Analysis of these data allows us

Table 6.1 Association between pollen characters and basic pollinator types based on living angiosperm (Modified from Taylor & Hu, 2010).

	Grain size	Surface feature	Pollen production	Dispersal method
Wind-pollinated pollen	25–40 μm	Smooth and dry	Large quantities	Individual
Possible wind or ambophilous	20–24 μm	Smooth and dry	Large quantities	Individual
	25–40 μm	Moderately sculptured, dry		
Animal-pollinated pollen	10–300 μm	Sculptured, may be sticky and oily	Variable quantities	Individual or clumped

to assess the ancestral pollen characters and formulate hypotheses on the nature of the pollination biology of early angiosperms.

6.2 Materials and methods

175 Early Cretaceous angiosperm pollen records were collected from the literature (Archangelsky and Gamerro 1967; Doyle et al. 1977; Doyle and Robbins 1977; Burger 1990, 1993; Brenner and Bickoff 1992; Hughes 1994; Brenner 1996; Schrank and Mahmoud 2002; Ibrahim 2002; Quattrocchio et al. 2006; Heimhofer et al. 2007). These pollen records span the Valanginian to the Aptian in which angiosperm floral fossils are rare. The ages of the majority of pollen taxa were based upon recent publications (Doyle 1992; Hughes 1994; Brenner 1996; Hochuli et al. 2006).

The majority of pollen data of extant basal angiosperms were collected from Erdtman (1966), the remaining data from other literature (Argue 1973; Mcconchie et al. 1982; Zavada 1984; Sampson 2000; Hesse 2001; Remizowa et al. 2008; Chaowasku et al. 2008). Family pollen size range data are based on information collected from Erdtman (1966) and placed in families based on current family conscriptions (Hu and Taylor 2010). Apertures and sculpturing are based on specific descriptions listed under the families. Family level pollination modes were collected from Mcconchie et al. (1982); Watson and Dallwitz (1992); Linder (1998, 2000); Judd et al. (2002); Lazaro and Traveset 2005; Hu et al. (2008); Thien et al. (2009); Rudall et al. (2009); Lander et al. (2009); and Taylor and Hu (2010). All data about ovary, gynoecium, and the majority of data about floral nectaries were from Watson and Dallwitz (1992). Additional floral nectaries data were from Smets (1986), Endress (1990), and Bernardello (2007).

Table 6.2 Pollen aperture categories.

Aperture categories	Aperture types
Inaperturate	Inaperturate, ulcerate
Monosulcate and derived	Monosulcate, trichotomosulcate, monoporate, dicolpate
Tricolpate and derived	Tricolpate, tricolporate, triporate, stephanocolpate, pericolpate

Table 6.3 Pollen sculpture categories.

Sculpture categories	Sculpture types
Sculptured	Reticulate
Moderately sculptured	Microreticulate, foveolate, echinate, gemmate, baculate, verrucate, striate, crotonoid
Minimally sculptured	Psilate, scabrate

Pollen terminology is from Traverse (2007). Three basic aperture categories were identified: inaperturate; monosulcate and derived; and tricolpate and derived. Table 6.2 shows the types assigned to each category.

Based upon the degree of roughness of the pollen ornamentation, three categories of sculpturing are proposed. Table 6.3 indicates the criteria to distinguish these categories. Five pollen size ranges are created starting with previous work (Hu and Taylor 2010). Taylor and Hu (2010) proposed four size ranges, < 20 μm, 20–24 μm, 25–40 μm, > 40 μm. Based on data on living plants, specifically the highly derived pollen from water-pollinated plants, we modified > 40 μm to 41–300 and added > 300 μm.

Determination of whether the fossil pollen clumps had a floral or coprolite origin was based on criteria from Taylor and Hu (2010). A variety of pollen and floral characters was mapped on the completely resolved tree (Stevens 2001) using the most-parsimonious reconstruction (MPR) method in MacClade (Maddison and Maddison, 1992).

6.3 Results

Four types of pollen, which are the oldest angiosperm pollen in fossil record, are from the Late Valanginian (Table 6.4) core 5 of Kokhav 2 well, Israel (Brenner and Bickoff 1992; Brenner 1996). All pollen grains have inaperturate apertures and reticulate sculpturing (sculptured type). Their sizes are between 14–24 μm, falling

Table 6.4 Pollen records from the Valanginian to the Barremian.

Taxa	Aperture	Sculpturing	Size (μm)	Predicted pollination mode	Geological age	Locality	Country	Continent	Reference
Inaperturate Pollen Type II	inaperturate	reticulate	14	animal	Late Valanginian to Early Hauterivian	core 5 in the Kokhav 2 well	Israel	Asia	Brenner and Bickoff 1992
Inaperturate circular form	inaperturate	reticulate	20	potentially ambophilous or wind pollinated	Late Valanginian to Early Hauterivian	Kokhav 2 well, core 5	Israel	Asia	Brenner 1996
Inaperturate elliptical grain	inaperturate	reticulate	19	animal	Late Valanginian to Early Hauterivian	Kokhav 2 well, core 5	Israel	Asia	Brenner 1996
larger Inaperturate circular grain	inaperturate	reticulate	25	potentially ambophilous or wind pollinated	Late Valanginian to Early Hauterivian	Kokhav 2 well, core 5	Israel	Asia	Brenner 1996
CfB Hauterivian-cactisulc (0401)	monosulcate	scabrate	15	animal	Hauterivian	Warlingham Borehole	England	Europe	Hughes 1994
Paleotaxon Hauterivian dimorph (0405)	monosulcate	rugulate	21	potentially ambophilous or wind pollinated	Hauterivian	Warlingham Borehole	England	Europe	Hughes 1994
Paleotaxon Hauterivian-lacebee (0409)	monosulcate	reticulate	19	animal	Hauterivian	Warlingham Borehole	England	Europe	Hughes 1994
CfB Retisulc-newling (0479)	monosulcate	reticulate	15	animal	Hauterivian	Warlingham Borehole	England	Europe	Hughes 1994
Paleotaxon Hauterivian-microtect (0232)	monosulcate	microreticulate	8	animal	Hauterivian	Warlingham Borehole	England	Europe	Hughes 1994
Paleotaxon Hauterivian-colthick (0162)	monosulcate (?)	reticulate	24	potentially ambophilous or wind pollinated	Hauterivian	Kingsclere Borehole	England	Europe	Hughes 1994

Taxon	Aperture	Sculpture	No.	Pollination	Age	Locality	Country	Region	Reference
Paleotaxon Hauterivian-colthick (0674)	monosulcate (?)	reticulate	22	potentially ambophilous or wind pollinated	Hauterivian	Kingsclere Borehole	England	Europe	Hughes 1994
Paleotaxon Hauterivian-colthick (0165)	monosulcate (?)	reticulate	22	potentially ambophilous or wind pollinated	Hauterivian	Kingsclere Borehole	England	Europe	Hughes 1994
Paleotaxon Retisulc-muriverm (0328)	monosulcate (?)	reticulate	18	animal	Hauterivian	Hunstanton Borehole	England	Europe	Hughes 1994
CfA Retisulc-muriverm (0432)	monosulcate	reticulate	18	animal	Hauterivian	Worbarrow	England	Europe	Hughes 1994
Paleotaxon Retisulc-muribeaded (0313)	monosulcate	reticulate	23	potentially ambophilous or wind pollinated	Hauterivian	Warlingham Borehole	England	Europe	Hughes 1994
CfA Retisulc-muribeaded (0318)	monosulcate (?)	reticulate	28	potentially ambophilous or wind pollinated	Hauterivian	Hunstanton Borehole	England	Europe	Hughes 1994
CfA Retisulc-newling (0343)	monosulcate	reticulate	18	animal	Hauterivian	HUN170	England	Europe	Hughes 1994
CfA Retisulc-newling (0339)	monosulcate	reticulate	25	potentially ambophilous or wind pollinated	Hauterivian	WOR18	England	Europe	Hughes 1994
CfA Hauterivian-cactisulc (0534)	monosulcate	gemmate	29	potentially ambophilous or wind pollinated	Hauterivian	Skegness Borehole	England	Europe	Hughes 1994
CfA Barremian-ring (0475)	monosulcate (?)	scabrate to rugulate	16	animal	Hauterivian	Skegness Borehole	England	Europe	Hughes 1994
Inaperturate Pollen Type III	inaperturate	reticulate	22	potentially ambophilous or wind pollinated	Late Hauterivian	Kokhav 2 well	Israel	Asia	Brenner and Bickoff 1992; Brenner 1996

Table 6.4 (cont.)

Taxa	Aperture	Sculpturing	Size (µm)	Predicted pollination mode	Geological age	Locality	Country	Continent	Reference
Inaperturate grain, Pre-*Afropollis* Group	inaperturate	reticulate	25	potentially ambophilous or wind pollinated	Late Hauterivian	Kokhav 2 well, core 2	Israel	Asia	Brenner 1996
monosulcate grain, Pre-*Afropollis* Group	inaperturate	reticulate	19	animal	Late Hauterivian	Kokhav 2 well, core 2	Israel	Asia	Brenner 1996
Inaperturate circular form, Spinatus Group	inaperturate	echinate (?)	19	animal	Late Hauterivian		Israel	Asia	Brenner 1996
Inaperturate elliptical grain, Spinatus Group	inaperturate	echinate	20	potentially ambophilous or wind pollinated	Late Hauterivian		Israel	Asia	Brenner 1996
Clavatipollenites, *Clavatipollenites* Group	monosulcate	reticulate	19	animal	Late Hauterivian	Kokhav 2 well, core	Israel	Asia	Brenner 1996
Retimonocolpites sp. A, *Liliacidites* Group	monosulcate	foveolate	15	animal	Late Hauterivian	Kokhav 2 well, core 2	Israel	Asia	Brenner 1996
Inaperturate Pollen Type I	inaperturate	microreticulate to scabrate	15	animal	Late Hauterivian	Kokhav 2 well	Israel	Asia	Brenner and Bickoff 1992; Brenner 1996
Monosulcites sp. D	monosulcate	reticulate	34	potentially ambophilous or wind pollinated	Late Hauterivian	Kokhav 2 well	Israel	Asia	Brenner and Bickoff 1992; Brenner 1996
Clavatipollenites sp. A	monosulcate	reticulate	14	animal	Late Hauterivian	Kokhav 2 well	Israel	Asia	Brenner and Bickoff 1992; Brenner 1996
Clavatipollenites sp. B	monosulcate	reticulate	15	animal	Late Hauterivian	Kokhav 2 well	Israel	Asia	Brenner and Bickoff 1992; Brenner 1996

Taxon	Aperture	Sculpture	No.	Pollination	Age	Locality	Country	Region	Reference
Echimonocolpites sp. cf. *E. tersus*	monosulcate	echinate	33	potentially ambophilous or wind pollinated	Late Hauterivian	Kokhav 2 well, Zohar 1 well	Israel	Asia	Brenner and Bickoff 1992; Brenner 1996
Clavatipollenites minutus	monosulcate	reticulate	17	animal	Late Hauterivian	Kokhav 2 well, Zohar 1 well	Israel	Asia	Brenner and Bickoff 1992; Brenner 1996
Paleotaxon Retisulc-dentat (0631)	monosulcate	reticulate	17	animal	Barremian	Warlingham Borehole	England	Europe	Hughes 1994
Paleotaxon Retisulc-dentat (0633)	monosulcate (?)	reticulate	20	potentially ambophilous or wind pollinated	Barremian	Warlingham Borehole	England	Europe	Hughes 1994
CfA Retisulc-dentat (0460)	monosulcate (?)	reticulate	23	potentially ambophilous or wind pollinated	Barremian	Isle of Wight	England	Europe	Hughes 1994
CfB Retisulc-dentat (0472)	monosulcate (?)	reticulate	23	potentially ambophilous or wind pollinated	Barremian	Skegness Borehole	England	Europe	Hughes 1994
CfA Retichot-baccat (0380)	trichotomosulcate (?)	reticulate	20	potentially ambophilous or wind pollinated	Barremian	Warlingham Borehole	England	Europe	Hughes 1994
CfA Retichot-baccat (0093)	trichotomosulcate (?)	reticulate	19	animal	Barremian	Isle of Wight	England	Europe	Hughes 1994
CfA Retisulc-dentat (0641)	monosulcate	reticulate	19	animal	Barremian	Warlingham Borehole	England	Europe	Hughes 1994
CfA Barremian-ring (0134)	monosulcate (?)	scabrate to rugulate	12	animal	Barremian	Isle of Wight	England	Europe	Hughes 1994
CfA Barremian-teebac (0104)	monosulcate	reticulate	21	potentially ambophilous or wind pollinated	Barremian	Alford Borehole	England	Europe	Hughes 1994

Table 6.4 (*cont.*)

Taxa	Aperture	Sculpturing	Size (μm)	Predicted pollination mode	Geological age	Locality	Country	Continent	Reference
Paleotaxon Aptian-dentsharp (0499)	monosulcate (?)	reticulate	20	potentially ambophilous or wind pollinated	Barremian	Hunstanton Borehole	England	Europe	Hughes 1994
Paleotaxon Aptian-dentsharp (0675)	monosulcate (?)	reticulate	19	animal	Barremian	Hunstanton Borehole	England	Europe	Hughes 1994
Paleotaxon Aptian-dentsharp (0615)	monosulcate	reticulate	31	potentially ambophilous or wind pollinated	Barremian	Alford Borehole	England	Europe	Hughes 1994
CfA retisulc-monbac (0116)	monosulcate	reticulate	19	animal	Barremian	Isle of Wight	England	Europe	Hughes 1994
CfA retisulc-monbac (0119)	monosulcate (?)	reticulate	16	animal	Barremian	Isle of Wight	England	Europe	Hughes 1994
CfB Retichot-baccat (0613)	monosulcate	reticulate	16	animal	Barremian	Warlingham Borehole	England	Europe	Hughes 1994
Paleotaxon Retisulc-crochet (0592)	monosulcate	reticulate	21	potentially ambophilous or wind pollinated	Barremian	Kingsclere Borehole	England	Europe	Hughes 1994
CfB Retisulc-crochet (0125)	monosulcate	reticulate	17	animal	Barremian	Kingsclere Borehole	England	Europe	Hughes 1994
CfA Retisulc-crochet (0589)	monosulcate	reticulate	13	animal	Barremian	Kingsclere Borehole	England	Europe	Hughes 1994
CfC Retisulc-crochet (0511)	monosulcate (?)	reticulate	17	animal	Barremian	Kingsclere Borehole	England	Europe	Hughes 1994
CfA Aptian-longcol (0485)	monosulcate (?)	reticulate	17	animal	Barremian	Isle of Wight	England	Europe	Hughes 1994

Taxon	Aperture	Sculpture	No.	Pollination	Stage	Locality	Country	Region	Reference
CfA Aptian-longcol (0488)	monosulcate (?)	reticulate	16	animal	Barremian	Kingsclere Borehole	England	Europe	Hughes 1994
CfA Aptian-perfotect (0502)	monosulcate	foveolate	17	animal	Barremian	Hunstanton Borehole	England	Europe	Hughes 1994
CfB Aptian-perfotect (0598)	trichotomosulcate	foveolate	20	potentially ambophilous or wind pollinated	Barremian	Isle of Wight	England	Europe	Hughes 1994
CfA Aptian-perfotect (0289)	monosulcate	foveolate	15	animal	Barremian	Isle of Wight	England	Europe	Hughes 1994
CfA Superret-croton (0076)	monosulcate	crotonoid	41	animal	Barremian	Isle of Wight	England	Europe	Hughes 1994
Paleotaxon Superret-croton (0082)	monosulcate	crotonoid	36	potentially ambophilous or wind pollinated	Barremian	Kingsclere Borehole	England	Europe	Hughes 1994
CfC Superret-croton (0278)	monosulcate (?)	crotonoid	31	potentially ambophilous or wind pollinated	Barremian	Isle of Wight	England	Europe	Hughes 1994
Paleotaxon Superret-subcrot (0676)	monosulcate	crotonoid	35	potentially ambophilous or wind pollinated	Barremian	Kingsclere Borehole	England	Europe	Hughes 1994
CfA Superret-subcrot (0575)	monosulcate (?)	crotonoid	22	potentially ambophilous or wind pollinated	Barremian	Kingsclere Borehole	England	Europe	Hughes 1994
CfB Aptian-longcol (0440)	monosulcate (?)	reticulate	17	animal	Barremian	Worbarrow	England	Europe	Hughes 1994
possible tricolpate pollen (0128)	tricolpate (?)	reticulate	16	animal	Barremian	Isle of Wight	England	Europe	Hughes 1994

Table 6.4 (cont.)

Taxa	Aperture	Sculpturing	Size (µm)	Predicted pollination mode	Geological age	Locality	Country	Continent	Reference
CfA Barremian-ring (0140)	monosulcate (?)	foveolate to scabrate	15	animal	Barremian	Isle of Wight	England	Europe	Hughes 1994
? Clavatipollenites sp.1	monosulcate	reticulate	22	potentially ambophilous or wind pollinated	Barremian	Well TB. 1	Congo	Africa	Doyle et al. 1977; Doyle 1992;
cf. Stellatopollis barghoornii	monosulcate	crotonoid	48	animal	Barremian	Well EMM.1	Congo	Africa	Doyle et al. 1977; Doyle 1992
Retimonocolpites sp.1	monosulcate	reticulate	25	potentially ambophilous or wind pollinated	Barremian	Well PN.1	Congo	Africa	Doyle et al. 1977; Doyle 1992
Retimonocolpites sp.2	monosulcate	reticulate	32	potentially ambophilous or wind pollinated	Barremian	Well TB.1	Congo	Africa	Doyle et al. 1977; Doyle 1992
cf. Clavatipollenites hughesii	monosulcate	reticulate	24	potentially ambophilous or wind pollinated	Barremian	Well TB.1	Congo	Africa	Doyle et al. 1977; Doyle 1992
Stellatopollis sp. 1	monosulcate	crotonoid	47	animal	Barremian	Well Pl.2b	Congo	Africa	Doyle et al. 1977; Doyle 1992
Tucanopollis annulatus	monoporate	rugulate	30	potentially ambophilous or wind pollinated	Late Barremian	Dakhla Oasis Area	Egypt	Africa	Schrank and mahmoud 2002;
Tucanopollis cf. crisopolensis	monoporate (?)	scabrate (?)	22	potentially ambophilous or wind pollinated	Late Barremian	Dakhla Oasis Area	Egypt	Africa	Schrank and mahmoud 2002

Taxon	Aperture	Sculpture	No.	Pollination	Age	Locality	Country	Continent	Reference
Harskutipollis cf. *robustus*	monosulcate	reticulate	27	potentially ambophilous or wind pollinated	Late Barremian	Dakhla Oasis Area	Egypt	Africa	Schrank and mahmoud 2002
Retimonocolpites pennyi	monosulcate	reticulate	25	potentially ambophilous or wind pollinated	Late Barremian	Dakhla Oasis Area	Egypt	Africa	Schrank and mahmoud 2002
Retimonocolpites matruhensis	monosulcate	reticulate	51	animal	Late Barremian	Dakhla Oasis Area	Egypt	Africa	Schrank and mahmoud 2002
Retimonocolpites sp.1	monosulcate	reticulate	29	potentially ambophilous or wind pollinated	Late Barremian	Dakhla Oasis Area	Egypt	Africa	Schrank and mahmoud 2002
Retimonocolpites sp.2	monosulcate	reticulate	33	potentially ambophilous or wind pollinated	Late Barremian	Dakhla Oasis Area	Egypt	Africa	Schrank and mahmoud 2002
Stellatopollis bituberensis	monosulcate	crotonoid	48	animal	Late Barremian	Dakhla Oasis Area	Egypt	Africa	Schrank and mahmoud 2002
Retiacolpites columellatus	inaperturate	reticulate	35	potentially ambophilous or wind pollinated	Late Barremian	Dakhla Oasis Area	Egypt	Africa	Schrank and mahmoud 2002
Retiacolpites sp.1	inaperturate	reticulate	28	potentially ambophilous or wind pollinated	Late Barremian	Dakhla Oasis Area	Egypt	Africa	Schrank and mahmoud 2002
Stellatopollis dejaxii	monosulcate	crotonoid	57	animal	Late Barremian	Western desert	Egypt	Africa	Ibrahim 2002

into two size ranges, i.e. < 20 μm and 20–24 μm. The majority (81 %) of these grains are less than 20 μm, and only 19 % of them are between 20–24 μm.

Hauterivian angiosperm pollen (Table 6.4) is from Warlingham Borehole, Kingsclere Borehole, Hunstanton Borehole, Skegness Borehole, Worbarrow, HUN170, WOR18 of England (Hughes 1994), and Kokhav 2 well, Zohar 1 well of Israel (Brenner and Bickoff 1992; Brenner 1996). Pollen aperture types include inaperturate and monosulcate, with monosulcate dominant (79 %) and inaperturate morphology accounting for only 21 %. All sculpturing categories were found including reticulate, scabrate, rugulate, microreticulate, gemmate, echinate, and foveolate types. The reticulate type is dominant (62 %) with grains having the minimally to moderately sculptured types accounting for only 38 %. Pollen sizes includes three ranges, i.e. < 20 μm, 20–24 μm, and 25–40 μm. Pollen grains less than 20 μm in diameter are dominant (55 %), and 20–24 μm and 25–40 μm ranges account for 24 % and 21 %, respectively.

Barremian pollen grains (Table 6.4) are from Warlingham Borehole, Kingsclere Borehole, Hunstanton Borehole, Alford Borehole, Isle of Wight, Worbarrow of England (Hughes 1994), Dakhla Oasis Area, Western Desert of Egypt (Schrank and Mahmoud 2002; Ibrahim 2002), and Well TB.1, Well EMM.1, Well PN.1, Well PI.2b of Congo (Doyle et al. 1977). Pollen aperture types include monosulcate, trichotomosulcate, monoporate, and tricolpate (?), with monosulcate dominant (88 %). All sculpturing categories are found including reticulate, scabrate, rugulate, foveolate, and crotonoid types, with reticulate dominant (67 %). Pollen size ranges include < 20 μm, 20–24 μm, 25–40 μm, 40–300 μm. Grains less than 20 μm in diameter account for 38 % of the flora. Grains in size ranges 20–24 μm, 25–40 μm, and 40–300 μm account for 24 %, 27 %, and 12 %, respectively.

Aptian pollen (Table 6.5) are from Zohar 1 well, core 7 of Israel (Brenner 1996), Santa Cruz Province of Argentina (Archangelsky and Gamerro 1967), Eromanga basin of Australia (Burger 1990; 1993), Hunstanton Borehole, Kingsclere Borehole, Isle of Wight, Worbarrow of England (Hughes 1994), Lusitanian Basin, Algarve Basin of Portugal (Heimhofer et al. 2007), Zohar 1 Well of Israel (Brenner and Bickoff 1992; Brenner 1996), Western Desert of Egypt (Ibrahim 2002), Well TM.1, Well CN.1, Well Gamba 2b, Gamba Formation Outcrop Locality M57 of Gabon (Doyle et al. 1977), Well K.38, Well K.8 of Congo (Doyle et al. 1977), Delaware City Well D12, D 13, Trent's Reach, Dutch Gap Canal of USA (Doyle and Robbins 1977), and Northeastern Tierra del Fuego of Argentina (Quattrocchio et al. 2006). Pollen aperture types include monosulcate, tricolpate, monoporate, inaperturate, and dicolpate, with monosulcate dominant (78 %). All sculpturing categories are found including the reticulate, crotonoid, scabrate, rugulate, foveolate, echinate, and striate types, with reticulate dominant (78 %). Pollen size ranges include < 20 μm, 20–24 μm, 25–40 μm, 40–300 μm. Pollen grains between 25–40 μm are dominant (46 %); pollen grains less than 20 μm in diameter account for

Table 6.5 Pollen records in the Aptian.

Taxa	Aperture	Sculpturing	Size (µm)	Predicted pollination mode	Geological age	Locality	Country	Continent	Reference
Liliacidites katangataensis	monosulcate	reticulate	46	animal	Barremian to Aptian	Zohar 1 well, core 7	Israel	Asia	Brenner 1996
Liliacidites sp.C	monosulcate	reticulate	40	potentially ambophilous or wind pollinated	Barremian to Aptian	Zohar 1 well, core 7	Israel	Asia	Brenner 1996
Clavatipollenites hughesii	monosulcate	reticulate	33	potentially ambophilous or wind pollinated	Barremian to Aptian	Santa Cruz Province	Argentina	South America	Archangelsky and Gamerro 1967
Clavatipollenites hughesii	monosulcate (?)	reticulate	22	potentially ambophilous or wind pollinated	Late Barremian to Aptian	Lusitanian basin	Portugal	Europe	Heimhofer et al. 2007
Clavatipollenites cf. *hughesii*	monosulcate	reticulate	23	potentially ambophilous or wind pollinated	Late Barremian to Aptian	Lusitanian basin	Portugal	Europe	Heimhofer et al. 2007
Clavatipollenites hughesii	monosulcate	reticulate	25	potentially ambophilous or wind pollinated	Late Barremian/ Early Aptian	Eromanga basin	Australia	Australia	Burger 1990; 1993
Retimonoporites operculatus	monoporate (?)	reticulate	19	animal	Early Aptian	Zohar 1 well, core 6	Israel	Asia	Brenner 1996
Retimonoporites operculatus	monoporate (?)	reticulate	20	potentially ambophilous or wind pollinated	Early Aptian	Zohar 1 well, core 6	Israel	Asia	Brenner 1996
Brenneripollis sp.	monocolpate (?)	reticulate	23	potentially ambophilous or wind pollinated	Early Aptian	Zohar 1 well, core 6	Israel	Asia	Brenner 1996

Table 6.5 (cont.)

Taxa	Aperture	Sculpturing	Size (μm)	Predicted pollination mode	Geological age	Locality	Country	Continent	Reference
Brenneripollis peroreticulatus	monosulcate	reticulate	21	potentially ambophilous or wind pollinated	Early Aptian	Core 6	Israel	Asia	Brenner and Bickoff 1992
Brenneripollis reticulatus	monosulcate	reticulate	18	animal	Early Aptian	Core 6	Israel	Asia	Brenner and Bickoff 1992
Liliacidites sp. A	monosulcate	reticulate	31	potentially ambophilous or wind pollinated	Early Aptian	Zohar 1 |well	Israel	Asia	Brenner and Bickoff 1992
Retimonocolpites sp. A	monosulcate	reticulate	24	potentially ambophilous or wind pollinated	Early Aptian	Zohar 1 well	Israel	Asia	Brenner and Bickoff, 1992
Retimonoporites operculatus	monoporate	reticulate	13	animal	Early Aptian	Core 6 in the Zohar 1 well	Israel	Asia	Brenner and Bickoff 1992
Tricolpites sp. A	tricolpate	reticulate	26	potentially ambophilous or wind pollinated	Early Aptian	Core 6 in the Zohar 1 well	Israel	Asia	Brenner and Bickoff 1992
Tricolpites sp. B	tricolpate	reticulate	24	potentially ambophilous or wind pollinated	Early Aptian	Core 6 in the Zohar 1 well	Israel	Asia	Brenner and Bickoff 1992
Tricolpites sp. C	tricolpate	reticulate	26	potentially ambophilous or wind pollinated	Early Aptian	Core 6 in the Zohar 1 well	Israel	Asia	Brenner and Bickoff 1992
Afropollis jardinus	inaperturate	reticulate	19	animal	Early Aptian	Core 6	Israel	Asia	Brenner and Bickoff 1992
tricolpate grain	tricolpate	reticulate	19	animal	Early Aptian	Zohar 1 well, core 6	Israel	Asia	Brenner 1996

Species	Aperture	Sculpture	Number	Pollination	Age	Locality	Region	Continent	Reference
Stellatopollis dejaxii	monosulcate	crotonoid	44	animal	Early Aptian	Western desert	Egypt	Africa	Ibrahim 2002
Dichastopollenites sp. cf. *D. ghazalatensis*	monosulcate (?)	reticulate	29	potentially ambophilous or wind pollinated	Early Aptian	Western desert	Egypt	Africa	Ibrahim 2002
Stellatopollis sp. cf. *S. hughesii*	monosulcate	crotonoid	31	potentially ambophilous or wind pollinated	Early Aptian	Western desert	Egypt	Africa	Ibrahim 2002
Retimonocolpites bueibensis	monosulcate	reticulate	24	potentially ambophilous or wind pollinated	Early Aptian	Western desert	Egypt	Africa	Ibrahim 2002
Arecipites microfoveolatus	monosulcate	foveolate	72	animal	Early Aptian	Western desert	Egypt	Africa	Ibrahim 2002
Tricolpites sp.	tricolpate	reticulate (?)	25	potentially ambophilous or wind pollinated	Early Aptian	Western desert	Egypt	Africa	Ibrahim 2002
Retimonocolpites ghazalii	monosulcate	reticulate	65	animal	Early Aptian	Western desert	Egypt	Africa	Ibrahim 2002
Stellatopollis limai	monosulcate	crotonoid to reticulate	52	animal	Early Aptian	Western desert	Egypt	Africa	Ibrahim 2002
Stellatopollis doylei	monosulcate	crotonoid	34	potentially ambophilous or wind pollinated	Early Aptian	Western desert	Egypt	Africa	Ibrahim 2002
Liliacidites sp.1	monosulcate	reticulate	23	potentially ambophilous or wind pollinated	Early Aptian	Well TM.1	Gabon	Africa	Doyle et al. 1977; Doyle 1992
aff. "*Clavatipollenites*" *minutus*	monosulcate	reticulate	16	animal	Early Aptian	Well TM.1	Gabon	Africa	Doyle et al. 1977; Doyle 1992

Table 6.5 (cont.)

Taxa	Aperture	Sculpturing	Size (µm)	Predicted pollination mode	Geological age	Locality	Country	Continent	Reference
aff. *Tricolpites micromurus*	tricolpate	reticulate	16	animal	Early Aptian	Well TM.1	Gabon	Africa	Doyle et al. 1977; Doyle 1992
aff. *Tricolpites crassimurus*	tricolpate	reticulate	30	potentially ambophilous or wind pollinated	Early Aptian	Well TM.1	Gabon	Africa	Doyle et al. 1977; Doyle 1992
"*Inaperturopollenites*" *crisopolensis*	inaperturate (?)	rugulate	28	potentially ambophilous or wind pollinated	Early Aptian	Well TM.1	Gabon	Africa	Doyle et al. 1977; Doyle 1992
"*Reticulatasporites*" *jardinus*	inaperturate (?)	reticulate	30	potentially ambophilous or wind pollinated	Early Aptian	Well TM.1	Gabon	Africa	Doyle et al. 1977; Doyle 1992
cf. *Clavatipollenites hughesii*	monosulcate	reticulate (?)	31	potentially ambophilous or wind pollinated	Early Aptian	northeastern Tierra del Fuego	Argentina	South America	Quattrocchio et al. 2006
Stellatopollis dejaxii	monosulcate	crotonoid	32	potentially ambophilous or wind pollinated	Middle Aptian	Western desert	Egypt	Africa	Ibrahim 2002
Stellatopollis barghoornii	monosulcate	crotonoid	38	potentially ambophilous or wind pollinated	Middle Aptian	Western desert	Egypt	Africa	Ibrahim 2002
Tricolpites sp.	tricolpate	reticulate (?)	26	potentially ambophilous or wind pollinated	Middle Aptian	Western desert	Egypt	Africa	Ibrahim 2002
Afropollis schrankii	monosulcate	reticulate	43	animal	Middle Aptian	Western desert	Egypt	Africa	Ibrahim 2002

Species	Aperture	Ornamentation	No.	Pollination	Age	Location	Country	Continent	Reference
Afropollis aff. *zonatus*	monosulcate (?)	reticulate	29	potentially ambophilous or wind pollinated	Middle Aptian	Western desert	Egypt	Africa	Ibrahim 2002
Afropollis operculatus subsp. *operculatus*	monosulcate (?)	reticulate	41	animal	Middle Aptian	Western desert	Egypt	Africa	Ibrahim 2002
Afropollis operculatus subsp. *microreticulatus*	monosulcate (?)	reticulate	45	animal	Middle Aptian	Western desert	Egypt	Africa	Ibrahim 2002
Echitricolpites sp.	tricolpate	echinate	35	potentially ambophilous or wind pollinated	Middle Aptian	Western desert	Egypt	Africa	Ibrahim 2002
Echimonocolpites sp.	monosulcate	echinate	37	potentially ambophilous or wind pollinated	Middle Aptian	Western desert	Egypt	Africa	Ibrahim 2002
Stellatopollis doylei	monosulcate	crotonoid	31	potentially ambophilous or wind pollinated	Middle Aptian	Western desert	Egypt	Africa	Ibrahim 2002
Stellatopollis limai	monosulcate	crotonoid to reticulate	36	potentially ambophilous or wind pollinated	Middle Aptian	Western desert	Egypt	Africa	Ibrahim 2002
Asteropollis sp. 1	monosulcate (?)	reticulate	27	potentially ambophilous or wind pollinated	Late Aptian	Algarve basin	Portugal	Europe	Heimhofer et al. 2007
Asteropollis sp.4	monosulcate (?)	reticulate	24	potentially ambophilous or wind pollinated	Late Aptian	Lusitanian basin	Portugal	Europe	Heimhofer et al. 2007
Pennipollis sp.1	monosulcate (?)	reticulate (?)	20	potentially ambophilous or wind pollinated	Late Aptian	Algarve basin	Portugal	Europe	Heimhofer et al. 2007

Table 6.5 (cont.)

Taxa	Aperture	Sculpturing	Size (µm)	Predicted pollination mode	Geological age	Locality	Country	Continent	Reference
Clavatipollenites cf. *tenellis*	monosulcate	reticulate	30	potentially ambophilous or wind pollinated	Late Aptian	Algarve basin	Portugal	Europe	Heimhofer et al. 2007
Clavatipollenites sp.1	monosulcate	reticulate	16	animal	Late Aptian	Algarve basin	Portugal	Europe	Heimhofer et al. 2007
Retimonocolpites sp.15	monosulcate	reticulate	30	potentially ambophilous or wind pollinated	Late Aptian	Lusitanian basin	Portugal	Europe	Heimhofer et al. 2007
Racemonocolpites cf. *exoticus*	monosulcate (?)	scabrate (?)	54	animal	Late Aptian	Algarve basin	Portugal	Europe	Heimhofer et al. 2007
cf. *Retimonocolpites peroreticulatus*	monosulcate	reticulate	18	animal	Late Aptian	Well K.38	Congo	Africa	Doyle et al. 1977; Doyle 1992
cf. *Retimonocolpites dividuus*	monosulcate	reticulate	33	potentially ambophilous or wind pollinated	Late Aptian	Well Gamba 2b	Gabon	Africa	Doyle et al. 1977; Doyle 1992
Clavatipollenites sp.2	monosulcate	reticulate	21	potentially ambophilous or wind pollinated	Late Aptian	Well K.38	Congo	Africa	Doyle et al. 1977; Doyle 1992
Striatopollis sp.2	tricolpate	striate to reticulate	25	potentially ambophilous or wind pollinated	Late Aptian	Gamba Fm. Outcrop locality M57	Gabon	Africa	Doyle et al. 1977; Doyle 1992
"Retitricolpites " geranioides	tricolpate	reticulate	52	animal	Late Aptian	Gamba Fm. Outcrop locality M57	Gabon	Africa	Doyle et al. 1977; Doyle 1992

Taxon	Aperture	Sculpture	No.	Pollination	Age	Locality	Country	Continent	Reference
cf. *Tricolpites georgensis*	tricolpate	reticulate	30	potentially ambophilous or wind pollinated	Late Aptian	Well K. 38	Congo	Africa	Doyle et al. 1977; Doyle 1992
Tricolpites sp.1	tricolpate	reticulate	18	animal	Late Aptian	Well K.8	Congo	Africa	Doyle et al. 1977; Doyle 1992
Tricolpites sp.2	tricolpate	reticulate	31	potentially ambophilous or wind pollinated	Late Aptian	Well K.38	Congo	Africa	Doyle et al. 1977; Doyle 1992
reticulate dicolpate	dicolpate	reticulate	29	potentially ambophilous or wind pollinated	Late Aptian	Well K.38	Congo	Africa	Doyle et al. 1977; Doyle 1992
monosulcate with extended sulcus	monosulcate	reticulate	23	potentially ambophilous or wind pollinated	Late Aptian	Well K.38	Congo	Africa	Doyle et al. 1977; Doyle 1992
CfA Retisulc-dubdent (0562)	monosulcate (?)	reticulate	14	animal	Aptian	Isle of Wight	England	Europe	Hughes 1994
CfB Retisulc-dubdent (0528)	monosulcate	reticulate	16	animal	Aptian	Isle of Wight	England	Europe	Hughes 1994
Paleotaxon retisulc-dident (0559)	monosulcate (?)	reticulate	17	animal	Aptian	Isle of Wight	England	Europe	Hughes 1994
Paleotaxon retisulc-dident (0373)	monosulcate (?)	reticulate	20	potentially ambophilous or wind pollinated	Aptian	Isle of Wight	England	Europe	Hughes 1994
Paleotaxon Afropol-murigroove (0540)	monosulcate (?)	reticulate	48	animal	Aptian	Hunstanton Borehole	England	Europe	Hughes 1994
Paleotaxon Afropol-murigroove (0542)	monosulcate (?)	reticulate	41	animal	Aptian	Hunstanton Borehole	England	Europe	Hughes 1994

Table 6.5 (cont.)

Taxa	Aperture	Sculpturing	Size (µm)	Predicted pollination mode	Geological age	Locality	Country	Continent	Reference
CfA Superret-croton (0363)	monosulcate (?)	crotonoid	32	potentially ambophilous or wind pollinated	Aptian	Isle of Wight	England	Europe	Hughes 1994
CfB Superret-croton (0604)	monosulcate (?)	crotonoid	38	potentially ambophilous or wind pollinated	Aptian	Hunstanton Borehole	England	Europe	Hughes 1994
CfA Superret-krinkel (0610)	monosulcate (?)	reticulate	34	potentially ambophilous or wind pollinated	Aptian	Worbarrow	England	Europe	Hughes 1994
CfB 'Tricolpopollenites' crassimurus (0130)	tricolpate	reticulate	27	potentially ambophilous or wind pollinated	Aptian	Isle of Wight	England	Europe	Hughes 1994
CfC Hauterivian-cactisulc (0570)	monosulcate (?)	scabrate to rugulate	18	animal	Aptian	Kingsclere Borehole	England	Europe	Hughes 1994
Clavatipollenites cf. minutus	monosulcate	reticulate	23	potentially ambophilous or wind pollinated	Aptian	Lusitanian basin	Portugal	Europe	Heimhofer et al. 2007
Asteropollis cf. asteroides	monosulcate (?)	reticulate	26	potentially ambophilous or wind pollinated	Aptian	Lusitanian basin	Portugal	Europe	Heimhofer et al. 2007
Retimonocolpites sp.2	monosulcate	reticulate	25	potentially ambophilous or wind pollinated	Aptian	Lusitanian basin	Portugal	Europe	Heimhofer et al. 2007
Retimonocolpites sp.4	monosulcate	reticulate	26	potentially ambophilous or wind pollinated	Aptian	Algarve basin	Portugal	Europe	Heimhofer et al. 2007

Taxon	Aperture	Ornamentation	Size	Pollination	Age	Locality	Country	Continent	Reference
Retimonocolpites sp.9	monosulcate	reticulate	25	potentially ambophilous or wind pollinated	Aptian	Algarve basin	Portugal	Europe	Heimhofer et al. 2007
Retimonocolpites sp.8	monosulcate	reticulate	15	animal	Aptian	Algarve basin	Portugal	Europe	Heimhofer et al. 2007
Pennipollis sp.2	monosulcate (?)	reticulate	18	animal	Aptian	Algarve basin	Portugal	Europe	Heimhofer et al. 2007
Retimonocolpites sp.6	monosulcate	reticulate	27	potentially ambophilous or wind pollinated	Aptian	Lusitanian basin	Portugal	Europe	Heimhofer et al. 2007
Stellatopollis barghoornii	monosulcate	crotonoid	75	animal	Aptian	Algarve basin	Portugal	Europe	Heimhofer et al. 2007
Tucanopollis aff. crisopolensis	monosulcate	scabrate (?)	15	animal	Aptian	Algarve basin	Portugal	Europe	Heimhofer et al. 2007
Monocolpites sp. 1	monosulcate	reticulate (?)	35	potentially ambophilous or wind pollinated	Aptian	Algarve basin	Portugal	Europe	Heimhofer et al. 2007
Striatopollis sp.1	tricolpate	striate	20	potentially ambophilous or wind pollinated	Aptian	Well CN.1	Gabon	Africa	Doyle et al. 1977; Doyle 1992
cf. *Clavatipollenites hughesii*	monosulcate	reticulate	26	potentially ambophilous or wind pollinated	Aptian	Delaware City Wells D12	USA	North America	Doyle and Robbins 1977; Hochuli et al. 2006;
Clavatipollenites sp.A	monosulcate	reticulate	31	potentially ambophilous or wind pollinated	Aptian	Delaware City Wells D12	USA	North America	Doyle and Robbins 1977; Hochuli et al. 2006

Table 6.5 (*cont.*)

Taxa	Aperture	Sculpturing	Size (μm)	Predicted pollination mode	Geological age	Locality	Country	Continent	Reference
aff. *Clavatipollenites minutus*	monosulcate	reticulate	19	animal	Aptian	Delaware City Wells D12	USA	North America	Doyle and Robbins 1977; Hochuli et al. 2006
Retimonocolpites peroreticulatus	monosulcate	reticulate	23	potentially ambophilous or wind pollinated	Aptian	Delaware City Wells D12	USA	North America	Doyle and Robbins 1977; Hochuli et al. 2006
Liliacidites sp.B	monosulcate	reticulate	24	potentially ambophilous or wind pollinated	Aptian	Delaware City Wells D13	USA	North America	Doyle and Robbins 1977; Hochuli et al. 2006
Liliacidites sp.A	monosulcate	reticulate	30	potentially ambophilous or wind pollinated	Aptian	Trent's Reach	USA	North America	Doyle and Robbins 1977; Hochuli et al. 2006
Stellatopollis sp. 1	monosulcate	crotonoid	41	animal	Aptian	Dutch Gap Canal	USA	North America	Doyle and Robbins 1977; Hochuli et al. 2006

19 %, and pollen grains in 20–24 μm and 40–300 μm account for 19 % and 14 %, respectively.

In summary, during the Valanginian, all pollen grains are inaperturate. From the Hauterivian through the Aptian, monosulcate type is dominant, although inaperturate type still exists. For sculpturing types, during the Valanginian, all grains are reticulate. From the Hauterivian through the Aptian, the reticulate type is still dominant. Lastly, during the Valanginian early angiosperm pollen are small, with two range sizes, i.e. < 20 μm and 20–24 μm, but, by the Aptian, pollen grains with the 25–40 μm size range become dominant (46 %).

To assess the ancestral pollen characters based on the basal-most angiosperms, we collected data on all the families of the ANITA grade, the basal-most monocots and the basal-most eudicots (as defined in Hu et al. 2008; Table 6.6). As with the fossils, the living taxa show a large range of aperture types, sculpturing and size ranges, although the most common aperture type is monosulcate (40 %), the most common sculpturing is sculptured morphology (39 %), and the most common size range is 25–40 μm (34 %).

The most parsimonious ancestral states were assessed at the family level, based on the topology of the most recent trees (APG III 2009; Stevens, 2001). Based on the parsimony analysis, the ancestral state is equivocal between inaperturate (that is scattered through basal angiosperms) and monosulcate (Fig 6.1). There are two clades with mostly inaperturate grades, the Laurales and a group of monocots with the Alismatales, and *Amborella* is inaperturate.

These parsimony data suggest the ancestral state for sculpturing is moderately sculptured (Fig 6.2) although the states are variable throughout the angiosperms. Sculptured grains are found in many of the families of the ANITA grade (but not Amborellaceae) and the ancestral states for the other basal dicots and the monocot–eudicot clades are equivocal.

Based on these analyses, the most ancestral and most frequent size range (Fig 6.3) appears to be 25–40 μm. Most other size types evolved multiple times. It also clearly shows that the > 300 μm size range is found in water-pollinated clade in the monocots. Note also that in the ANITA grade, Trimeniaceae, Illiciaceae, and Schisandraceae have smaller pollen.

To better understand the pollination modes of basal angiosperms, we updated the known pollinator for the basal families (Table 6.6) and made a tentative enhanced criteria table for determining pollinators based on pollen characteristics (Table 6.7) used in the past (Hu et al, 2008; Taylor and Hu, 2010). Based on the literature, which was unfortunately mostly based on the study of derived angiosperms in which sticky pollen can be found, we suggest animal pollination is found in small grains (< 20 μm), large grains (41–300 μm), and grains that are sculptured or moderately sculptured with size ranges 20–24 μm and 25–40 μm. Grains are from wind-pollinated plants if minimally sculptured, non-sticky, and

Table 6.6 Pollen data on basal angiosperms at family and species level with references. For pollination mode, C = Coleoptera, D = Diptera, H = Hemiptera, Ho = Homoptera, Hy = Hymenoptera, L = Lepidoptera, M = Micropterigidae, T = thrips, W = Wind, Wa = Water.

Family	Species	Sculpturing	Grain size (µm)	Aperture type	Pollen morphology references
Amborellaceae		moderately sculptured	29	inaperturate	
Amborellaceae	*Amborella trichopoda*	gemmate	20 (17–24)	ulcerate	Sampson 2000; Hesse 2001
Amborellaceae	*Amborella trichopoda*	gemmate	29	ulcerate	Sampson 2000; Hesse 2001; Erdtman 1966
Hydatellaceae		moderately sculptured	16–23	monosulcate and derived	
Hydatellaceae	*Trithuria inconspicua*	foveolate, echinate	23 (20–27)	monosulcate	Remizowa et al. 2008
Hydatellaceae	*Trithuria bibrateata*	foveolate, echinate	16 (13–19)	monosulcate	Remizowa et al. 2008
Hydatellaceae	*Trithuria laterna*	foveolate, echinate	17 (15–21)	monosulcate	Remizowa et al. 2008
Cabombaceae		minimally sculptured, moderately sculptured	42–100	monosulcate and derived	
Cabombaceae	*Brasenia schreberi*	scabrate	42	monosulcate	Remizowa et al. 2008
Cabombaceae	*Brasenia purpurea*	scabrate	58	monosulcate	Erdtman 1966
Cabombaceae	*Cabomba aquatica*	striate	81	monosulcate	Remizowa et al., 2008
Cabombaceae	*Cabomba caroliniana*	striate	100	monosulcate	Erdtman 1966
Nymphaeaceae		minimally sculptured, moderately sculptured	28–75	inaperturate, monosulcate and derived, tricolpate and derived	

Predicted pollinator	Pollination mode	Pollination references	Ovary (family level) (Watson and Dallwitz (1992 onwards))	Gynoecium (family level) (Watson and Dallwitz (1992 onwards))	Floral nectaries (family level)	Nectary references
	C H Ho Hy LW	Thien et al. 2009	apocarpous	superior	absent	Bernardello 2007
animal						
animal						
	?W Wa	Rudall et al. 2009	apocarpous	superior	absent	Watson and Dallwitz (1992 onwards)
animal						
animal						
animal						
	D Hy W	Thien et al. 2009; Rudall et al. 2009	apocarpous	superior	absent, present	Thien et al. 2009
animal	W	Rudall et al. 2009				
animal	W	Rudall et al. 2009				
animal	D Hy	Thien et al. 2009				
animal	D Hy	Thien et al. 2009				
	C D Hy	Thien et al. 2009; Hu et al. 2008	syncarpous	superior to partly inferior	most primitive nectaries	Bernardello 2007

Table 6.6 (*cont.*)

Family	Species	Sculpturing	Grain size (µm)	Aperture type	Pollen morphology references
Nymphaeaceae	*Nuphar lutea*	echinate	51 (47–55)	monosulcate	Halbritter and Svojtk 2000 onwards; Erdtman 1966
Nymphaeaceae	*Nuphar advena*	echinate	66	monosulcate	Erdtman 1966
Nymphaeaceae	*Euryale ferox*	echinate	42	variable, monosulcate, dicolpate, tricolpate	Erdtman 1966
Nymphaeaceae	*Victoria regia*	echinate	75	variable, monosulcate, dicolpate, tricolpate	Erdtman 1966
Nymphaeaceae	*Barclaya spp.*	psilate	48	inaperturate	Erdtman 1966
Nymphaeaceae	*Nymphaea alba*	baculate, verrucate	36 (32–39)	ulcerate	Halbritter and Svojtk 2000 onwards
Nymphaeaceae	*Nymphaea zanzibariensis*	baculate	42	monosulcate	Erdtman 1966
Austrobaileyaceae		sculptured	30 (28–32)	monosulcate and derived	
Austrobaileyaceae	*Austrobaileya maculata*	reticulate	30 (28–32)	monosulcate	Zavada, 1984
Trimeniaceae		minimally sculptured	20–24	inaperturate, monosulcate and derived	
Trimeniaceae	*Trimenia papuana*	scabrate	20	inaperaturate, periporate	Sampson 2000
Trimeniaceae	*Trimenia weinmannieaefolia*	scabrate	23	periporate	Erdtman 1966
Trimeniaceae	*Trimenia neocaledonica*	scabrate	24	bicolpate	Sampson 2000
Illiciaceae		sculptured	19–33	monosulcate and derived	
Illiciaceae	*Illicium parviflorum*	reticulate	19	trichomosulcate	Sampson 2000

Predicted pollinator	Pollination mode	Pollination references	Ovary (family level) (Watson and Dallwitz (1992 onwards))	Gynoecium (family level) (Watson and Dallwitz (1992 onwards))	Floral nectaries (family level)	Nectary references
animal						
animal						
animal						
animal						
animal						
animal						
animal						
	C D	Thien et al. 2009; Hu et al. 2008	apocarpous	superior	present	Endress 1990
animal						
	D Hy W	Thien et al. 2009; Hu et al. 2008	monomerous, apocarpous	superior	absent	Endress 1990
animal/ wind						
animal/ wind						
animal/ wind						
	C D H	Thien et al. 2009	apocarpous	superior	present	Endress 1990
animal						

Table 6.6 (*cont.*)

Family	Species	Sculpturing	Grain size (µm)	Aperture type	Pollen morphology references
Illiciaceae	*Illicium floridanum*	reticulate	33	trichomosulcate	Erdtman 1966
Schisandraceae		sculptured	18–32	monosulcate and derived	
Schisandraceae	*Schisandra arisanensis*	reticulate	27	trichomosulcate	Erdtman 1966
Schisandraceae	*Schisandra chinesis*	reticulate	32	trichomosulcate	Erdtman 1966
Schisandraceae	*Kadsura japonica*	reticulate	18 (17–19)	trichomosulcate	Sampson 2000
Schisandraceae	*Kadsura coccinea*	reticulate	31	trichomosulcate	Erdtman 1966
Ceratophyllaceae		minimally sculptured	40 (35–45)	inaperturate	
Ceratophyllaceae	*Ceratophyllum demersum*	psilate	40 (35–45)	inaperturate	Erdtman 1966
Chloranthaceae		sculptured	20–41	monosulcate and derived	
Chloranthaceae	*Ascarina lucida*	reticulate	24	monosulcate	Erdtman 1966
Chloranthaceae	*Chloranthus inconspicuus*	reticulate	20	6-colpate	Erdtman 1966
Chloranthaceae	*Hedyosmum brasiliense*	reticulate	41	monosulcate	Erdtman 1966
Myristicaceae		moderately sculptured	20–50	monosulcate and derived	
Myristicaceae	*Myristica sebifera*	foveolate	33	monosulcate	Erdtman 1966
Magnoliaceae		sculptured	40–110	monosulcate and derived	
Magnoliaceae	*Magnolia hamorii*	reticulate	100	monosulcate	Erdtman 1966
Degeneriaceae		minimally sculptured	80	monosulcate and derived	
Degeneriaceae	*Degeneria vitiensis*	psilate	80	monosulcate	Erdtman 1966; Sampson 2000

Predicted pollinator	Pollination mode	Pollination references	Ovary (family level) (Watson and Dallwitz (1992 onwards))	Gynoecium (family level) (Watson and Dallwitz (1992 onwards))	Floral nectaries (family level)	Nectary references
animal						
	C D	Thien et al. 2009	apocarpous	superior	present	Bernardello 2007
animal						
animal						
animal						
animal						
	Wa	Hu et al. 2008	monomerous	superior	absent	Endress 1990
wind						
	T W	Hu et al. 2008	monomerous	superior, or partly inferior	absent	Bernardello 2007
animal						
animal						
animal						
	C T	Hu et al. 2008	monomerous	superior	absent	Bernardello 2007
animal						
	C D T Hy	Hu et al. 2008	apocarpous	superior	petal nectaries in some	Bernardello 2007
animal						
	C	Hu et al. 2008	monomerous	superior	absent	Bernardello 2007
animal						

Table 6.6 (*cont.*)

Family	Species	Sculpturing	Grain size (μm)	Aperture type	Pollen morphology references
Himantandraceae		minimally sculptured	30–38	monosulcate and derived	
Himantandraceae	*Galbulimima belgraveana*	scabrate	30	monosulcate	Erdtman 1966
Eupomatiaceae		minimally sculptured	34–35	monosulcate and derived	
Eupomatiaceae	*Eupomatia laurina*	psilate	35	monosulcate	Erdtman 1966
Annonaceae		moderately sculptured	30–160	inaperturate, monosulcate and derived	
Annonaceae	*General*		30–160		Erdtman 1966
Annonaceae	*Alphonsea siamensis*	rugulate	41	inaperturate	Chaowasku et al. 2008
Annonaceae	*Miliusa mollis*	rugulate	46	inaperturate	Chaowasku et al. 2008
Annonaceae	*Miliusa brahei*	verrucate	36	inaperturate	Chaowasku et al. 2008
Annonaceae	*Miliusa macropoda*	verrucate	41	inaperturate	Chaowasku et al. 2008
Annonaceae	*Orophea polycarpa*	rugulate	37	inaperturate	Chaowasku et al. 2008
Annonaceae	*Platymitra macrocarpa*	verrucate	27	inaperturate	Chaowasku et al. 2008
Annonaceae	*Polyalthia cerasoides*	verrucate	42	inaperturate	Chaowasku et al. 2008
Atherospermataceae		sculptured	41–50	monosulcate and derived	
Atherospermataceae	*Atherosperma moschata*	reticulate	41	dicolpate	Erdtman 1966
Calycanthaceae		sculptured	42–60	monosulcate and derived	
Calycanthaceae	*Calycanthus floridus*	reticulate	42	dicolpate	Erdtman 1966, Sampson 2000

Predicted pollinator	Pollination mode	Pollination references	Ovary (family level) (Watson and Dallwitz (1992 onwards))	Gynoecium (family level) (Watson and Dallwitz (1992 onwards))	Floral nectaries (family level)	Nectary references
	insect	Watson and Dallwitz (1992 onwards)	apocarpous	superior	absent	Bernardello 2007
wind						
	C	Hu et al. 2008	apocarpous	partly inferior	absent	Bernardello 2007
wind	C					
	insect	Judd et al. 2002	apocarpous (syncarpous)	superior	petal nectaries in some	Bernardello 2007
animal						
animal						
animal						
animal						
animal						
animal						
animal						
animal						
	?	Hu et al. 2008	apocarpous	superior to inferior	present	Bernardello 2007
animal						
	C	Hu et al. 2008	apocarpous	superior	absent	Bernardello 2007
animal						

Table 6.6 (*cont.*)

Family	Species	Sculpturing	Grain size (µm)	Aperture type	Pollen morphology references
Gomortegaceae		minimally sculptured	31	inaperturate	
Gomortegaceae	*Gomortega nitida*	scabrate	31	inaperturate	Erdtman 1966
Hernandiaceae		moderately sculptured	18–115	inaperturate	
Hernandiaceae	*Sparattanthelium tarapotanum*	echinate	19	inaperature	Erdtman 1966
Hernandiaceae	*Hernandia moerenhoutiana*	echinate	115	inaperature	Erdtman 1966
Lauraceae		moderately sculptured	24–70	inaperturate	
Lauraceae	*Cinnamonmum camphora*	echinate	34	inaperturate	Erdtman 1966
Lauraceae	*Beilschmiedia tartaire*	echinate	23	inaperturate	Sampson 2000
Lauraceae	*Cassytha filiformis*	echinate	24–30	inaperturate	Erdtman 1966
Monimiaceae		minimallly to moderately sculptured	15–50	inaperturate	
Monimiaceae	*Hedycara arborea*	scabrate	39 (tetrad)	inaperturate	Erdtman 1966
Monimiaceae	*Peumus boldus*	echinate	33	inaperturate	Erdtman 1966
Siparunaceae		moderately sculptured	16–20	inaperturate	
Siparunaceae	*Siparuna cujabana*	rugulate ?	20	inaperturate	Erdtman 1966
Canellaceae		moderately sculptured	32–48	monosulcate and derived	
Canellaceae	*Canella abla*	microreticulate/ foveolate	48	monosulcate	Erdtman 1966
Winteraceae		sculptured	50 -74 (tetrad)	inaperturate	
Winteraceae	*Drimys winteria*	reticulate	50 (tetrad)	ulcerate	Erdtman 1966

Predicted pollinator	Pollination mode	Pollination references	Ovary (family level) (Watson and Dallwitz (1992 onwards))	Gynoecium (family level) (Watson and Dallwitz (1992 onwards))	Floral nectaries (family level)	Nectary references
	D	Lander et al. 2009	syncarpous	inferior	present	Bernardello 2007
wind						
	?	Hu et al. 2008	monomerous	inferior	present	Bernardello 2007
animal						
animal						
	C D T Hy	Hu et al. 2008	monomerous	superior (usually)	present	Smets 1986
animal						
animal						
animal						
	C D T Hy	Hu et al. 2008	monomerous, or apocarpous	superior to partly inferior	present	Bernardello 2007
wind						
animal						
	?	Hu et al. 2008				
animal			apocarpous	superior to partly inferior	absent	Bernardello 2007
	C T	Hu et al. 2008	syncarpous	superior	absent	Endress 1990
animal						
	C D M T	Hu et al. 2008	monomerous, or apocarpous, or syncarpous	superior	present in some	Bernardello 2007
animal						

Table 6.6 (*cont.*)

Family	Species	Sculpturing	Grain size (μm)	Aperture type	Pollen morphology references
Winteraceae	*Takhtajania perrieri*	reticulate	74 (tetrad)	ulcerate	Sampson 2000
Aristolochiaceae		sculptured	27–73	monosulcate and derived	
Aristolochiaceae	*Saruma henryi*	reticulate	28	monosulcate	Erdtman 1966
Hydnoraceae		minimally sculptured	44	monosulcate and derived	
Hydnoraceae	*Hydnora afriacana*	scabrate	44	dicolpate	Erdtman 1966
Lactoridaceae		minimally sculptured	35 (tetrad)	inaperturate	
Lactoridaceae	*Lactoris fernandeziana*	scabrate	35 (tetrad)	ulcerate	Erdtman 1966
Piperaceae		moderately sculptured	10–17	inaperturate, monosulcate and derived	
Piperaceae	*Peperomia tithymaloides*	rugulate ?	11	inaperturate to monosulcate	Erdtman 1966
Piperaceae	*Heckeria subpeltata*	rugulate ?	13	monosulcate	Erdtman 1966
Piperaceae	*Piper majusculum*	rugulate ?	13	monosulcate	Erdtman 1966
Saururaceae		moderately sculptured	11–18	monosulcate and derived	
Saururaceae	*Anemopsis californica*	rugulate ?	14	monosulcate	Erdtman 1966
Acoraceae		moderately sculptured	18–21	monosulcate and derived	
Acoraceae	*Acornus calamus*	microreticulate	21	monosulcate	Erdtman 1966
Toefieldiaceae		moderately sculptured	30	monosulcate and derived	
Toefieldiaceae	*Tofieldia calycullata*	rugulate ?	30	dicolpate	Erdtman 1966
Araceae		moderately sculptured	11–75	monosulcate and derived	

Predicted pollinator	Pollination mode	Pollination references	Ovary (family level) (Watson and Dallwitz (1992 onwards))	Gynoecium (family level) (Watson and Dallwitz (1992 onwards))	Floral nectaries (family level)	Nectary references
animal						
	D	Hu et al. 2008	syncarpous	inferior (usually)	present	Smets 1986
animal						
	C	Hu et al. 2008	syncarpous	inferior	absent	Bernardello 2007
animal						
	W	Hu et al. 2008	apocarpous to syncarpous	superior	absent	Bernardello 2007
wind						
	C D Hy	Hu et al. 2008	syncarpous	superior	absent	Endress 1990
animal						
animal						
animal						
	C D Hy T W	Hu et al. 2008	apocarpous, or syncarpous	superior (mostly)	absent	Bernardello 2007
animal						
	?	Hu et al. 2008	syncarpous	superior	absent	Bernardello 2007
animal						
	?	Hu et al. 2008			present	Bernardello 2007
animal						
	C D Hy	Hu et al. 2008	monomerous, or syncarpous	superior	absent	Bernardello 2007

Table 6.6 (*cont.*)

Family	Species	Sculpturing	Grain size (µm)	Aperture type	Pollen morphology references
Araceae	*Lemna gibba*	echinate	22	monosulcate	Erdtman 1966
Araceae	*Afroraphidophora africana*	microreticulate	27	monosulcate	Erdtman 1966
Araceae	*Pothos lonipes*	microreticulate	21	monosulcate	Erdtman 1966
Alismataceae		minimally sculptured to sculptured	19–38	monosulcate and derived	
Alismataceae	*Limnophyton obtusifolim*	reticulate	36	periporate	Erdtman 1966
Alismataceae	*Limnocharis flava*	scabrate	30	periporate	Argue 1973
Alismataceae	*Hydrocleis nymphoides*	echinate	28	periporate	Argue 1973
Alismataceae	*Tenagocharis latifolia*	echinate	31	periporate	Argue 1973
Aponogetonaceae		sculptured	21–45	monosulcate and derived	
Aponogetonaceae	*Aponogeton abyssinicum*	reticulate (?)	37	monosulcate	Erdtman 1966
Aponogetonaceae	*A. dinteri*	reticulate (?)	30	monosulcate	Erdtman 1966
Aponogetonaceae	*A. guillotii*	reticulate (?)	45	monosulcate	Erdtman 1966
Butomaceae		sculptured	35	monosulcate and derived	
Butomaceae	*Butomus umbellatus*	reticulate	35	monosulcate	
Cymodoceaceae		minimally sculptured	2840–3000	inaperturate	
Cymodoceaceae	*Amphibolis griffith*	scabrate	3000	inaperturate?	Mcconchie et al. 1982
Cymodoceaceae	*A. antarctica*	scabrate	2840	inaperturate?	Mcconchie et al. 1982

Predicted pollinator	Pollination mode	Pollination references	Ovary (family level) (Watson and Dallwitz (1992 onwards))	Gynoecium (family level) (Watson and Dallwitz (1992 onwards))	Floral nectaries (family level)	Nectary references
animal						
animal						
animal						
	D Hy	Hu et al. 2008	apocarpous	superior	present	Watson and Dallwitz (1992 onwards)
animal						
wind						
animal						
animal						
	C Hy	Watson and Dallwitz (1992 onwards)	apocarpous	superior	present	Watson and Dallwitz (1992 onwards)
animal						
animal						
animal						
	D Hy	Hu et al. 2008	apocarpous	superior	present	Watson and Dallwitz (1992 onwards)
animal						
	Wa	Mcconchie et al. 1982	apocarpous	superior	absent	Bernardello 2007
water	Wa					
water	Wa					

Table 6.6 (*cont.*)

Family	Species	Sculpturing	Grain size (µm)	Aperture type	Pollen morphology references
Juncaginaceae		sculptured	20–30	inaperturate	
Juncaginaceae	*Tetroncium magellanicum*	reticulate	28	inaperaturate	Erdtman 1966
Hydrocharitaceae		minimally sculptured	25–130 (long strands >300)	inaperaturate	
Hydrocharitaceae	*Limnobium stoloniferum*	psilate	27	inaperaturate	Erdtman 1966
Posidoniaceae		?	450	inaperturate	
Posidoniaceae	*Posidonia australis*	?	450 (thread-like)	inaperturate	
Potamogetonaceae		moderately sculptured to sculptured	20–35	inaperaturate	
Potamogetonaceae	*Potamogeton natans*	reticulate	24	inaperaturate	Erdtman 1966
Potamogetonaceae	*Potamogeton perfoliatus*	microreticulate	31	inaperaturate	Erdtman 1966
Ruppiaceae		sculptured	70	inaperturate	
Ruppiaceae	*Ruppia maritima*	reticulate	70	inaperturate?	Erdtman 1966
Scheuchzeriaceae		sculptured	44 (dyads)	inaperturate	
Scheuchzeriaceae	*Scheuchzeria palustris*	reticulate	44 (dyads)	inaperturate	Erdtman 1966
Zosteraceae		without exine	2550		
Zosteraceae	*Zostera marina*	?	2250	exine missing	Erdtman 1966
Berberidaceae		sculptured	30–65	tricolpate and derived	

Predicted pollinator	Pollination mode	Pollination references	Ovary (family level) (Watson and Dallwitz (1992 onwards))	Gynoecium (family level) (Watson and Dallwitz (1992 onwards))	Floral nectaries (family level)	Nectary references
	W	Hu et al. 2008	apocarpous, or syncarpous, or monomerous	superior	absent	Bernardello 2007
animal						
	C D W Wa	Hu et al. 2008	syncarpous	inferior	present, or absent	Watson and Dallwitz (1992 onwards)
wind						
	Wa	Hu et al. 2008	monomerous	superior	absent	Bernardello 2007
water						
	W Wa	Hu et al. 2008	apocarpous	superior	absent	Bernardello 2007
animal						
animal						
	Wa	Hu et al. 2008	apocarpous	superior	absent	Bernardello 2007
animal						
	W	Hu et al. 2008	apocarpous to syncarpous	superior	absent	Bernardello 2007
animal						
	Wa	Hu et al. 2008	monomerous	superior	absent	Bernardello 2007
water						
	D Hy	Hu et al. 2008	monomerous	superior	present	Watson and Dallwitz (1992 onwards)

Table 6.6 (*cont.*)

Family	Species	Sculpturing	Grain size (µm)	Aperture type	Pollen morphology references
Berberidaceae	*Berberis dictyophylla*	reticulate	46	tricolpate	Erdtman 1966
Berberidaceae	*Leontice smirnowii*	reticulate	62	tricolpate	Erdtman 1966
Eupteleaceae		minimally sculptured	30–39	tricolpate and derived	
Eupteleaceae	*Euptelea francheti*	scabrate?	31	6-colpate	Erdtman 1966
Eupteleaceae	*Euptelea pleiosperma*	scabrate?	34	tricopate	Erdtman 1966
Eupteleaceae	*Euptelea polyandra*	scabrate?	30	tricopate or 6-colpate	Erdtman 1966
Circaeasteraceae		moderately sculptured	23	tricolpate and derived	
Circaeasteraceae	*Circasaster agrestis*	finely striate	23	tricolpate	Erdtman 1966
Lardizabalaceae		sculptured	20–40	tricolpate and derived	
Lardizabalaceae	*Sinofranchetia chinensis*	reticulate	30	tricolporate	Erdtman 1966
Menispermaceae		moderately sculptured to sculptured	14–45	inaperturate, tricolpate and derived	
Menispermaceae	*Cocculus trilobus*	reticulate	22	tricolporate	Erdtman 1966
Menispermaceae	*Legnephora moorii*	striate	45	tricolporate	Erdtman 1966
Menispermaceae	*Tiliacora funifera*	reticulate	15	inaperturate	Erdtman 1966
Papaveraceae (inc. Fumarioideae Papaveroideae, Pteridophylloideae)		sculptured	13–55	tricolpate and derived	

Predicted pollinator	Pollination mode	Pollination references	Ovary (family level) (Watson and Dallwitz (1992 onwards))	Gynoecium (family level) (Watson and Dallwitz (1992 onwards))	Floral nectaries (family level)	Nectary references
animal						
animal						
	W	Hu et al. 2008	apocarpous	superior	absent	Watson and Dallwitz (1992 onwards)
wind						
wind						
wind						
	?	Hu et al. 2008	monomerous to apocarpous	superior	absent	Watson and Dallwitz (1992 onwards)
animal						
	D Hy	Hu et al. 2008	apocarpous	superior	present, or absent	Watson and Dallwitz (1992 onwards)
animal						
	Hy W	Hu et al. 2008	monomerous, or apocarpous	superior	present	Endress 1990
animal						
animal						
animal						
	C D Hy	Hu et al. 2008	syncarpous	superior	present, or absent	Bernardello 2007

Table 6.6 (*cont.*)

Family	Species	Sculpturing	Grain size (µm)	Aperture type	Pollen morphology references
Papaveraceae (inc. Fumarioideae Papaveroideae, Pteridophylloideae)	*Arctomecon californicum*	reticulate	35	tricolpate	Erdtman 1966
Papaveraceae (inc. Fumarioideae Papaveroideae, Pteridophylloideae)	*Canbya candida*	reticulate	14	tricolpate	Erdtman 1966
Papaveraceae (inc. Fumarioideae Papaveroideae, Pteridophylloideae)	*Chelidonium majus*	reticulate	31	tricolpate	Erdtman 1966
Papaveraceae (inc. Fumarioideae Papaveroideae, Pteridophylloideae)	*Dendromecon rigida*	reticulate	40	4(-5)-colpate	Erdtman 1966
Papaveraceae (inc. Fumarioideae Papaveroideae, Pteridophylloideae)	*Eschscholtzia pulchella*	reticulate	25	6–7-colpate	Erdtman 1966
Papaveraceae (inc. Fumarioideae Papaveroideae, Pteridophylloideae)	*Hylomecon japonica*	reticulate	34	tricolpate	Erdtman 1966
Papaveraceae (inc. Fumarioideae Papaveroideae, Pteridophylloideae)	*Papaver somniferum*	reticulate	34	tricolpate	Erdtman 1966
Papaveraceae (inc. Fumarioideae Papaveroideae, Pteridophylloideae)	*Romneya coulteri*	reticulate	33	tricolpate	Erdtman 1966
Ranunculaceae		moderately sculptured to sculptured	17–60	tricolpate and derived	
Ranunculaceae	*Coptis asplenifolia*	echinate	25	periporate (?)	Erdtman 1966
Ranunculaceae	*Helleborus atrorubens*	reticulate	42	tricolpate	Erdtman 1966
Sabiaceae		sculptured	20–33	tricolpate and derived	

Predicted pollinator	Pollination mode	Pollination references	Ovary (family level) (Watson and Dallwitz (1992 onwards))	Gynoecium (family level) (Watson and Dallwitz (1992 onwards))	Floral nectaries (family level)	Nectary references
animal						
animal						
animal						
animal						
animal						
animal						
animal						
animal						
	insect, W	Watson and Dallwitz (1992 onwards)	monomerous, or apocarpous, or syncarpous	superior	present (usually), or absent	Watson and Dallwitz (1992 onwards)
animal						
animal						
	insect, birds	Taylor and Hu 2010	syncarpous	superior	present	Bernardello 2007

Table 6.6 (*cont.*)

Family	Species	Sculpturing	Grain size (μm)	Aperture type	Pollen morphology references
Sabiaceae	*Meliosma arnottiana*	reticulate	20	tricolporate	Erdtman 1966
Sabiaceae	*Meliosma myriantha*	reticulate	24	tricolporate	Erdtman 1966
Sabiaceae	*Ophiocaryum heterophyllum*	reticulate	26	tricolporate	Erdtman 1966
Sabiaceae	*Sabia dumicola*	reticulate	23	tricolporate	Erdtman 1966
Sabiaceae	*Sabia gracilis*	reticulate	24	tricolporate	Erdtman 1966
Sabiaceae	*Sabia paniculata*	reticulate	26	tricolporate	Erdtman 1966
Nelumbonaceae		sculptured	65–79	tricolpate and derived	
Nelumbonaceae	*Nelumbo nucifera*	rugulate ?	79	tricolpate	Erdtman 1966
Platanaceae		sculptured	23–25	tricolpate and derived	
Platanaceae	*Platanus orientalis*	reticulate	23	tricolpate	Erdtman 1966
Proteaceae		moderately sculptured to sculptured	18–100	tricolpate and derived	
Proteaceae	*Agastachys odorata*	reticulate	29	triporate	Erdtman 1966
Proteaceae	*Beauprea elegans*	reticulate	48	triporate	Erdtman 1966
Proteaceae	*Dilobeia thouarsii*	reticulate ?	26	triporate	Erdtman 1966
Proteaceae	*Franklandia fucifolia*	baculate	92	triporate ?	Erdtman 1966
Proteaceae	*Serruria acrocarpa*	reticulate ?	42	triporate	Erdtman 1966

Predicted pollinator	Pollination mode	Pollination references	Ovary (family level) (Watson and Dallwitz (1992 onwards))	Gynoecium (family level) (Watson and Dallwitz (1992 onwards))	Floral nectaries (family level)	Nectary references
animal						
animal						
animal						
animal						
animal						
animal						
	C	Watson and Dallwitz (1992 onwards)	apocarpous	superior	absent	Bernardello 2007
animal						
	W	Watson and Dallwitz (1992 onwards)	apocarpous	superior	absent	Watson and Dallwitz (1992 onwards)
animal		Linder 1998, 2000	apocarpous	superior		
	insect, birds, bats, rodents	Watson and Dallwitz (1992 onwards)	monomerous	superior	present	Bernardello 2007
animal						
animal						
animal						
animal						
animal						

Table 6.6 (*cont.*)

Family	Species	Sculpturing	Grain size (µm)	Aperture type	Pollen morphology references
Proteaceae	*Spatalla curvifolia*	reticulate	26	triporate	Erdtman 1966
Proteaceae	*Symphyonema montanum*	reticulate	25	triporate	Erdtman 1966
Proteaceae	*Embothrium wickhamii*	reticulate	54	triporate	Erdtman 1966
Proteaceae	*Euplassa inaequalis*	reticulate	31	triporate	Erdtman 1966
Proteaceae	*Guevina avellana*	reticulate	37	triporate	Erdtman 1966
Proteaceae	*Hakea ivoryi*	reticulate	55	triporate	Erdtman 1966
Proteaceae	*Hakea laurina*	baculate	100	triporate	Erdtman 1966
Proteaceae	*Kermadecia vitiensis*	reticulate	32	triporate	Erdtman 1966
Proteaceae	*Lambertia ericifolia*	reticulate	58	triporate	Erdtman 1966
Proteaceae	*Panopsis sessilifolia*	reticulate	33	triporate	Erdtman 1966
Proteaceae	*Stenocarpus elegans*	reticulate	39	triporate	Erdtman 1966
Proteaceae	*Xylomelum salicinum*	echinate	39	triporate	Erdtman 1966
Trochodendraceae		sculptured	16–24	tricolpate and derived	
Trochodendraceae	*Tetracentron sinense*	reticulate	16	tricolpate	Erdtman 1966
Trochodendraceae	*Trochodendron aralioides*	reticulate	24	tricolpate	Erdtman 1966
Buxaceae		moderately sculptured to sculptured	21–45	tricolpate and derived	
Buxaceae	*Buxus balearica*	reticulate	31	periporate	Erdtman 1966

Predicted pollinator	Pollination mode	Pollination references	Ovary (family level) (Watson and Dallwitz (1992 onwards))	Gynoecium (family level) (Watson and Dallwitz (1992 onwards))	Floral nectaries (family level)	Nectary references
animal						
animal						
animal						
animal						
animal						
animal						
animal						
animal						
animal						
animal						
animal						
animal						
	D W	Hu et al. 2008	apocarpous to syncarpous	partly inferior	present	Watson and Dallwitz (1992 onwards)
animal						
animal						
	C D Hy L W	Lazaro & Traveset 2005	syncarpous	superior	present	Bernardello 2007
animal						

Table 6.6 (*cont.*)

Family	Species	Sculpturing	Grain size (µm)	Aperture type	Pollen morphology references
Buxaceae	*Buxus japonica*	reticulate	33	periporate	Erdtman 1966
Buxaceae	*Buxus sempervirens*	reticulate	30	periporate	Erdtman 1966
Buxaceae	*Notobuxus obtusifolius*	rugulate ?	34	tricolporate	Erdtman 1966
Buxaceae	*Pachysandra procumbens*	crotonoid	45	periporate	Erdtman 1966
Buxaceae	*Pachysandra stylosa*	crotonoid	38	periporate	Erdtman 1966
Buxaceae	*Sarcococca hookeriana*	crotonoid	32	periporate	Erdtman 1966
Buxaceae	*Simmondsia californica*	foveolate	35	tricolpate	Erdtman 1966
Buxaceae	*Styloceras laurifolium*	microreticulate	35	periporate	Erdtman 1966
Buxaceae	*Styloceras parvifolium*	echinate	21	tricolporate	Erdtman 1966
Didymelaceae		sculptured	23	tricolpate and derived	
Didymelaceae	*Didymeles madagascariensis*	reticulate	23	Tricolporiorate	Erdtman 1966
Gunneraceae		minimally sculptured	36	tricolpate and derived	
Gunneraceae	*Gunnera petaloidea*	scabrate?	36	tricolpate	Erdtman 1966
Myrothamnaceae		minimally sculptured	31 (tetrad)	tricolpate and derived	
Myrothamnaceae	*Myrothamnus flabellifolia*	psilate	31 (Tetrad)	faintly tricolpate	Erdtman 1966

Predicted pollinator	Pollination mode	Pollination references	Ovary (family level) (Watson and Dallwitz (1992 onwards))	Gynoecium (family level) (Watson and Dallwitz (1992 onwards))	Floral nectaries (family level)	Nectary references
animal						
animal						
animal						
animal						
animal						
animal						
animal						
animal						
animal						
	W	Hu et al. 2008	monomerous	superior	?	Watson and Dallwitz (1992 onwards)
animal						
	W	Hu et al. 2008	syncarpous	inferior	absent	Bernardello 2007
wind						
	W	Hu et al. 2008	syncarpous	superior	absent	Watson and Dallwitz (1992 onwards)
wind						

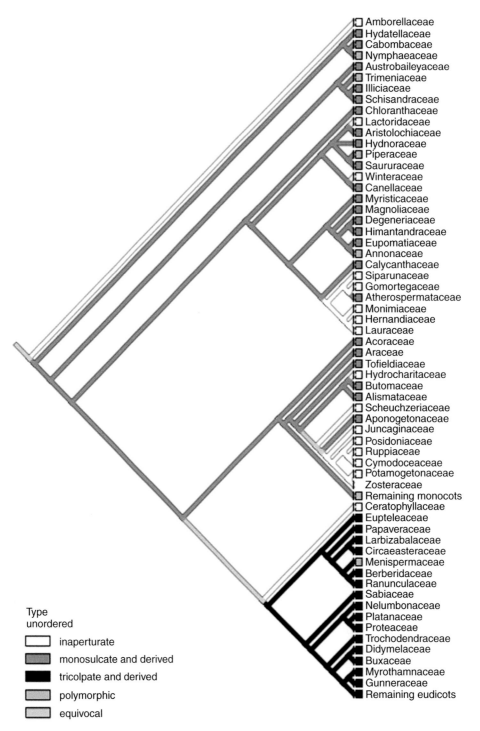

Fig 6.1 MacClade reconstruction of aperture types based on molecular topology (Stevens 2001).

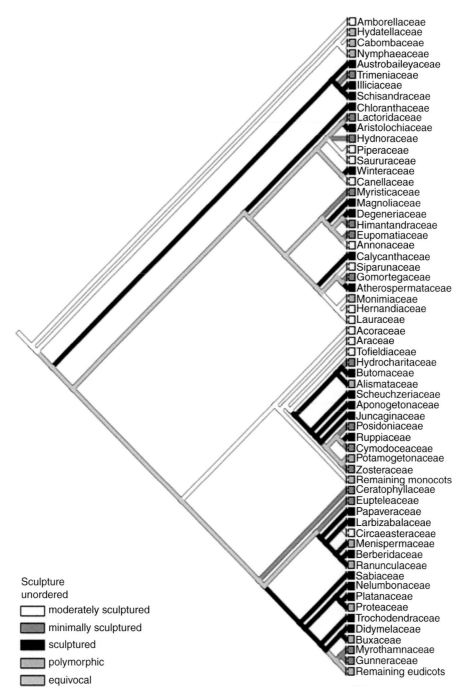

Sculpture
unordered

☐ moderately sculptured
⬛ minimally sculptured
■ sculptured
▨ polymorphic
▨ equivocal

Fig 6.2 MacClade reconstruction of sculpturing based on molecular topology (Stevens 2001).

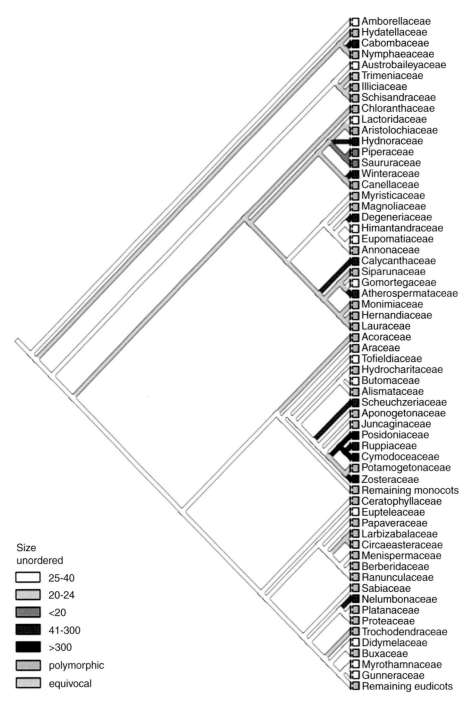

Fig 6.3 MacClade reconstruction of size ranges based on molecular topology (Stevens 2001).

Table 6.7 Proposed pollen criteria to predict pollinators based on pollen type based on living angiosperms including the basal angiosperms.

	< 20 μm	20–24 μm	25–40 μm	41–300 μm	> 300 μm
Sculptured	animal	animal	animal (?ambophilous if not sticky)	animal	water
Moderately sculptured	animal	animal	animal (?ambophilous if not sticky)	animal	water
Minimally sculptured	animal	?ambophilous if not sticky	wind	animal	water

size range between 25 and 40 μm. We note that ambophilous pollination may be possible for 20–24 μm grains that are minimally sculptured and not sticky, and that ambophilous pollination is possible for any type of sculpturing in 25–40 μm range if the grains are not sticky. Grains greater than 300 μm are advanced water pollinated and are based upon living angiosperm pollen data.

Family-level pollination modes were assigned to three categories: insect pollination includes those pollinated by Coleoptera, Diptera, and Thysanoptera; specialized pollination includes water and Hymenoptera; and wind pollination. Based on parsimony analysis (Fig 6.4), the ancestral angiosperm state is insect pollination with scattered occurrences of specialized pollination in the basal dicots, and specialized pollination becomes important in the monocot–eudicot clades. Wind pollination becomes particularly important in the basal eudicots.

In addition to examining pollination biology, we also examine if there was an association between pollen characters, pollination biology, and other floral characters. We examined the family distribution of nectaries (Fig 6.5). It supports Bernardello's (2007) suggestion that the absence of nectaries is plesiomorphic and they have evolved multiple times, as shown by their diverse morphology. In the basal angiosperms ovary, connation is rare and the apocarpous state is ancestral (Fig 6.6). Only the Piperales have many members with syncarpous ovaries. Finally, hypogynous flowers are ancestral and common with epigynous flowers evolving multiple times (Fig 6.7). There are no clear associations between the various characters and pollination syndromes.

6.4 Discussion

Based upon worldwide fossil pollen records, Valanginian pollen is small, inaperturate, and sculptured. There are only < 20 μm and 20–24 μm size classes, with

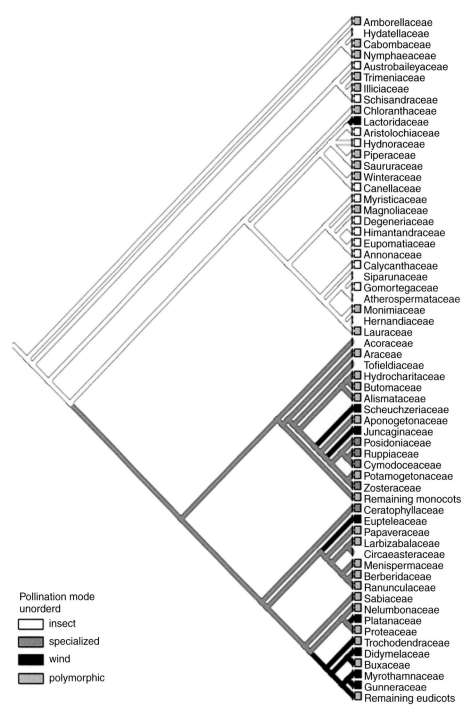

Fig 6.4 MacClade reconstruction of pollination modes based on molecular topology (Stevens 2001).

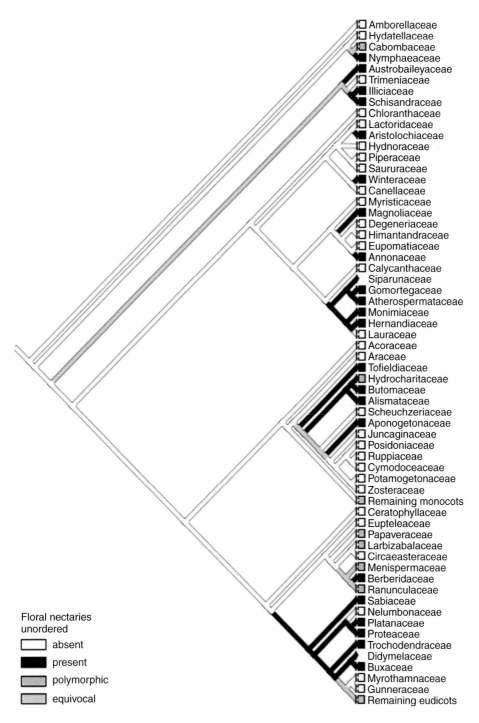

Fig 6.5 MacClade reconstruction of presence of floral nectaries based on molecular topology (Stevens 2001).

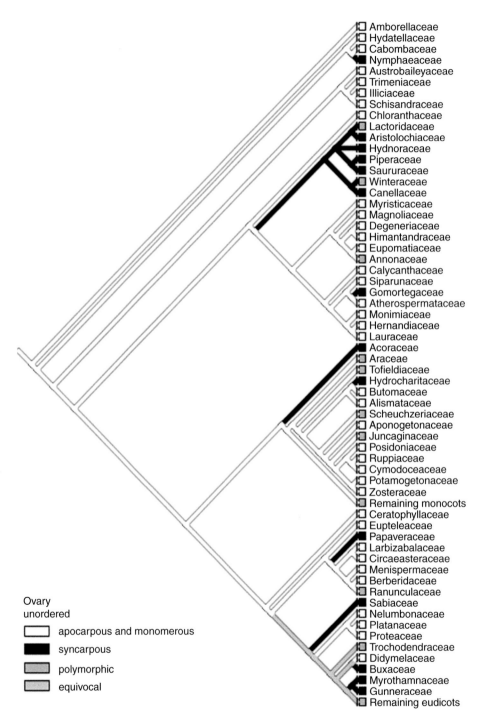

Fig 6.6 MacClade reconstruction of ovary connation based on molecular topology (Stevens 2001).

Fig 6.7 MacClade reconstruction of gynoecial position based on molecular topology (Stevens 2001).

the < 20 μm class dominant (81 %). Judging from our criteria, the majority of early angiosperms should have been pollinated by insects. However, we speculate, in contrast to these criteria, that the size class 20–24 μm may represent the ambophilous class since early angiosperm pollen appears not to be sticky (Taylor and Hu 2010). So ambophily could be present during the Valanginian. Because of the absence of grains with a diameter of 25–40 μm, the probability of early wind pollination is very low. According to Furness and Rudall (2004), an increase in the pollen aperture number is a general trend in angiosperms, and inaperturate morphology is fairly common among basal angiosperms. The lack of a clearly defined aperture for the Valanginian pollen may indicate the ancestral trait of angiosperm pollen grains (Brenner, 1996). Interestingly, all pollen grains in the Valanginian were reticulate. Linder (1998) mentioned that smooth pollen evolved later as wind pollination evolved from ambophily. Therefore, sculptured ornamentation may represent the ancestral character. We suggest that it is not the smoothness that is important for wind pollination, but rather having non-sticky pollen. Considering that early angiosperm pollen probably was not sticky (Taylor and Hu 2010), the sculpturing on the pollen surface may be useful for pollinators to collect pollen grains. Friis et al. (2006) suggested that early pollinators of angiosperms are probably pollen collectors due to the lack of nectaries in early angiosperm flowers.

During the Hauterivian, pollen aperture types included inaperturate and monosulcate, and monosulcate became dominant (79 %). Although common pollen sculpturing such as scabrate, rugulate, microreticulate, gemmate, echinate, and foveolate appeared at this time, reticulate was still dominant (62 %). The minimally sculptured sculpturing may be associated with the evolution of wind pollination at this time. Pollen size includes three classes, i.e. < 20 μm, 20–24 μm, and 25–40 μm. The size class of 25–40 μm appeared first during the Hauterivian. Since pollen grains were not sticky at this time, pollen with a diameter of 25–40 μm may be considered a proxy of wind-pollinated flowers. Wind pollination could have evolved during the Hauterivian, although ambophilous pollination cannot be excluded. However, pollen grains less than 20 μm in diameter are still dominant (55 %), and pollen with a 20–24 μm range accounts for 21 %. So, insect pollination was still dominant and ambophily may have been important as a transitional state.

During the Barremian, the aperture types of trichotomosulcate, monoporate, and possibly tricolpate first appeared. But the monosulcate state was still dominant (88 %). Sculptured ornamentation was still dominant and accounted for 67 %. Pollen size classes include < 20 μm (dominant at 38 %), 20–24 μm, 25–40 μm, and 40–300 μm. The size class of 40–300 μm first appeared and may indicate the appearance of specialized pollination because some pollinators such as Lepidoptera tend to collect large pollen grains (Harder 1998). However, angiosperm pollen grains with size ranges of 20–24 μm and 25–40 μm accounted for 51 %. This situation may indicate that ambophily and possibly wind pollination became important in the

Barremian. These grains most closely match the ancestral states suggested based on the parsimony of living angiosperms. This indicates that the living angiosperms are part of a crown group and early fossils reflect ancestral characteristics not found in living angiosperms.

During the Aptian, unequivocal tricolpate aperture first appeared, but the monosulcate aperture was still dominant (78 %). This suggests that eudicots were in place already by the Aptian, but in an early stage of diversification. A small staminate flower, *Teixeiria lusitanica* from Portugal, also provided evidence of the presence of eudicots in the Late Aptian (von Balthazar et al. 2005). Major pollen sculpturing types appeared, but the reticulate type was still dominant (78 %). Interestingly, the pollen of size range 25–40 µm becomes dominant (48 %) at this time. Considering the dry pollen grains and nectarless flowers, ambophily and wind pollination could have been common during the Aptian.

In contrast, based on parsimony analysis of basal living angiosperms, the ancestral states were equivocally inaperturate or monosulcate, moderately sculptured and 25–40µm in size. Doyle's (2005) analyses of basal angiosperms indicated that the ancestral pollen was monosulcate with a continuous tectum. This suggests that living basal angiosperms do not reflect the earliest pollen morphologies found in the fossil record and probably do not reflect all the ancestral states. Yet these morphological data also suggest animal or ambophilous pollination syndromes were present very early. If we look at the predicted pollinators for all the basal angiosperms that we examined, in six families (mostly in the Magnoliales) we predict wind pollination but that has not been reported in these families. We also have eight families (most often monocots but with some scattered basal dicots and eudicots) with cases where we predict animal pollinators but they are wind pollinated. The main problem is that they are sculptured, not smooth. This phenomenon may indicate that the evolution of pollen morphology temporally lags behind the evolution of floral morphology, as Crepet (2008) suggested. We hypothesize that these inconsistencies are due to the lack of pollen stickiness that appears to have evolved later in angiosperms. Previous reports show that early angiosperm pollen appears not to be sticky as shown by the absence of clumps (Hu et al. 2008, Taylor and Hu, 2010). In addition, the pollen of the ANITA grade also lacks the lipid matrix that can cause stickiness (Thien et al. 2009), although *Austrobaileya* pollen initially forms sticky clumps but becomes powdery when older (J. Williams, personal communications, 2010). Thus pollen adapted for wind pollination does not have to be smooth to be dry, and animal-pollinated pollen does not have to be sculptured because they are not sticky. The lack of clumping results in the possibility that pollen in the 20–40 µm size range could be ambophilous. Every family in Table 6.6 that has both wind and animal pollination, have species that fall in that size range except for Cabombaceae (42–100 µm) and Saururaceae (11–18 µm). Work on the presence and nature of pollen stickiness is needed for other basal angiosperms.

It was clear to Dilcher (1979) that the late Albian Dakota Flora demonstrated a full range of diverse pollination strategies as he diagramed the presence of both wind and insect pollination at that time. He illustrated fossils of large flowers, showy medium-sized flowers with nectaries, and small flowers. Dilcher (1996, 2010) also suggested that the positioning of the floral organs were related to pollinators. The large showy flowers were borne singly while the small flowers were positioned in an attractive cluster of flowers, which allows the pollinators to crawl over the surface of an umbel. Thus by the end of the Lower Cretaceous, pendulous catkins, clusters of small flowers and large showy flowers were all present to accommodate maximum pollination potential. Basinger and Dilcher (1984) also demonstrated the very early presence of a well-developed stigma terminating an elongate style. This suggests that pollen clumping was also involved in insect transport that then deposited many pollen grains at one time and thus allowed for pollen tube competition. This provided a further step in the enhancement of the genetic potential of plants through the male gametophyte contained in their pollen.

The results of the parsimony analysis of pollen grains from living basal angiosperms also show some interesting pollination trends (Fig 6.4). They suggest that the early and normal size range is 25–40 μm. If the grains are dry, they could be ambophilous with transition to wind pollination accomplished early. Most other types of pollination syndromes appear to have evolved multiple times. Our data also clearly show mega-size pollen, water-pollinated clade in the monocots (Fig 6.3). Note that within the ANITA grade, Trimeniaceae, Illicaceae, and Schisandraceae have smaller pollen (Fig 6.3), and that appears partly correlated with the presence of nectaries (Fig 6.5). This suggests that pollen grains smaller than the 20 μm size range is probably associated with animal pollination, which includes nectar rewards, and does not appear to be associated to wind pollination or strictly pollen-collecting pollinators as previously proposed (Lupia et al. 2002; Hu et al. 2008).

We note that there are not obvious associations between pollen characteristics and other floral characters that we examined. But there are some associations in specific clades. In the Laurales clade their pollen are mostly inaperturate and the size range is 20–24 μm, while the flowers generally have inferior ovaries and nectaries. The monocot clade starting with Juncaginaceae is inaperturate and also includes exineless water-pollinated pollen types. Although Amborellaceae is moderately sculptured, most of the Austrobaileyales are sculptured. We do not see any associations between pollen size and gynoecium types, although some have suggested pollen size is associated with pollen tube length (e.g. Anderson and Barrett 1986; Fernandez et al. 2009).

The parsimony analysis of floral nectaries (Fig 6.5) shows that absence of nectaries is ancestral. This is consistent with early Cretaceous flower fossil records (Friis et al. 2006). Recently, Taylor and Hu (2010) described seven insect coprolite types from the Campanian (Late Cretaceous) and two types were probably from insects with the ability of obtaining nutrients by osmotic shock or penetration of digestive enzymes through the wall. These coprolite records would predict that nectar was present because methods of osmotic shock or penetrating of digestive enzymes through the wall are dependent on mixing nectar and pollen to achieve nutrient extraction (Roulston and Cane 2000). The predicted presence of nectaries in the Campanian is consistent with the fossil flowers with nectaries in the Late Albian (Basinger and Dilcher 1984). Interestingly, nectaries are very scattered in basal angiosperms, but are rare in the ANITA group (Thien et al. 2009). The pattern supports the hypothesis that nectaries have independently evolved several times through the geological history (Bernardello 2007).

There are no pollen clumps of floral origin from the late Valanginian through the Aptian. The few pollen clumps reported from the Late Barremian through the Aptian (Friis et al. 1999, 2000, 2004; Hughes 1994) have insect coprolite origin based upon coprolite morphology and pollen preservation (Taylor and Hu 2010). This information may indicate that pollen grains of early angiosperms are not sticky enough to hold the grains together. In extant angiosperms, pollenkitt is present in all angiosperms investigated (Hesse 1981b) and is the most common adhesive material and responsible to the stickiness of pollen grains from angiosperms pollinated by animals (Hesse 1981b; Pacini and Hesse 2005). However, the degree of the stickiness of pollenkitt is different. Hesse (1981a, 1981b) mentioned that highly viscous pollenkitt in angiosperms pollinated by animals is mainly located on the surface of the exine. Therefore, pollen from animal-pollinated angiosperms usually is sticky. In contrast, less viscous pollenkitt in wind-pollinated angiosperms can flow rapidly into the exine cavities (Hesse 1981a). As a result, pollen grains from wind-pollinated angiosperms are dry and powdery. Interestingly, pollenkitt in ambophilous taxa shows intermediate characters of structure (Pacini and Hesse 2005). Usually, the pollen grains from ambophilous taxa are dry and not sticky due to pollenkitt preferentially filling the cavities of the exine (Hesse 1981b). So, at least, the highly viscous pollenkitt probably did not evolve during the Aptian and before. Coincidently thus far, fossil records of pollenkitt, or pollenkitt-like substances are all from the Albian and younger sediments in North America and Sweden (Friis 1985; Friis et al. 1988; Crane et al. 1989; Friis and Pedersen 1996; Pedersen et al. 1991; Zetter et al. 2002).

At the same time, the size range of 49 % pollen from the Late Valanginian through the Barremian is between 20 and 40 µm, which is an optimal size range for wind-dispersed pollen grains (Whitehead 1983; Proctor et al. 1996). Also there is the rare occurrence of pollen that has a maximum of 51 µm diameter in the Late Barremian. In the Aptian, pollen grains with the size range of 20–40 µm accounted for 65 % of total angiosperm pollen. However, the insect coprolites and rare pollen occurrence in the sediments of the late Valanginian through the Aptian may suggest insect pollination was also important (Friis et al. 2006). Friis et al. (2006)

indicated that all flowers recovered from Torres Vedras locality of Portugal may be pollinated by both insect and wind, considering little sterile tissue in stamens and the presence of potentially wind-dispersed pollen grains in coprolites. Linder (1998) proposed a model of the evolution of anemophily and pointed out that plants with small open flowers, a small perianth, and dry pollen may be ambophilous. Ambophily is an unspecialized basic condition, from which either insect or wind pollination can evolve. Ambophily might have been an important pollination strategy of early angiosperms in the newly invaded habitats because the pollen grains of early angiosperms were probably not sticky (Taylor and Hu 2010) and the flowers illustrated by Friis et al. (2006) were small, unshowy and lacked nectarines.

Moreover, Endress (2010) noticed that nearly all bisexual flowers of basal angiosperms are dichogamous (having stamens and pistils that mature at different time) and protogynous. Interestingly, Sargent and Otto (2004) found strong support for correlated evolution between protogyny and abiotic pollination, and transitions from biotic to abiotic pollination were more likely among protogynous species. Endress (1997) hypothesized that protogyny is an ancestral trait in flowering plants. Sun et al. (2002) noted that in *Archaefructus*, which dates from the Barremian–Aptian boundary of 125 million years, the stamens matured at a different time than the carpels and Ji et al. (2004) showed that in at least one species, the stigmas have developed first. So dichogamy and protogyny could be common in angiosperm flowers by the earliest Aptian.

Based upon Thien et al. (2009), the majority of pollinators of basal angiosperms are generalists, such as flies, beetles, and small pollen-collecting bees. Wind pollination evolves more often in species that are pollinated by generalists and have small pale-colored flowers (Friedman and Barrett, 2008). Also most ambophilous species are pollinated by generalists (Friedman and Barrett, 2008). So the potential pollinators during the Late Valanginian through the Aptian could be generalists. The absence of floral nectaries (Friis et al. 2006) and dry pollen (Taylor and Hu 2010) in early angiosperms during the Late Valanginian through the Aptian provide the indirect evidence of pollen collecting behavior. Actually, the fossil records of insects suggest that major groups of pollinators such as Coleoptera (beetles), Diptera (flies), Thysanoptera (thrips), Hymenoptera (mostly bees and wasps), and Lepidoptera (butterflies and moths), were present by the Early Cretaceous (Labandeira 1997, 2000, 2002; Ren 1998; Ren et al. 2009; Grimaldi 1999; Grimaldi and Engel 2005, Poinar and Danforth 2006).

Based upon a parsimony analysis of basal angiosperm pollination vectors, insect pollination is ancestral (Fig 6.4). Worldwide fossil pollen records from the Late Valanginian through the Aptian support this conclusion from parsimony analysis, but ambophily could have been present as early as the Late Valanginian. Based upon this research and the literature, three stages of pollen diversification and floral pollination could be distinguished during the Early Cretaceous.

Stage 1 (Valanginian) – all pollen grains are less than 25 μm, sculptured, and inaperturate. Generalized pollinators, such as beetles, flies, and small bees, could have been the major pollinators. But ambophily probably was present.

Stage 2 (Hauterivian through the Barremian) – pollen grains include sizes larger than 40 μm, 20–40 μm, and smaller than 20 μm. Pollen sculpturing and aperture became diverse, but unequivocal tricolpate aperture was still not present. Generalized insect pollination and ambophily were dominant, but specialized wind pollination and specialized insect pollination could have been in place already.

Stage 3 (Aptian through the Albian) – more specialized pollen types appeared including tricolpates, and pollen clumping is found (multiple times) (Hu et al. 2008; Taylor and Hu 2010), suggesting specialization in monocot and dicot pollination including wind pollination.

6.5 Conclusions

Generalized insect pollination, which includes beetles, flies, and small bees, is probably ancestral to early angiosperms, but ambophily could have been a very important strategy of pollination from the Valanginian through the Aptian before sticky pollen evolved. Ambophily and wind pollination could have been common in the Aptian. Ambophily probably was the prerequisite for the evolution of wind pollination. Sticky pollen and some specialized pollination probably appeared in the Albian, while nectaries evolved by the end of the Early Cretaceous.

Acknowledgements

We would like to thank Dr Sébastien Patiny for his invitation to contribute this chapter to this book, and Paul Zeigler for assistance with the manuscript. We also appreciate the comments of the anonymous reviewers.

Reference

Ackerman, J. D. (2000). Abiotic pollen and pollination: ecological, functional and evolutionary perspectives. *Plant Systematic and Evolution*, **222**, 167–85.

Aizen, M. A., Garibaldi, L. A., Cunningham, S. A. and Klein, A. M. (2009). How much does agriculture depend on pollinators? Lessons from long-term trends in crop production and diversity deficits. *Annals of Botany*, **103**, 1579–88.

Anderson, J. M. and Barrett, S. C. H. (1986). Pollen tube growth in tristylous Pontederia cordata (Pontederiaceae). *Canadian Journal of Botany*, **64**, 2602–7.

APG III (2009). An update of the Angiosperm Phylogeny Group classification for the orders and families of flowering plants: APG III. *Botanical Journal of the Linnean Society*, **161**, 105–21.

Archangelsky, S. and Gamerro, J. C. (1967). Spore and pollen types of the Lower Cretaceous in Patagonia (Argentina). *Review of Palaeobotany and Palynology*, **1**, 211–7.

Argue, C. L. (1973). The pollen of Limnocharis flava Buchen., Hydrocleis nymplloides (Willd.) Buchen., and Tenagocltaris latifolia (Don.) Buchen. (Limnocharitaceae). *Grana*, **14**, 108–12.

Basinger, J. F. and Dilcher, D. L. (1984). Ancient bisexual flowers. *Science*, **224**, 511–3.

von Balthazar, M., Pedersen, K. R. and Friis, E. M., (2005). Teixeiria lusitanica gen. et nov., a ranunculalean flower from the Early Cretaceous of Portugal. *Plant Systematics and Evolution*, **255**, 55–75.

Batten, D. J. (1986). *Possible functional implications of exine sculpture and architecture in some Late Cretaceous Normapolles pollen. In Pollen and Spores: Form and Function*, 1st edn. London, UK: Academic Press, pp. 219–232.

Bernardello, G. (2007). *A systematic survey of floral nectaries*. In *Nectaries and Nectar*, 1st edn. Dordrecht, Netherlands: Springer, pp. 19–128.

Brenner, G. J. (1996). Evidence for the earliest stage of angiosperms pollen evolution: a paleoequatorial section from Israel. In *Flowering Plant Origin, Evolution and Phylogeny*, 1st edn. New York, NY: Chapman & Hall, pp. 91–115.

Brenner, G. J. and Bickoff, I. (1992). Palynology and age of the Lower Cretaceous basal Kurnub Group from the coastal plain to the northern Negev of Israel. *Palynology*, **16**, 137–85.

Burger, D. (1990). Early Cretaceous angiosperms from Queensland, Australia. *Review of Palaeobotany and Palynology*, **65**, 153–63.

Burger, D. (1993). Early and middle Cretaceous angiosperm pollen grains from Australia. *Review of Palaeobotany and Palynology*, **78**, 183–234.

Chaowasku, T., Mols, J. and van der Ham, R. W. J. M. (2008). Pollen morphology of Miliusa and relatives (Annonaceae). *Grana*, **47**, 175–84.

Cox, P. A. (1991). Abiotic pollination: an evolutionary escape for animal-pollinated angiosperms. *Philosophical Transactions of the Royal Society London*, **333**, 217–24.

Crane, P. R. (1986). *Form and function in wind dispersed pollen*. In *Pollen and Spores: Form and Function*, 1st edn. London, UK: Academic Press, pp. 179–202.

Crane, P. R., Friis, E. M. and Pedersen, K. R. (1986). Lower Cretaceous angiosperm flowers: fossil evidence on early radiation of Dicotyledons. *Science*, **232**, 852–4.

Crane, P.R., Friis, E. M. and Pedersen, K. R. (1989). Reproductive structure and function in Cretaceous Chloranthaceae. *Plant Systematic and Evolution*, **165**, 211–26.

Crepet, W. L. (2008). The fossil record of angiosperms: requiem or renaissance? *Annals of the Missouri Botanical Garden*, **95**, 3–33.

Crepet, W. L. and Friis, E. M. (1987). The evolution of insect pollination in angiosperms. In *The Origin of Angiosperms and Their Biological Consequences*, 1st edn. Cambridge, UK: Cambridge University Press, pp. 181–201.

Crepet, W. L. and Nixon, K. C. (1996). The fossil history of stamens. In *The Anther: Form, Function and Phylogeny*, 1st edn. Cambridge, UK: Cambridge University Press, pp. 25–57.

Crepet, W. L., Friis, E. M. and Nixon, K. C. (1991). Fossil evidence for the evolution of biotic pollination. *Philosophical Transactions of the Royal Society London*, **333**, 187–95.

Culley, T. M., Weller, S. G. and Sakai, A. K. (2002). The evolution of wind pollination in angiosperms. *Trends in Ecology & Evolution*, **17**, 361–9.

Dilcher (1979). Early angiosperm reproduction: an introductory report. *Review of Palaeobotany and Palynology*, **27**, 291–328.

Dilcher D. L. (1996). La importanicia del origen de las angiospermas y como formaron el mundo alrededor de ellas. In *Copnferendias VI Congreso Latinoamericano De Botanica, Mar Del Plata, Argentia*, 1st edn. Kew, UK: The Trustees of the Royal Botanical Garden, pp. 29–48.

Dilcher, D. L. (2010). Major innovations in angiosperm evolution. In *Plants in Mesozoic Time: Morphological Innovations, Phylogeny, Ecosystems*, 1st edn. Bloomington, IN: Indiana University Press, pp. 97–116.

Dilcher, D.L., Sun, G., Ji, Q. and Li, H. (2007). An early infructescence Hyrcantha decussata (comb. nov.) from the Yixian Formation in northeastern China. *Proceedings of the National Academy of Sciences*, **104**, 9370–4.

Doyle, J. A. (1992). Revised palynological correlations of the lower Potomac Group (USA) and the Cocobeach sequence of Gabon (Barremian-Aptian). *Cretaceous Research*, **13**, 337–49.

Doyle, J. A. (2005). Early evolution of angiosperm pollen as inferred from molecular and morphological phylogenetic analyses. *Grana*, **44**, 227–51.

Doyle, J. A. and Hickey, L. J. (1976). Pollen and leaves from the mid Cretaceous Potomac Group and their bearing on early angiosperm evolution. In *Origin and Early Evolution of Angiosperms*, 1st edn. New York, NY: Columbia University Press, pp. 139–206.

Doyle, J. A. and Robbins, E. I. (1977). Angiosperm pollen zonation of the continental Cretaceous of the Atlantic coastal plain and its application to deep wells in the Salisbury Embayment. *Palynology*, **1**, 43–78.

Doyle, J. A., Biens, P., Doerenkamp, A. and Jardine, S. (1977). Angiosperm pollen from the pre-Albian Lower Cretaceous of equatorial Africa. *Bulletin des Centres de Recherches Exploration-Production Elf-Aquitaine*, **1**, 451–73.

Endress, P. K. (1990). Evolution of reproductive structures and functions in primitive angiosperms (Magnoliidae). *Memoirs of the New York Botanical Garden*, **55**, 5–34.

Endress, P. K. (1997). Evolutionary biology of flowers: prospects for the next century. In *Evolution and Diversification of Land Plants*, 1st edn. New York, NY: Springer, pp. 99–119.

Endress, P. K. (2010). The evolution of floral biology in basal angiosperms. *Philosophical Transactions of the Royal Society London*, **365**, 411–21.

Erdtman, G. (1966). *Pollen Morphology and Plant Taxonomy*, 1st edn. New York, NY: Hafner Publishing Company.

Faegri, K. and van der Pijl, L. (1979). *The Principles of Pollination Ecology*, 3rd edn. New York, NY: Pergamon Press.

Feild, T. S. and Arens, N. C. (2005). Form, function and environments of the early angiosperms: merging extant phylogeny and ecophysiology with fossils. *New Phytologist*, **166**, 383–408.

Fernandez, V. A., Galetto, L. and Astegiano, J. (2009). Influence of flower functionality and pollination system on the pollen size–pistil length relationship. *Organisms Diversity and Evolution*, **9**, 75–82.

Friedman, J. and Barrett, S. C. H. (2008). A phylogenetic analysis of the evolution of wind pollination in the angiosperms. *International Journal of Plant Sciences*, **169**, 49–58.

Friedman, J. and Barrett, S. C. H. (2009). Wind of change: new insights on the ecology and evolution of pollination and mating in wind-pollinated plants. *Annals of Botany*, **103**, 1515–27.

Friis, E. M. (1985). Actinocalyx gen. nov., sympetalous angiosperm flowers from the Upper Cretaceous of southern Sweden. *Review of Palaeobotany and Palynology*, **45**, 171–83.

Friis, E. M. and Pedersen, K. R. (1996). Angiosperm pollen in situ in Cretaceous reproductive organs. In Vol. 1 of *Palynology: principles and applications*, 1st edn. College Station, TX: Texas A&M University, pp. 409–426.

Friis, E. M., Crane, P. R. and Pedersen, K. R. (1988). Reproductive structures of Cretaceous Platanaceae. *Biologiska Skrifter K. Danske Videnskabernes Selskab*, **31**, 1-56.

Friis, E. M., Pedersen, K. R. and Crane, P. R. (1999). Early angiosperm diversification: the diversity of pollen associated with angiosperm reproductive structures in early Cretaceous floras from Portugal. *Annals of the Missouri Botanical Garden*, **86**, 259–96.

Friis, E. M., Pedersen, K. R. and Crane, P. R. (2000). Fossil floral structures of a basal angiosperm with monocolpate, reticulate–acolumellate pollen from the Early Cretaceous of Portugal. *Grana*, **39**, 226–39.

Friis, E. M., Pedersen, K. R. and Crane, P. R. (2004). Araceae from the Early Cretaceous of Portugal: Evidence on the emergence of monocotyledons. *Proceedings of the National Academy of Sciences*, **101**, 16565–70.

Friis, E. M., Pedersen, K. R. and Crane, P. R. (2006). Cretaceous angiosperm flowers: Innovation and evolution in plant reproduction. *Palaeogeography, Palaeoclimatology, Palaeoecology*, **232**, 251–93.

Furness, C. A. and Rudall, P. J. (2004). Pollen aperture evolution: a crucial factor for eudicot success? *Trends in Plant Sciences*, **9**, 154–8.

Grimaldi, D. (1999). The coradiations of pollinating insects and angiosperms in the Cretaceous. *Annals of the Missouri Botanical Garden*, **86**, 373–406.

Grimaldi, D. and Engel, M. S. (2005). *Evolution of the Insects*, 1st edn. Cambridge, UK: Cambridge University Press.

Halbritter, H. and Svojtka, M. (2000 onwards). Nuphar lutea. In PalDat – a palynological database: Descriptions, illustrations, identification, and information retrieval. from http://www.paldat.org/

Harder, L. D. (1998). Pollen-size comparisons among animal-pollinated angiosperms with different pollination characteristics. *Biological Journal of the Linnean Society*, **64**, 513–25.

Heimhofer, U., Hochuli, P. A., Burla, S. and Weissert, H. (2007). New records of Early Cretaceous angiosperm pollen from Portuguese coastal deposits:

implications for the timing of the early angiosperm radiation. *Review of Palaeobotany and Palynology*, **144**, 39–76.

Herendeen, P. S., Crane, P. R. and Drinnan, A. N. (1995). Fagaceous flowers, fruits and cupules from the Campanian (Late Cretaceous) of central Georgia, USA. *International Journal of Plant Sciences*, **156**, 93–116.

Hesse, M. (1979a). Entwicklungsgeschichte und Ultrastruktur von Pollenkitt und Exine bei nahe verwandten entomophilen und anemophilen Sippen der Oleaceae, Scrophulariaceae, Plantaginaceae und Asteraceae. *Plant Systematic and Evolution*, **132**, 107–39.

Hesse, M. (1979b). Ultrastructure and distribution of pollenkitt in the insect- and wind-pollinated genus *Acer* (Aceraceae). *Plant Systematic and Evolution*, **131**, 277–89.

Hesse, M. (1981a). The fine structure of the exine in relation to the stickiness of angiosperm pollen. *Review of Palaeobotany and Palynology*, **35**, 81–92.

Hesse, M. (1981b). Pollenkitt and viscin threads: their role in cementing pollen grains. *Grana*, **20**, 145–152.

Hesse, M. (2001). Pollen characters of Amborella trichopoda (Amborellaceae): a reinvestigation. *International Journal of Plant Sciences*, **162**, 201–8.

Hochuli, P. A., Heimhofer, U. and Weissert, H. (2006). Timing of early angiosperm radiation: recalibrating the classical succession. *Journal of the Geological Society, London*, **163**, 587–94.

Hu, S. (2006). Palynomorphs and selected mesofossils from the Cretaceous Dakota Formation, Minnesota, USA.

Ph.D. dissertation, Department of Geology. Gainesville, University of Florida, p. 217.

Hu, S. and Taylor, D. W. (2010). *Predictions of early Cretaceous floral structure and pollination biology by fossil pollen proxy*. Abstract 2010 Annual Meeting of Botanical Society of America. p. 71.

Hu, S., Dilcher, D. L., Jarzen, D. M. and Taylor, D. W. (2008). Early steps of angiosperm-pollinator coevolution. *Proceedings of the National Academy of Sciences*, **105**, 240–5.

Hughes, N. F. (1994). *The Enigma of Angiosperm Origin*, 1st edn. Cambridge, UK: Cambridge University Press.

Ibrahim, M. I. A. (2002). New angiosperm pollen from the Upper Barremian–Aptian of the western desert, Egypt. *Palynology*, **26**, 107–33.

Ji, Q., Li, H., Bowe, L. M., Liu, Y., and Taylor, D. W. (2004). Early Cretaceous Archaefructus eoflora sp. nov. with bisexual flowers from Beipiao, Western Liaoning, China. *Acta Geologica Sinica*, **78**, 883–96.

Judd, W. S., Campbell, C. S., Kellogg, E. A., Stevens, P. F. and Donoghue, M. J. (2002). *Plant Systematics, a Phylogenetic Approach*, 2nd edn. Sunderland, MA: Sinauer Associates, Inc.

Labandeira, C. C. (1997). Insect mouthparts: ascertaining the paleobiology of insect feeding strategies. *Annual Review of Ecology, Evolution and Systematic*, **28**, 153–93.

Labandeira, C. C. (2000). The paleobiology of pollination and its precursors. In *Phanerozoic terrestrial ecosystems*. 1st edn. Boulder, CO: Paleontological Society Papers, pp. 233–269.

Labandeira, C. C. (2002). The history of associations between plants and animals. In *Plant-animal interactions: an evolutionary approach*. 1st edn. Oxford, UK: Blackwell Science, pp. 26–74.

Lander, T. A., Harris, S. A. and Boshier, D. H. (2009). Flower and fruit production and insect pollination of the endangered Chilean tree, Gomortega keule in native forest, exotic pine plantation and agricultural environments. *Revista Chilena de Historia Natural*, **82**, 403–12.

Lazaro, A. and Traveset, A. (2005). Spatio-temporal variation in the pollination mode of Buxus balearica (Buxaceae), an ambophilous and sefing species: mainland–island comparison. *Ecography*, **28**, 640–52.

Linder, H. P. (1998). Morphology and the evolution of wind pollination. In *Reproductive Biology*, 1st edn. Kew, UK: The Royal Botanic Gardens, pp. 123–35.

Linder, H. P. (2000). Pollen morphology and wind pollination in angiosperms. In *Pollen and Spores: Morphology and Biology*, 1st edn. Kew, UK: The Royal Botanic Gardens, pp. 73–88.

Lonsdorf, E., Kremen, C., Ricketts, T. H., Winfree, R., Williams, N. M. and Greenleaf, S. S. (2009). Modeling pollination services across agricultural landscapes. *Annals of Botany*, **103**, 1589–600.

Lupia, R., Herendeen, P. S. and Keller, J. A. (2002). A new fossil flower and associated coprolites: evidence for angiosperm-insect interactions in the Santonian (Late Cretaceous) of Georgia, USA. *International Journal of Plant Science*, **163**, 675–86.

Maddison, W. P. and Maddison, D. R. (1992). *MacClade: Analysis of Phylogeny and Character Evolution*, 3rd edn. Sunderland, MA: Sinauer Associates, Inc.

Mcconchie, C. A., Knox, R. B., Ducker, S. C. and Pettitt, J. M. (1982). Pollen wall structure and cytochemistry in the seagrass Amphibolis griffithii (Cymodoceaceae). *Annals of Botany*, **50**, 729–32.

Mitchell, R. J., Flanagan, R. J., Brown, B. J., Waser, N. M. and Karron, J. D. (2009). New frontiers in competition for pollination. *Annals of Botany*, **103**, 1403–13.

Pacini, E. (2000). From anther and pollen ripening to pollen presentation. *Plant Systematic and Evolution*, **222**, 19–43.

Pacini, E. and Hesse, M. (2005). Pollenkitt: its composition, forms and functions. *Flora*, **200**, 399–415.

Pedersen, K.R., Crane, P. R., Drinnan, A. N. and Friis, E. M. (1991). Fruits from the mid Cretaceous of North America with pollen grains of the Clavatipollenites type. *Grana*, **30**, 577–90.

Poinar, G. O. and Danforth, B. N. (2006). A fossil bee from Early Cretaceous Burmese amber. *Science*, **314**, 614.

Proctor, M., Yeo, P. and Lack, A. (1996). *The Natural History of Pollination*, 1st edn. Portland, OR: Timber Press.

Proctor, M. C. F. (1978). Insect pollination syndromes in an evolutionary and systematic context. In *The Pollination of Flowers by Insects*, vol.6, London: Symposium Series Linnaean Society. pp. 105–116.

Quattrocchio, M. E., Martınez, M. A., Pavisich, A. C. and Volkheimer, W. (2006). Early Cretaceous palynostratigraphy, palynofacies and palaeoenvironments of well sections in northeastern Tierra del Fuego, Argentina. *Cretaceous Research*, **27**, 584–602.

Remizowa, M. V., Sokoloff, D. D., Macfarlane, T. D., Yadav, S. R., Prychid, C. J. and Rudall, P. J. (2008).

Comparative pollen morphology in the early-divergent angiosperm family Hydatellaceae reveals variation at the infraspecific level. *Grana*, **47**, 81–100.

Ren, D. (1998). Flower-associated Brachycera flies as fossil evidence for Jurassic angiosperm origins. *Science*, **280**, 85–8.

Ren, D., Labandeira, C. C., Santiago-Blay, J. A., Rasitsyn, A., Shih, C., Bashuev, A., Logan, M. A. V., Hotton, C. L. and Dilcher, D. (2009). A probable pollination mode before angiosperms: eurasian, long-proboscid scorpionflies. *Science*, **326**, 840–7.

Retallack, G. and Dilcher, D. L. (1981). A coastal hypothesis for the dispersal and rise to dominance of flowering plants. In *Paleobotany, Paleoecology and Evolution*, 1st edn. New York, NY: Praeger, pp. 27–77.

Roulston, T. H. and Cane, J. H. (2000). Pollen nutritional content and digestibility for animals. *Plant Systematic and Evolution*, **222**, 187–209.

Rudall, P. J., Remizowa, M. V., Prenner, G., Prychid, C. J., Tuckett, R. E. and Sokoloff, D. D. (2009). Non-flowers near the base of extant angiosperms? Spatio–temporal arrangement of organs in reproductive units of Hydatellaceae and its bearing on the origin of the flower. *American Journal of Botany*, **96**, 67–82.

Sampson, F. B. (2000). Pollen diversity in some modern Magnoliids. *International Journal of Plant Sciences*, **161**, S193–S210.

Sannier, J., Baker, W. J., Anstett, M.-C. and Nadot, S. (2009). A comparative analysis of pollinator type and pollen ornamentation in the Araceae and the Arecaceae, two unrelated families of the monocots. *BMC Research Notes*, **2**, 145.

Sargent, R. D. and Otto, S. P. (2004). A phylogenetic analysis of pollination mode and the evolution of dichogamy in angiosperms. *Evolutionary Ecology Research*, **6**, 1183–99.

Schrank, E. and Mahmoud, M. S. (2002). Barremian angiosperm pollen and associated palynomorphs from the Dakhla oasis area, Egypt. *Palaeontology*, **45**, 33–56.

Smets, E. (1986). Localization and systematic importance of the floral nectaries in the Magnoliatae (Dicotyledons). *Bulletin du Jardin botanique national de Belgique*, **56**, 51–76.

Soltis, P. S. and Soltis, D. E. (2004). The origin and diversification of angiosperms. *American Journal of Botany*, **91**, 1614–26.

Stevens, P. F. (2001 onwards). Angiosperm Phylogeny Website. 9th. 2010. From http://www.mobot.org/MOBOT/research/APweb/

Sun, G., Ji, Q., Dilcher, D. L., Zheng, S., Nixon, K. C. and Wang, X. (2002). Archaefructaceae, a new basal angiosperm family. *Science*, **296**, 899–904.

Tanaka, N., Uehara, K. and Murata, J. (2004). Correlation between pollen morphology and pollination mechanisms in the Hydrocharitaceae. *Journal of Plant Research*, **117**, 265–76.

Taylor, D. W. (2010). Implications of Fossil Floral Data on Understanding the Early Evolution of Molecular Developmental Controls of Flowers. In *Plants in Mesozoic Time: Morphological Innovations, Phylogeny, Ecosystems*, 1st edn. Bloomington, IN: Indiana University Press, pp. 118–69.

Taylor, D. W. and Hickey, L. J. (1996). Evidence for and implications of an herbaceous origin for angiosperms. In *Flowering Plant Origin, Evolution and Phylogeny*, 1st edn. New York, NY: Chapman & Hall, pp. 232–66.

Taylor, D. W. and Hu, S. (2010). Coevolution of early angiosperms and their pollinators: Evidence from pollen. *Palaeontographica, Abteilung B: Palaeobotany -Paleophytology*, **283**, 103–35.

Theissen, G. and Melzer, R. (2007). Molecular mechanisms underlying origin and diversification of the angiosperm flower. *Annals of Botany*, **100**, 603–19.

Thien, L. B., Bernhardt, P., Devall, M. S., Chen, Z. D., Luo, Y. B., Fan, J. H., Yuan, L. C. and Williams, J. H. (2009). Pollination biology of basal angiosperms (ANITA grade). *American Journal of Botany*, **96**, 166–82.

Traverse, A. (2007). *Paleopalynology*, 2nd edn. Dordrecht, Netherlands: Springer.

Watson, L. and Dallwitz, M. J. (1992 onwards). *The families of flowering plants: descriptions, illustrations, identification, and information retrieval*. 20th. 2010, from http://delta-intkey.com

Whitehead, D. R. (1969). Wind pollination in the angiosperms: evolutionary and environmental considerations. *Evolution*, **23**, 28–35.

Whitehead, D. R. (1983). Wind pollination: Some ecological and evolutionary perspectives. In *Pollination Biology*. 1st edn. Orlando, FL: Academic Press, Inc., pp. 97–108.

Wing, S. L. and Boucher, L. D. (1998). Ecological aspects of the Cretaceous flowering plant radiation. *Annual Review of Earth Planet Science*, **26**, 379–421.

Zavada, M. S. (1984). Pollen wall development of Austrobaileya maculata. *Botanical Gazette*, **145**, 11–21.

Zetter, R., Weber, M., Hess, M. and Pingen, M. (2002). Pollen, pollenkitt, and orbicules in Craigia bronnii flower buds (Tilioideae, Malvaceae) from the Miocene of Hambach, Germany. *International Journal of Plant Sciences*, **163**, 1067–71.

7

Pollinator mediated floral divergence in the absence of pollinator shifts

ALLAN G. ELLIS AND BRUCE ANDERSON

7.1 Introduction

The remarkable diversity of the angiosperms is often attributed to their specialized reproductive associations with diverse pollinating vectors (Crepet 1984; Eriksson and Bremer 1992; Grimaldi 1999). The appeal of this argument lies in the premise that specializing on particular pollinating groups can directly result in reproductive isolation from related plant species, even in sympatry, thus generating diversity (Grant 1949; Stebbins 1970). This pollinator shift model is the dominant paradigm explaining floral diversification and is often invoked as an important plant speciation mechanism (reviewed in Kay and Sargent 2009). However, pollinator shifts, which are usually quantitative, not qualitative, often result only in weak reproductive isolation (Armbruster and Muchhala 2009). In addition, sympatric species that use different pollinators are rarely sister species, and most often also exhibit substantial postmating isolation. These observations question the validity of the link between pollinator shifts and reproductive isolation, which underlies the pollinator shift paradigm, and suggest that floral divergence associated with pollinator shifts is unlikely to be a product of selection for reproductive isolation in sympatry, except perhaps upon secondary contact. Instead, floral diversity might result largely from

Evolution of Plant–Pollinator Relationships, ed S. Patiny. Published by Cambridge University Press. © The Systematics Association 2012.

spatially variable influences on efficient gamete transfer (Johnson 2006), a per-spective that opens the possibility of numerous mechanisms influencing the divergence of floral traits.

A hallmark of this perspective should be substantial variation in floral traits between closely related allopatrically distributed species, but also between pop-ulations of species. Numerous studies document such floral diversity (Herrera et al. 2006; Ellis and Johnson 2009; Schlumpberger et al. 2009), suggesting that geographical variation in floral traits is ubiquitous. Approximately 20 % of British plant species (Warren and Mckenzie 2001), 40 % of the Polemoniaceae (Schemske and Bierzychudek 2007), 38 % of Cape Erica species (Rebelo and Siegfried 1985), and 40 % of Protea species (Carlson and Holsinger 2010) exhibit flower color polymorphisms. Although these numbers confound intra- and interpopulation variation, they do attest to the extent of variation in just a sin-gle floral trait.

Geographic variation in floral traits can arise through three predominant processes: phenotypic plasticity, genetic drift, and selection (Herrera et al. 2006). Although some traits undoubtedly exhibit plasticity (e.g. inflorescence size asso-ciations with water availability, Caruso 2006), most floral traits (except nectar vol-umes) that have been investigated have high heritability (Ashman and Majetic 2006) and are under selection (Harder and Johnson 2009). Random fixation of flo-ral characters through genetic drift could play a role, even in the face of strong selection (Wright 1943). Although it is possible that much of the floral variation observed in nature is due to genetic drift, little is known about the importance of this process for floral evolution (Rausher 2008). In contrast, many studies have shown that selection regimes are geographically variable (see Conner 2006) and some show that selection is the cause of spatially structured floral variation (e.g. Herrera et al. 2006; Harder and Johnson 2009).

Because the ranges of plants and their pollinators are not perfectly overlapping (e.g. Johnson and Steiner 1997), a single plant species might be visited by differ-ent pollinator species across its range. Functionally different pollinators may then exert contrasting selective pressures in different parts of the plants' range, lead-ing to morphological divergence and even speciation (Grant 1949). This method of divergence can even work for species with multiple floral visitors (e.g. Gomez et al. 2009; Herrera et al. 2006; Anderson et al. 2010a) if plants are adapted to the most effective pollinator (*sensu* Stebbins 1970) because shifts in pollinator composition at different sites will generate contrasting patterns of selection. Several lines of evidence support the pollinator shift model of floral divergence including: pollin-ation ecotypes (Grant and Grant 1965; Armbruster 1985; Johnson 1997; Johnson and Steiner 1997; Anderson et al. 2010a), the propensity of unrelated plants to form guilds of morphologically similar flowers that are pollinated by similar pollinators

(Vogel 1954; van der Pijl 1961; Fenster et al. 2004; Pauw 2006), and the association between lineage divergence and pollinator transitions on plant phylogenies (e.g. Johnson et al. 1998; Beardsley et al. 2003; Perez et al. 2006; Okuyama et al. 2008). More direct evidence for the effect of pollinators on flowers can be found in experiments documenting pollinator preferences for different floral types (Hodges and Arnold 1994; Schemske and Bradshaw 1999; Ramsey et al. 2003; Aldridge and Campbell 2007) and studies that measure selection by pollinators on floral traits (Schemske and Horvitz 1989; Alexandersson and Johnson 2002; Maad 2000; Herrera et al. 2006).

Although evidence suggests that selection on plant traits through pollinator shifts is an important driver of floral variation, there are also key selective roles to be played by other biotic forces, such as floral herbivory (Strauss et al. 1996; Galen 1999; Ehrlen et al. 2002; Vanhoenacker et al. 2006) and pollen theft (Hargreaves et al. 2009), which are potentially geographically variable and can sometimes have stronger effects on fitness than pollination (Cariveau et al. 2004; Gomez 2003). The abiotic environment (e.g. rainfall, soil pH) has also been shown to affect floral traits directly, especially since flowers are great contributors to water stress (e.g. Galen et al. 1987; Galen 1999, 2000). In addition, environmentally imposed selection on non-floral traits could pleiotropically affect floral traits, leading to variation when these indirect selective processes are geographically variable (see reviews by Strauss and Whittall 2006; Rausher 2008).

The influence of non-pollinator agents on floral traits has been extensively reviewed (e.g. Strauss and Whittall 2006), as has floral variation as a result of pollinator shifts (e.g. Johnson 2006; Kay and Sargent 2009), and thus these topics will not be the focus of this chapter. Instead, we explore an additional, largely overlooked, suite of pollinator-driven mechanisms with the potential to generate geographically structured variation in floral traits in species that utilize the same pollinating vector. These include: coevolutionary dynamics; morphological, and behavioral variation within pollinator species; exploitation of different behaviors in the same pollinating species; the balance of selection operating through both genders in hermaphrodites; and various consequences of geographic variation in plant (as opposed to pollinator) community composition. In each case we describe the selective pressures driving evolution of floral traits and the mechanisms by which this translates into geographically structured floral divergence (Table 7.1). Although mechanisms generating spatial variation are frequently interlinked, we deal first with those predominantly associated with spatially structured variation in pollinator traits, then those arising through geographic variation in the plant community context, and finally those linked to spatial variation in the degree of pollen limitation experienced by plant populations.

Table 7.1 The mechanisms by which geographically structured variation in floral traits can arise. We list both previously considered mechanisms (in white) and those that we highlight in this chapter (in grey).

Mechanism dictating floral phenotype of individuals	Mechanisms generating spatial structure	Sample references
Genetic drift	Spatial isolation of (small) populations	Wright 1943
Phenotypic plasticity	Variation in the abiotic environment	Caruso 2006
Selection imposed by the abiotic environment (either directly or indirectly via pleiotropic effects)	Variation in the abiotic environment	Reviewed in Strauss and Whittall 2006.
Selection imposed by non-pollinating biotic agents (florivores, seed predators etc)	Spatial variation in composition and densities of communities of enemies	Reviewed in Strauss and Whittall 2006.
Selection imposed by the most effective pollinator	Spatial variation in composition of available pollinator communities (qualitative or quantitative) – Pollinator shift model	Reviewed in Johnson 2006, Kay and Sargent 2009
	Spatial variation in the morphology / behavior of focal pollinator	Ings et al. 2009
	Spatial variation in plant community context	Moeller 2005
Selection for resemblance to co-flowering species (mimicry)	Spatial variation in plant community context (available models)	Johnson 1994, Combs and Pauw 2009
Selection through pollinator mediated positive (facilitative) or negative (competitive) interactions with coflowering species	Spatial variation in the plant community context	Armbruster 1985, Moeller 2004, Mucchala and Potts 2007, Smith and Rausher 2008, De Jager et al. 2010
Coevolution of plants and pollinators	Spatially variable constraints on traits of either interacting partner (abiotic or plant community context)	Anderson and Johnson 2008, Pauw et al. 2009

Table 7.1 (*cont.*)

Mechanism dictating floral phenotype of individuals	Mechanisms generating spatial structure	Sample references
Contrasting selection imposed by multiple behaviors in the same pollinator	Spatial variation in the strength of selection operating through different behaviors (pollen limitation and plant community context)	Temeles and Kress 2003, Ellis and Johnson 2010a, Alarcon et al. 2010.
Contrasting gender specific selection on floral traits	Spatial variation in the strength (or direction) of selection operating through male and female function (pollen limitation)	Ellis and Johnson 2010b
Selection for reproductive assurance	Spatial variation in pollinator services and thus in the importance of selfing.	Moeller 2006, Fishman and Willis 2008

7.2 Pollinator-driven mechanisms generating floral variation without pollinator shifts

7.2.1 Spatially structured intraspecific variation in pollinators

If we accept the premise that selection on floral phenotype is primarily imposed by the most effective pollinator, then subtle geographically structured variation in floral traits could result from intraspecific variation in pollinator morphological traits, dictating the fit between pollinator and flower or in behaviors dictating floral preferences. Insect morphologies can vary spatially within a species for many reasons. Perhaps one of the most well-studied patterns are latitudinal (and altitudinal) body size gradients (Bergmann clines – Chown and Gaston 1999, 2010). Plant populations occurring along these gradients would experience contrasting selection for the fit between pollinators and flowers. In one of few studies that have investigated this possibility, Malo and Baonza (2002) demonstrated that pollinator body size and *Cytisus scoparius* flower size show correlated responses across altitudinal gradients. Demonstration of this mechanism is complicated by the possibility that insect traits are in turn influenced by plant traits, i.e. coevolution, which is discussed in more detail later.

Like morphology, intrinsic or learned preferences of pollinators might also vary across their range, which could lead to divergence in attractive traits between plant populations. Although numerous studies have examined the floral preferences and learning ability of pollinators (e.g. Giurfa et al. 1995; Weiss 2001; Riffell et al. 2008), very few have explored variation in these traits within species (Chittka et al. 2004; Raine et al. 2006; Ings et al. 2009). Ings et al. (2009) demonstrated that bumblebee populations exhibit significant variation in both the strength and persistence of blue color preference in learning trials. Such variation in floral preferences between pollinator populations would potentially impose divergent selection on flower color between plant populations, although this has not been demonstrated. Even in the absence of variation in innate preference or the ability to learn, geographical variation in conditioned preferences is likely because this is to some extent dictated by the plant community context, i.e. variation in both the flower color of favored forage plants and the background of flower colors against which learning occurs (e.g. Forrest and Thomson 2009). For example, the color preferences of pollinators foraging in two populations of a low density plant species might differ because the colors of alternate or complementary food sources in the surrounding communities of coflowering species differ.

Although geographically structured morphological and behavioral variation in pollinators is an extremely obvious potential source of geographic variation in floral traits, there has been remarkably little work on variation in pollinator traits across their range (Chittka et al. 2004; Ings et al. 2009; Herrera et al. 2006). The majority of studies reporting pollinator variation investigate coevolved systems (e.g. Anderson and Johnson 2008; Pauw et al. 2009), which differ from the mechanism discussed here because variation in measured insect traits is determined by variation in floral traits and vice versa.

7.2.2 Coevolution of plants and pollinators

Several studies have now emerged suggesting that coevolutionary races can give rise to divergence in the traits of the interacting partners, and that this divergence does not necessarily require any shifts in pollinator (or plant) community context (Pauw et al. 2009; Toju and Sota 2006, 2009; Anderson and Johnson 2008, 2009; Laine 2009). Darwin (1862) used the Madagascar star orchid (*Angraecum sesquipedale*) to illustrate his idea of coevolution (see Johnson and Anderson 2010 for a historical account). He hypothesized that this orchid, with an extraordinary long spur, is only effectively pollinated by a moth with shorter mouth parts than the floral spur. This forces the moth to push up against the reproductive parts of the flower in order to get the nectar found at the spur's base, making pollen transfer more efficient. As a result, he and Wallace hypothesized that there is a strong selective pressure on flowers to evolve longer tubes than the mouthparts of their pollinators (experimentally shown by Nilsson 1988; Johnson and Steiner 1997; Anderson and Johnson 2008; Muchhala and Thomson 2009; Pauw et al. 2009; Anderson et al. 2010b). However, they also realized that longer proboscid pollinators were able to access more nectar than shorter proboscid pollinators (experimentally shown by Pauw et al. 2009 for long-proboscid flies). Consequently, the evolution of longer floral tubes forced the evolution of longer insect proboscides, which in turn forced the selection for even longer floral tubes.

Wallace (1867) noted that this positive feedback system would continue generating longer and longer traits until it is balanced by an opposing selective pressure. Although he did not elaborate much on opposing selective pressures, Wallace (1867) implied that proboscis and tube lengthening would only be advantageous to a point, after which increased length may become a liability (e.g. Harder 1983; Kunte 2007). Insects with excessively long proboscides may have difficulty maneuvering them and inserting them accurately into the narrow gullets of flowers (e.g. Harder 1983). In particular, wind may be an important factor affecting flight maneuverability, and on windy days long-proboscid flies frequently miss the floral openings with their proboscis and often stop foraging entirely when winds are strong (Anderson personal Observation 2008, Pauw personal Communication 2007). It is possible that in very windy places, insect proboscides would not be able to evolve to the same lengths as in windless places and so, the coevolutionary race between plants and insects would proceed to different end-points depending on environmental factors. If the ranges of a plant and its long-proboscid pollinator encompassed a heterogeneous abiotic environment, then one would expect coevolution to have different end-points in different localities, but that plant and insect traits would match at each site. This is what Anderson and Johnson (2008, 2009) and Pauw et al. (2009) found when they examined proboscis and tube length covariation between specialist flowers and their long-proboscid fly pollinators (Fig 7.1). They found that floral tubes were not very variable within populations, but varied several-fold in length, even between geographically close populations. Similar covariation has also been found between the spurs of *Diascia* flowers and the forelegs of their oil-collecting bee pollinators (Steiner and Whitehead 1990, 1991). Although Anderson and Johnson (2008) showed that coevolutionary end-points also correlated with latitude (also see Toju and Sota 2006), there has been no experimental evidence to confirm that it is indeed the interaction between the abiotic environment and coevolution that is causing the divergence of floral traits, such as spur and tube length.

Alternatively, coevolution may generate trait divergence through changes in plant community composition (see review by Laine 2009). For example most pollinators forage from multiple plant species, even though some of the plant species that they forage from can be highly specialized (Jordano 1987; Bascompte et al. 2006). Consider a long-proboscid pollinator feeding in a community comprising a long-tubed plant species, which it pollinates effectively, and an abundant short-tubed species from which it robs nectar. The short-tubed flowers may never experience selection for longer tubes because the mismatch between the long-proboscid

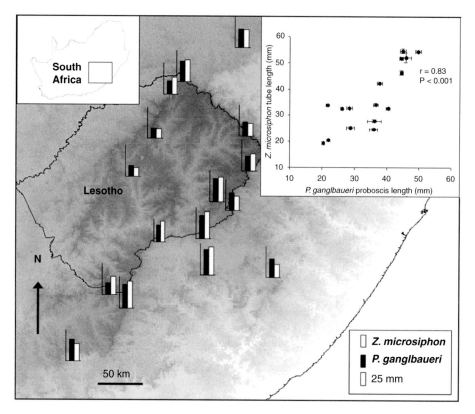

Fig 7.1 Geographic variability in the outcomes of coevolutionary races leading to divergence of floral tube lengths in *Zaluzianskya microsiphon*. Bars represent tube and proboscis lengths of *Z. microsiphon* and the long-proboscid fly pollinator, *Prosoeca ganglbaueri*. These are closely matched at each site and covary strongly (inset). Data redrawn and modified from Anderson and Johnson (2008).

insect and the flower may be too large (Pauw et al. 2009). However, if the long-proboscid pollinator receives much of its energetic requirements from robbing the short-tubed species, it would experience selection for shorter proboscides to improve maneuverability and foraging efficiency on the short-tubed plants. This may then exert selective pressure on the long-tubed flowers to become shorter. In contrast, in a community where insects only forage from highly specialized flowers or if there is only a single plant species to forage from, one may expect an evolutionary race selecting for longer traits as predicted by Darwin (1862) and Wallace (1867). The result is that plant community structure can affect the strengths of reciprocal selection, making it equally reciprocal in some instances but unbalanced in others and this can lead to trait divergence. For example, Anderson and Johnson (2008) showed that the tubes and proboscides of the specialized plant *Zaluzianskya microsiphon* and its long-proboscid fly pollinator (*Prosoeca ganglbaueri*) are probably involved in a coevolutionary race. However, they also showed that there are numerous other specialist flowers that are pollinated by *P. ganglbaueri*, many of which are uncommon or rewardless, with their traits tracking the coevolutionary race through a process of unilateral evolution (Anderson and Johnson 2009). However, in some populations, long-proboscid flies rob very short-tubed, common generalist species of nectar and it has recently been shown that these generalist species are potentially selecting for flies with shorter proboscides than the tube lengths of the specialist community (Anderson et al. 2010a). Thus both abiotic constraints and plant community context can generate a geographic mosaic of coevolution (Thompson 1994, 2005), resulting in contrasting selection on floral traits across a species' range.

7.2.3 Facilitation and competition between plant species for pollinators

The composition of plant communities can also influence visitation frequency, preferences, and floral constancy of pollinators, which in turn can influence the extent of heterospecific pollen transfer, and the quantity and quality of pollen loads arriving on stigmas (Hersch and Roy 2007; Geber and Moeller 2006). These indirect interactions between plant community members, mediated by shared pollinators are likely to exert strong selection for floral traits, which either reduce negative interactions (competition) or facilitate positive interactions between plant species (Rathcke 1983; Feinsinger 1987).

Coflowering plant species may compete directly for a limited resource, i.e. the pollinator, or indirectly through fitness reductions associated with interspecific pollen transfer (Robertson 1895; Feinsinger 1978; Rathcke 1983; Waser 1983). Numerous studies have demonstrated that indirect competition reduces seed production by clogging stigmas and styles, or by the production of infertile or inviable hybrid seed, and reduces pollen export through loss of pollen to heterospecific stigmas (Campbell and Motten 1985; Morales and Traveset 2008; Mitchell et al. 2009). This implies that competition for pollination might exert strong selection for floral traits that either reduce pollinator sharing or prevent heterospecific pollen transfers by shared pollinators. In the only study experimentally testing this (but see Fishman and Wyatt 1999; Caruso 2000), Smith and Rausher (2008) found that the presence of a competitor selected for floral traits that reduced heterospecific pollen receipt in *Ipomoea hederacea*. In addition, a number of studies comparing species' floral traits in populations sympatric and allopatric with a competitor have demonstrated character displacement of floral traits in sympatry, especially for traits involved in pollen placement (Armbruster 1985; Armbruster et al. 1994; Muchhala and Potts 2007). Thus negative reproductive interactions between plant species, mediated by common pollinators, do exert selection on floral traits, particularly those that reduce the incidence of interspecific pollen transfer.

Alternatively, co-occuring plant species could facilitate each other's reproduction by together maintaining larger or more persistent pollinator populations (Macior 1971; Brown and Kodric-Brown 1979; Rathcke 1983; Feinsinger 1987), although evidence for this idea is surprisingly rare (see Ghazoul 2006). Moeller (2004) presents evidence suggesting facilitation of pollination in communities of coexisting *Clarkia* species, and other studies have shown that non-rewarding species benefit from the proximity of coflowering rewarding species ("Magnet effect" – Laverty 1992; Johnson et al. 2003; Peter and Johnson 2008). In these systems, the fitness advantage arises from pollinator sharing, and thus selection, particularly in low density species (Feinsinger 1987), might favor floral traits that enhance similarity to congeneric species (Brown and Kodric-Brown 1979; Waser 1986; Chittka et al. 1997).

Geographic variation in floral traits under this mechanism is generated when the plant community context, and thus the indirect pollinator mediated interactions that a plant species encounters, varies in space. This requirement is no doubt ubiquitously met as the composition of plant communities varies enormously, but very few studies have addressed its importance. For example, Muchhala and Potts (2007) showed that differences between sympatric *Burmeistera* species in the degree of exertion of the staminal column were greater than expected under random community assembly, and that exaggerated differences were likely due to trait evolution (character displacement) in the most widespread species in response to pollinator mediated competition with congeners. Similarly, De Jager et al. (2010), in an analysis of the distribution of flower colors within *Oxalis* communities, found that color was significantly clustered within communities and that this was driven by the convergence of low abundance species on the colors of more common species, which varied in space, suggesting a possible role for facilitation of pollinator visits by the shared pollinator in driving the pattern. In both these systems, geographic variation in floral traits of widespread species arises due to differences in selection imposed by interactions with co-occurring plant species in different community contexts.

7.2.4 Floral mimicry and variation in models

Geographic variation in plant communities can also affect floral features of non-rewarding species (especially floral Batesian mimics) that rely on resemblance to co-occurring rewarding species for pollinator visitation. It has been shown that visitation to floral mimics is promoted by close resemblance in color (Benitez-Vieyra et al. 2007; Peter and Johnson 2008), size (Anderson et al. 2005), and shape (Johnson et al. 2003) to their models. Thus, it is likely that selection on mimic floral traits occurs through the ability of pollinators to discriminate between mimic and model, and thus ultimately through selection on the floral traits of the model. Populations of mimics should unilaterally track the floral phenotypes of their local models (e.g. Anderson et al. 2005). Furthermore, if widespread mimics traverse the ranges of several different looking models, mimic populations may diverge in accordance with floral traits of those models.

Disa nervosa, for example, mimics two geographically isolated *Watsonia* species pollinated by the same species of long-proboscid fly, *Philoliche aethiopica* (Johnson and Morita 2006). In morphology, the two *Watsonia* species are so similar that it has not led to morphological differences between the different mimic populations. However, because the two model species are geographically discrete and flower at different times, the mimic populations that flower at the same time as their sympatric models probably do not exchange any genetic material and are evolving on separate trajectories.

The orchid *Disa ferruginea* takes this process one step further and is pollinated by the butterfly *Aeropetes tulbaghia* across its entire range, (Johnson 1994). In the southern parts of its range, *D. ferruginea* occurs sympatrically with several red flowers, which *A. tulbaghia* visits for nectar and looks strikingly similar to one of these flowers, (*Tritoniopsis triticea*, Fig 7.2). However, the mountains in the extreme eastern parts of the *D. ferruginea* range do not have any red-flowered taxa that co-occur and flower at the same time as *D. ferruginea*. Instead, the butterflies in this part of the range tend to visit orange flowers and here, *D. ferruginea* is orange instead of red, closely resembling the orange inflorescences of a different model, *Kniphoffia uvaria* (Fig 7.2). This evidence for divergence due to different model floras is, however, still rather anecdotal and requires selection and translocation experiments to determine whether the geographic color variation is in fact an adaptation to different models.

In contrast to specialized Batesian mimicry that selects for uniformity of models within a population, generalized mimicry systems may give rise to floral polymorphisms within populations. Polymorphisms are maintained within populations because they retard pollinator learning through negative frequency dependent selection (e.g. Gigord et al. 2001). However, they may also become fixed in different populations due to genetic drift.

7.2.5 Balance of selection imposed by different behaviors in the same pollinating species

Although the majority of non-rewarding plant species, such as *Disa ferruginea*, exploit the sensory biases of feeding insects to achieve pollination (Batesian mimics, generalized food deception), a number of deceptive species exploit other behavioral modalities of flower-visiting insects (brood site mimicry, shelter mimicry, and sexual deception; Dafni 1984). These alternate strategies are made possible by the fact that animals exhibit a range of different behaviors whilst visiting flowers. They forage for diverse nutritional rewards including nectar, pollen, resins, and oils. They use flowers to shelter from inclement weather or as sleeping

Fig 7.2 Color divergence in *Disa ferruginea*, which mimics different color models in the southern and eastern parts of its range. Models and mimics are visited by the butterfly *Aeropetes tulbachia* in all parts of the range. Photos by Johnson and Anderson. See plate section for color version.

chambers. Flowers provide convenient arenas for aggregation and social inter-action. Males search for mates, gather compounds necessary for sexual signalling, and mark territories. Females search for oviposition sites or gather nest-building

Fig 3.1 Some representative flowers offering specialized rewards. A. *Yucca* cf. *schedigera*, brood-site reward flower pollinated by *Tegiticula* moths. B. *Krameria* sp. (Krameriaceae) offering oil to attract pollinating centridine bees (Hymenoptera: Apidae). C. *Dalechampia spathulata* being polinated by a fragrance-collecting male *Eulaema* sp. (Apidae: Euglossini) in Costa Rica (photo by W. Hallwachs, reproduced with permission). D. *Euglossa viridissima* (Euglossini) collecting eugenol from a blotter-paper bait, Quintana Roo, Mexico. E. *Dalechampia tiliifolia*, which offers a bright orange resin as the pollinator reward. F. *Trigona* (sensu lato) sp. (Apidae: Meliponini) collecting resin from a staminate *Clusia* sp. (Clusiaceae) in Suriname. The resin is secreted from staminodia in the center of the flower. Note balls of resin already placed in the bee's corbiculae (pollen baskets).

Fig 3.2 *Euglossa* nest constructed entirely of plant resins, with "front door" open, hanging from branch of cf. *Lycopodium*, Parque Nacional Soberanía, Panama. The resin used in the hard outer wall is most likely to be a different chemical composition to that of the internal resins and the "door" resin, which need to stay liquid and malleable. Note female *Euglossa* sp. in the entrance. This opening is closed with a resin "door" by night.

Fig 10.4 *Melocactus macracanthus*, spine-less cephalium as adaptation to ornithophily.

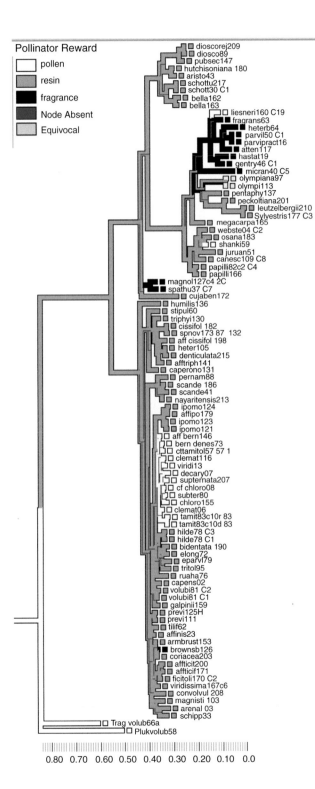

Pollinator Reward

- pollen
- resin
- fragrance
- Node Absent
- Equivocal

dioscorej209
diosco89
pubsec147
hutchisoniana 180
aristo43
schottu217
schott30 C1
bella162
bella163
liesneri160 C19
fragrans63
heterb64
parvil50 C1
parvipract16
atten117
hastat19
gentry46 C1
micran40 C5
olympiana97
olympi113
pentaphy137
peckoltiana201
leutzelbergii210
Sylvestris177 C3
megacarpa165
webste04 C2
osana183
shanki59
juruan51
canesc109 C8
papilli82c2 C4
papilli166
magnol127c4 2C
spathu37 C7
cujaben172
humilis136
stipul60
triphyi130
cissifol 182
spnov173 87 132
aff cissifol 198
heter105
denticulata215
afftriph141
caperono131
pernam88
scande 186
scande41
nayaritensis213
ipomo124
affipo179
ipomo123
ipomo121
aff bern146
bern denes73
cttamitol57 57 1
clemat116
viridi13
decary07
supternata207
cf chloro08
subter80
chloro155
clemat06
tamit83c10r 83
tamit83c10d 83
hilde78 C3
hilde78 C1
bidentata 190
elong72
eparvl79
tritol95
ruaha76
capens02
volubi81 C2
volubi81 C1
galpinii159
previ125H
previ111
tilif62
affinis23
armbrust153
brownsb126
coriacea203
affticit200
affticif171
ficitoli170 C2
viridissima167c6
convolvul 208
magnisti 103
arenal 03
schipp33
Trag volub66a
Plukvolub58

0.80 0.70 0.60 0.50 0.40 0.30 0.20 0.10 0.0

Fig 3.3 Bayesian estimate of *Dalechampia* (Euphorbiaceae) phylogeny, showing evolution of pollinator rewards. The branch lengths are proportional to divergence in the ITS sequence. See Armbruster et al. (2009) for estimation details. Pollinator rewards are mapped onto the tree using parsimony, trace-across-all-trees, and MPRs modes in Mesquite (Maddison and Maddison, 2009). The width of each color on each branch indicates the proportion reconstructions with that character state, across 35 trees sampled from the posterior distribution of 5×10^6 retained trees. The proportion of trees lacking that branch and node is indicated in red.

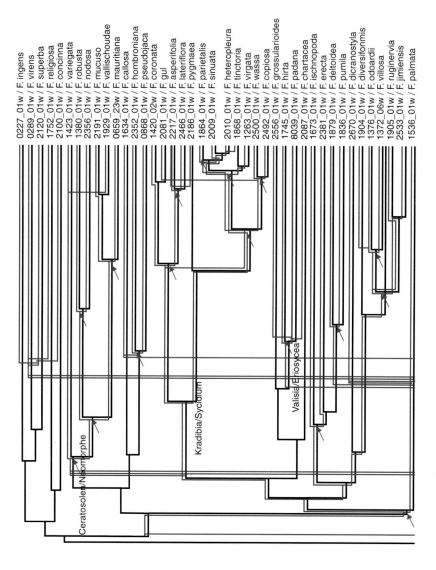

Fig 4.4 Hypothetical codivergence scenario inferred by Jane. Arrows indicate putative cospeciation event.

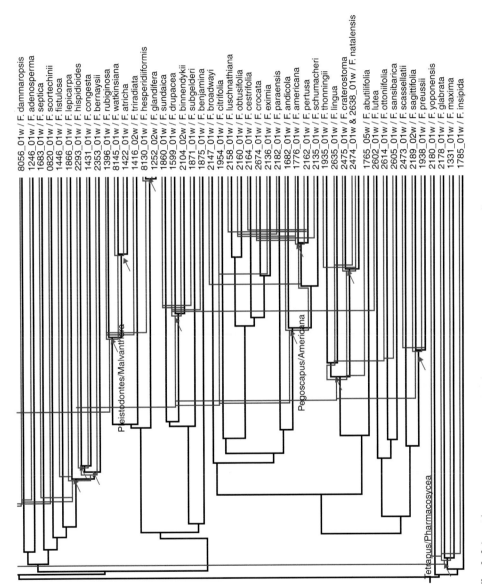

Fig 4.4 (cont.)

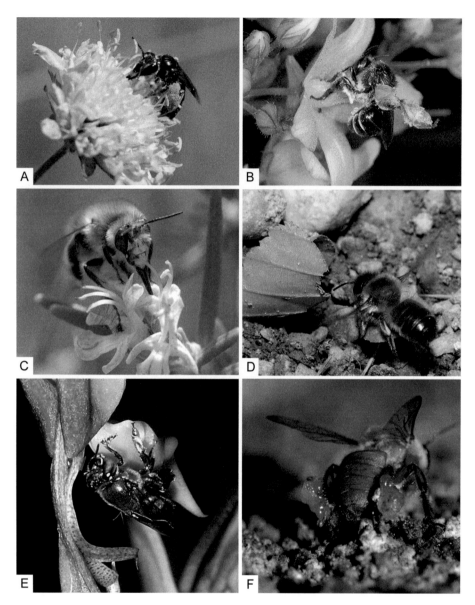

Fig 5.1 Modern bee–plant interactions. A. *Andrena hattorfiana* female foraging
on pollen of *Scabiosa* sp. (Schrophulariaceae) (picture Edith Tempez). B. *Macropis
europaea* female foraging on oil and pollen of *Lysimachia vulgaris* (Myrsinaceae)
(picture Yvan Barbier). C. *Anthophora plumipes* male foraging on nectar of Lamiaceae
(picture Jean-Marc Michalowski). D. *Megachile circumcincta* with peace of leaf
for cell linning (picture Nicolas J. Vereecken). E. Male of orchid bee collecting
fragrances (picture Günter Gerlach). F. Worker of stingless bee *Melipona* cf. *rufiventris*
(Meliponini) carrying resin (picture Claus Rasmussen).

Fig 7.2 Color divergence in *Disa ferruginea*, which mimics different color models in the southern and eastern parts of its range. Models and mimics are visited by the butterfly *Aeropetes tulbachia* in all parts of the range. Photos by Johnson and Anderson.

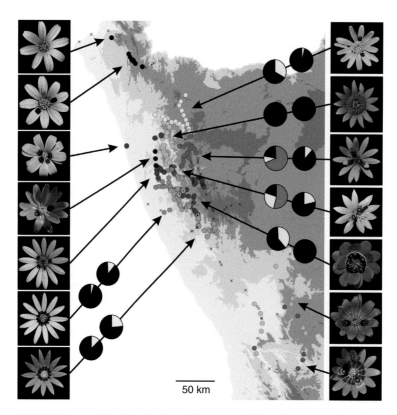

Fig 7.3 Geographic distribution of *Gorteria diffusa* floral morphotypes all visited predominantly by the bee fly *Megapalpus capensis*. Pie charts show the frequency of behaviors (feeding: black; short inspection visits: yellow; copulation attempts: orange) exhibited by female (near the base of the arrow) and male flies on each floral morphotype. Data redrawn and modified from Ellis and Johnson 2009, 2010a.

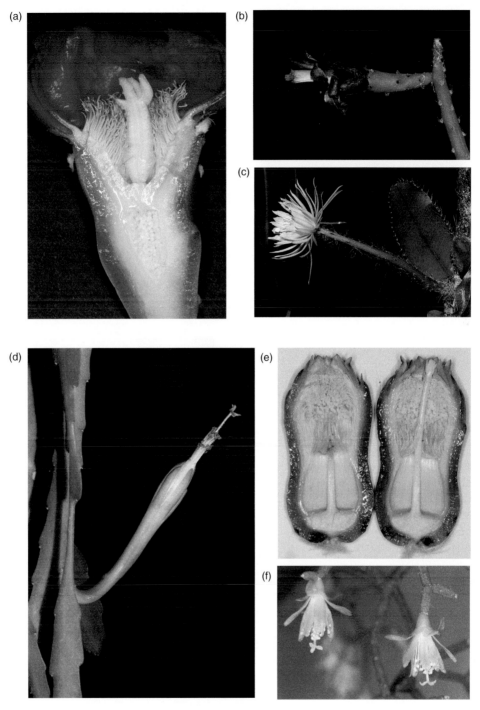

Fig 10.2 A. *Opuntia quimilo*, male-sterile flower: incipient ornithophily. B. *Tacinga funalis*, hummingbird flower without hypanthium. C. *Selenicereus wittii*, sphingid flower with elongated hypanthium. D. *Disocactus quetzaltecus*, hummingbird flower with elongated, colorful hypanthium. E. *Pilosocereus pachycladus* ssp. *pernambucensis*, bat flower with hypanthium, forming a massive nectar chamber. F. *Rhipsalis cereuscula*, member of the Cactoideae lacking a hypanthium.

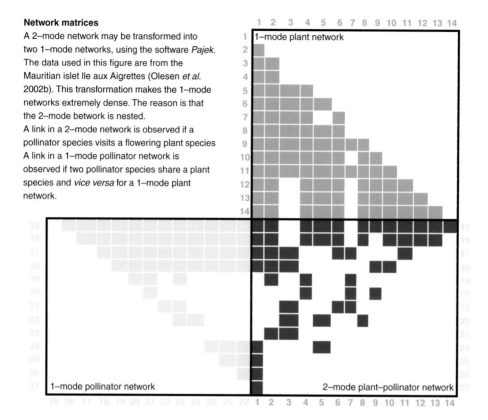

Network matrices

A 2–mode network may be transformed into two 1–mode networks, using the software *Pajek*. The data used in this figure are from the Mauritian islet Ile aux Aigrettes (Olesen *et al.* 2002b). This transformation makes the 1–mode networks extremely dense. The reason is that the 2–mode betwork is nested.

A link in a 2–mode network is observed if a pollinator species visits a flowering plant species A link in a 1–mode pollinator network is observed if two pollinator species share a plant species and *vice versa* for a 1–mode plant network.

Fig 13.1 Descriptive network parameters: P, no. plant species in the study network; A, no. pollinator species in the study network. AP, matrix size or total no. potential links in the 2-mode network. $0.5P(P\text{-}1)$, total no. potential links in the 1-mode pollinator network. I, total no. observed links in the network. C, 2-mode network connectance = $I/(AP)$ or 1-mode network connectance = $I/(0.5A(A\text{-}1))$ L_1, linkage level of species 1 = 1 link, i.e. no. links from species 1 to other species. $<L_A>$, average linkage level for all pollinator species in either 1- or 2-mode networks. The following two parameters are only calculated for 1-mode networks. $l_{1\text{-}2}$, shortest path length between species 1 and 2, i.e. no. links between the two species. $<l>$, average shortest path length among all species. c_3, clustering coefficient of species 3, i.e. link density among neighbors of species 3. Species 3 interacts with three species. Max. no. links among these species is 3. Observed no. links among these three species is 1 (between species 2 and 4). c = observed no. links among neighbors/potential no. links among neighbors. NB. If a species has only one link, it has no clustering coefficient. The software *Pajek* calculates these parameters.

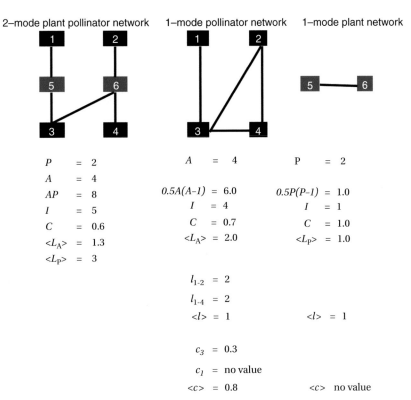

2–mode plant pollinator network   1–mode pollinator network   1–mode plant network

P = 2	A = 4	P = 2
A = 4		
AP = 8	$0.5A(A–1)$ = 6.0	$0.5P(P–1)$ = 1.0
I = 5	I = 4	I = 1
C = 0.6	C = 0.7	C = 1.0
$\langle L_A \rangle$ = 1.3	$\langle L_A \rangle$ = 2.0	$\langle L_p \rangle$ = 1.0
$\langle L_P \rangle$ = 3		

$l_{1\text{-}2}$ = 2

$l_{1\text{-}4}$ = 2

$\langle l \rangle$ = 1 $\langle l \rangle$ = 1

c_3 = 0.3

c_1 = no value

$\langle c \rangle$ = 0.8 $\langle c \rangle$ no value

Fig 13.1 (*cont.*)

Fig 14.2 Proportion of bilabiate (top) and disk blossoms (bottom) along the first axis of an environmental PCA corresponding to elevation. To better visualize the pattern, a boxplot of the vegetation plots belonging to each interval with a length of 1 ranging from –4 to 4 in the first PCA axis were used instead of drawing points. The dashed line represents the GLMs applied independently on the proportion of the two blossom types with the first axis of the environmental PCA as explaining factor (see Pellissier et al. 2010c, pictures: L. Pellissier).

materials. Any flower-visiting species likely exhibits multiple behaviors on the flowers of a particular plant species and the floral cues that elicit (or prevent) each behavioral response might differ.

Importantly different behaviors within the same pollinator can impose contrasting selection on floral traits. For example, whereas nectar feeding and pollen-collecting behavior in a bee species might contribute to plant seed set equivalently, they are likely to have contrasting effects on fitness through male function because a large portion of the pollen removed by pollen-collecting bees is not available for export. Thus, selection might favor traits that attract nectar-feeding individuals but deter pollen-collecting ones, such as hidden, cryptic, or chemically defended pollen (Lunau 2000; Hargreaves et al. 2009). Alternatively, under some circumstances (e.g. pollen limited seed set), the female fitness benefit of attracting more visitors (i.e. pollen collectors) might outweigh the male function cost, imposing selection for traits that promote visitation by pollen collectors without deterring nectar feeders, such as increased quantities and visibility of pollen.

In the beetle daisy, *Gorteria diffusa*, many allopatric floral variants exist that all provide nectar and pollen rewards to their bee–fly pollinators, but which differ in the degree to which they elicit mate-seeking and even copulatory responses from male flies (Fig 7.3, Ellis and Johnson 2009; 2010a). Traits that elicit copulation responses by flies enhance pollen export (Ellis and Johnson 2010a), but seed set is not significantly influenced by visitation from mate-seeking flies, presumably because visits from feeding flies are adequate to set all fruit (Johnson and Midgley 1997). Intriguingly, sexually deceptive floral forms are less attractive to female food-seeking flies (Ellis and Johnson 2010a), and male and female flies have contrasting preferences for components of floral phenotype (De Jager and Ellis, unpublished data). Thus, selection on floral traits by male (mate-seeking) and female (food-seeking) flies likely differs, with male-attracting traits being favored through the male component of plant fitness (i.e. pollen export). Ne'eman et al. (2006) found similar gender differences in pollination efficiency and potential for pollen export in solitary bees, and Alarcon et al. (2010) demonstrated gender differences in flower preference in hawk moths. Many flower-visiting insects (and other animals) exhibit gender dimorphism in a range of traits that might influence their interactions with flowers, including body size (e.g. Fischer and Fiedler 2000) and visual pigments (e.g. Arikawa et al. 2005), suggesting that differences in selection imposed by male and female individuals of the same pollinating species might be common (see Temeles and Kress 2003).

Contrasting selection by different behaviors or by differences between male and female insects could generate geographic variation in floral traits when coupled with mechanisms generating spatial variation in the strength of selection operating through alternate plant gender pathways. This could result directly from variation in plant community context (Temeles and Kress 2003). For example, the presence (or not) of a congeneric plant preferred as a mating site by a pollinating

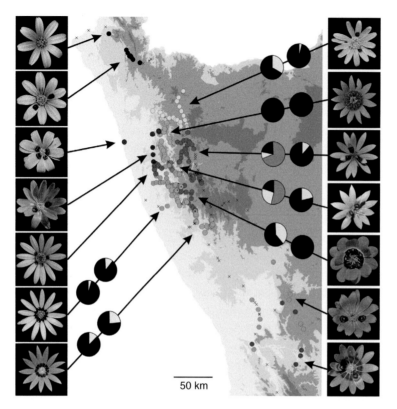

Fig 7.3 Geographic distribution of *Gorteria diffusa* floral morphotypes all visited predominantly by the bee fly *Megapalpus capensis*. Pie charts show the frequency of behaviors (feeding: black; short inspection visits: yellow; copulation attempts: orange) exhibited by female (near the base of the arrow) and male flies on each floral morphotype. Data redrawn and modified from Ellis and Johnson 2009, 2010a. See plate section for color version.

species, would dictate the strength of selection imposed by the mate-seeking sensory biases of male insects on floral traits of the focal plant species. As mentioned above, another important source of spatial variation in strengths of contrasting selection pressures is pollen limitation (Harder and Aizen, 2010). The degree of pollen limitation is dictated by the availability and abundance of pollinators, by the abiotic environment (i.e. resource limitation), and by the plant community context, which affects both the quantity and quality of pollen transfer (Caruso et al. 2005; Aizen and Harder 2007; Harder and Aizen 2010). Pollen limitation influences the intensity (and sometimes direction) of selection on floral traits, but can also influence the balance of selection operating through alternate gender functions (i.e. seeds set or sired) in hermaphrodite plants (Morgan 1992; Ashman and Morgan 2004; Lankinen and Larsson 2009; Sandring and Agren 2009; Harder and Aizen 2010). In the *Gorteria* system for instance, exploitation of mate-seeking male flies improves pollen export relative to feeding female flies but seed set is not influenced (Ellis and Johnson 2010a). We would thus expect to see strong selection for traits eliciting copulation only when selection operating through male function is strong relative to that operating through female function, a situation which is most likely when seed set is not pollen limited. Persistent spatial variation in the degree of pollen limitation, coupled with contrasting selection through plant gender functions imposed by different pollinator behaviors or genders, could lead to geographically structured floral variation. No studies that we are aware of have tested this mechanism, although many have demonstrated substantial spatial variation in pollen limitation of plant reproduction.

7.2.6 Gender specific selection in plants

Contrasting selection on floral traits through alternate plant gender functions does not require different pollinator behaviors. Inherent differences in the opportunity for and intensity of selection on floral traits result from asymmetry in the reproductive capacities of the male and female plant fitness pathways (Bateman 1948). For example, no selection on attractive floral traits can occur through female function if seed production is limited by the availability of resources or ovules and not pollinator visits. In contrast, male function should continue benefiting from attracting more pollinators until all pollen is removed. Under pollen limitation, both gender functions benefit from increased pollinator attraction, thus potentially intensifying selection on attractive floral traits, or alternatively traits facilitating selfing might evolve to alleviate pollen limitation (Ashman and Morgan 2004; Fishman and Willis 2008; Harder and Aizen 2010). As a result, persistent spatial variation in pollen limitation could generate varying intensities of selection on traits through gender functions, and ultimately geographically structured floral variation. Population divergence would be strongest when selection on floral traits through alternate sex roles is in conflict (Lankinen and Larsson 2009). Morgan (1992) in his "gender-balance hypothesis" suggested that floral traits might often represent compromises between optimization of female and male fitness. Instead, the direction of selection on most traits doesn't seem to vary between gender functions (Ashman and Morgan 2004; Delph and Ashman 2006; Harder and Johnson 2009), although most studies use pollen removal as a proxy of male fitness, which may often only be weakly related to seed siring success (Ellis and Johnson 2010b).

In Darwin's classic description of coevolution of moths' proboscides and orchid spurs described previously, increased elongation of the plant nectar spurs is favored by both plant genders. Flowers with spurs longer than moth proboscides will both successfully receive pollen and place it on the head of the moth for export, whereas flowers with spurs shorter than moth proboscides will neither receive nor export pollen (Fig 7.4). Under this scenario, variation in pollen limitation between

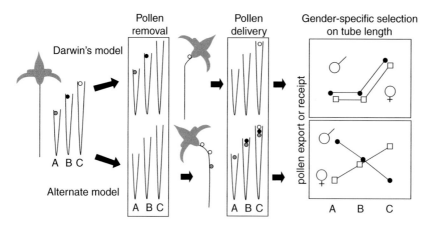

Fig 7.4 Schematic demonstrating the effects of pollen placement position on gender-specific selection on nectar tube length. Pollen placement on the head (i.e. at maximum proboscis extension) results in Darwin's classical model of tube length coevolution because the longest tubed flowers both export and receive most pollen. However, if pollen is placed on the proboscis (the alternate model) short-tubed flowers may have an export advantage over longer tubed ones, but will always have lower relative fitness through female function. Floral tubes of variable lengths are represented by A, B and C. Model redrawn and modified from Ellis and Johnson 2010b.

populations (and thus the intensity of selection through female function) might simply affect the rate at which the coevolutionary race proceeds in each population. However, as Ellis and Johnson (2010b) point out, if pollen placement occurs on the proboscis and not the head of the moth, placement will occur independently of the match between proboscis and spurs (within the limits imposed by the proboscis morphology). In fact, pollen from short-spurred flowers will potentially have an export advantage over that from longer spurred flowers because it will contact stigmas of a greater proportion of the population (i.e. all flowers with spurs equal to or longer than its own), and its position lower on the proboscis reduces the potential for burial. In contrast, potential female fitness is still highest in the longest spurred flowers, which can receive pollen from flowers of any spur length, resulting in conflicting selection on nectar spurs through gender functions (Ellis and Johnson 2010b). Under this scenario, variation in pollen limitation between populations might switch the balance between selection operating through alternate genders imposing variable contraints on the endpoints of coevolution. Interestingly, the few studies that have measured selection using equivalent estimates of both male and female fitness have found contrasting patterns of selection (see references in Conner 2006). Contrasting or asymmetric gender-specific selection coupled with ubiquitous geographically variable pollination contexts could thus commonly generate substantial variation in floral traits (Lankinen and Larsson 2009; Harder and Aizen 2010; Ellis and Johnson 2010b).

7.3 Concluding remarks

In this chapter, we have outlined a series of mechanisms by which substantial floral divergence can accrue between plant populations using the same pollinating vector. Many recent papers have cautioned against a pollinator-biased perspective on floral evolution (Galen 1999, 2000; Galen and Cuba 2001; Gomez 2003; Strauss and Whittall 2006; Rausher 2008). Here, we caution against dismissing pollinator-mediated selection as the driver of floral divergence when no differences in pollinator identity can be detected between plant populations. Herrera et al. (2006) found that in 20 % of published studies that found variation in floral traits, there was no significant geographic difference in pollinator communities. These were taken to be cases in which non-pollinator agents determine floral divergence, but as we have shown here, in many of these cases divergence could have resulted due to variation in selection imposed by the same pollinating species (or assemblage). Thus Herrera et al.'s (2006) estimate that pollinators account for floral trait divergence in 80 % of systems investigated could well be an underestimate.

Because the mechanisms we review here have received so little empirical attention, it is impossible at this stage to gauge their importance relative to the dominant pollinator shift paradigm. In all cases though, the selective pressures driving floral phenotype evolution within populations have been established empirically, and the spatial variation in pollinator traits, plant community context, and pollen limitation required to translate this into population divergence is ubiquitous. Also plant community context is confounded with the pollinator community context, and both in turn influence the degree of pollen limitation a plant population encounters. The view that floral divergence resulting from shifts in flower visiting assemblages (especially when differences are quantitative and not qualitative), results from a switch in selective agent from one effective pollinator to another, may be too simplistic.

Perhaps the largest unknown, for the mechanisms we propose and for the pollinator shift paradigm, is the temporal stability of the biotic community context. Feinsinger (1987) and Waser et al. (1996) suggest that the context is too transient in an evolutionary sense to allow adaptation to any single pollinating species, and should instead favor traits that promote generalization. In some environments, at least, this is not the case, as illustrated by the numerous examples of pollination ecotypes that have evolved within species and attest to the stability of the geographical mosaic of pollination (e.g. Johnson 1997). Or alternatively, floral divergence could reflect periods of strong consistent selection on floral traits interspersed with long periods of weak or inconsistent selection (Harder and Johnson 2009).

Either way, we need more information on the nature and stability of geographic pollination mosaics. One approach is to examine areas of contrasting landscape and evolutionary stability. Do more stable environments support higher levels of intraspecific floral variation or a higher incidence of variation without pollinator shifts by the mechanisms discussed here? Johnson and Steiner (2000) suggest that stability might influence levels of pollinator specialization, which in turn might influence plant speciation. Interestingly, the accumulation of diversity itself strongly influences the plant (and pollinator) community context and perhaps the extent of pollen limitation (Vamosi et al. 2006; Armbruster and Muchhala 2009; Harder and Aizen 2010) – all factors which underlie the generation of further floral diversity by the mechanisms we have discussed to establish a positive feedback loop by which diversity begets diversity.

Acknowledgments

We thank the South African National Research Foundation, Stellenbosch University and the University of Kwazulu-Natal for funding much of the research we refer to in this chapter, and an anonymous reviewer for comments on the manuscript.

References

Aizen, M. A. and Harder, L. D. (2007). Expanding the limits of the pollen-limitation concept: Effects of pollen quantity and quality. *Ecology*, **88**, 271–81.

Alarcon, R., Riffell, J. A., Davidowitz, G., Hildebrand, J. G. and Bronstein, J. L. (2010). Sex-dependent variation in the floral preferences of the hawkmoth *Manduca sexta*. *Animal Behaviour*, **80**, 289–96.

Aldridge, G. and Campbell, D. R. (2007). Variation in pollinator preference between two *Ipomopsis* contact sites that differ in hybridization rate. *Evolution*, **61**, 99–110.

Alexandersson, R. and Johnson, S. D. (2002). Pollinator-mediated selection on flower-tube length in a hawkmoth-pollinated *Gladiolus* (Iridaceae). *Proceedings of the Royal Society of London Series B-Biological Sciences*, **269**, 631–6.

Anderson, B. and Johnson, S. D. (2008). The geographical mosaic of coevolution in a plant–pollinator mutualism. *Evolution*, **62**, 220–5.

Anderson, B. and Johnson, S. D. (2009). Geographical covariation and local convergence of flower depth in a guild of fly-pollinated plants. *New Phytologist*, **182**, 533–40.

Anderson, B., Alexandersson, R. and Johnson, S. D. (2010a). Evolution and coexistence of pollination ecotypes in an African *Gladiolus* (Iridaceae). *Evolution*, **64**, 960–72.

Anderson, B., Johnson, S. D. and Carbutt, C. (2005). Exploitation of a specialized mutualism by a deceptive orchid. *American Journal of Botany*, **92**, 1342–9.

Anderson, B., Terblanche, J. S. and Ellis, A. G. (2010b). Predictable patterns of trait mismatches between interacting plants and insects. *BMC Evolutionary Biology*, **10**, 204.

Arikawa, K., Wakakuwa, M., Qiu, X. D., Kurasawa, M. and Stavenga, D. G. (2005). Sexual dimorphism of short-wavelength photoreceptors in the small white butterfly, *Pieris rapae crucivora*. *Journal of Neuroscience*, **25**, 5935–42.

Armbruster, W. S. (1985). Patterns of character divergence and the evolution of reproductive ecotypes of *Dalechampia scandens* (Euphorbiaceae). *Evolution*, **39**, 733–52.

Armbruster, W. S. and Muchhala, N. (2009). Associations between floral specialization and species diversity: cause, effect, or correlation? *Evolutionary Ecology*, **23**, 159–79.

Armbruster, W. S., Edwards, M. E. and Debevec, E. M. (1994). Floral character displacement generates assemblage structure of Western Australian triggerplants (*Stylidium*). *Ecology*, **75**, 315–29.

Ashman, T. L. and Majetic, C. J. (2006). Genetic constraints on floral evolution: a review and evaluation of patterns. *Heredity*, **96**, 343–52.

Ashman, T. L. and Morgan, M. T. (2004). Explaining phenotypic selection on plant attractive characters: male function, gender balance or ecological context? *Proceedings of the Royal Society of London Series B, Biological Sciences*, **271**, 3–559.

Bascompte, J., Jordano, P. and Olesen, J. M. (2006). Asymmetric coevolutionary networks facilitate biodiversity maintenance. *Science*, **312**, 431–3.

Bateman, A. J. (1948). Intra-sexual selection in *Drosophila*. *Heredity*, **23**, 349–68.

Beardsley, P. M., Yen, A. and Olmstead, R. G. (2003). AFLP phylogeny of *Mimulus* section Erythranthe and the evolution of hummingbird pollination. *Evolution*, **57**, 1397–410.

Benitez-Vieyra, S., de Ibarra, N. H., Wertlen, A. M. and Cocucci, A. A. (2007). How to look like a mallow: evidence of floral mimicry between Turneraceae and Malvaceae. *Proceedings of the Royal Society B, Biological Sciences*, **274**, 2239–48.

Brown, J. H. and Kodric-Brown, A. (1979). Convergence, competition, and mimicry in a temperate community of hummingbird-pollinated flowers. *Ecology*, **60**, 1022–35.

Campbell, D. R. and Motten, A. F. (1985). The mechanism of competition for pollination between two forest herbs. *Ecology*, **66**, 554–65.

Cariveau, D., Irwin, R. E., Brody, A. K., Garcia-Mayeya, L. S. and von der Ohe, A. (2004). Direct and indirect effects of pollinators and seed predators to selection on plant and floral traits. *Oikos*, **104**, 15–26.

Carlson, J. E. and Holsinger, K. E. (2010). Natural selection on inflorescence color polymorphisms in wild *Protea* populations: The role of pollinators, seed predators, and intertrait correlations. *American Journal of Botany*, **97**, 934–44.

Caruso, C. M. (2000). Competition for pollination influences selection on floral traits of *Ipomopsis aggregata*. *Evolution*, **54**, 1546–57.

Caruso, C. M. (2006). Plasticity of inflorescence traits in *Lobelia siphilitica* (Lobeliaceae) in response to soil water availability. *American Journal of Botany*, **93**, 531–8.

Caruso, C. M., Remington, D. L. D. and Ostergren, K. E. (2005). Variation in resource limitation of plant reproduction influences natural selection on floral traits of *Asclepias syriaca*. *Oecologia*, **146**, 68–76.

Chittka, L., Gumbert, A. and Kunze, J. (1997). Foraging dynamics of bumble bees: correlates of movements within and between plant species. *Behavioral Ecology*, **8**, 239–49.

Chittka, L., Ings, T. C. and Raine, N. E. (2004). Chance and adaptation in the evolution of island bumblebee behaviour. *Population Ecology*, **46**, 243–51.

Chown, S. L. and Gaston, K. J. (1999). Exploring links between physiology and ecology at macroscales: the role of respiratory metabolism in insects. *Biological Reviews*, **74**, 87–120.

Chown, S. L. and Gaston, K. J. (2010). Body size variation in insects: a macroecological perspective. *Biological Reviews*, **85**, 139–69.

Combs, J. K. and Pauw, A. (2009). Preliminary evidence that the long-proboscid fly, *Philoliche gulosa*, pollinates *Disa karooica* and its proposed Batesian model *Pelargonium stipulaceum*. *South African Journal of Botany*, **75**, 757–61.

Conner, J. K. (2006). Ecological genetics of floral evolution. In *Ecology and Evolution of Flowers*, ed. L. D. Harder and S. C. H. Barrett. Oxford, UK: Oxford University Press, pp. 260–274.

Crepet, W. L. (1984). Advanced (constant) insect pollination mechanisms: pattern of evolution and implications vis-à-vis angiosperm diversity. *Annals of the Missouri Botanical Garden*, **71**, 607–30.

Dafni, A. (1984). Mimicry and deception in pollination. *Annual Review of Ecology and Systematics*, **15**, 259–78.

Darwin, C. R. (1862). *On the Various Contrivances by which British and Foreign Orchids Are Fertilized by Insects*. London, UK: John Murray.

De Jager, M. L., Dreyer, L. D. and Ellis, A. G. (2010). Do pollinators influence the assembly of flower colours within plant communities? *Oecologia*, in press.

Delph, L. F. and Ashman, T. L. (2006). Trait selection in flowering plants: how does sexual selection contribute? *Integrative and Comparative Biology*, **46**, 465–72.

Ehrlen, J., Kack, S. and Agren, J. (2002). Pollen limitation, seed predation and scape length in *Primula farinosa*. *Oikos*, **97**, 45–51.

Ellis, A. G. and Johnson, S. D. (2009). The evolution of floral variation without pollinator shifts in *Gorteria diffusa* (Asteraceae). *American Journal of Botany*, **96**, 793–801.

Ellis, A. G. and Johnson, S. D. (2010a). Floral mimicry enhances pollen export: the evolution of pollination by sexual deceit outside of the Orchidaceae. *American Naturalist*, **176**, E143–E151.

Ellis, A. G. and Johnson, S. D. (2010b). Gender differences in in the effects of floral spur length manipulation on fitness in a hermaphrodite orchid. *International Journal of Plant Sciences*, **171**, 1010–19.

Eriksson, O. and Bremer, B. (1992). Pollination systems, dispersal modes, life forms, and diversification rates in Angiosperm families. *Evolution*, **46**, 258–66.

Feinsinger, P. (1978). Ecological interactions between plants and hummingbirds in a successional tropical community. *Ecological Monographs*, **48**, 269–87.

Feinsinger, P. (1987). Effects of plant species on each other's pollination: is community structure influenced? *Trends in Ecology and Evolution*, **2**, 123–6.

Fenster, C. B., Armbruster, W. S., Wilson, P., Dudash, M. R. and Thomson, J. D. (2004). Pollination syndromes and floral specialization. *Annual Review of Ecology Evolution and Systematics*, **35**, 375–403.

Fischer, K. and Fiedler, K. (2000). Sex-related differences in reaction norms in the butterfly *Lycaena tityrus* (Lepidoptera:Lycaenidae). *Oikos*, **90**, 372–80.

Fishman, L. and Willis, J. H. (2008). Pollen limitation and natural selection on floral characters in the yellow monkeyflower, *Mimulus guttatus*. *New Phytologist*, **177**, 802–10.

Fishman, L. and Wyatt, R. (1999). Pollinator-mediated competition, reproductive character displacement, and the evolution of selfing in *Arenaria uniflora* (Caryophyllaceae). *Evolution*, **53**, 1723–33.

Forrest, J. and Thomson, J. D. (2009). Background complexity affects colour preference in bumblebees. *Naturwissenschaften*, **96**, 921–25.

Galen, C. (1999). Why do flowers vary? The functional ecology of variation in flower size and form within natural plant populations. *Bioscience*, **49**, 631–40.

Galen, C. (2000). High and dry: Drought stress, sex-allocation trade-offs, and selection on flower size in the alpine wildflower *Polemonium viscosum* (Polemoniaceae). *American Naturalist*, **156**, 72–83.

Galen, C. and Cuba, J. (2001). Down the tube: Pollinators, predators, and the evolution of flower shape in the Alpine skypilot, *Polemonium viscosum*. *Evolution*, **55**, 1963–71.

Galen, C., Zimmer, K. A. and Newport, M. E. (1987). Pollination in floral scent morphs of *Polemonium viscosum* – a mechanism for disruptive selection on flower size. *Evolution*, **41**, 599–606.

Geber, M. A. and Moeller, D. A. (2006). Pollinator responses to plant communities and implications for reproductive character evolution. In *Ecology and Evolution of Flowers*, ed. L. D. Harder and S. C. H. Barrett. Oxford, UK: Oxford University Press, pp. 102–19.

Ghazoul, J. (2006). Floral diversity and the facilitation of pollination. *Journal of Ecology*, **94**, 295–304.

Gigord, L. D. B., Macnair, M. R. and Smithson, A. (2001). Negative frequency-dependent selection maintains a dramatic flower color polymorphism in the rewardless orchid *Dactylorhiza sambucina* (L.) Soo. *Proceedings of the National Academy of Sciences USA*, **98**, 6253–5.

Giurfa, M., Núñez, J., Chittka, L. and Menzel, R. (1995). Colour preferences of flower-naive honeybees. *Journal of Comparative Physiology A (Sensory, Neural, and Behavioral Physiology)*, **177**, 247–59.

Gomez, J. M. (2003). Herbivory reduces the strength of pollinator-mediated selection in the Mediterranean herb *Erysimum mediohispanicum*: Consequences for plant specialization. *American Naturalist*, **162**, 242–56.

Gomez, J. M., Perfectti, F., Bosch, J. and Camacho, J. P. M. (2009). A geographic selection mosaic in a generalized plant–pollinator–herbivore system. *Ecological Monographs*, **79**, 245–63.

Grant, V. (1949). Pollination systems as isolating mechanisms in angiosperms. *Evolution*, **3**, 82–97.

Grant, V. and Grant, K. A. (1965). *Flower pollination in the Phlox family*, Columbia.

Grimaldi, D. (1999). The coradiations of pollinating insects and angiosperms in the Cretaceous. *Annals of the Missouri Botanical Garden*, **86**, 373–406.

Harder, L. D. (1983). Flower handling effeciency of bumble bees:

morphological aspects of probing time. *Oecologia*, **57**, 274–80.

Harder, L. D. and Aizen, M. A. (2010). Floral adaptation and diversification under pollen limitation. *Philosophical Transactions of the Royal Society B-Biological Sciences*, **365**, 529–43.

Harder, L. D. and Johnson, S. D. (2009). Darwin's beautiful contrivances: evolutionary and functional evidence for floral adaptation. *New Phytologist*, **183**, 530–45.

Hargreaves, A. L., Harder, L. D. and Johnson, S. D. (2009). Consumptive emasculation: the ecological and evolutionary consequences of pollen theft. *Biological Reviews*, **84**, 259–76.

Herrera, C. M., Castellanos, M. C. and Medrano, M. (2006). Geographical context of floral evolution: towards an improved research programme in floral diversification. In *Ecology and Evolution of Flowers*, ed. L. D. Harder and S. C. H. Barrett. Oxford, UK: Oxford University Press, pp. 278–294.

Hersch, E. I. and Roy, B. A. (2007). Context-dependent pollinator behavior: an explanation for patterns of hybridization among three species of Indian paintbrush. *Evolution*, **61**, 111–24.

Hodges, S. A. and Arnold, M. L. (1994). Floral and ecological isolation between *Aquilegia formosa* and *Aquilegia pubescens*. *Proceedings of the National Academy of Sciences USA*, **91**, 2493–6.

Ings, T. C., Raine, N. E. and Chittka, L. (2009). A population comparison of the strength and persistence of innate colour preference and learning speed in the bumblebee *Bombus terrestris*. *Behavioral Ecology and Sociobiology*, **63**, 1207–18.

Johnson, S. D. (1994). Evidence for Batesian mimicry in a butterfly-pollinated orchid. *Biological Journal of the Linnean Society*, **53**, 91–104.

Johnson, S. D. (1997). Pollination ecotypes of *Satyrium hallackii* (Orchidaceae) in South Africa. *Botanical Journal of the Linnean Society*, **123**, 225–35.

Johnson, S. D. (2006). Pollinator-driven speciation in plants. In *Ecology and Evolution of Flowers*, ed. L. D. Harder and S. C. H. Barrett. Oxford, UK: Oxford University Press, pp. 295–310.

Johnson, S. D. and Anderson, B. (2010). Coevolution between food-rewarding flowers and their pollinators. *Evolution Education Outreach*, **3**, 32–9.

Johnson, S. D. and Midgley, J. J. (1997). Fly pollination of *Gorteria diffusa* (Asteraceae), and a possible mimetic function for dark spots on the capitulum. *American Journal of Botany*, **84**, 429–36.

Johnson, S. D. and Morita, S. (2006). Lying to Pinocchio: floral deception in an orchid pollinated by long-proboscid flies. *Botanical Journal of the Linnean Society*, **152**, 271–8.

Johnson, S. D. and Steiner, K. E. (1997). Long-tongued fly pollination and evolution of floral spur length in the *Disa draconis* complex (Orchidaceae). *Evolution*, **51**, 45–53.

Johnson, S. D. and Steiner, K. E. (2000). Generalisation versus specialisation in plant pollination systems. *Trends in Ecology and Evolution*, **15**, 140–3.

Johnson, S. D., Linder, H. P. and Steiner, K. E. (1998). Phylogeny and radiation of pollination systems in *Disa* (Orchidaceae). *American Journal of Botany*, **85**, 402–11.

Johnson, S. D., Peter, C. I., Nilsson, L. A. and Agren, J. (2003). Pollination success in a deceptive orchid is enhanced by co-occurring rewarding magnet plants. *Ecology*, **84**, 2919–27.

Jordano, P. (1987). Patterns of mutualistic interactions in pollination and seed dispersal: connectance, dependence asymmetries and coevolution. *The American Naturalist*, **129**, 657–77.

Kay, K. M. and Sargent, R. D. (2009). The role of animal pollination in plant speciation: integrating ecology, geography, and genetics. *Annual Review of Ecology Evolution and Systematics*, **40**, 637–56.

Kunte, K. (2007). Allometry and functional constraints on proboscis lengths in butterflies. *Functional Ecology*, **21**, 982–7.

Laine, A. L. (2009). Role of coevolution in generating biological diversity: spatially divergent selection trajectories. *Journal of Experimental Botany*, **60**, 2957–70.

Lankinen, A. and Larsson, M. C. (2009). Conflicting selection pressures on reproductive functions and speciation in plants. *Evolutionary Ecology*, **23**, 147–57.

Laverty, T. M. (1992). Plant interactions for pollinator visits: a test of the magnet species effect. *Oecologia*, **89**, 502–8.

Lunau, K. (2000). The ecology and evolution of visual pollen signals. *Plant Systematics and Evolution*, **222**, 89–111.

Maad, J. (2000). Phenotypic selection in hawkmoth-pollinated *Platanthera bifolia*: Targets and fitness surfaces. *Evolution*, **54**, 112–23.

Macior, L. W. (1971). Coevolution of plants and animals: systematic insights from plant-insect interactions. *Taxon*, **20**, 17–28.

Malo, J. E. and Baonza, J. (2002). Are there predictable clines in plant-pollinator interactions along altitudinal gradients? The example of *Cytisus scoparius* (L.) link in the Sierra de Guadarrama (Central Spain). *Diversity and Distributions*, **8**, 365–71.

Mitchell, R. J., Flanagan, R. J., Brown, B. J., Waser, N. M. and Karron, J. D. (2009). New frontiers in competition for pollination. *Annals of Botany*, **103**, 1403–13.

Moeller, D. A. (2004). Facilitative interactions among plants via shared pollinators. *Ecology*, **85**, 3289–301.

Moeller, D. A. (2005). Pollinator community structure and sources of spatial variation in plant–pollinator interactions in *Clarkia xantiana* ssp xantiana. *Oecologia*, **142**, 28–37.

Moeller, D. A. (2006). Geographic structure of pollinator communities, reproductive assurance, and the evolution of self-pollination. *Ecology*, **87**, 1510–1522.

Morales, C. L. and Traveset, A. (2008). Interspecific pollen transfer: magnitude, prevalence and consequences for plant fitness. *Critical Reviews in Plant Sciences*, **27**, 221–38.

Morgan, M. T. (1992). The evolution of traits influencing male and female fertility in outcrossing plants. *American Naturalist*, **139**, 1002–51.

Muchhala, N. and Potts, M. D. (2007). Character displacement among bat-pollinated flowers of the genus *Burmeistera*: analysis of mechanism, process and pattern. *Proceedings of the Royal Society B, Biological Sciences*, **274**, 2731–7.

Muchhala, N. and Thomson, J. D. (2009). Going to great lengths: selection for long corolla tubes in an extremely specialized bat-flower mutualism. *Proceedings of the Royal Society B, Biological Sciences*, **276**, 2147–52.

Ne'eman, G., Shavit, O., Shaltiel, L. and Shmida, A. (2006). Foraging by male and female solitary bees with implications for pollination. *Journal of Insect Behavior*, **19**, 383–401.

Nilsson, L. A. (1988). The evolution of flowers with long corolla tubes. *Nature*, **334**, 147–9.

Okuyama, Y., Pellmyr, O. and Kato, M. (2008). Parallel floral adaptations to pollination by fungus gnats within the genus *Mitella* (Saxifragaceae). *Molecular Phylogenetics and Evolution*, **46**, 560–75.

Pauw, A. (2006). Floral syndromes accurately predict pollination by a specialized oil-collecting bee (*Rediviva peringueyi*, Melittidae) in a guild of South African orchids (Coryciinae). *American Journal of Botany*, **93**, 917–26.

Pauw, A., Stofberg, J. and Waterman, R. J. (2009). Flies and flowers in Darwin's race. *Evolution*, **63**, 268–79.

Perez, F., Arroyo, M. T. K., Medel, R. and Hershkovitz, M. A. (2006). Ancestral reconstruction of flower morphology and pollination systems in *Schizanthus* (Solanaceae). *American Journal of Botany*, **93**, 1029–38.

Peter, C. I. and Johnson, S. D. (2008). Mimics and magnets: The importance of color and ecological facilitation in floral deception. *Ecology*, **89**, 1583–1595.

Raine, N. E., Ings, T. C., Ramos-Rodriguez, O. and Chittka, L. (2006). Intercolony variation in learning performance of a wild British bumblebee population (Hymenoptera:Apidae:*Bombus terrestris audax*). *Entomologia Generalis*, **28**, 241–56.

Ramsey, J., Bradshaw, H. D. and Schemske, D. W. (2003). Components of reproductive isolation between the monkeyflowers *Mimulus lewisii* and *M. cardinalis* (Phrymaceae). *Evolution*, **57**, 1520–34.

Rathcke, B. (1983). Competition and facilitation among plants for pollination. In *Pollination Biology*, ed.

L. Real. Orlando, FL: Academic Press, pp. 305–29.

Rausher, M. D. (2008). Evolutionary transitions in floral color. *International Journal of Plant Sciences*, **169**, 7–21.

Rebelo, A. G. and Siegfried W. R. (1985). Colour and size of flowers in relation to pollination of *Erica* species. *Oecologia*, **65**, 584–590.

Riffell, J. A., Alarcon, R., Abrell, L., Davidowitz, G., Bronstein, J. L. and Hildebrand, J. G. (2008). Behavioral consequences of innate preferences and olfactory learning in hawkmoth–flower interactions. *Proceedings of the National Academy of Sciences USA*, **105**, 3404–9.

Robertson, C. (1895). The philosophy of flower seasons, and the phaenological relations of the entomophilous flora and anthophilous insect fauna. *American Naturalist*, **29**, 97–117.

Sandring, S. and Agren, J. (2009). Pollinator-mediated selection on floral display and flowering time in the perennial herb *Arabidopsis lyrata*. *Evolution*, **63**, 1292–1300.

Schemske, D. W. and Bierzychudek, P. (2007). Spatial differentiation for flower color in the desert annual *Linanthus parryae*: Was Wright right? *Evolution*, **61**, 2528–43.

Schemske, D. W. and Bradshaw, H. D. (1999). Pollinator preference and the evolution of floral traits in monkeyflowers (*Mimulus*). *Proceedings of the National Academy of Sciences of the United States of America*, **96**, 11910–5.

Schemske, D. W. and Horvitz, C. C. (1989). Temporal variation in selection on a floral character. *Evolution*, **43**, 461–5.

Schlumpberger, B. O., Cocucci, A. A., More, M., Sersic, A. N. and Raguso, R. A. (2009). Extreme variation in floral characters and its consequences for pollinator attraction among populations of an Andean cactus. *Annals of Botany*, **103**, 1489–1500.

Smith, R. A. and Rausher, M. D. (2008). Experimental evidence that selection favors character displacement in the ivyleaf morning glory. *American Naturalist*, **171**, 1–9.

Stebbins, G. L. (1970). Adaptive radiation of reproductive characteristics in angiosperms. I. Pollination mechanisms. *Annual Review of Ecology and Systematics*, **1**, 307–26.

Steiner, K. E. and Whitehead, V. B. (1990). Pollinator adaptation to oil-secreting flowers: *Rediviva* and *Diascia*. *Evolution*, **44**, 1701–7.

Steiner, K. E. and Whitehead, V. B. (1991). Oil flowers and oil bees: further evidence for pollinator adaptation. *Evolution*, **45**, 1493–501.

Strauss, S. Y. and Whittall, J. B. (2006). Non-pollinator agents of selection on floral traits. In *Ecology and Evolution of Flowers*, ed. L. D. Harder and S. C. H. Barrett. Oxford, UK: Oxford University Press, pp. 120–35.

Strauss, S. Y., Conner, J. K. and Rush, S. L. (1996). Foliar herbivory affects floral characters and plant attractiveness to pollinators: implications for male and female plant fitness. *American Naturalist*, **147**, 1098–107.

Temeles, E. J. and Kress, W. J. (2003). Adaptation in a plant-hummingbird association. *Science*, **300**, 630–3.

Thompson, J. N. (1994). *The Coevolutionary Process*. Chicago, MI: Chicago University Press.

Thompson, J. N. (2005). *The Geographic Mosaic of Coevolution*. Chicago, MI: University of Chicago Press.

Toju, H. and Sota, T. (2006). Imbalance of predator and prey armament:

Geographic clines in phenotypic interface and natural selection. *American Naturalist*, **167**, 105–17.

Toju, H. and Sota, T. (2009). Do arms races punctuate evolutionary stasis? Unified insights from phylogeny, phylogeography and microevolutionary processes. *Molecular Ecology*, **18**, 3940–54.

Vamosi, J. C., Knight, T. M., Steets, J. A., Mazer, S. J., Burd, M. and Ashman, T. L. (2006). Pollination decays in biodiversity hotspots. *Proceedings of the National Academy of Sciences USA*, **103**, 956–61.

Van der Pijl, L. (1961). Ecological aspects of flower evolution. II. Zoophilous flower classes. *Evolution*, **15**, 44–59.

Vanhoenacker, D., Agren, J. and Ehrlen, J. (2006). Spatio-temporal variation in pollen limitation and reproductive success of two scape morphs in *Primula farinosa*. *New Phytologist*, **169**, 615–21.

Vogel, S. (1954). Blutenbiologische Typen als Elemente der Sippengliederung: dargestellt anhand der Flora Sudafrikas. *Botanische Studien*, **1**.

Wallace, A. R. (1867). Creation by Law. *Quarterly Journal Science*, **4**, 470–88.

Warren, J. and Mackenzie, S. (2001). Why are all colour combinations not equally represented as flower-colour polymorphisms? *New Phytologist*, **151**, 237–41.

Waser, N. M. (1983). Competition for pollination and floral character differences among sympatric species: a review of evidence. In *Handbook of Experimental Pollination Biology*, ed. C. E. Jones and R. J. Little. New York, NY: Van Nostrand Reinhold Co., pp. 277–93.

Waser, N. M. (1986). Flower constancy: definition, cause, and measurement. *American Naturalist*, **127**, 593–603.

Waser, N. M., Chittka, L., Price, M. V., Williams, N. M. and Ollerton, J. (1996). Generalization in pollination systems, and why it matters. *Ecology*, **77**, 1043–60.

Weiss, M. R. (2001). Vision and learning in some neglected pollinators: beetles, flies, moths, and butterflies. In *Cognitive Ecology of Pollination; Animal Behaviour and Floral Evolution*, ed. L. Chittka and J. D. Thomson. Cambridge, UK: Cambridge University Press, pp. 171–90.

Wright, S. (1943). An analysis of local variability of flower color in *Linanthus parryae*. *Genetics*, **28**, 139–56.

8

Animal pollination and speciation in plants: general mechanisms and examples from the orchids

FLORIAN P. SCHIESTL

8.1 Introduction

Orchids have served as a model system for pollinator-driven evolution since Darwin's milestone contribution on the fertilization of this plant family (Darwin 1862; van der Pijl 1966; Johnson 2006; Peakall 2007). Orchids represent a major component of angiosperm diversity, and besides their famous and often highly specialized pollination systems (van der Pijl 1966; Schiestl and Schlüter 2009; Tremblay 1992), they have evolved strikingly different lifestyles and thrive in various different habitats (Dressler 1981). Many different mechanisms have been invoked as drivers for orchid-species richness (van der Pijl 1966; Dressler 1981; Peakall 2007; Gravendeel et al. 2004). For example, Cozzolino and Widmer (2005b) suggested the evolution of deceptive pollination as a key trait for orchid diversity. Another study inferred epiphytic lifestyle as a main factor for orchid diversity (Gravendeel et al. 2004). For epiphytic taxa, which indeed represent a major component of orchid diversity, a combination of fine, dust-like seeds and specialized pollination has been attributed as the key factor for species richness (Gentry and Dodson 1987). Although my review focuses on pollination and its link to species richness, it is not the purpose to advocate pollination as the main mechanism driving speciation.

Evolution of Plant–Pollinator Relationships, ed S. Patiny. Published by Cambridge University Press. © The Systematics Association 2012.

Rather, I aim to better clarify its role among the undoubtedly many mechanisms that shape the incredible diversity in orchids and other plants.

Despite little agreement on the definition of a species in general (Coyne and Orr 2004; Johnson 2006), most researchers acknowledge the importance of some degree of reproductive isolation between populations of incipient species (Rieseberg and Willis 2007; Grant 1981; Widmer et al. 2009). In plants, however, reproductive isolation can depend on various intrinsic and extrinsic factors that act at different levels of reproduction. For example, in two sister species of *Mimulus*, habitat and pollinator differences, as well as post-pollination and post-zygotic isolation work together to prevent interspecific geneflow almost totally (Ramsey et al. 2003). Early acting barriers (habitat, pollinators), however, contribute more to total isolation in this system, which is a common situation among angiosperm taxa (Rieseberg and Willis 2007; Widmer et al. 2009). Pollinator or floral isolation can be the by-product of adaptation to a pollination niche, i.e. the differential use of pollinator resources by plants (Grant 1949; Grant 1994; Johnson 2010; Levin 2004). Floral isolation can work through floral morphology (morphological isolation), allowing only certain pollinators access to rewards (e.g. through long floral spurs or tubes) or placement of pollen on different body parts of a pollinator (group). In ethological floral isolation, a specific (group of) pollinator(s) are attracted through innate preferences to given floral signals, or establishes floral constancy through learning of floral signals. Thus, floral isolation is the consequence of specialization to pollinators with specific morphology (e.g. long proboscis) or behavior (e.g. preferences for specific floral signals).

Although specialization in pollination is a common phenomenon among angiosperms (Johnson and Steiner 2000; Johnson 2010), its ultimate causes are currently little understood. A recent study by Scopece et al. (2010) provides support for the assumption that specialized pollination leads to more efficient pollen transfer. Specialized pollination does not, however, automatically lead to strong floral isolation, although this link has often been implied (Waser 2001; van der Pijl 1966; Gravendeel et al. 2004). Examples for specialized pollination systems without floral isolation are the South African *Eucomis autumnalis* and *E. comosa*, which are specialized on pollination by pompilid wasps but broadly share the wasp species visiting the plants (Shuttleworth and Johnson 2009). An even more extreme example is the five Australian species of the sexually deceptive orchid genus *Cryptostylis*, which are all pollinated by the same single species of ichneumonid wasp (Schiestl et al. 2004; Gaskett and Herberstein 2010). In these two examples, which contain at least partly sympatric species, specialized plant lineages may have diverged during temporal allopatry evolving postzygotic isolation; alternatively, already diverged lineages may have adapted convergently to the same pollinators. On the other hand, plants fairly generalized in pollination can still have strong floral isolation, e.g. in the orchid genera *Gymnadenia* or *Earina* (Schiestl and Schlüter 2009). Indeed and perhaps surprising, it has been shown that there is no correlation between the degree of specialization in pollination (number of pollinators) and floral isolation (pollinator sharing) in orchids (Schiestl and Schlüter 2009). Thus, a distinction should be made between floral isolation and specialization in pollination, and the two phenomena investigated individually.

8.2 Adaptation to pollinators and speciation

Although there is little doubt that adaptation to pollinators can drive floral diversification, its role in the speciation process is more contentious (Johnson 2006). Because shifts in pollination systems among species are often accompanied by shifts in habitat or soil types (Goldblatt and Manning 2006; Van der Niet et al. 2006), it is sometimes not clear whether adaptation to pollinators is the primary cause of divergence, or a side effect of allopatric divergence or habitat adaptation (Johnson 2010). Because pollinators are not evenly distributed among the landscape and different habitats (a phenomenon called the "pollination climate" *sensu* Grant and Grant 1965), different geographic distribution or habitats may often lead to selection for pollinator switches in different plant lineages (Grant 1993, Grant and Grant 1965, Goldblatt and Manning 2006, Johnson 2006). In addition, gene flow between populations adapted to different habitats will automatically select for pre-zygotic barriers such as floral isolation, to reduce the production of maladapted hybrids. The same is true after the evolution of genetic incompatibility, for example through polyploidisation or BDM incompatibilities, where floral isolation may be selected for through reinforcement (Van der Niet et al. 2006; but see Coyne and Orr (2004, pp 369–75) for controversies around reinforcement).

In certain plant-pollinator systems, however, it seems plausible that adaptation to different pollinators can also be the primary cause of divergence (Xu et al. 2011). Such adaptation to different "pollination niches" (Johnson 2010; Levin 2004) without any preceding differences in geographic distribution can lead to sympatric speciation because initial populations have the opportunity for gene flow (for a discussion of geographic versus demographic definition of sympatry see Fitzpatrick et al. 2008). Importantly, gene flow is mediated by pollinators, thus foraging range and distribution of pollinators are more important factors for gene-flow than small-scale differences in the distribution and habitat of plant populations. It seems likely, for example, that highly mobile male euglossine bees, which have been shown to move within forest areas of over 100 ha on a daily basis (Wikelski et al. 2010), will find and visit flowers even if they grow in quite distinct habitats. Male thynnine wasps that pollinate sexually deceptive orchids have been shown to move over 100 m during search for females (Peakall 1990), however, average mate-search distances of male solitary bees may be shorter (Peakall and Schiestl 2004). Female solitary bees collecting pollen have been shown to fly well over 500 m, and natural "barriers" like forests and rivers can be readily crossed (Zurbuchen et al. 2010), and the same is likely true for large foraging moths. Because pollinators can mediate gene flow even among distant plant populations, a switch to a new pollinator is only possible if selection for differential pollinator adaptation is strong, and traits mediating adaptation are linked to reproductive isolation (Coyne and Orr 2004; Waser and Campbell 2004). Selection for such a switch in pollination system may be fuelled by low pollination success (Knight et al. 2005), which may be improved by the utilization of more abundant or efficient pollinators. The switch to an, as yet, unexploited pollinator niche may also relieve competition for pollination service, however, intraspecific competition of pollination has rarely been demonstrated (Spigler and Chang 2009). A switch in pollinators can be mediated by few adaptive genes, which are, for example, responsible for floral color (Bradshaw and Schemske 2003; Hoballah et al. 2007) or floral scent (Schluter and Schiestl 2008). Thus, changes in pollinator attraction can evolve quickly, and adaptive traits do not necessarily face the challenge of being maintained despite recombination with ongoing gene-flow (Coyne and Orr 2004). Initially, a new genotype will be rare, possibly comprising a single individual. For a successful establishment, clonal reproduction, self-pollination, or geitonogamy are thus necessary. A pollinator switch will automatically induce some degree of assortative mating, even if some pollinator sharing remains because pollinators mediate gene flow among and between populations (i.e. the dual role of pollinators; Waser and Campbell 2004). If floral isolation resulting from a switch in pollinators is strong enough to allow adaptive alleles to become fixed, i.e. geneflow does not override selection (Rieseberg et al. 2004), two (incipient) species will persist. In the sexually deceptive genus *Ophrys*, recent investigations have shown that pollinator differences can lead to strong floral isolation preventing gene-flow almost totally (Xu et al. 2011).

Collectively, in situations without any geographical barriers to gene flow, pollinator adaptation as a driving force for plant speciation becomes likely when:

specialization to different pollination niches, i.e. a switch in pollinator conveys a selective advantage, for example through higher pollination success or efficiency;

there are no constraints in key factors mediating adaptations to pollinators, for example in the basal floral structure (radial symmetry), or a lack of variability in floral morphology or signals (Johnson et al. 1998; Barrett and Schluter 2008);

reproductive isolation is sufficiently strong to prevent or at least reduce gene flow in sympatry. Obviously, besides floral isolation, this may be achieved by other mechanisms like shift in flowering time or autogamy.

Under these assumptions, species richness will be determined by:

(1) the number of "realizable" pollination niches, i.e. number of potential pollinator species in a given habitat, or the potential of different placement of pollinia on a single pollinator (a further sub-division of the available pollination niches)

(2) the degree of specialization by the plant

(3) the strength of floral isolation in the pollination system

(4) intrinsic developmental factors like floral bauplan and variability in adaptive genes (adaptability).

The second and third factor seems readily testable by comparing species numbers of sister taxa showing variation in them. Such an analysis within the Orchidaceae is presented below. In the next section, I discuss possible examples of pollination systems with pollinator adaptation primarily driving speciation.

8.3 Pollination systems where pollinator adaptation may drive speciation

One of the more investigated pollination systems with high specialization in pollination and strong floral isolation is sexual deception (Xu et al. 2011). Details of this pollination mode, its evolutionary consequences and a comparison to a system with weak floral isolation are outlined in the next paragraph. Another prominent example involving reproductive behavior of the pollinators is plants pollinated by male euglossine bees, which collect fragrances from flowers (Dodson et al. 1969). In this pollination system, pollination is highly specialized and floral isolation usually strong (Dressler 1968; Schiestl and Schlüter 2009). Pollination by male euglossines occurs most prominently in the Orchidaceae (Williams 1982; Vogel 1963; Dodson et al. 1969), but also among several other angiosperms families like Araceae (Hentrich et al. 2010; Williams and Dressler 1976), Gesneriaceae (Dressler 1968; Vogel 1966), Euphorbiaceae (Armbruster and Webster 1979), Solanaceae (Sazima et al. 1993) and Annonaceae (Teichert et al. 2009). Interestingly, within the Araceae, the euglossine-pollinated *Anthurium* and *Spatiphyllum* are also among the largest genera (Hentrich et al. 2010). This family, however, is also rich in other specific pollination mechanisms with likely strong floral isolation, e.g. involving Dynastine (Scarabaeidae) beetles in several genera (Gibernau et al. 2003, Gottsberger and Silberbauer-Gottsberger 1991, Maia and Schlindwein 2006). Other well-known pollination systems with highly specific pollination and low pollinator sharing are obligatory nursery pollination systems in *Ficus* and *Yucca* (Dufay and Anstett 2003) and the gymnosperms group of cycads.

In the latter group, despite its ancient origin, recent radiations are hypothesized in both the pants (Treutlein and Wink 2002) and their pollinators (Downie et al. 2008). In pollination systems with nectar reward, pollination can be highly specialized, too, and the best-known examples come from the flora of South Africa (Johnson and Steiner 2003), within the plant families Orchidaceae and Iridaceae (Johnson 1997; Johnson et al. 1998; Goldblatt and Manning 2006; Goldblatt and Manning 1996). However, in these pollination systems, the degree of floral isolation is often unknown, and pollinator switches are often accompanied by shifts in habitat, making pollinator-driven speciation difficult to assess (Goldblatt and Manning 2002; Johnson 2010; Goldblatt and Manning 1996; but see van der Niet and Johnson 2009). Among other rewarding systems, switches from bee to hummingbird or sphingid pollination seem to usually convey rather strong floral isolation, e.g. in the genera *Mimulus, Ipomopsis, Aquilegia, Costus* (Campbell 2008; Ramsey et al. 2003; Grant 1993; Hodges and Arnold 1994; Schemske and Bradshaw 1999; Fulton and Hodges 1999; Kay and Schemske 2003). Again, differences in pollinators are often accompanied by some degree of habitat differences in these pollination systems.

8.4 Patterns of speciation in food- and sexually deceptive orchids

Mediterranean orchids in the subtribe Orchidinae provide a nice opportunity to compare pollinator adaptation and floral isolation in related taxa with similar ecology (Fig 8.1). Some of the main genera representing (generalized) food deception (*Orchis, Anacamptis*) and sexual deception (*Ophrys*) are closely related; in fact *Ophrys* is likely sister to *Anacamptis* and *Serapias* (Bateman et al. 2003; Aceto et al. 1999). General ecology and pollination in both systems is relatively well investigated. Plants in both groups are terrestrial, perennial herbs with a tuber, perferring dry meadows or open woodland on poor, calcareous soils. Bees are the primary pollinators and both groups are almost exclusively deceptive, producing no food reward for the pollinators. However, the main difference is the primary use of female bees pollinating through food search behavior in the food-deceptive species, contrasting with the exclusive exploitation of male bees (and few other taxa) pollinating through mating behavior in the sexually deceptive *Ophrys* (Schiestl 2005). Flowers in the food-deceptive group are often colorful and arranged on large inflorescences with a spur commonly present. Flowers of the sexually deceptive *Ophrys* are highly drived, with dark-red to black labella, spurless, and coated with fine hairs. In sexual deception, pollination is specific with strong floral isolation (Schiestl and Schlüter 2009; Xu et al. 2011), whereas food deception leads to fairly generalized pollinator attraction with high pollinator

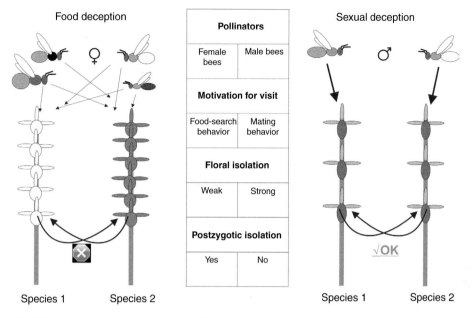

Fig 8.1 Schematic comparison of pollination and reproductive isolation in (generalized) food deception (*Anacamptis, Orchis, Dactylorhiza*) and sexual deception (*Ophrys*) in mediterranean Orchidinae.

sharing (Cozzolino et al. 2005). This difference in specialisation is due to the behavior of the pollinators utilised. Female bees often switch to different flower types after unrewarding visits during foraging (Smithson and Macnair 1997), whereas male bees during mate search are more or less specifically tuned towards their specific mating singals (Mant et al. 2005a). In addition to different figures of floral isolation, interspecific genetic divergence differs between the two groups of orchids. Despite weak floral isolation, food-deceptive genera typically show higher levels of neutral genetic divergence compared with sexually deceptive orchids (Cozzolino and Widmer 2005b). This is explained by the evolution of karyotype differences leading to postzygotic isolation in food-deceptive orchids (Cozzolino et al. 2004, Cozzolino and Widmer 2005a), whereas postzygotic isolation is typically absent among *Ophrys* species (Scopece et al. 2007; Xu et al. 2011). In general, such postzygotic barriers in combination with pollinator sharing are unlikely to evolve in sympatry, since pollinator sharing in initially compatible popuations should lead to homogenisation of genomes and thus prevent the evolution of karytype differences. Thus, in food-deceptive genera, divergence in allopatry seems to be a prerequsite for incipient speciation, and adaptation to different pollinators is unliklely to play any significant role in the speciation process. In *Ophrys*, however, the apparent lack of postzygotic isolation and strong floral isolation among closely related taxa suggests floral isolation evolves during speciation as a by-product of pollinator adaptation. Such adaptive speciation may in principle happen both in sympatry and allopatry (Grant 1993), however, the sympatric occurrence of closely related *Ophrys* species often suggests a prominent role for sympatric speciation in this genus. The most important trait for pollinator adaptation, which also mediates floral isolation, is the scent of *Ophrys* flowers, being a mimicry of the sex pheromones of the pollinator species and attracting pollinator males on a highly specific basis (Schiestl and Ayasse 2002; Schiestl and Schlüter 2009; Schiestl 2005; Stökl et al. 2009). Despite behaviorally active scent, bouquets in *Ophrys* usually consist of several different compounds in specific proportions (Mant et al. 2005b) and the molecular mechanism controlling the production of such different blends may be based on modification of common precursor molecules through few genes of large effect (Schluter and Schiestl 2008; Schluter et al. 2011).

8.5 Does strong floral isolation and specialization lead to higher species numbers in orchids? A meta-analysis

I have suggested four factors to impact on species diversity when pollinator adaptation drives speciation:

(1) Niche abundance

(2) Specialization in pollination

(3) Floral isolation

(4) Adaptability

Of those, (1) and (4) are difficult to test for the lack of data on, e.g. bee species diversity in given habitats, as well as population variability in genes coding for adaptive traits. Thus, I focus on (2) and (3) which typically differ in pollination systems involving food-seeking behavior (specialization usually weak) and reproductive behavior (specialization usually strong; Schiestl and Schlüter 2009). An exception seems to be the sexually deceptive genus *Cryptostylis*, with high pollinator sharing in the Australian species, however, nothing is known about the pollinators in the majority of species distributed in the oriental region. To test whether specialization in pollination and floral isolation contributes to species richness, a sister-clade comparison was done. The sister clades analyzed usually showed similar ecology but different pollination system, with one clade representing reproductive behavior (scent reward or sexual deception) and the other a system with food seeking behavior (food deception, nectar reward, and others, see Table 8.1). Among published molecular orchid phylogenies at the subribe/

Table 8.1 Orchid sister clades with different strengths of floral isolation. The first two columns give the respective sister clades and their mean number of species used in the analysis.

Genera/clades with strong floral isolation (mean) no. of species	Genera/clades with weak floral isolation (mean) no. of species	Tribe/subtribe	Ref. Phylogeny
Sexual deception	*Other*		
Ophrys (126)	Anacamptis[1], Serapias[2] (12.5)	Orchideae/Orchidinae	Bateman et al. 2003
Chiloglottis, Drakea, Spiculea, Paracaleana, Caleana (9.6)	Megastylis[1,3], Rimacola[5] (4)	Diurideae/Drakaeinae	Kores et al. 2001
Caladenia (84)	Cyanicula[1], Glossodia[1], Elythranthera[1] (6.3)	Diuridae/Caladeniinae	Kores et al. 2001
Cryptostylis (23)	Coilochilus[4] (1)	Diuridae/Cryptostylidinae	Kores et al. 2001
Calochilus[4] (23)	Thelymitra[1] (50)	Diuridae/Thelymitrinae	Kores et al. 2001
Cyrtidiorchis (5)	Maxillaria sect. trigonae[1] (10)	Cymbidieae/Maxillariinae	Whitten et al. 2007
Scent reward[6]	*Other*		
Zygopetalum (15)	Batemannia[3] (5)	Cymbidieae/Zygopetalinae	Whitten et al. 2000
Trichopilia (42)	Psychopis[5] (5)	CymbidieaeOncidiinae	Whitten et al. 2000
Macroclinium, Notylia, Macradenia, Warmingia (27.75)	Ionopsis[3], Trizeuxis[3] (3.5)	Cymbidieae/Oncidiinae	Chase and Palmer 1997

[1] food deception; [2] sleeping hole pollination [3] nectar [4] autogamy [5] pollination system unknown [6] pollination by male euglossine bees

genus level, clear sister genera/clades with known pollination systems were identified. Only phylogenic studies with extensive species sampling were included to avoid incorrect sister-clade assignment. When choosing sister groups for the analysis, only groups with little intrageneric variation in pollination system were included. In the genus *Caladenia*, comprising both food- and sexually deceptive species, only sexually deceptive species were included. Clades including too many (> 5) genera were avoided, as they would have incorporated too much variation in factors other than pollination. Several very large genera with likely strong floral isolation, e.g. *Lepanthes* (sexual deception), *Pleurothallis*, and *Bulbophyllum* (brood-site mimicry, and other pollination systems) could not be included in this analysis because their sister groups are currently unresolved. Species numbers of the genera included were compiled using the online database "world checklist of selected plant families" (http://www.kew.org/) and the encyclopedia "Genera Orchidacearum" (Pridgeon et al. 2001). For clades with more than one genus, mean species numbers were calculated. For genera with unclear taxonomy, like *Ophrys*, an intermediate species number (126 for *Ophrys*) was assumed. Moreover, the exact number of species was not of key importance in the statistical analysis because all variables were ln(1 + x) transformed before comparison, to assume homogeneity of variances. A dependent students t-test for paired samples was calculated to compare species numbers in sister clades.

In the analysis, I included a total of nine sister groups, spread over two subfamilies (Orchidoideae, Epidendroideae), three tribes and eight subtribes were included in the analyses (Table 8.1). The specialized group had a mean number of 39.48 ± 13.35 species, whereas the unspecialized group had 10.81 ± 5.03 species ($t_8 = 3.02$; P = 0.017). In seven pairs, the specialized groups had a higher species number than the unspecialized group, and only in two pairs was it reversed. The result of significantly higher species numbers in clades with strong specialization supports the assumption that specialization in pollination and strong floral isolation can drive speciation in orchids. This does not imply that pollinator-driven speciation is the only mechanism of speciation in those orchids, but it can work on top of other mechanisms like allopatric divergence or habitat adaptation. For example, in the Oncidiinae, the primary radiation was based on habitat and life history specialization, followed by adaptation to different pollinators (Chase and Palmer 1997). Within this subtribe, all species of the *Rodriguezia* clade are twig epiphytes, adapted to different pollinators. In this situation, the above factors (1–4) should determine species richness within genera, and thus highly specialized species with strong floral isolation should be richer in species than those lacking these traits. Indeed, the highly specialized, euglossine-pollinated genera *Macroclinium, Notylia, Macradenia, Warmingia* contain a mean of 27.7 species, whereas the sister clade, encompassing the nectar reward-producing, butterfly-pollinated sister groups *Ionopsis* and *Trizeuxis*, comprise only a mean of 3.5 species.

8.6 Conclusion

My analysis supports the assumption that pollinator adaptation *per se* does drive plant speciation when floral isolation is strong. This is the case in pollination systems involving reproductive behavior in pollinators, however, several other pollination systems may fulfil these requirements, too. As highlighted in Kay and Sargent, (2009) and Schiestl and Schlüter, (2009) more scrutiny is needed in investigations assessing the role of pollinator adaptation in plant speciation. First of all, floral isolation must not be confused with absolute specialization in pollination, i.e. the absolute number of pollinators utilized by a plant. Secondly, floral isolation should be assessed in phylogenetic sister species, rather than communities of unrelated plants, to highlight its role in the speciation process. Especially powerful are investigations combining assessment of floral isolation with postpollination barriers and habitat differences in a phylogenetic framework. In sister species with no postpollination or postzyotic barriers and an overlap in distribution (of pollinator ranges), pollinator adaptation is a likely driving force in plant species diversification.

Acknowledgements

I would like to thank Peter Linder and one anonymous reviewer for commenting on an earlier version of this manuscript, and Sébastien Patiny for inviting me to write a chapter for this book. The author is funded by the Swiss National Science Foundation (SNF) grant No. 31003A_125340.

References

Aceto, S., Caputo, P., Cozzolino, S., Gaudio, L. and Moretti, A. (1999). Phylogeny and evolution of Orchis and allied genera based on ITS DNA variation: morphological gaps and molecular continuity. *Molecular Phylogenetics & Evolution*, **13**, 67–76.

Armbruster, W. S. and Webster, G. L. (1979). Pollination of 2 species of Dalechampia (Euphorbiaceae) in Mexico by euglossine bees. *Biotropica*, **11**, 278–83.

Barrett, R. D. H. and Schluter, D. (2008). Adaptation from standing genetic variation. *Trends in Ecology & Evolution*, **23**, 38–44.

Bateman, R. M., Hollingsworth, P. M., Preston, J., Yi-Bo, L., Pridgeon, A. M. and Chase, M. W. (2003). Molecular phylogenetics and evolution of Orchidinae and eelected Habenariinae (Orchidaceae). *Botanical Journal of the Linnean Society*, **142**, 1–40.

Bradshaw, H. D. and Schemske, D. W. (2003). Allele substitution at a flower colour locus produces a pollinator shift in Monkeyflowers. *Nature*, **426**, 176–178.

Campbell, D. R. (2008). Pollinator shifts and the origin and loss of plant species. *Annals of the Missouri Botanical Garden*, **95**, 264–274.

Chase, M. W. and Palmer, J. D. (1997). Leapfrog radiation in floral and vegetative traits among twig epiphytes in the orchid subtribe Oncidiinae. In *Molecular Evolution and Adaptive Radiation*, ed. T. J. Givnish and K. J. Sytsma. Cambridge, UK: Cambridge University Press, pp. 331–52.

Coyne, J. A. and Orr, H. A. (2004). *Speciation*. Sunderland, MA: Sinauer Associates.

Cozzolino, S. and Widmer, A. (2005a). The evolutionary basis of reproductive isolation in Mediterranean orchids. *Taxon*, **54**, 977–85.

Cozzolino, S. and Widmer, A. (2005b). Orchid diversity: an evolutionary consequence of deception? *Trends in Ecology & Evolution*, **20**, 487–94.

Cozzolino, S., D'Emerico, S. and Widmer, A. (2004). Evidence for reproductive isolate selection in Mediterranean orchids: karyotype differences compensate for the lack of pollinator specificity. *Proceedings of the Royal Society of London Series B, Biological Sciences*, **271**, S259–S262.

Cozzolino, S., Schiestl, F. P., Muller, A., De Castro, O., Nardella, A. M. and Widmer, A. (2005). Evidence for pollinator sharing in Mediterranean nectar-mimic orchids: absence of premating barriers? *Proceedings of the Royal Society B, Biological Sciences*, **272**, 1271–8.

Darwin, C. (1862). *On the Various Contrivances by which British and Foreign Orchids Are Fertilised by Insects*. London, UK: John Murray.

Dodson, C. H., Dressler, R. L., Hills, H. G., Adams, R. M. and Williams, N. H. (1969). Biologically active compounds in orchid fragrances. *Science*, **164**, 1243–9.

Downie, D. A., Donaldson, J. S. and Oberprieler, R. G. (2008). Molecular systematics and evolution in an African cycad–weevil interaction: Amorphocerini (Coleoptera:Curculionidae:Molytinae) weevils on Encephalartos. *Molecular Phylogenetics and Evolution*, **47**, 102–16.

Dressler, R. L. (1968). Pollination by euglossine bees. *Evolution*, **22**, 202–10.

Dressler, R. L. (1981). *The Orchids: Natural History and Classification*. Cambridge, MA: Harvard University Press.

Dufay, M. and Anstett, M. C. (2003). Conflicts between plants and pollinators that reproduce within inflorescences: evolutionary variations on a theme. *Oikos*, **100**, 3–14.

Fitzpatrick, B. M., Fordyce, J. A. and Gavrilets, S. (2008). What, if anything, is sympatric speciation? *Journal of Evolutionary Biology*, **21**, 1452–9.

Fulton, M. and Hodges, S. A. (1999). Floral isolation between Aquilegia formosa and Aquilegia pubescens. *Proceedings of the Royal Society of London Series B, Biological Sciences*, **266**, 2247–52.

Gaskett, A. C. and Herberstein, M. E. (2010). Colour mimicry and sexual deception by Tongue orchids (Cryptostylis). *Naturwissenschaften*, **97**, 97–102.

Gentry, A. H. and Dodson, C. H. (1987). Diversity and biogeography of neotropical vascular epiphytes. *Annals of the Missouri Botanical Garden*, **74**, 205–33.

Gibernau, M., Barabe, D., Labat, D., Cerdan, P. and Dejean, A. (2003). Reproductive biology of Montrichardia arborescens (Araceae) in French

Guiana. *Journal of Tropical Ecology*, **19**, 103–7.

Goldblatt, P. and Manning, J. C. (1996). Phylogeny and speciation in Lapeirousia subgenus Lapeirousia (Iridaceae: Ixioideae). *Annals of the Missouri Botanical Garden*, **83**, 346–61.

Goldblatt, P. and Manning, J. C. (2002). Plant diversity of the Cape Region of southern Africa. *Annals of the Missouri Botanical Garden*, **89**, 281–302.

Goldblatt, P. and Manning, J. C. (2006). Radiation of pollination systems in the iridaceae of sub-Saharan Africa. *Annals of Botany*, **97**, 317–44.

Gottsberger, G. and Silberbauer-Gottsberger, I. (1991). Olfactory and visual attraction of *Erioscelis emarginata* (Cyclocephalini, Dynastinae) to the inflorescences of *Philodendron selloum* (Araceae). *Biotropica*, **23**, 23–8.

Grant, V. (1949). Pollination systems as isolating mechanisms in angiosperms. *Evolution*, **3**, 82–97.

Grant, V. (1981). *Plant Speciation*. New York, NY: Columbia University Press.

Grant, V. (1993). Origin of floral isolation between ornithophilous and sphingophilous plant species. *Proceedings of the National Academy of Sciences USA*, **90**, 7729–33.

Grant, V. (1994). Modes and origins of mechanical and ethological isolation in angiosperms. *Proceedings of the National Academy of Sciences USA*, **91**, 3–10.

Grant, V. and Grant, K. A. (1965). *Flower Pollination in the Phlox Family*. New York, NY: Columbia University Press.

Gravendeel, B., Smithson, A., Slik, F. J. W. and Schuiteman, A. (2004). Epiphytism and pollinator specialization: drivers for orchid diversity? *Philosophical Transactions of the Royal Society of London Series B-Biological Sciences*, **359**, 1523–35.

Hentrich, H., Kaiser, R. and Gottsberger, G. (2010). Floral biology and reproductive isolation by floral scent in three sympatric aroid species in French Guiana. *Plant Biology*, **12**, 587–96.

Hoballah, M. E., Gubitz, T., Stuurman, J., Broger, L., Barone, M., Mandel, T., Dell'Olivo, A., Arnold, M. and Kuhlemeier, C. (2007). Single gene-mediated shift in pollinator attraction in Petunia. *Plant Cell*, **19**, 779–90.

Hodges, S. A. and Arnold, M. L. (1994). Floral and ecological isolation between *Aquilegia formosa* and *Aquilegia pubescens*. *Proceedings of the National Academy of Sciences USA*, **91**, 2493–6.

Johnson, S. D. (1997). Insect pollination and floral mechanisms in South African species of *Satyrium* (Orchidaceae). *Plant Systematic and Ecology*, **204**, 195–206.

Johnson, S. D. (2006) Pollinator driven speciation in plants. In *Ecology and Evolution of flowers*, ed. L. D. Harder and Barrett, S. C. H. Oxford, UK: Oxford University Press, pp. 295–310.

Johnson, S. D. (2010). The pollination niche and its role in the diversification and maintenance of the southern African flora. *Philosophical Transactions of the Royal Society B, Biological Sciences*, **365**, 499–516.

Johnson, S. D. and Steiner, K. E. (2000). Generalization versus specialization in plant pollination systems. *TREE*, **15**, 140–3.

Johnson, S. D. and Steiner, K.E. (2003). Specialized pollination systems in southern Africa. *South African Journal of Science*, **99**, 345–8.

Johnson, S. D., Linder, H. P. and Steiner, K. E. (1998). Phylogeny and Radiation of Pollination Systems in Disa (Orchidaceae). *American Journal of Botany*, **85**, 402–11.

Kay, K. M. and Sargent, R. D. (2009). The role of animal pollination in plant speciation: integrating ecology, geography, and genetics. *Annual Review of Ecology Evolution and Systematics*, **40**, 637–56.

Kay, K. M. and Schemske, D. W. (2003). Pollinator assemblages and visitation rates for 11 species of neotropical Costus (Costaceae). *Biotropica*, **35**, 198–207.

Kores, P. J., Molvray, M., Weston, P. H., Hopper, S. D., Brown, A. P., Cameron, K. M. and Chase, M. W. (2001). A phylogenetic analysis of Diurideae (Orchidaceae) based on plastid DNA sequence data. *American Journal of Botany*, **88**, 1903–14.

Knight, T. M., Steets, J. A., Vamosi, J. C., Mazer, S. J., Burd, M., Campbell, D. R., Dudash, M. R., Johnston, M. O., Mitchell, R. J. and Ashman, T. L. (2005). Pollen limitation of plant reproduction: pattern and process. *Annual Review of Ecology Evolution and Systematics*, **36**, 467–97.

Levin, D. A. (2004). Ecological speciation: crossing the divide. *Systematic Botany*, **29**, 807–16.

Maia, A. C. D. and Schlindwein, C. (2006). *Caladium bicolor* (Araceae) and *Cyclocephata celata* (Coleoptera, Dynastinae): a well-established pollination system in the northern Atlantic Rainforest of Pernambuco, Brazil. *Plant Biology*, **8**, 529–34.

Mant, J., Peakall, R. and Schiestl, F. P. (2005a). Does selection on floral odor promote differentiation among populations and species of the sexually deceptive orchid genus *Ophrys*? *Evolution*, **59**, 1449–63.

Mant, J. G., Brändli, C., Vereecken, N. J., Schulz, C. M., Francke, W. and Schiestl, F. P. (2005b). Cuticular hydrocarbons as sex pheromone of the bee *Colletes cunicularius* and the key to its mimicry by the sexually deceptive orchid, *Ophrys exaltata*. *Journal of Chemical Ecology*, **31**, 1765–87.

Peakall, R. (1990). Responses of male *Zaspilothynnus trilobatus* Turner wasps to females and the sexually deceptive orchid it pollinates. *Functional Ecology*, **4**, 159–167.

Peakall, R. (2007). Speciation in the Orchidaceae: confronting the challenges. *Molecular Ecology*, **16**, 2834–7.

Peakall, R. and Schiestl, F. P. (2004). A mark-recapture study of male Colletes cunicularius bees: implications for pollination by sexual deception. *Behavioral Ecology and Sociobiology*, **56**, 579–84.

Pridgeon, A. M., Cribb, P. J., Chase, M. W. and Rasmussen, F. N. (2001). *Genera Orchidacearum*. Oxford, UK: Oxford University press,.

Ramsey, J., Bradshaw, H. D. and Schemske, D. W. (2003). Components of reproductive isolation between the monkeyflowers *Mimulus lewisii* and *M. cardinalis* (Phrymaceae). *Evolution*, **57**, 1520–34.

Rieseberg, L. H. and Willis, J. H. (2007). Plant speciation. *Science*, **317**, 910–4.

Rieseberg, L. H., Church, S. A. and Morjan, C. L. (2004). Integration of populations and differentiation of species. *New Phytologist*, **161**, 59–69.

Sazima, M., Vogel, S., Cocucci, A. and Hausner, G. 1993. The perfume flowers of Cyphomandra (Solanaceae): pollination by euglossine bees, bellow mechanism, osmophors, and volatiles. *Plant Systematic and Evolution*, **187**, 51–88.

Schemske, D. W. and Bradshaw, H. D. (1999). Pollinator Preference and the Evolution of Floral Traits in Monkeyflowers (Mimulus). *Proceedings of the National Academy of Sciences USA*, **96**, 11910–5.

Schiestl, F. P. (2005). On the success of a swindle: pollination by deception in orchids. *Naturwissenschaften*, **92**, 255–64.

Schiestl, F. P. and Ayasse, M. (2002). Do changes in floral odor cause speciation in sexually deceptive orchids? *Plant Systematics and Evolution*, **234**, 111–9.

Schiestl, F. P. and Schlüter, P. M. (2009). Floral isolation, specialized pollination, and pollinator behavior in orchids. *Annual Review of Entomology*, **54**, 425–46.

Schiestl, F. P., Peakall, R. and Mant, J. G. (2004). Chemical communication in the sexually deceptive orchid genus *Cryptostylis*. *Botanical Journal of the Linnean Society*, **144**, 199–205.

Schluter, P. M. and Schiestl, F. P. (2008). Molecular mechanisms of floral mimicry in orchids. *Trends in Plant Science*, **13**, 228–35.

Schluter, P. M., Xu, S., Gagliardini, V., Whittle, E. J., Shanklin, J., Grossniklaus, U. and Schiestl, F. P. (2011) Stearoyl-ACP desaturases are associated with floral isolation in sexually deceptive orchids. *Proceedings of the National Academy of Sciences*, **108**, 5696–701.

Scopece, G., Cozzolino, S., Johnson, S. D. and Schiestl, F. P. (2010). Pollination efficiency and the evolution of specialized deceptive pollination systems. *The American Naturalist*, **175**, 98–105.

Scopece, G., Musacchio, A., Widmer, A. and Cozzolino, S. (2007). Patterns of reproductive isolation in Mediterranean deceptive orchids. *Evolution*, **61**, 2623–42.

Shuttleworth, A. and Johnson, S. D. (2009). A key role for floral scent in a wasp-pollination system in Eucomis (Hyacinthaceae). *Annals of Botany*, **103**, 715–25.

Smithson, A. and Macnair, M. R. (1997). Negative frequency-dependent selection by pollinators on artificial flowers without rewards. *Evolution*, **51**, 715–23.

Spigler, R. B. and Chang, S. M. (2009). Pollen limitation and reproduction varies with population size in experimental populations of Sabatia angularis (Gentianaceae). *Botany-Botanique*, **87**, 330–338.

Stökl, J., Schluter, P. M., Stuessy, T. F., Paulus, H. F., Fraberger, R., Erdmann, D., Schulz, C., Francke, W., Assum, G. and Ayasse, M. (2009). Speciation in sexually deceptive orchids: pollinator-driven selection maintains discrete odour phenotypes in hybridizing species. *Biological Journal of the Linnean Society*, **98**, 439–51.

Teichert, H., Dotterl, S., Zimma, B., Ayasse, M. and Gottsberger, G. (2009). Perfume-collecting male euglossine bees as pollinators of a basal angiosperm: the case of Unonopsis stipitata (Annonaceae). *Plant Biology*, **11**, 29–37.

Tremblay, R. L. (1992). Trends in the pollination ecology of the orchidaceae: evolution and systematics. *Canadian Journal of Botany*, **70**: 642–50.

Treutlein, J. and Wink, M. (2002). Molecular phylogeny of cycads inferred from rbcL sequences. *Naturwissenschaften*, **89**, 221–5.

van der Niet, T. and Johnson, S. D. (2009). Patterns of plant speciation in the Cape floristic region. *Molecular Phylogenetics and Evolution*, **51**, 85–93.

van der Niet, T., Johnson, S. D. and Linder, H. P. (2006). Macroevolutionary data suggest a role for reinforcement in pollination system shifts. *Evolution*, **60**, 1596–601.

van der Pijl, L. D. D. H. (1966). *Orchid Flowers: Their Pollination and Evolution*. Coral Gables, FL: University of Miami Press.

Vogel, S. (1963). Das sexuelle Anlockungsprinzip der Catasetinen- und Stanhopeen- Blüten und die wahre Funktion ihres sogenannten Futtergewebes. *Österreichische botanische Zeitschrift*, **110**, 308–37.

Vogel, S. (1966). Parfümsammelnde Bienen als Bestäuber von Orchidaceen und Gloxinia. *Österreichische botanische Zeitschrift*, **113**, 302–61.

Waser, N. M. (2001) Pollinator behavior and plant speciation: looking beyond the "ethological isolation" paradigm. In *Cognitive Ecology of Pollination*, ed. L. Chittka, L and J. D. Thomson. Cambridge, UK: Cambridge University Press, pp. 318–35.

Waser, N. M. and Campbell, D. R. (2004) Ecological speciation in flowering plants. In *Adaptive Speciation*, ed. U. Dieckmann, M. Doebeli, M. J. Metz, and D. Tautz. Cambridge, UK: Cambridge University Press, pp. 264–277.

Widmer, A., Lexer, C. and Cozzolino, S. (2009). Evolution of reproductive isolation in plants. *Heredity*, **102**, 31–8.

Wikelski, M., Moxley, J., Eaton-Mordas, A., Lopez-Uribe, M. M., Holland, R., Moskowitz, D., Roubik, D. W. and Kays, R. (2010). Large-range movements of neotropical orchid bees observed via radio telemetry. *PLoS ONE* **5**.

Williams, N. H. (1982) The biology of orchids and euglossine bees. In *Orchid Biology: Reviews and Perspectives, 2.*, ed. J. Arditti. New York, NY: Cornell University Press, pp. 119–72.

Williams, N. H. and Dressler, R. L. (1976). Euglossine pollination of Spathiphyllum (Araceae). *Selbyana*, **1**, 349–56.

Whitten, W. M., Williams, N. H. and Chase, N.W. (2000). Subtribal and generic relationship of Maxillarieae (Orchidaceae) with emphasis on Stanhopeinae: combined molecular evidence. *American Journal of Botany*, **87**, 1842–56.

Whitten, W. M., Blanco, M. A., Williams, N. H., Koehler, S., Carnevali, G., Singer, R. B., Endara, L. and Neubig, K. M. (2007). Molecular phylogenetics of Maxillaria and related genera (Orchidaceae:Cymbidieae) based on combined molecular data sets. *American Journal of Botany*, **94**, 1860–89.

Xu, S., Schluter, P. M., Scopece, G., Breitkopf, H., Cozzolino, S. and Schiestl, F. P. (2011). Floral isolation is the main reproductive isolation barrier among sexually deceptive orchids. *Evolution*, DOI: 10.1111/j.1558-5646.2011.01323.x

Zurbuchen, A., Bachofen, C., Muller, A., Hein, S. and Dorn, S. (2010). Are landscape structures insurmountable barriers for foraging bees? A mark-recapture study with two solitary pollen specialist species. *Apidologie*, **41**, 497–508.

9

Why are floral signals complex?
An outline of functional hypotheses

ANNE S. LEONARD, ANNA DORNHAUS
AND DANIEL R. PAPAJ

9.1 Introduction

Plants produce a remarkable variety of displays to attract animals that transfer pollen. These floral displays are usually complex, broadcasting various combinations of visual, olfactory, gustatory, tactile, and thermal stimuli (Raguso 2004a). Even acoustic stimuli may be involved, as in the case of structural nectar guides used by echolocating flower-feeding bats (von Helversen and von Helversen 1999). Yet these sensorially complex advertisements likely evolved from an ancestor that primarily transmitted only chemicals, serving a defensive function (Pellmyr and Thein 1986). The subsequent amplification and elaboration of floral stimuli therefore offers an intriguing opportunity to study signal evolution. However, at present, we know surprisingly little about why floral displays consist of so many elements. This contrasts with progress in other areas: recently, researchers studying topics as diverse as sexual selection, warning displays, animal learning, and parent–offspring communication have explored the function of signal complexity (Rowe 1999; Candolin 2003; Hebets and Papaj 2005; Partan and Marler 2005).

Researchers studying plant–pollinator interactions, however, have not to date shown a comparable degree of interest in the topic of complex signals, as judged

Evolution of Plant–Pollinator Relationships, ed S. Patiny. Published by Cambridge University Press. © The Systematics Association 2012.

by an analysis of the research literature. An August 2010 search on the ISI Web of Science® database on journal articles published since 1995 returned only two on plant–pollinator topics containing the words "multimodal" and "signal-" in their titles, abstracts, or keywords (those articles being Raguso and Willis 2002; Kulahci et al. 2008). In comparison, the same search returned 59 articles on sexual selection topics.

A related search on bee learning provides further evidence that our understanding of how pollinators process and learn floral signals is based upon single sensory modalities studied in isolation. A search for articles published since 1995 that contain the words "bee" and "learning" returned 268 articles on how bees learn or process unimodal stimuli (olfactory = 142, visual = 112, tactile = 14), whereas only 12 focus on multimodal stimuli. This disparity is striking given that the great ethologist Karl von Frisch first showed that honeybees learn both colors and scents nearly a century ago (von Frisch 1914, 1919).

Not only would pollination biologists benefit from a better understanding of how pollinators interact with the complex floral signal, but such knowledge could also contribute significantly to our understanding of signal complexity in general. The study of plant–pollinator interactions integrates research from many disciplines, and thus is well-positioned to tackle fundamental questions regarding the function of signal complexity. From a proximate perspective, not only is there a wealth of information regarding the sensory and cognitive systems of pollinators (Chittka and Thomson, 2001; Giurfa, 2007), but the means by which flowers produce stimuli used by pollinators are relatively well-understood (Dudareva and Pichersky, 2006; Grotewold, 2006). Moreover, in comparison to animal signalers, plants offer opportunities for manipulative experiments that are often difficult, if not impossible, to run in other systems. For example, it is straightforward to change a display though use of artificial flowers (Makino and Sakai, 2007), through minor alteration of real flowers (Waser and Price, 1985), or through modification of individual components using both selective breeding (Odell et al. 1999) and molecular techniques (Hoballah et al. 2007). In contrast, students of animal communication may be limited to use of fewer and relatively sophisticated techniques, such as the use of robots (Taylor et al. 2008).

Although we are increasingly informed about *how* complex signals work in various plant–pollinator systems, we still know little about *why* they work as they do. We believe that a conceptual framework for the function of floral displays could stimulate research in that area. Here we present such a framework, in the form of testable hypotheses addressing the function of floral signal complexity. First, we describe hypotheses that propose benefits both to the sender (plant) and the receiver (pollinator). We highlight new research that points towards the role of uncertainty reduction in floral trait evolution. Later, we consider instances in which the benefit of multicomponent signaling accrues mainly to the plant, including situations in which the interests of plant and pollinator are in explicit conflict.

9.2 The multicomponent nature of floral stimuli

The suite of stimuli emitted by a typical flower constitutes a complex signal. Although definitions of both terms can vary, we use "signal" and "complex" as described in Hebets and Papaj (2005). Specifically, we follow Markl's (1983) definition of a signal as "a packet of energy or matter generated by a display or action of one organism (the signaler) that is selected for its effects in influencing the probability pattern of behavior of another organism (the receiver) via its sensory-nervous system in a fashion that is adaptive either to one or both parties." Much of the literature relevant to complex signaling refers to "multimodal", "bimodal", "multiple" "multicomponent" or "composite" signals. We use "complex" as a general term including signals that are multimodal (e.g. color + scent) or generally multicomponent (e.g. color + visual pattern). Aspects of visual complexity include flower color, iridescence, color contrast, intensity contrast, photoreceptor contrast, pattern, shape, size, symmetry, and the architecture of an inflorescence (e.g. Giurfa and Lehrer 2001; Glover and Whitney 2010). Likewise, floral scents are complex blends of volatile organic compounds (Knudsen et al. 2006). Flowers may even transmit olfactory "patterns," as scents often show a spatial gradient in concentration (Bergström et al. 1995) or vary in composition across floral structures (Dötterl and Jürgens 2005). Additional forms of olfactory complexity relevant to pollinators involve differences in the identity, abundance, and ratio of chemical components (Raguso 2008). Although less commonly studied, both the microtexture of petals (Kevan and Lane 1985) and the 3D morphology of the flower (Heinrich 1979) contribute to tactile complexity; and within the thermal modality, pollinators may perceive and discriminate among different floral temperatures (Whitney et al. 2008; Hammer et al. 2009).

9.3 Why are floral signals complex? Functional hypotheses

Producing complex displays probably entails costs for plants. Although the metabolic costs of adding a signal to a floral display are largely unknown, several traits that contribute to signal complexity are thought to incur these costs, such as flower size (Galen 1999), and to some extent, floral scent (Helsper et al. 1998; but see Grison-Pigé et al. 2001). Floral display complexity may also attract the attention of unintended receivers, such as herbivores drawn to floral scent (Theis 2006). What benefits offset these costs to the sender and thereby drive the evolution of complex signals?

Functional hypotheses for signals generally fall into two groups: content-based and efficacy-based (Guilford and Dawkins 1991; Hebets and Papaj 2005). Content-based hypotheses refer to the "what" of a signal. The proposition that different components of a complex signal convey different "messages" is an example of a content-based hypothesis. For example, studies of avian sexual signaling (Candolin 2003) commonly test the hypothesis that different male plumage traits provide females with different kinds of information used in mate choice, e.g. age, nutrition, parasite load, immunocompetence. In contrast, efficacy-based hypotheses refer to the "how" of a signal; that is, how a message might be more effectively transmitted, detected and/or processed using multiple components. Such a function likely accounts for the transmission of both visual and vibratory signals during courtship in wolf spiders: by transmitting both signals, a male is able to maintain a similar rate of courtship success even where transmission in one modality is blocked by darkness or vibration-impeding substrates (Fig 2 in Hebets and Papaj 2005).

Any particular signal is under selection for both content and efficacy; and some explanations for signal function have elements of both content and efficacy. Thus, hypotheses from each of these perspectives are not necessarily mutually exclusive, and an emphasis on the content versus efficacy of a floral signal is ultimately a matter of individual preference. The distinction can be especially subtle when the meaning of a signal is not separable from its contribution to efficacy, as occurs when a signal conveys a quality such as "location." Yet, even in this case, the content/efficacy framework can still help guide thinking about signal function. For example, consider the hypotheses that a floral pattern (1) conveys information about the location of nectar (signal content) and (2) facilitates close-range detection because of strong color contrast (signal efficacy). Both may be true, but a researcher interested in the evolution of honest floral signaling likely finds 1 most relevant, whereas a researcher studying the overlap between floral signals and pollinator visual systems might frame an experiment around 2. Table 9.1 organizes functional hypotheses likely to be relevant to interactions between plants and their pollinators into these categories.

9.3.1 Content-based hypotheses

One commonly-cited hypothesis for complex signals is the **multiple messages hypothesis**, which states that different components of the complex signal convey different kinds of information (Møller and Pomiankowski 1993; Johnstone 1996). What messages might the plant convey to pollinators? One component of a floral signal may permit pollinators to distinguish it from competitors ("species

Table 9.1 A framework of functional explanations for why floral signals consist of multiple components, adapted from reviews of animal communication by Hebets and Papaj (2005) and Rowe (1999).

Content-based hypotheses

Multiple messages		*Multiple signal components convey different information*
	Floral types	
	Species identity	
	Reward status	
	Reward quality	
	Reward type	
	Location	
	Location of patch within habitat	
	Location of flowers within patch	
	Location of reward within flower	
Redundant signals		*Multiple signal components improve accuracy of information*

Efficacy-based hypotheses

Signal transmission		
	Efficacy backup	*Multiple signal components facilitate transmission in variable environments*
	Efficacy tradeoff	*Multiple signal components overcome transmission constraints faced by each component independently*
Signal detection		*Multiple signal components are detected more successfully or quickly*
Signal processing		
	Parallel processing	*Multimodal signals processed more quickly along parallel neural pathways*
	Perceptual variability	*Multiple signal components reach pollinators with varying sensory systems*

Table 9.1 (*cont.*)

Inter-signal interaction hypotheses		
	Attention-altering	*One signal focuses pollinators' selective attention on a second signal*
	Context	*One signal component provides a context for pollinators' response to a second component*

identity"), as Wright and Schiestl (2009) have argued for the function of floral scent. Floral identity may allow pollinators to predict handling efficiency – bees may select particular species because they have learned how to extract nectar effectively (Chittka et al. 1999). Another display component could signal the presence of nectar ("reward status"). This information might be conveyed by a different chemical component (Howell and Alarcon 2007; Goyret et al. 2008) or by a visual cue (Thorp et al. 1975; Weiss 1991). Pollinators might also use a different display component to assess the value of the floral reward ("reward quality") – as in, for example, Raine and Chittka's (2007a) finding that bumblebees' (*Bombus terrestris*) innate preference for the color violet corresponds with a higher rate of nectar production by violet flowers. Yet another display component could indicate the kind of reward available ("reward type"), as in the visual or olfactory stimuli associated with nectar (Raguso 2004b) or pollen (Dobson and Bergström 1999). Finally, other display components may be useful in finding a patch of flowers from a distance (Williams and Dodson 1972), a flower within a patch (Hurley et al. 2009), or a reward within a flower, as in the case of floral patterns that function as nectar guides ("location" on different spatial scales) (Waser and Price 1985).

Another content-based hypothesis is the **redundant signals hypothesis** (Hebets and Papaj 2005). The redundant signals hypothesis proposes that floral signals are complex because from the standpoint of signal production (i.e. independent of environmental transmission or receiver processing), any one signal encodes information about the sender imperfectly; producing multiple, redundant, signals, which improves the overall accuracy of the (single) message (Bradbury and Vehrencamp 1998). Redundant signals may thus function as a tactical check on signaler honesty; in mate choice, for example, females may assess multiple male ornaments because faking the production of several quality indicators is thought to be difficult (Candolin 2003).

The equivalent of this kind of "quality control" in plant–pollinator relationships is perhaps best considered in the context of rewardless mimic flowers. These flowers exploit an animal for pollination service but provide nothing in return (Renner 2006). Some rewardless orchids, for example, mimic another flower species that does offer a reward; others mimic a female insect, luring males who attempt to mate with the flower but succeed only in picking up pollen and transferring it to the next mimic (Schiestl 2005). Transmitting additional signal components that provide pollinators with more information regarding floral identity could facilitate discrimination between rewarding flowers and unrewarding mimics, benefitting both the pollinator and the rewarding plant species. Support for such facilitation in discrimination is found in the bee learning literature. For example, Kunze and Gumbert (2001) found that *B. terrestris* learned more quickly to distinguish between two similar colors of artificial flower (one rewarding, one unrewarding) when they transmitted different scents than when they were unscented or transmitted the same scent. Additionally, in a discrimination learning test, Kulahci et al. (2008) found that *B. impatiens* showed the highest visitation rate to the rewarding flower type when flowers differed in two features (shape and scent) versus a single feature (shape only or scent only). Although these results are consistent with other explanations, one function of transmitting both a visual as well as an olfactory signal may be to provide pollinators with redundant information about floral identity.

9.3.2 Efficacy-based hypotheses

Efficacy-based hypotheses for complex signals propose that multiple components allow a plant's message to be more effectively transmitted through the environment and/or more effectively detected or processed by the pollinator.

All signals tend to degrade as they propagate through the environment (Bradbury and Vehrencamp 1998). Transmission-based hypotheses propose that multicomponent signals reduce the effects of environmental degradation of information produced by the sender. One such hypothesis, the **efficacy backup hypothesis**, states that flowers emit multiple stimuli so that under any given set of environmental conditions, at least one will convey information (cf. "robustness" in Ay et al. 2007). For example, flowers may produce both scent and visual stimuli so as to ensure that at least one kind of stimulus is useful, regardless of environmental conditions (Kaczorowski et al. unpublished data). On windy days, when scent is less localizable, visual components may be more useful; on overcast days or in deep shade, when visual stimuli are difficult to discern, scent may function more effectively. Although this explanation for multimodal signals seems highly intuitive, we know of no evidence even in a controlled semi-field situation to support this hypothesis.

The **efficacy tradeoff hypothesis** proposes that different components of the complex floral display solve different challenges in signal efficacy related to transmission or detection. Perhaps no single component can maximize efficacy on all counts, regardless of environmental variability. For example, scent may be more detectable at a distance than visual cues, while a visual cue may better allow a flower to be localized precisely once the pollinator is in the vicinity of the plant.

A recent study by Streinzer et al. (2009) supports such a scenario for the display of the sexually deceptive orchid *Ophrys heldreichii*. *O. heldreichii* transmits an olfactory signal that mimics the scent of female solitary bees and is detected by males at long distances (Kullenberg and Bergström 1976). However, the flower also has a conspicuously colored perianth. By comparing the responses of male long-horned bees (*Tetralonia berlandi*) at different distances from intact flowers versus flowers with the perianth removed, Streinzer et al. 2009 showed that presence of the color signal reduced search time when the bee was within 30 cm of the flower; and at this close range, disruption of the olfactory signal (increasing wind speed) had no effect on the bees' ability to locate the flower. At greater distances, however, removal of the perianth did not affect searching behavior, but search time slowed with increasing wind speed. Ultimately, the range at which signals in each modality operate may depend greatly upon pollinator ecology and sensory physiology (Giurfa et al. 1996; Spaethe et al. 2001; Balkenius et al. 2006) as well as plant species, as some flowers transmit more scent than others, and some are less visually detectable than others.

The environment not only degrades a signal but is also a source of competing signals and stimuli that can obscure a floral display. **Detection-based hypotheses** for floral complexity propose that multicomponent signals facilitate detection of the signal against this background noise. In this case, the benefit of signal complexity relates to the enhanced efficacy of detection by the sensory system of the pollinators, rather than enhanced transmission through the environment. Within the visual modality, floral size, color, and brightness are all likely to influence detectability (Chittka and Spaethe 2007) and thus a combination of these may convey additional benefits. Adding a signal in a different sensory modality can also increase detectability: human-based psychophysical research suggests that multimodal stimuli are detected both more quickly and successfully than unimodal stimuli (Stein and Meredith 1993; Rowe 1999). The benefit of increased detection is perhaps self-explanatory; additionally, even small increases in speed of detection may contribute to a higher nectar collection rate (Burns 2005), a factor that can directly affect reproductive success in species that make countless foraging decisions daily, such as bumblebees (Heinrich 1979; Pelletier and McNeil 2003).

Once a signal has been successfully transmitted and detected, it is processed by the receiver's nervous system. Could complex floral signals be processed more effectively than simple signals? The **parallel processing hypothesis** proposes that a complex signal, whose components are processed in parallel, conveys information more effectively than a signal that attempts to transmit the same amount of information in a single channel (modality or component). Parallel processing refers to the capacity of a modularized nervous system to process multiple streams of information more or less simultaneously, rather than sequentially. If complex stimuli are processed along parallel neural pathways, then use of multiple components may allow receivers to process a greater amount of information without sacrificing processing speed (Hebets and Papaj 2005). Even within one sensory modality, aspects of signals may be processed along parallel circuits, as is the case for visual processing of movement and color in mammals (Livingstone and Hubel 1988) and olfactory processing in many insect species (Galizia and Rossler 2010). Of course, information in different modalities may be processed at different speeds: humans for example, process an auditory stimulus 40–60 ms more quickly than a visual stimulus (Stein and Meredith 1993). A test of this hypothesis would thus require studying in more detail the decision times of receivers in response to multi-component and single-component signals. If all components of a signal have to be processed in order to make a correct decision, then total decision time when components are processed in parallel should be similar to the slowest of the individual components (Thomas 1996; Kulahci et al. 2008). Alternatively, when components are processed in series, total decision time would approximate the sum of the decision times for each component or modality separately. On the other hand, if processing only one component of the complex signal is sufficient to make a correct decision, then total decision time under parallel processing should approximate that of the component that can be processed fastest, whereas under serial processing decision, time would be determined by which component is analyzed first (and thus not necessarily the fastest component).

Finally, complex floral displays might be a response to **perceptual variability** among receivers (Hebets and Papaj 2005). It is well-established that pollinators show variability in sensory acuity both within and across species. For example, bumblebees of different sizes (Spaethe and Chittka 2003; Spaethe et al. 2007) and *M. sexta* reared on different quality larval diets (Goyret et al. 2009) show different sensitivities to visual and olfactory floral stimuli. Production of a multi-component signal might thus allow a flower to attract a wider range of pollinators. A recent experiment suggests that this function could apply as well across species: Muchhala et al. (2008) found that even though bats transfer the most pollen to the flowers of the tropical shrub *Aphelandra acanthus*, its flowers transmit a sweet scent attractive to hawkmoths, and also remain open during the day, attracting (visually-oriented) hummingbirds with a bright yellow color. In this case, the ability to attract hummingbirds as well as bats may reflect the value of pollen quality as well as quantity: 73 % of pollen transferred by bats was heterospecific, compared to only 6 % of the pollen brought by hummingbirds (Muchhala et al. 2008).

9.3.3 Inter-signal interactions

Often, the function of one signal may include altering the receiver's response to another signal (Hebets and Papaj 2005). For example, Kunze and Gumbert (2001) also reported that bumblebees learned to distinguish between two similar colors of flower more quickly when the flowers transmitted the same scent than when they were both unscented. Thus, beyond transmitting information about floral identity, the mere presence of floral scent affected the bees' response to visual stimuli. Such inter-signal interactions can be understood within either an efficacy framework (e.g. one signal enhances the detection or processing of another), or a content framework (e.g. the meaning of one signal depends upon the presence of a second signal).

One efficacy-based hypothesis that involves an inter-signal interaction relates directly to a cognitive process known as selective attention (Smith 1996; Dukas 2002). The **attention-altering hypothesis** proposes that one component facilitates detection of a second component by directing a receiver's attention selectively to that component. As with any explanation based on attention, this hypothesis implies that there are constraints on the pollinator's sensitivity to particular stimuli; in the absence of such constraints, the pollinator would always be in a state of maximal sensitivity to all floral stimuli at once. For example, a floral scent may focus a pollinator's attention on visual floral stimuli specifically, such as color or shape, or trigger a search image associated with a particular flower type (Goulson 2000). A testable prediction of this hypothesis is that when a floral scent is detected, a pollinator trades sensitivity to visually based predatory stimuli off against sensitivity to visually based floral stimuli. Or, one could test this prediction in nocturnally foraging moths, asking whether floral scent affects the ability to detect the ultrasound of insectivorous echolocating bats (Skals et al. 2003).

Some evidence that complex signals are better at capturing the attention of pollinators comes from the literature on flower constancy. Flower constancy refers to the tendency of a pollinator to visit one floral type even when other equally rewarding flower types are available. One explanation for flower constancy involves a constraint on the capacity of pollinators' working memory to contain multiple flower types (Chittka et al. 1999). Working memory (information stored for a short duration), and attention (information processed moment-to-moment) are inexorably linked (Dukas 2002). In support of the connection between limited working memory and constancy, field observations show that bees are more likely to be constant (choose a similar flower type) within a few seconds of leaving the preceding flower (Chittka et al. 1997), a timespan during which the last-visited flower type would be stored in working memory. Interestingly, bees show increased constancy as floral signal complexity increases (Gegear 2005; Gegear and Laverty 2005), a finding that would be consistent with a complex floral signal occupying more of the bees' working memory capacity than a relatively simpler signal.

A second, content-based, form of interaction is the **context hypothesis** (Hebets and Papaj 2005), which specifies that one component of a display provides the context in which the pollinator interprets another signal. Research in experimental psychology has shown that, in addition to learning to associate a stimulus with reward, subjects also learn "background" stimuli that provide context; subsequent removal of these cues can impair performance (Shettleworth 1998; Skow and Jakob 2005). For example, all pollinators encounter or learn stimuli in situations other than foraging – when locating and selecting host plants (e.g. Weiss and Papaj 2003; Goyret et al. 2008), home sites (Fauria et al. 2002), or mates. It is possible that one component of a floral display helps pollinators to distinguish between different contexts, triggering them to interpret and learn other floral signals. For example, bumblebees land more frequently on artificial flowers that transmit scent (Leonard et al. 2011), and Giurfa et al. (1995) found that flower-naïve honeybees would not land on unscented artificial flowers. These findings suggest that scent may provide a foraging context, priming pollinators to learn or recall floral stimuli (Raguso and Willis 2002; Goyret et al. 2007). A test of this hypothesis might include comparing the performance of individuals trained to learn colors in two contexts, for example, at the colony entrance versus at feeders (as in Worden et al. 2005) when a scent is present in one context versus when both are unscented.

Beyond helping pollinators to identify stimuli as belonging to a "foraging" context, floral stimuli may provide a context for learning and remembering other stimuli associated with floral identity. Psychological research suggests that forgetting may be caused in part by changes in background stimuli (i.e. the "context-change account of forgetting," Bouton et al. 1999); in an ecological setting, the background stimuli experienced by pollinators will vary substantially across time and space. If one component of the floral display (e.g. scent) was transmitted relatively constantly and consistently across different environments, then this component might facilitate recall of a second signal (e.g. color) by recreating the context in which that signal was first learned.

9.4 Uncertainty reduction and the complex floral signal

A number of the hypotheses in our framework suggest that the floral signal is an uncertain one from the pollinator's perspective. Pollinators searching for floral rewards may experience uncertainty at several levels, such as in locating a flower against a background of distracting stimuli (Goulson 2000) or in distinguishing among flowers of similar species. For example, in their survey of flowers visited by bumblebees (*B. terrestris*) in Würzburg (Germany), Raine and Chittka (2007b) report an average nectar collection rate (microliters/24 hours) of 2120 for *Salvia pratensis* and 520 for *S. verticillata*. A bee might therefore benefit by selectively visiting the more profitable species, *S. pratensis*. However, flowers of these two species present rather similar visual and tactile stimuli: both transmit strongly in bee UV–blue color space (Raine and Chittka 2007b), both are bilaterally symmetrical flowers with landing platforms, and both are arranged on vertical inflorescences. After gaining experience with the two species, what happens when a bee enters a new patch and locates a vertical inflorescence of UV-blue, bilaterally symmetrical flowers?

In a general sense, the answer depends upon the bee's level of uncertainty in discriminating between the two floral types, as well as the costs and benefits of landing on each. Uncertainty can be influenced by several factors: for example, the degree of signal overlap between the two flower types, environmental stimuli that obscure transmission of the floral signal, and internal "noise" in the bee's sensory processing system. Signal Detection Theory (SDT) (Green and Swets 1966; Wiley 2006) provides a framework for predicting pollinators' behavior in the face of this signal uncertainty. SDT can also be invoked to help explain why floral displays are complex signals.

A first assumption is that, from the perspective of the pollinator, uncertainty in any form is likely to be costly. It may cause pollinators to take longer to make decisions; for example, bees take longer to make landing decisions as the difficulty of a discrimination task increases (Chittka et al. 2009). In addition to time costs, there may be costs associated with errors in choice. These may be errors both of commission (e.g. "false alarm:" visiting an unprofitable flower type) and omission ("missed detection:" failing to land on a profitable flower type). In an SDT framework, we can model two similar flower types as transmitting overlapping distributions of stimuli along some perceptual axis, and assume that pollinators use a threshold-based rule to decide which stimuli to land upon (Fig 9.1a). In that case, the probability of false alarm and correct detection (=1–p[missed detection]) are related: a pollinator may be conservative, landing on few flowers and thus experiencing a low rate of false alarms but low rate of correct detections (Fig 9.1b); alternatively, a "cavalier" pollinator might land on almost all stimuli encountered, yielding a high rate of correct detections but also a high rate of false alarms (Fig 9.1c). Both false alarms and missed detections are not only potentially costly to the pollinator but may be costly to the plant as well. In general, any factor that causes pollinators to be less likely to locate, contact, and transport pollen to conspecific plants represents a potential loss of the plant's reproductive success.

Recently, researchers interested in SDT and floral signal evolution have utilized a classical psychophysical "peak shift" approach (Hanson 1959; Shettleworth 1998) to explore how uncertainty in distinguishing between flowers that differ in reward value affects where a pollinator should optimally locate its decision threshold. In these experiments, the optimal decision threshold minimizes the probability of false alarms while maximizing the probability of correct detections. In a peak shift experiment, subjects gain experience with two similar stimuli that differ in reward value. One stimulus, the "S+" provides a reward; the other stimulus, "S–" is unrewarding or punishing. During a test phase, subjects' responses are measured across a wide range of stimulus values. Rather than responding most strongly to the S+ value, subjects' strongest ("peak") response to test stimuli

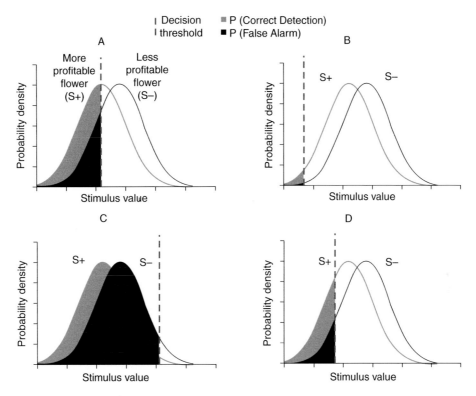

Fig 9.1 Pollinators discriminating between similar flowers that differ in reward value (S+ versus S–) face a classic signal detection problem. (A) When flower types transmit overlapping distributions of stimuli, pollinators may use a decision threshold to decide which stimuli to land upon. Regardless of where the decision threshold is located, pollinators will face a probability of making a false alarm (landing on S–, the less profitable flower) as well as a probability of correct detection (landing on S+, the more profitable flower, =1–p[missed detection]). (B) A pollinator might use a conservative decision threshold, not landing on most stimuli that it encounters. In this case, false alarms are reduced but correct detections are reduced as well. (C) A cavalier pollinator might land upon almost all stimuli it encounters, ensuring a high probability of correct detection, but a high probability of false alarms as well. (D) When false alarms are costly, relative to missed detections, a slightly conservative decision threshold minimizes false alarms while maximizing correct detections. This threshold is not located at the most common value of the rewarding flower (as in A) but is shifted in a direction away from the S–.

is often observed to be a novel stimulus value that is shifted in a direction away from the S– value. Lynn et al. (2005) used SDT to develop a functional account of the peak shift phenomenon: given uncertainty in distinguishing between S+ and S–, subjects' observed preference for a novel stimulus value (that is more distinct from S– than S+) can be interpreted as a strategy adopted to reduce the risk of false alarm (incorrectly responding to S–) (Fig 9.1d).

These expectations have relevance for plant–pollinator interactions: Lynn et al. (2005) showed that bumblebees (*B. impatiens*) trained to discriminate between a rewarding S+ and punishing S– that are similar colors of artificial flowers show peak shift when offered a wide array of floral colors. Moreover, the degree of the shift varied in relation to the nature of the relative costs of false alarms and missed detections. Wright et al. (2009) similarly demonstrated that honeybees show peak shift in response to olfactory stimuli (S+ and S– were two scents presented as blends in two ratios). In both these experiments, rather than responding most strongly to a floral signal they had learned was rewarding, bees instead preferred a novel stimulus value that was more different than S–. In ecological terms, this bias suggests that bees who experience uncertainty in discriminating between two similar floral signals associated with different payoffs (e.g. *S. pratensis* versus *S. verticillata*, or a model and its Batesian mimic) become conservative in their decision-making, seeking out not the average (or most common) signal value of the model (S+) but other, rarer values of the floral signal, in order to minimize the chances of visiting the less profitable flower. Of course, these rarer signal values may not even be the same species as S+; if not, then pollen transported from the model species is wasted.

In both Lynn et al. (2005) and Wright et al. (2009), the stimuli that bees encountered differed in only a single aspect (color or blend ratio). One way to explore whether more complex floral signals function to reduce pollinators' uncertainty would be to compare the magnitude of bees' peak shift in response to one stimulus type (e.g. color) when another stimulus (e.g. odor) is added to the floral signal. To this end, Leonard et al. (2011) performed a peak shift experiment on two groups of bumblebees. For one group, floral stimuli differed only in color; for another group, floral stimuli differed in both color and scent. While bees trained and tested on floral stimuli that differed in color showed a color preference shift away from the S– color, bees trained and tested on stimuli that differed in both color and scent did not. Bees thus behaved as though more certain about the color of the rewarding flower type when in the presence of floral scent. Intriguingly, bees showed this enhanced ability to identify the color of S+ without showing evidence that they learned the identity of the odor associated with it. These findings suggest an inter-signal interaction, whereby bees acquire better information about color in the presence of floral scent. While the process underlying this inter-signal interaction has yet to be determined, it is so far consistent with either the attention-altering or the context hypothesis.

9.5 Sender–receiver conflict: a third axis of explanation for multicomponent floral signals

Uncertainty reduction hypotheses generally assume that the complexity of a signal is mutually beneficial to sender and receiver. However, this is not always the case. In reality, plant–pollinator interactions are distributed along a continuum from mutualism to exploitation (Bronstein, 1994). In the gray area between the two extremes, we may find that different components of the signal have different patterns of joint fitness consequences for pollinator and plant. For example, flowers often transmit stimuli that attract pollinators in other contexts, as reviewed recently by Schaefer and Ruxton (2010) and Schiestl et al. (2010). Such sensory exploitation is thought to be the case with floral scent. Many components of floral scent also play a role in within- (pollinator) species communication (e.g. benzaldehyde, geraniol, linalool), and Schiestl (2010) has argued that this duality in function is not coincidence but, in many cases, a plant's exploitation of a pre-existing pollinator sensory bias for a chemical compound. Along a similar vein, Biesmeijer et al. (2005) have suggested that several characteristics of floral patterns (dark central spots, radiating lines) exploit a visual preference that evolved in the context of locating an entrance to a nest or burrow. These hypotheses suggest that plants may benefit by adding a signal that pollinators already find attractive; such a signal could improve the detectability of the flower, but potentially reduce pollinator fitness if attraction to the signal is strong enough to allow the plant to limit rewards.

Yet another example of potential sender-receiver conflict was put forward recently by Kessler and colleagues for the *Nicotiana*-hawkmoth interactions. Kessler et al. (2008) used transgenic *Nicotiana* plants to show that one component of the floral signal (benzyl acetone) served as an attractant for hawkmoth pollinators. This component is presumably of benefit both to hawkmoths, which receive a nectar reward, and to *Nicotiana*, which obtains pollination services. However, another chemical component of the floral signal, nicotine contained in the nectar, acts as a repellent. In field assays, flowers of transgenic plants lacking this component experienced a higher visitation rate than those of control plants. The authors argue that the repellent effect of nicotine thereby enhances plant fitness by promoting outcrossing. At the same time, the repellent probably reduces pollinator fitness by reducing the rate at which nectar is collected. Thus, one floral component (benzyl acetone) may be mutualistic, while the other (nicotine) is exploitative. This example implies that the complexity of the floral signal can effectively be a consequence of the complexity of the evolutionary game played between plant and pollinator. We thus propose that, in addition to content and efficacy, a third "manipulation" perspective on signal complexity that considers the sometimes-coincident, sometimes-conflicting interests of plant and pollinator.

9.6 Conclusions

Floral signals act in concert to influence pollinator behavior, yet pollinators' responses to these signals are usually studied in isolation. Without more research on complex, multimodal signaling by flowers, our understanding of floral evolution is incomplete: in many cases the function of one signal cannot be fully understood independently from another signal. Moreover, the study of floral complexity offers a new perspective on both the maintenance of pollination mutualisms and the relationship between signal complexity and receiver uncertainty. With the goal of spurring new research, we end this review with some open questions for plant–pollinator researchers intrigued by the complexity of floral displays to consider.

What are the production costs of floral complexity? Most research to date focuses on the benefits of floral signal complexity in terms of pollinator learning and decision-making. Yet floral displays are costly to produce and possible costs of complexity have not been quantified. Although Bradbury and Verhrencamp (1998) provide a framework describing the costs associated with signal production in different sensory modalities by animals, we know of no equivalent review covering plant signalers. Such an overview could provide interesting points of comparison. For example, what are the relative costs of a simple versus complex display, or of increasing the complexity of a display relative to changing the quality or quantity of reward? How can we measure the cost of adding a component/modality, relative to increasing signal strength in the same modality?

What advantages do multimodal signals offer plants in terms of dynamic signaling? Signals are not static entities, but vary over time and space. Among animal signalers, signal components of different modalities vary at different scales (Bradbury and Vehrencamp 1998), for example, a visual signal is generally modulated more rapidly than a chemical signal. But do these modality-specific expectations hold true for complex floral signals? In flowers, at least, both visual and olfactory signals can be changed quite rapidly (Weiss 1991; Knudsen et al. 2006). Moreover, could the use of a multimodal signal allow plants to adjust their signaling depending on environmental conditions, such as producing more scent under low light levels? Recent research suggests this kind of functional flexibility in animal signalers (Cheroske et al. 2009), but to our knowledge this possibility has not been directly explored in plants, although data exist regarding environmental influences on signal production (Jakobson and Olsen 1994). Given the recent interest in plants' induced responses to herbivory (Kessler et al. 2010), the potential for multimodal signals to modulate pollinator attraction seems to us to be a wide-open area of research.

References

Ay, N., Flack J. and Krakauer, D. C. (2007). Robustness and complexity coconstructed in multimodal signaling networks. *Philosophical Transactions of the Royal Society of London, Series B, Biological Sciences*, **362**, 441–7.

Balkenius, A., Rosen, W. and Kelber, A. (2006). The relative importance of olfaction and vision in a diurnal and nocturnal hawkmoth. *Journal of Comparative Physiology, A*, **192**, 431–7.

Bergström, G., Dobson, H. E. M. and Groth, I. (1995). Spatial fragrance patterns within the flowers of *Ranunculus acris* (Ranunculaceae). *Plant Systematics and Evolution*, **195**, 221–42.

Biesmeijer, J. C., Giurfa, M., Koedam, D. Potts, S. G., Joel, D. M. and Dafni, A. (2005). Convergent evolution: floral guides, stingless bee nest entrances, and insectivorous pitchers. *Naturwissenschaften*, **92**, 444–50.

Bouton, M. E., Nelson, J. B. and Rosas, J. M. (1999). Stimulus generalization, context change, and forgetting. *Psychological Bulletin*, **125**, 171–86.

Bradbury, J. W. and Vehrencamp, S. L. (1998). *Principles of Animal Communication*. Sunderland, MA: Sinauer Associates.

Bronstein, J. L. (1994). Conditional outcomes in mutualistic interactions. *Trends in Ecology and Evolution*, **9**, 214–7.

Burns, J. G. (2005). Impulsive bees forage better: the advantage of quick, sometimes inaccurate foraging decisions. *Animal Behaviour*, **70**, e1–e5.

Candolin, U. (2003). The use of multiple cues in mate choice. *Biological Reviews of the Cambridge Philosophical Society*, **78**, 575–95.

Cheroske, A. G., Cronin, T. W., Durham, M. F. and Caldwell, R. L. (2009). Adaptive signaling behavior in stomatopods under varying light conditions. *Marine and Freshwater Behaviour and Physiology*, **42**, 219–32.

Chittka, L. and Spaethe, J. (2007). Visual search and the importance of time in complex decision making by bees. *Arthropod–Plant Interactions*, **1**, 7–44.

Chittka, L. and Thomson, J. D. (2001). *Cognitive Ecology of Pollination*. Cambridge, UK: Cambridge University Press.

Chittka, L., Gumbert, A. and Kunze, J. (1997). Foraging dynamics of bumble bees: correlates of movements within and between plant species. *Behavioral Ecology*, **8**, 239–49.

Chittka, L., Skorupski, P. and Raine, N. E. (2009). Speed–accuracy tradeoffs in animal decision-making. *Trends in Ecology and Evolution*, **24**, 400–7.

Chittka, L., Thomson, J. D. and Waser, N. M. (1999). Flower constancy, insect psychology, and plant evolution. *Naturwissenschaften*, **86**, 361–77.

Dobson, H. E. M. and Bergström G. (1999). The ecology and evolution of pollen odors. *Plant Systematics and Evolution*, **222**, 63–87.

Dötterl, S. and Jürgens, A. (2005). Spatial fragrance patterns in flowers of *Silene latifolia*: Lilac compounds as olfactory nectar guides? *Plant Systematics and Evolution*, **255**, 99–109.

Dudareva, N. and Pichersky, E. (2006). *Biology of Floral Scent*. Boca Raton: CRC Press.

Dukas, R. D. (2002). Behavioural and ecological consequences of limited attention. *Philosophical Transactions of the Royal Society of London, Series B*, **357**, 1539–47.

Fauria, K., Dale, K., Colborn, M. and Collett, T. S. (2002). Learning speed and contextual isolation in bumblebees. *Journal of Experimental Biology*, **205** (7), 1009–18.

von Frisch, K. (1914). Der Farben und Formensinn der Bienen. *Zoologische Jarbücher (Physiologie)*, **35**, 1–188.

von Frisch, K. (1919). Über den Geruchssinn der Bienen und seine blütenbiologische Bedeutung. *Zoologische Jahrbücher (Physiologie)*, **37**, 2–238.

Galen, C. (1999). Why do flowers vary? *Bioscience*, **49**, 631–40.

Galizia, C. G. and Rossler, W. (2010). Parallel olfactory systems in insects: anatomy and function. *Annual Review of Entomology*, **55**, 399–420.

Gegear, R. J. (2005). Multicomponent floral signals elicit selective foraging in bumblebees. *Naturwissenschaften*, **92**, 269–271.

Gegear, R. J. and Laverty, T. M. (2005). Flower constancy in bumblebees: a test of the trait variability hypothesis. *Animal Behaviour*, **69**, 939–49.

Giurfa, M. (2007). Behavioral and neural analysis of associative learning in the honeybee: a taste from the magic well. *Journal of Comparative Physiology, A*, **193**, 801–24.

Giurfa, M. and Lehrer, M. (2001). Honeybee vision and floral displays: from detection to close-up recognition. In *Cognitive Ecology of Pollination*, ed. L. Chittka and J. D. Thomson. Cambridge, UK: Cambridge University Press, pp. 61–82.

Giurfa, M., Nunez, J., Chittka, L. and Menzel, R. (1995). Color preferences of flower-naive honeybees. *Journal of Comparative Physiology, A*, **177**, 247–259.

Giurfa, M., Vorobyev, M., Kevan, P. and Menzel, R. (1996). Detection of coloured stimuli by honeybees: minimum visual angles and receptor specific contrasts. *Journal of Comparative Physiology, A*, **178**, 699–709.

Glover, B. J. and Whitney, H. M. (2010). Structural colour and iridescence in plants: the poorly studied relations of pigment colour. *Annals of Botany*, **105**, 505–11.

Goulson, D. (2000). Are insects flower constant because they use search images to find flowers? *Oikos*, **88**, 547–552.

Goyret, J., Markwell, P. M. and Raguso, R. A. (2007). The effect of decoupling olfactory and visual stimuli on the foraging behavior of *Manduca sexta*. *Journal of Experimental Biology*, **210**, 1398–1405.

Goyret, J., Markwell, P. M. and Raguso, R. A. (2008). Context- and scale-dependent effects of floral CO_2 on nectar foraging by *Manduca sexta*. *Proceedings of the National Academy of Sciences USA*, **105**, 4565–70.

Goyret, J., Kelber, A., Pfaff, M. and Raguso, R. A. (2009). Flexible responses to visual and olfactory stimuli by foraging *Manduca sexta*: larval nutrition affects adult behaviour. *Proceedings of the Royal Society of London, Series B, Biological Sciences*, **276**, 2739–45.

Green, D. M. and Swets, J. A. (1966). *Signal Detection Theory and Psychophysics*. New York, NY: Wiley & Sons.

Grison-Pigé, L., Salager, J., Hossaert-McKey, M. M. and Roy, J. (2001). Carbon allocation to volatiles and other reproductive components in male *Ficus carica* (Moraceae). *American Journal of Botany*, **88**, 2214–20.

Grotewold, E. (2006). The genetics and biochemistry of floral pigments. *Annual Review of Plant Biology*, **57**, 761–80.

Guilford, T. and Dawkins, M. S. (1991). Receiver psychology and the evolution of animal signals. *Animal Behaviour*, **42**, 1–14.

Hammer, T. J., Hata, C. and Nieh, J. C. (2009). Thermal learning in the honeybee, *Apils mellifera*. *Journal of Experimental Biology*, **212**, 3928–23.

Hanson, H. M. (1959). Effects of discrimination training on stimulus generalization. *Journal of Experimental Psychology*, **58**, 321–34.

Hebets, E. A. and Papaj, D. R. (2005). Complex signal function: developing a framework of testable hypotheses. *Behavioral Ecology and Sociobiology*, **57**, 197–214.

Heinrich, B. (1979). *Bumblebee Economics*. Cambridge, MA: Harvard University Press.

Helsper, J. P. F. G., Davies, J. A., Bouwmeester, H. J., Krol, A. F. and van Kampen, M. H. (1998). Circadian rhythmicity in emission of volatile compounds by flowers of *Rosa hybrida* L. cv. Honesty. *Planta* **207**, 88–95.

von Helversen, D. and von Helverson, O. (1999). Acoustic guide in a bat-pollinated flower. *Nature* **398**, 759–60.

Hoballah, T. M., Gübitz, T., Stuurman, J., Broger, L., Barone, M., Mandel, T., Dell'Olivo, A., Arnold M., and Kulemeier, C. (2007). Single gene-mediated shift in pollinator attraction in *Petunia*. *The Plant Cell*, **19**, 779–90.

Howell, A. D. and Alarcon, R. (2007). *Osmia* bees (Hymenoptera:Megachilidae) can detect nectar-rewarding flowers using olfactory cues. *Animal Behaviour*, **74**, 199–205.

Hurley, T. A., Franz, S. and Healy, S. D. (2009). Do rufous hummingbirds (*Selasphorus rufus*) use visual beacons? *Animal Cognition*, **13**, 377–83.

Jakobsen, H. B. and Olsen, C. E. (1994). Influence of climatic factors on emission of flower volatiles. *Planta*, **192**, 365–71.

Johnstone, R. A. (1996). Multiple displays in animal communication: "backup signals" and "multiple messages". *Philosophical Transactions of the Royal Society of London, Series B, Biological Sciences*, **351**, 329–38.

Kessler, D., Diezel, C. and Baldwin, I. T. (2010). Changing pollinators as a means of escaping herbivores. *Current Biology*, **20**, 1–6.

Kessler, D., Gase, K. and Baldwin, I. T. (2008). Field experiments with transformed plants reveal the sense of floral scents. *Science*, **321**, 1200–2.

Kevan, P. G. and Lane, M. A. (1985). Flower petal microtexture is a tactile cue for bees. *Proceedings of the National Academy of Sciences USA*, **82**, 4750–2.

Knudsen, J. T., Eriksson, R., Gershenzon, J. and Ståhl, B. (2006). Diversity and distribution of floral scent. *Botanical Review*, **72**, 1–120.

Kulahci, I. G., Dornhaus, A. and Papaj, D. R. (2008). Multimodal signals enhance decision making in foraging bumble-bees. *Proceedings of the Royal Society of London, Series B, Biological Sciences*, **275**, 797–802.

Kullenberg, B. and Bergström, G. (1976). *Hymenoptera aculeata* males as pollinators of *Ophrys* orchids. *Zoologica Scripta*, **5**, 13–23.

Kunze, J. and Gumbert, A. (2001). The combined effect of color and odor on flower choice behavior of bumble bees in flower mimicry systems. *Behavioral Ecology*, **12**, 447–56.

Leonard, A. S., Dornhaus, A. and Papaj, D. R. (2011) Flowers help bees cope with uncertainty: signal detection and the function of floral complexity. *Journal of Experimental Biology*, **214**, 113–21 .

Livingstone, M. and Hubel, D. (1988). Segregation of form, color, movement, and depth: anatomy, physiology, and perception. *Science*, **240**, 740–9.

Lynn, S. K., Cnaai, J. and Papaj, D. R. (2005). Peak shift discrimination learning as a mechanism of signal evolution. *Evolution*, **59**, 1300–5.

Makino, T. T. and Sakai, S. (2007). Experience changes pollinator responses to floral display size: from size-based to reward-based foraging. *Functional Ecology*, **21**, 854–63.

Markl, H. (1983). Vibrational communication. In *Neuroethology and Behavioral Physiology*, ed. R. Huber and H. Markl. New York, NY: Springer, pp. 332–353.

Møller, A. P. and Pomiankowski, A. (1993). Why have birds got multiple sexual ornaments? *Behavioral Ecology and Sociobiology*, **32**, 167–76.

Muchhala, N., Caiza, A., Vizuete, J. C. and Thomson, J. D. (2008). A generalized pollination system in the tropics: bats, birds and *Aphelandra acanthus*. *Annals of Botany*, **103**, 1481–7.

Odell, E., Raguso, R. A. and Jones, K. N. (1999). Bumblebee foraging responses to variation in floral scent and color in snapdragons. *American Midland Naturalist*, **142**, 257–65.

Partan, S. and Marler, P. (2005). Issues in the classification of multimodal communication signals. *American Naturalist*, **166**, 231–45.

Pelletier, L. and McNeil, J. N. (2003). The effect of food supplementation on reproductive success in bumblebee field colonies. *Oikos*, **103**, 688–94.

Pellmyr, O. and Thien, L. B. (1986). Insect reproduction and floral fragrances: keys to the evolution of the angiosperms? *Taxon*, **35**, 76–85.

Raguso, R. A. (2004a). Flowers as sensory billboards: progress towards an integrated understanding of floral advertisement. *Current Opinion in Plant Biology* **7**, 434–40.

Raguso, R. A. (2004b). Why are some floral nectars scented? *Ecology*, **85**, 1486–94.

Raguso, R. A. (2008). Wake up and smell the roses: the ecology and evolution of floral scent. *Annual Review of Ecology, Evolution and Systematics*, **39**, 549–69.

Raguso, R. A. and Willis, M. A. (2002). Synergy between visual and olfactory cues in nectar feeding by naive hawkmoths, *Manduca sexta*. *Animal Behaviour*, **64**, 685–95.

Raine, N. E. and Chittka, L. (2007a). The adaptive significance of sensory bias in a foraging context: floral colour preferences in the bumblebee *Bombus terrestris*. *PLoS ONE*, **2** (6): e556.

Raine, N. E. and Chittka, L. (2007b). Nectar production rates of 75 bumblebee-visited flower species in a German flora (Hymenoptera:Apidae:*Bombus terrestris*). *Entomologia Generalis*, **30**, 191–2.

Renner, S. S. (2006). Rewardless flowers in the angiosperms and the role of insect cognition in their evolution. In *Plant–Pollinator Interactions*, ed. N. M. Waser and J. Ollerton. Chicago MI: University of Chicago Press, pp. 123–144.

Rowe, C. (1999). Receiver psychology and the evolution of multicomponent signals. *Animal Behaviour*, **58**, 921–31.

Schaefer, H. M. and Ruxton, G. D. (2010). Deception in plants: mimicry or perceptual exploitation? *Trends in Ecology & Evolution*, **24**, 676–84.

Schiestl, F. P. (2005). On the success of a swindle, pollination by deception in orchids. *Naturwissenschaften*, **92**, 255–64.

Schiestl, F. P. (2010). The evolution of floral scent and insect chemical communication. *Ecology Letters*, **13**, 643–56.

Schiestl, F. P., Johnson, S. D. and Raguso, R. A. (2010). Floral evolution as a figment of the imagination of pollinators. *Trends in Ecology & Evolution*, **25**, 382–3.

Shettleworth, S. J. (1998). *Cognition, Evolution and Behavior*. New York, NY: Oxford University Press.

Skals, N., Plepys, D. and Löfstedt, C. (2003). Foraging and mate-finding in the silver Y moth, *Autographa gamma* (Lepidoptera:Noctuidae) under risk of predation. *Oikos*, **102**, 351–7.

Skow, C. D. and Jakob, E. J. (2005). Jumping spiders attend to context during learned avoidance of aposematic prey. *Behavioral Ecology*, **17**, 34–40.

Smith, B. H. (1996). The role of attention in learning about odorants. *Biological Bulletin, Marine Biological Laboratory, Woods Hole*, **191**, 76–83.

Spaethe, J. and Chittka, L. (2003). Interindividual variation of eye optics and single object resolution in bumblebees. *Journal of Experimental Biology*, **206**, 3447–53.

Spaethe, J., Tautz, J. and Chittka, L. (2001). Visual constraints in foraging bumblebees, Flower size and color affect search time and flight behavior. *Proceedings of the National Academy of Sciences USA*, **98**, 3898–903.

Spaethe, J., Brockmann, A., Halbig, C., and Tautz, J. (2007). Size determines antennal sensitivity and behavioral threshold to odors in bumblebee workers. *Naturwissenschaften*, **94**, 733–9.

Stein, B. E. and Meredith, M. A. (1993). *The Merging of the Senses*. Cambridge: The MIT Press.

Streinzer, M., Paulus, H. G. and Spaethe, J. (2009). Floral colour signal increases short-range detectability of a sexually deceptive orchid to its bee pollinator. *Journal of Experimental Biology*, **212**, 1365–70.

Taylor, R. C., Klein, B. A., Stein, J. and Ryan, M. (2008). Faux frogs: multimodal signalling and the value of robotics in animal behaviour. *Animal Behaviour*, **76**, 1089–97.

Theis, N. (2006). Fragrance of Canada thistle (*Cirsium arvense*) attracts both floral herbivores and pollinators. *Journal of Chemical Ecology*, **32**, 1573–61.

Thomas, R. D. (1996). Separability and independence of dimensions within the same-different judgment task. *Journal of Mathematical Psychology*, **40**, 318–341.

Thorp, R. W., Briggs, D. L., Estes, J. R. and Erickson, E. H. (1975). Nectar fluorescence under ultraviolet irradiation. *Science*, **189**, 476–8.

Waser, N. M. and Price, M. V. (1985). The effect of nectar guides on pollinator preference: experimental studies with a montane herb. *Oecologia*, **67**, 121–6.

Weiss, M. R. (1991). Floral color changes as cues for pollinators. *Nature*, **354**, 227–9.

Weiss, M. R. and Papaj, D. R. (2003). Colour learning in two behavioural contexts: how much can a butterfly keep in mind? *Animal Behaviour*, **65**, 425–34.

Whitney, H. M., Dyer, A. G., Chittka, L., Rands, S. A., and Glover, B. J. (2008). The interaction of temperature and sucrose concentration on foraging preferences in bumblebees. *Naturwissenschaften*, **95**, 845–50.

Williams, N. H. and Dodson, C. H. (1972). Selective attraction of male Euglossine bees to orchid floral fragrances and its importance in long distance pollen flow. *Evolution*, **26**, 84–95.

Wiley, R. H. (2006). Signal detection and animal communication. *Advances in the Study of Behaviour*, **36**, 217–47.

Worden, B. D., Skemp, A. K. and Papaj, D. R. (2005). Learning in two contexts: the effects of interference and body size in bumblebees. *Journal of Experimental Biology*, **208**, 2045–53.

Wright, G. A. and Schiestl, F. P. (2009). The evolution of floral scent: the influence of olfactory learning by insect pollinators on the honest signaling of floral rewards. *Functional Ecology*, **23**, 841–51.

Wright, G. A., Choudhary, A. F. and Bentley, M. A. (2009). Reward quality influences the development of learned olfactory biases in honeybees. *Proceedings of the Royal Society Series B*, **276**, 2597–2604.

10

A survey on pollination modes in cacti and a potential key innovation

BORIS O. SCHLUMPBERGER

10.1 Introduction

Flower–pollinator interactions are among the most fascinating research areas in ecology. The ways in which such interactions evolve are influenced by a number of biotic and abiotic factors, such as habitat and climate, plant and animal morphology, phylogenetic background etc. The increasing amount of data on pollination ecology combined with more and more detailed and precise phylogenetic reconstructions based on (not only) molecular evidence, allows improvement of our understanding of how flower–pollinator relationships evolve. Plant lineages especially well-suited for the study of factors influencing the evolution of flower–pollinator interactions would contain pollination by various groups of pollinators, preferably both diurnal and nocturnal, including both vertebrates and invertebrates, and sufficient evolutionary shifts among pollination modes.

Cactaceae are ideal for this type of study for a number of reasons:

(1)   manageable number of species

(2)   great diversity of habitats

(3)   diverse growth forms

Evolution of Plant–Pollinator Relationships, ed S. Patiny. Published by Cambridge University Press. © The Systematics Association 2012.

(4) adaptation to several pollinator guilds, including bimodal and generalized pollination

(5) an increasingly solid phylogenetic base of data

Cacti inhabit a majority of the latitudinal range of the Americas, from 56° northern latitude in Canada to 50° south in Patagonia, plus the entire east–west extension of the continent (Anderson 2001). Their distribution also covers most of the altitudinal range of vascular plants, from sea level to about 4500 m (Anderson 2001). Cacti grow in a diverse range of temperate to tropical habitats with at least seasonal or/and local aridity, including such extremes as the Atacama Desert and Amazonian inundation forests.

This article aims to evaluate the current state of knowledge about pollination modes in cacti and the correlation of pollination modes with phylogenetics, growth form and distribution. For this purpose, known or proposed pollinators, growth form and distribution ranges were assigned to all cactus species (1433 species sensu Hunt et al. 2006). Furthermore, all phylogenetic data available were screened for indications of evolutionary shifts in pollination mode.

10.2 Cactus flowers and their pollinators

With relatively simple floral adaptations, cacti are mostly pollinated by four groups of animals: bees (Grant and Grant 1979; Simpson and Neff 1987; Schlindwein and Wittmann 1997; Reyes-Agüero et al. 2006; Roig-Alsina and Schlumpberger 2008), birds (Locatelli and Machado 1999a; Nassar et al. 2007), bats (Valiente-Banuet et al. 1996; Fleming and Valiente-Banuet 2002; Tschapka et al. 2008), and sphingid moths (Sazima and Sazima 1995; Locatelli and Machado 1999b; Raguso et al. 2003; Schlumpberger and Raguso 2008; Schlumpberger et al. 2009), or combinations thereof (Sahley 1996; Fleming et al. 2001; Diaz and Cocucci 2003; Schlumpberger and Badano 2005; Walter 2010). One species from the Sonoran desert, *Pachycereus schottii*, is pollinated by the moth *Upiga virescens* (Crambidae), an example of nursery pollination (Fleming and Holland 1998) similar to the *Yucca-Yucca* moth mutualism (Pellmyr 2003). Future field studies may reveal more cases of mixed pollination modes. For instance, in several species of *Echinopsis* s.l. with clearly sphingid-adapted flowers, bees play a significant role as backup pollinators (Schlumpberger, unpublished data).

10.3 The ancestral pollination syndrome of Cactaceae

Molecular phylogenetic data indicate that basal cacti evolved from a grade of the former Portulacaceae, i.e. the newly defined Anacampserotaceae, Portulacaceae

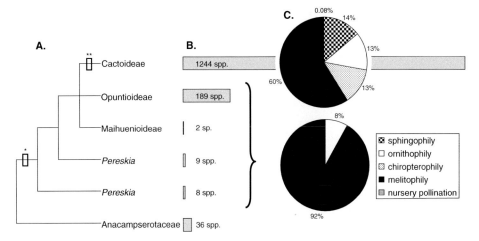

Fig 10.1 A. Simplified phylogeny of major lineages of the Cactaceae and their sister family Anacampserotaceae based on Nyffeler and Eggli 2010a and 2010b. B. Number of species in the respective lineages. C. Relative abundance of pollinator guilds associated with Cactoideae and basal lineages respectively. * = pericarp; ** = hypanthium.

and Talinaceae (Nyffeler and Eggli 2010a). These weedy, occasionally woody plants with moderate (sometimes no) succulence, share a relatively uniform flower type, with free tepals and few to many free anthers (4 to rarely 100). The ovary is superior (Talinaceae and Anacampserotaceae) to half-inferior or inferior (Portulacaceae), and the fruits are capsules. Flowers appear to be adapted to bees, which is supported by scattered reports (e.g. Valerio and Ramírez 2003). In addition, all species seem to be autogamous (Geesink 1969), with some reports on cleistogamy (Vengris et al. 1972) and rarely apomixis (Lombello and Pinto-Maglio 2009). In a study on Hawaiian *Portulaca* species, no nectaries were found (Kim and Carr 1990). Honeybees visiting cultivated *Portulaca grandiflora* foraged for pollen (Mogford 1980). Phylogenetic reconstruction suggests a sister group position of the Cactaceae to the Anacampserotaceae (Fig 10.1), with *Pereskia* as a basal grade (Nyffeler and Eggli 2010a). *Pereskia* is thought to resemble early cacti, with limited succulence and deciduous leaves. Flowers are reminiscent of those of the previously mentioned families, i.e. simple, open, bowl-shaped, choripetalous, and diurnal. However, the flowers already possess the typical fleshy pericarpel of the Cactaceae with half-inferior to inferior ovaries, and are, as most cacti, obligate outcrossers (Leuenberger 1986). Another synapomorphy of cacti are the fleshy, mostly non-dehiscent fruits. Although detailed studies on the spectrum of flower visitors and pollinators are lacking for *Pereskia*, these flowers are clearly adapted to pollination by bees (Leuenberger 1986). Bee pollination is also prevalent in other early-branching lineages of Cactaceae, such as the Maihuenioideae and the Opuntioideae (Fig 10.1; Reyes-Agüero et al. 2006).

10.4 A key innovation in floral morphology of cacti

In plant lineages with nectar as a major reward, modifications of the nectar compartment allow adaptation to different pollinator guilds. Such compartments can be formed by corolla tubes in sympetalous flowers, sepal tubes as in *Silene*, or petal or sepal spurs as in *Aquilegia* or *Disa*, respectively. In the case of floral nectar spurs, it was hypothesized that they may have been key innovations that promoted the diversification of certain plant lineages (Hodges 1997). The morphological preconditions in the lineages ancestral to Cactaceae, however, are limited, lacking spurs, tubes, or fused floral organs that may function as evolutionarily modifiable nectar compartments. Therefore, these plants are restricted to relatively unspecialized bee pollination with pollen reward, and a high amount of autogamy (see citations above).

In early diverging cactus lineages, bee pollination prevails by far (Fig 10.1), and the few shifts to new pollinator guilds are limited to hummingbird pollination (a few cases in the Opuntioideae and one possible case in the Pereskioideae). Some species like *Opuntia quimilo* may be incipient ornithophilous, showing pollination by bees and hummingbirds and lacking a number of typical adaptations to hummingbirds (Fig 10.2A; Diaz and Cocucci 2003). Adaptation to hummingbird pollination in these basal lineages is achieved by relatively simple floral modifications, such as limitation of access to the nectar resources by tubular arrangement of the filaments (e.g. *Tacinga*, Fig 10.2B) or tepals (e.g. *Nopalea*), loss of landing opportunities for insects, and a switch to red corollas. These modifications can be mainly understood as anti-bee adaptations, i.e. reducing the loss of pollen to less-efficiently pollinating bees (see Castellanos et al. 2004). In the species-rich subfamily Cactoideae, the fleshy pericarpel extends to form a hypanthium, which acts as a floral filter, generating a greater diversity of bird flowers (Fig 10.2D), and allowing adaptation to a wider array of pollinator guilds such as sphingids (Fig 10.2C), and bats (Fig 10.2E). The extensive possibilities of floral modification via a hypanthium are an important factor for diversification of pollination modes in Cactoideae, possibly a key innovation like nectar spurs in *Aquilegia* (Hodges 1997) that led to the diversification of Cactoideae: while the early diverging lineages comprise 206 species (two pollinator guilds, Fig 10.1), the Cactoideae comprise 1244 species (five pollinator guilds, Fig 10.1). Members of the Cactoideae lacking a hypanthium, such as in the genera *Rhipsalis* (Fig 10.2F) and *Blossfeldia*, are confined to bee pollination and are often capable of selfing.

10.5 Evolution of pollination modes and directionality of syndrome evolution

The evolution of pollination modes is considered biased by phylogenetic preconditions. For example, specialization may result in evolutionary dead-ends, and

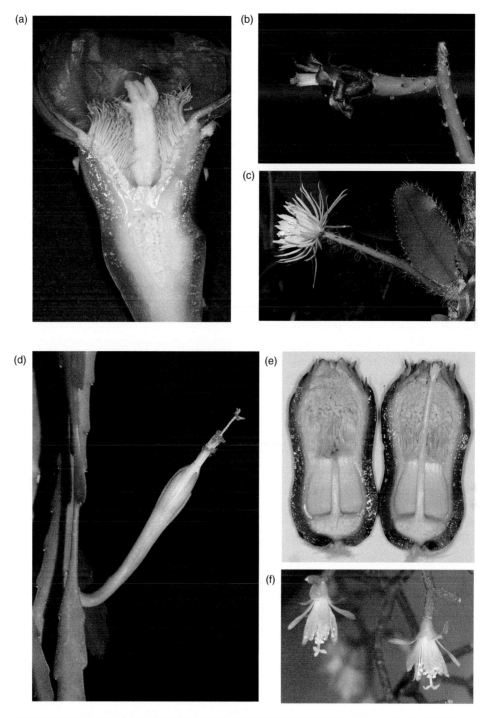

Fig 10.2 A. *Opuntia quimilo*, male-sterile flower: incipient ornithophily. B. *Tacinga funalis*, hummingbird flower without hypanthium. C. *Selenicereus wittii*, sphingid flower with elongated hypanthium. D. *Disocactus quetzaltecus*, hummingbird flower with elongated, colorful hypanthium. E. *Pilosocereus pachycladus* ssp. *pernambucensis*, bat flower with hypanthium, forming a massive nectar chamber. F. *Rhipsalis cereuscula*, member of the Cactoideae lacking a hypanthium. See plate section for color version.

the loss of certain traits such as flower pigments is usually irreversible (see Tripp and Manos 2008 and citations therein). In fact, certain shifts in pollination mode are often documented, as for example, bee to hummingbird (e.g. in *Penstemon*; Wilson et al. 2004; Wolfe et al. 2006), while other shifts, such as from hummingbird to bee pollination, are rarely observed (e.g. in Sinningieae, Gesneriaceae; Perret et al. 2003). In fact, adaptation to pollination by sphingids and bats was considered to be potentially non-reversable to other pollinator guilds (Tripp and Manos 2008). However, Fleming et al. (2009) list two examples of bird-pollination originating from chiropterophily, and two examples (among them the Saguaro cactus, *Carnegiea gigantea*) of evolution towards generalized pollination.

To assess the situation in cacti, the currently available phylogenetic data for the family were screened and 21 shifts in pollinator guilds were inferred (Table 10.1). With 48 % of all observed shifts, bee to hummingbird pollination is by far the most common case, followed by shifts from bee to sphingid pollination (23 %). In contrast, only one or two shifts were found for sphingid to bee or hummingbird adaptation, bat to crambid moth pollination (nursery pollination in *Pachycereus schottii*), bat to combined bat-hummingbird pollination, and bat to mixed pollination (*Carnegiea gigantea*). (Table 10.1).

Pollination by bees is the ancestral pollination mode in cacti, as is the case in many plant families (Tripp and Manos 2008). Pollination by sphingids, bats, and hummingbirds are derived conditions, and only a few shifts away from these pollination modes are deducable from current phylogenetic studies. For a better understanding of evolutionary transitions in pollination modes in cacti, more phylogenetic data are needed on the species level. The documentation of shifts away from bat and sphingid pollination contrasts data from previous publications dealing with other angiosperm families (cited above). However, certain shifts may result from hybridization or introgression, which may, for example, reverse the loss of floral pigments, thus facilitating the reversal from sphingid or bat pollination. In addition, white flowers may be exclusively or, in part, pollinated by bees or/and hummingbirds, as in *Echinopsis ancistrophora* (Schlumpberger et al. 2009), *E. atacamensis* ssp. *pasacana* (Schlumpberger and Badano 2005), *E. chiloensis* (Walter 2010), *E. thionantha* (Schlumpberger, unpublished data), or *Weberbauerocereus weberbaueri* (Sahley 1996). Interestingly, generalized pollination modes evolved several times from more specialized ones, e.g. in the genera *Echinopsis* s.l., *Pachycereus*, and *Carnegiea*.

10.6 Correlation of growth form and pollination syndromes

Pollination mode may depend on the plant's growth form to a certain degree, as the position of flower presentation influences their visibility and accessability

Table 10.1 Shifts in pollination modes in the Cactaceae, inferred from phylogenetic data (except the assumed shifts in *Echinocereus*).

Direction and number of shifts		Genera	References
Bee to hummingbird	10	*Pereskia*, *Opuntia* s.l., *Mammillaria*, *Echinopsis* (*Lobivia* clade), *Echinocereus**	Edwards et al. 2005; Griffith and Porter 2009; Butterworth and Wallace 2004
Bee to sphingid[1]	5	*Echinopsis* (*Lobivia* clade, *Echinopsis* s.str. clade, *Helianthocereus* clade, *Acanthocalycium* clade)	Schlumpberger & Renner in review
Bee to bat	1	Pachycereeae	Arias et al. 2005
Sphingid to bee	2	*Echinopsis* (*Lobivia* clade, *Echinopsis* s.str. clade)	Schlumpberger and Renner, in review
Sphingid to hummingbird	1	*Peniocereus*	Arias et al. 2005
Bat to hummingbird and bat[2]	1	*Pachycereus*	Hartmann et al. 2002; Arias et al. 2003
Bat to crambid moth	1	*Pachycereus*	Hartmann et al. 2002; Arias et al. 2003
Bat to mixed pollination[3]	1	*Carnegiea*	Hartmann et al. 2002; Arias et al. 2003

[1] Several cases of additional pollination by bees
[2] *Pachycereus marginatus* is pollinated by bats and hummingbirds
[3] *Carnegiea gigantea* is pollinated by doves, bats and bees
* no phylogenetic data for transitions in *Echinocereus*

for pollinators (and flower predators), and the evaporation of floral volatiles. Growth forms of cacti range from low, discoid, globular, or creeping habits to erect, shrubby, or tree-like habit, and, in the case of epiphytic or epilithic taxa, pendular growth. To grossly assess the correlation of growth form and pollination mode, cacti species are divided into two groups: species with flowers close to the surface of the substrate, i.e. globular, discoid or creeping; and species with exposed flower position, i.e. erect to columnar or pendular growth. The diversity of growth forms was evaluated for the four major guilds of pollinators (bees, birds, sphingids, bats) in the subfamily Cactoideae (1244 species). The strongest correlation with growth form (or rather flower exposition) was found in bat pollination, with 100 % of the respective species exhibiting columnar (94 %) or pendular (6 %) growth forms (Fig 10.3). To a lesser degree, about 80 % of observed

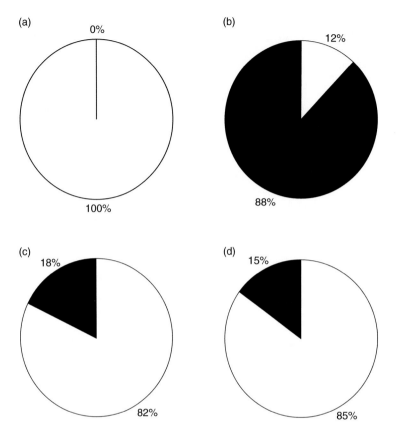

Fig 10.3 Pollination mode and flower position, i.e. growth form. The four major pollination modes in cacti: A. bat, B. bee, C. hummingbird, and D. sphingid pollination, and the percentages of low versus exposed flower position.

sphingid and bird pollinated flowers were exposed. In contrast, pollination by bees was the only syndrome in which low, mostly globular growth prevailed (Fig 10.3). In fact, genera confined to mostly globular growth like *Gymnocalycium* and *Parodia* are entirely adapted to bee pollination. In *Mammillaria* and the tribe Cacteae (with mostly globular growth) as a whole, two shifts to hummingbird pollination resulted in only four ornithophilous species (three of which evolved elongated growth).

These observations mainly reflect the pollinators' behavior: while bees usually forage in a relatively small area close to their nesting sites, birds and bats are capable of patrolling much larger areas, and sphingids are long-distance flyers not confined to nesting or roosting areas. Thus, it is important to present the flowers in positions that allow easy detection by these animals, which are, in contrast to many cactus-pollinating bees, mostly opportunistic flower

Fig 10.4 *Melocactus macracanthus*, spine-less cephalium as adaptation to ornithophily. See plate section for color version.

visitors. For scent-guided bats and sphingids, exposed flowers are advantageous because volatiles are freely released and odor plumes can develop with less disturbance than would be possible close to the ground. Similarly, elevated, uncovered flowers are needed for bats that detect their nectar resources with the help of echolocation (von Helversen et al. 2003). For both bats and birds, i.e. mostly hummingbirds, exposed flowers facilitate hovering and landing. In *Melocactus*, with globular growth and hummingbird-pollination, the spineless, in older plants elongated, terminal cephalia can be seen as a unique adaptation to this pollination mode (Fig 10.4).

Epiphytism, usually with pendant but also upright or climbing habit, evolved independently in several lineages of cacti (Anderson 2001; Korotkova et al. 2010). In an evaluation of the 128 currently accepted species (*sensu* Hunt et al. 2006), all four major pollination modes are represented. However, compared to the family average, bee and bat pollination are under-represented, while ornithophily and especially sphingophily are more common (Fig 10.5).

10.7 Pollination modes and geography

The latitudinal and altitudinal distribution of floral syndromes is biased by the distribution of the respective pollinator guilds (Ollerton et al. 2006 and citations therein). In particular, pollination by vertebrates seems to be centered around the tropics, with decreasing diversity towards the north, south, and higher elevations. For example, bat pollination in cacti is centered in tropical areas, with increasingly

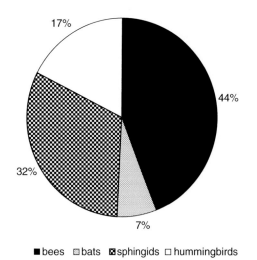

■ bees ⊞ bats ⊠ sphingids □ hummingbirds

Fig 10.5 Pollination modes in epiphytic cacti.

generalized pollination modes towards the northern limits of vertebrate-pollinated columnar cacti (Fleming et al. 2001). Similar observations were made in South America, towards the southern limits of columnar cacti distribution, where species are pollinated by representatives of different nocturnal and diurnal visitor guilds (Sahley 1996; Schlumpberger and Badano 2005; Walter 2010). The situation with elevation is similar: for flower-visiting New World bats, a strong correlation of species diversity and altitude was found, and bat pollination was considered to be primarily a lowland phenomenon (Fleming et al. 2009). In contrast, the distribution of both hummingbirds and sphingids stretches far into temperate zones, though their centers of diversity are the tropics (Schreiber 1978; Griggs 1997). Among the four relevant pollinator groups, bees (Apoidea) are the only guild with higher diversity in temperate zones (Michener 2000).

The known or putative pollination modes of all cacti species, i.e. the above mentioned four guilds plus nursery pollination, were plotted onto maps for a family-wide analysis. The most obvious result is that the cacti from northern- and southernmost habitats are exclusively bee-pollinated, reaching to about 57° north and to about 50° in the south (Fig 10.6). In contrast, hummingbird and sphingid-pollinated cacti do not extend beyond 37° and 40° south, respectively, and 40° and 33° north, respectively, despite their pollinators' larger natural distribution. The reason for this may be phylogenetic constraints: the cacti which inhabit the extreme southern and northern regions belong to the subfamily Opuntioideae and to the tribes Cacteae (*Coryphantha*) and Notocacteae (*Austrocactus*), all of which are almost exclusively adapted to bee pollination. The likely rather recent range extension towards the north and south respectively (after the last glaciation period), combined with the

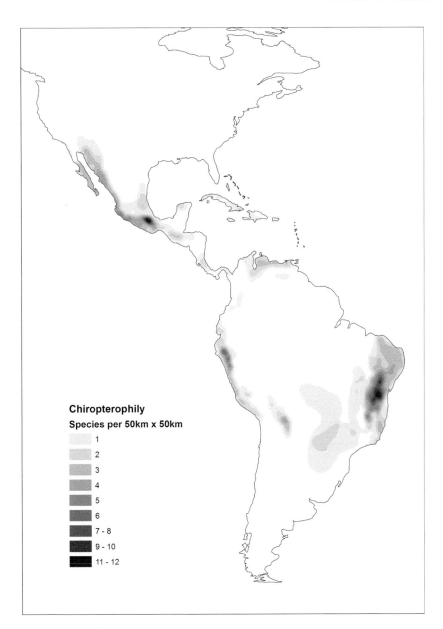

Fig 10.6 Distribution of pollination modes in cacti.

ancestral adaptations to bee pollination in these lineages, and both the dominance and efficiency of bees as pollen vectors in these habitats made shifts to sphingid or bird pollination unlikely. Another striking observation is the unequal distribution of sphingid pollination: while widespread in South America, sphingid pollination is rare north of a local diversity center in the Mexican Veracruz state. Sphingid-

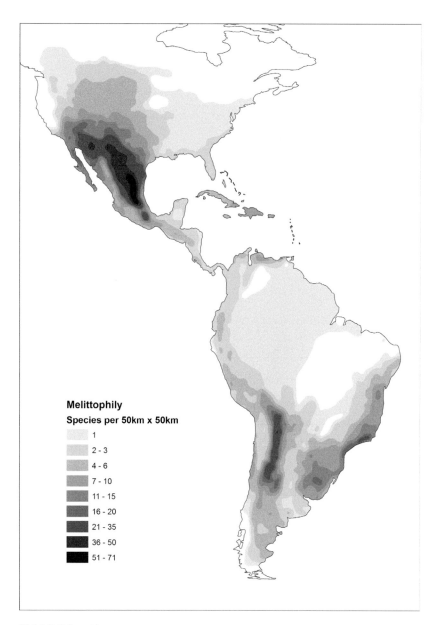

Fig 10.6 (cont.)

adapted cacti in South America are globular (e.g. *Echinopsis* s.l. and *Discocactus*) to columnar (e.g. *Cereus*) and those from Central and southern North America are often epiphytic (e.g. *Epiphyllum*). Globular, sphingophilous cacti are restricted to South America. Bat pollination in cacti is confirmed to be the pollination mode most restricted to the tropics, barely extending into subtropical regions (Fig 10.6). Centers of bat pollination in cacti are the Mexican Tehuacan Valley region

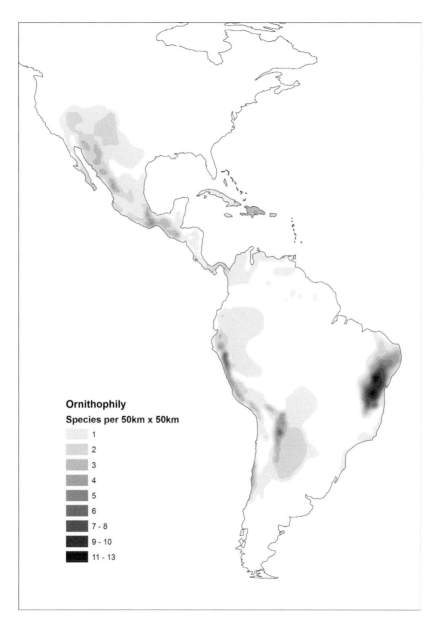

Fig 10.6 (*cont.*)

and eastern Brazil. While these cacti have exclusively columnar growth, Central American bat-pollinated cacti are epiphytic.

Altitudinal patterns are similar to latitudinal patterns. In low elevations all pollination modes occur, while melitophily is the most common one at high altitudes. For instance, bee pollination is the common syndrome among species from the high Andes, reaching more than 4500 m (Fig 10.6, Table 10.2; Anderson 2001

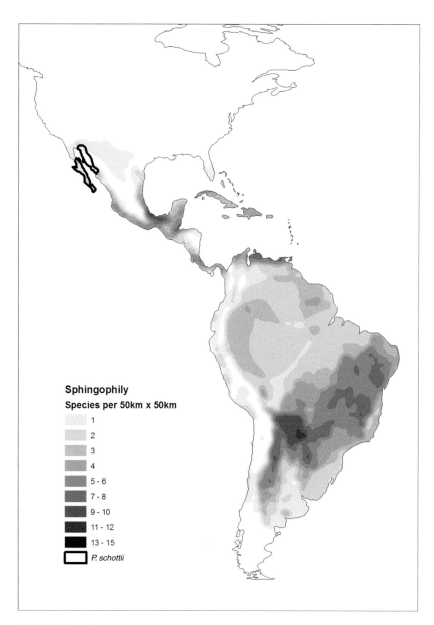

Fig 10.6 (*cont.*)

and personal observation). Although less widespread in the highest mountain regions, hummingbird-pollination can reach similar altitudes, with more than 4000 m in the case of *Echinopsis maximiliana* (syn. *Lobivia*) (Schlumpberger, unpublished data). Sphingophily in cacti reaches about 3600 m (Schlumpberger, unpublished data), and bat pollination only about 2800 m (Table 10.2; Sahley 1996).

Table 10.2 Altitudinal limits for pollination modes in cacti, inferred from literature and own observations.

Pollination mode	Altitudinal limit	Examples
Bee pollination	4500–4700	*Austrocylindropuntia lagopus*, *Cumulopuntia boliviana*, *Echinopsis chrysochete*
Hummingbird pollination	4000–4100	*Echinopsis maximiliana*, *Oreocereus celsianus*, *Oroya peruviana*
Sphingid pollination	3400–3600	*Echinopsis lageniformis*, *E. tacaquirensis*
Bat pollination	2500–2800	*Espostoa melanostele*, *Lasiocereus rupicola*, *Weberbauerocereus weberbaueri*

10.8 Conclusions

Cactus flowers are adapted to pollination by bees, birds, sphingids, bats, crambid moths (one species only), and combinations thereof. Phylogenetic data from early diverging cactus lineages and sister groups indicate that bee pollination is the ancestral pollination mode. Bee pollination is also the most common pollination mode in basal cacti lineages, with few shifts to hummingbird pollination, mostly achieved by modification of tepal or/and stamen arrangement. In contrast, in the subfamily Cactoideae, a hypanthium allows for modification of the nectar-containing flower structure and therefore the adaptation to a wider array of pollinators. Similar to flower spurs in *Aquilegia*, the formation of a hypanthium may be a key innovation for the Cactoideae, which contain more than 85 % of all cacti species. The current state of knowledge allows inference of 22 shifts in pollination modes, with almost 50 % from bee to hummingbird pollination. Several cases of shifts towards more generalized pollination were found. Thus far, no shift away from hummingbird pollination is known. The evaluation of data for the subfamily Cactoideae shows a strong association of pollinator guild and flower exposition: while flowers pollinated by birds, sphingids, and bats tend to have exposed positions, e.g. on columnar or pendular cacti, bee-pollinated cacti are mostly low-growing, i.e. globular. The distribution of pollination modes along latitudinal gradients is similar to altitudinal distribution: bat pollination is most restricted to tropical and lowland habitats, and pollination by bees is most widespread, reaching 57° north and 50° south, and altitudes of more than 4500 m.

Although some clear patterns and interesting observations can be drawn from the present body of knowledge, this can only be a first approach. Much more fieldwork is needed to comprehensively understand flower–pollinator associations in cacti, and more phylogenetic data, especially at the species level, are necessary to infer evolutionary processes in flower–pollinator interactions.

Acknowledgements

The author is grateful to W. Barthlott and associates (J. L. Geffert, J. Mutke, D. Rafiqpoor and N. Korotkova) who created the distribution maps. Kate Goodrich and Robert Raguso kindly revised the manuscript. Mats Winberg provided the photo of *Melocactus macracanthus*.

References

Anderson, E. F. (2001). *The Cactus Family*. Portland, OR: Timber Press.

Arias, S., Terrazas, T. and Cameron, K. M. (2003). Phylogenetic analysis of *Pachycereus* (Cactaceae, Pachycereeae) based on chloroplast and nuclear DNA sequences. *Systematic Botany*, **28**, 547–57.

Arias, S., Terrazas, T., Arreola-Nava, H. J., Vázquez-Sánchez, M. and Cameron, K. M. (2005). Phylogenetic relationships in *Peniocereus* (Cactaceae) inferred from plastid DNA sequence data. *Journal of Plant Research*, **118**, 317–28.

Butterworth, C. A. and Wallace, R. S. (2004). Phylogenetic studies of *Mammillaria* (Cactaceae): insights from chloroplast sequence variation and hypothesis testing using the parametric bootstrap. *American Journal of Botany*, **91**, 1086–98.

Castellanos, M. C., Wilson, P. and Thomson, J. D. (2004). "Anti-bee" and "pro-bird" changes during the evolution of hummingbird pollination in *Penstemon* flowers. *Journal of Evolutionary Biology*, **17**, 876–85.

Diaz, L. and Cocucci, A. A. (2003). Functional gynodioecy in *Opuntia quimilo* (Cactaceae), a tree cactus pollinated by bees and hummingbirds. *Plant Biology*, **5**, 1–9.

Edwards, E. J., Nyffeler, R. and Donoghue, M. J. (2005). Basal cactus phylogeny: implications of *Pereskia* (Cactaceae) paraphyly for the transition to the cactus life form. *American Journal of Botany*, **92**, 1177–88.

Fleming, T. H. and Holland, J. N. (1998). The evolution of obligate pollination mutualisms: senita cactus and senita moth. *Oecologia*, **114**, 368–75.

Fleming, T. H. and Valiente-Banuet, A. (2002). *Columnar Cacti and their Mutualists: Evolution, Ecology, and Conservation*. Tucson, AZ: University of Arizona Press.

Fleming, T. H., Geiselman, C. and Kress, W. J. (2009). The evolution of bat pollination: a phylogenetic perspective. *Annals of Botany*, **104**, 1017–43.

Fleming, T. H., Sahley, C. T., Holland, J. N., Nason, J. D. and Hamrick, J. L. (2001). Sonoran Desert columnar cacti and the evolution of generalized pollination systems. *Ecological Monographs*, **71**, 511–30.

Geesink, R. (1969). An account of the genus *Portulaca* in Indo–Australia and the Pacific (Portulacaceae). *Blumea*, **17**, 274–301.

Grant, V. and Grant, K. A. (1979). The pollination spectrum in the southwestern American cactus flora. *Plant Systematics and Evolution*, **133**, 29-37.

Griffith, M.P. and Porter, J.M. (2009). Phylogeny of Opuntioideae (Cactaceae). *International Journal of Plant Sciences*, **170**, 107-16.

Griggs, J. (1997). *All the Birds of North America*. London, UK: Harper Collins Publishers.

Hartmann, S., Nason, J. D. and Bhattacharya, D. (2002). Phylogenetic origins of *Lophocereus* (Cactaceae) and the Senita cactus: Senita moth pollination mutualism. *American Journal of Botany*, **89**, 1085-92.

Helversen, D. v., Holderied, M. W. and Helversen, O. v. (2003). Echoes of bat-pollinated, bell-shaped flowers: conspicuous for nectar-feeding bats? *Journal of Experimental Biology*, **206**, 1025-34.

Hodges, S. A. (1997). Floral nectar spurs and diversification. *International Journal of Plant Sciences*, **158**, 81-8.

Hunt, D., Taylor, N. and Charles, G. (2006). *The New Cactus Lexicon*. Milborne Port, UK: DH Books.

Kim, I. and Carr, G. D. (1990). Reproductive biology and uniform culture of *Portulaca* in Hawaii. *Pacific Science*, **44**, 123-9.

Korotkova, N., Zabel, L., Quandt, D. and Barthlott, W. (2010). A phylogenetic analysis of *Pfeiffera* and the reinstatement of *Lymanbensonia* as an independently evolved lineage of epiphytic Cactaceae within a new tribe Lymanbensonieae. *Willdenowia*, **40**, 151-72.

Leuenberger, B. (1986). *Pereskia* (Cactaceae). *Memoirs of the New York Botanical Garden*, **41**, 1-141.

Locatelli E. and Machado, I. C. S. (1999a). Comparative study of the floral biology in two ornithophilous species of Cactaceae: *Melocactus zehntneri* and *Opuntia palmadora*. *Bradleya*, **17**, 75-85.

Locatelli E. and Machado I. C. S. (1999b). Floral biology of *Cereus fernambucensis*: a sphingophilous cactus of restinga. *Bradleya*, **17**, 86-94.

Lombello, R. A. and Pinto-Maglio, C. A. F. (2009). Cytogenetic and reproductive biology of *Talinum triangulare* (Portulacaceae), an invasive plant with medicinal properties. Resumos do 55° Congresso Brasileiro de Genética.

Michener, C. D. (2000). *The Bees of the World*. Baltimore, MD: Johns Hopkins University Press.

Mogford, D. J. (1980). Colour discrimination in the pollination of *Portulaca grandiflora* Hook., by *Apis mellifera* L. *Journal of South African Botany*, **46**, 121-5.

Nassar, J. M., Ramírez, N., Lampo, M., González, J. A., Casado, R. and Navaz, F. (2007). Reproductive biology and mating system estimates of two Andean Melocacti, *Melocactus schatzlii* and *M. andinus* (Cactaceae). *Annals of Botany*, **99**, 29-38.

Nyffeler, R. and Eggli, U. (2010a). Disintegrating Portulacaceae: a new familial classification of the suborder Portulacineae (Caryophyllales) based on molecular and morphological data. *Taxon*, **59**, 227-40.

Nyffeler, R. and Eggli, U. (2010b). A farewell to dated ideas and concepts: molecular phylogenetics and a revised suprageneric classification of the family Cactaceae. *Schumannia*, **6**, 109-49.

Ollerton, J., Johnson, S. D. and Hingston, A. B. (2006). Geographical variation in diversity and specificity of pollination systems. In *Plant–Pollinator Interactions: from Specialization to Generalization*, ed. N. M. Waser and J. Ollerton. Chicago, MI: University of Chicago Press.

Pellmyr, O. (2003). Yuccas, yucca moths, and coevolution: a review. *Annals of the Missouri Botanical Garden*, **90**, 35–55.

Perret, M., Chautems, A., Spichiger, R., Kite, G. and Savolainen, V. (2003). Systematics and evolution of tribe Sinningieae (Gesneriaceae): evidence from phylogenetic analyses of six plastid DNA regions and nuclear ncpGS. *American Journal of Botany*, **90**, 445–60.

Raguso R. A., Henzel C., Buchmann S. L. and Nabhan G. P. (2003). Trumpet flowers of the Sonoran desert: floral biology of *Peniocereus* cacti and Sacred *Datura*. *International Journal of Plant Sciences*, **164**, 877–92.

Reyes-Agüero, J. A., Aguirre, J. R. and Valiente-Banuet, A. (2006). Reproductive biology of *Opuntia*: a review. *Journal of Arid Environments*, **64**, 549–85.

Roig-Alsina, A. and Schlumpberger, B. O. (2008). The cactus-specialist bees of the genus *Brachyglossula* Hedicke (Hymenoptera, Colletidae): notes on host associations and description of a new species. *Journal of the Kansas Entomological Society*, **81**, 84–91.

Sahley, C. T. (1996). Bat and hummingbird pollination of an autotetraploid columnar cactus, *Weberbauerocereus weberbaueri* (Cactaceae). *American Journal of Botany*, **83**, 1329–36.

Sazima, W. R. and Sazima M. (1995). Hawkmoth pollination in *Cereus peruvianus*, a columnar cactus from southeastern Brazil. *Flora*, **190**, 339–43.

Schlindwein, C. and Wittmann, D. (1997). Stamen movements in flowers of *Opuntia* (Cactaceae) favour oligolectic pollinators. *Plant Systematics and Evolution*, **204**, 179–94.

Schlumpberger, B. O. and Badano, E. I. (2005). Diversity of floral visitors to *Echinopsis atacamensis* ssp. *pasacana* (Cactaceae). *Haseltonia*, **11**, 18–26.

Schlumpberger, B. O. and Raguso, R. A. (2008). Geographic variation in floral scent of *Echinopsis ancistrophora* (Cactaceae); evidence for constraints on hawkmoth attraction. *Oikos*, **117**, 801–14.

Schlumpberger, B. O., Cocucci, A. A., Moré, M., Sérsic, A. N. and Raguso, R. A. (2009). Extreme variation in floral characters and its consequences for pollinator attraction among populations of an Andean cactus. *Annals of Botany*, **103**, 1489–1500.

Schreiber, H. (1978). Dispersal centres of Sphingidae (Lepidoptera) in the neotropical region. *Biogeographica*, **10**, 1–195.

Simpson, B. B. and Neff, J. L. (1987). Pollination ecology in the arid Southwest. *Aliso*, **11**, 417–40.

Tripp, E. A. and Manos, P. S. (2008). Is floral specialization an evolutionary dead-end? Pollination system transitions in *Ruellia* (Acanthaceae). *Evolution*, **62**, 1712–36.

Tschapka, M., Helversen, O. v. and Barthlott, W. (2008). Bat pollination of *Weberocereus tunilla*, an epiphytic rain forest cactus with functional flagelliflory. *Plant Biology*, **1**, 554–559.

Valerio, R. and Ramírez, N. (2003). Depresión exogámica y biología reproductiva de *Talinum Paniculatum* (Jacq.) Gaertner (Portulacaceae). *Acta Botánica Venezuelica*, **26**, 111–24.

Valiente-Banuet, A., Arizmendi, M. C., Rojas-Martinez, A., and Dominguez-Canseco, L. (1996). Ecological relationships between columnar cacti and nectar feeding bats in México. *Journal Tropicla Ecology*, **12**, 103–19.

Vengris, J., Dunn, S., and Stacewicz-Sapuncakis, M. (1972). Life history studies as related to weed control in the northeast. 7. Common purslane. *Massachusetts Agricultural Experiment Station Research Bulletin*, **598**, 1–44.

Walter, H. (2010). Floral biology of *Echinopsis chiloensis* ssp. *chiloensis* (Cactaceae): evidence for a mixed pollination syndrome. *Flora*, **210**, 757–63.

Wilson, P., Castellanos, M. C., Hogue, J. N., Thomson, J. D. and Armbruster, W. S. (2004). A multivariate search for pollination syndromes among penstemons. *Oikos*, **104**, 345–61.

Wolfe, A. D., Randle, C. P., Datwyler, S. L., Morawetz, J. J., Argueas, N. and Diaz, J. (2006). Phylogeny, taxonomic affinities, and biogeography of *Penstemon* (Plantaginaceae) based on ITS and cpDNA sequence data. *American Journal of Botany*, **93**, 1699–713.

11

Zygomorphy, area, and the latitudinal biodiversity gradient in angiosperms

JANA C. VAMOSI AND STEVEN M. VAMOSI

11.1 Introduction

In angiosperms, floral symmetry (Sargent 2004) is one of several traits, including mating system (Vamosi and Vamosi 2004), fruit type (Smith 2001), growth form (Verdú 2002), and distribution (Ricklefs and Renner 1994; Jansson and Davies 2008), that have been investigated in terms of their roles as key innovations leading to differential diversification rates in angiosperms. When a key innovation evolves, either through the evolution of a new structure or a transition from one state to another, it is thought that this will lead to what is known as phylogenetic tree imbalance (Fig 11.1). The existence and putative mode of action of key innovations have a long history in evolutionary biology (Farrell et al. 1991; Heard and Cox 2007). Darwin was initially reluctant to embrace the idea of key innovations because they did not adhere to the ideas of gradualism. Yet, the rapid diversification of flowering plants presented what he coined as an "abominable mystery" that could only be reconciled with theories of gradualism if one imagined that certain traits conferred higher levels of diversification. Whether any of the aforementioned traits presents a consistent overarching key to the abominable mystery has remained elusive (Davies et al. 2004a). Although previous studies have found associations between particular states of various traits and diversification, most of these have focused on individual traits and rather limited numbers of sister

Evolution of Plant–Pollinator Relationships, ed S. Patiny. Published by Cambridge University Press. © The Systematics Association 2012.

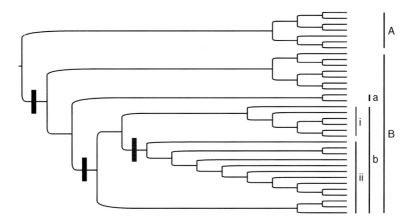

Fig 11.1 Hypothetical effects of key innovations (extrinsic: colonization of new ecozone; intrinsic: evolution of zygomorphic flowers) on diversification. Colonization of a new ecozone (indicated with left-most brown dash) occurs twice, initially leading to greater diversification in clade *B* than in its sister group *A*. Within clade *B*, further variation in diversification rates follows from subsequent colonization of a second new ecozone (indicated with second brown dash; partially leading to the difference in species richness between sisters groups *a* and *b*), and the evolution of zygomorphy (indicated with blue dash; boosting the total species richness of lineage *b*, and leading to the difference in species richness between sister groups *i* and *ii*). Figure modified from Vamosi and Vamosi (2011).

groups. Furthermore, geographical aspects of particular lineages have been relatively neglected, despite their inherent influences on speciation and extinction rates (Gaston 1998; Orme et al. 2006).

A recent multivariate phylogenetic investigation revealed that area was the predominant predictor of species richness (Vamosi and Vamosi 2010). Although herbaceousness was associated with larger area and was thus a possible indirect trigger of diversification, zygomorphy was the single most important intrinsic trait with an additive effect upon species richness (Table 11.1). That is, for a given area zygomorphic families generally had higher species richness than actinomorphic ones. We thus begin our consideration of these possible explanations of diversification with a brief introduction to the theoretical frameworks proposed to explain how the evolution of zygomorphy may affect plant–pollinator relationships and the geographical scale of speciation. Secondly, we assess the frequency of transitions from actinomorphic to zygomorphic flowers and what effect it may have on species richness in descendant lineages. Finally, we ask whether there are important interactions to consider, specifically whether zygomorphy has differing effects on diversification within tropical versus temperate landmasses.

Table 11.1 The effect of geographical versus intrinsic traits on species richness (SR) and geographical extent (GE) in angiosperms. Note that the large effect of available area overrides many of the effects of other traits and some traits may appear to be key innovations through their effects on increasing the available area (e.g. herbaceousness) while others (e.g. zygomorphy) appear to operate additively by increasing SR per unit area. Other traits (e.g. breeding system, fleshy fruits) do not appear to be very important and may have hitchhiked upon other more important traits.

Trait	Effect on SR	Effect on GE	Phylogenetic signal?
Available area (AA)	Y	Y	N
Age	N	N	—
Zygomorphy	Y	N	Y
Fleshy fruits	N	N	N
Herbaceousness	N	Y	Y
Tropicality (T)	Y	Y	N
Breeding System	N	N	N
AA × T	N	Y	—

11.2 Theoretical background of diversification dynamics

11.2.1 Area versus zygomorphy in diversification patterns

The tremendous variation among clades in species richness sparked widespread support for the idea that traits influence speciation and/or extinction rates. In plants, putative evolutionarily successful traits include, among other morphological attributes, many floral traits involved in sexuality, e.g. self-incompatibility (Igic et al. 2008) and pollinator guilds (biotic pollination versus abiotic pollination (Ricklefs and Renner 1994)), as well as more specific surrogates of specialization in pollination, e.g. floral asymmetry (Sargent 2004). Conversely, there have been some floral traits associated with evolutionary dead-ends, including dioecy (Heilbuth 2000; Vamosi and Vamosi 2004) and selfing (Igic et al. 2008).

The reasons why certain traits are associated with increased or decreased diversification are often intuitively appealing: some traits inherently encourage speciation via increased genetic diversity (self-incompatibility), or an association with increased opportunities for character displacement (floral asymmetry and pollinator specialization). Other traits may be affiliated with less stochastic

Table 11.2 Correlations between zygomorphy and other traits ($df = 407$ in all cases). Although the magnitude of the correlations is never particularly strong, they are generally higher for other intrinsic traits than for extrinsic ones. *P*-values are provided for reference and are not adjusted for multiple comparisons.

Trait	r	P-value
(i) Intrinsic		
Herbaceousness	0.237	< 0.0001
Fleshy fruits	−0.159	0.001
Cosexuality	0.127	0.010
(ii) Extrinsic/geographical		
Available area	0.104	0.035
Tropicality	0.097	0.050

pollen receipt (hermaphroditism versus dioecy (Vamosi and Otto 2002)). All of the traits have underlying correlations with one another (Vamosi and Vamosi 2004; Vamosi et al. 2003; Vamosi and Vamosi 2005a), although the strength of these correlations varies for different traits (Table 11.2), making it difficult to tease apart the effect of any single trait. In fact, whether any single trait provides a consistent cue to the long-term stability of certain traits has received rather equivocal support (Davies et al. 2004a).

Despite these doubts regarding the ability of any particular floral trait to consistently spur increased diversification, at least in isolation, one intrinsic trait of relevance here – zygomorphic flowers – has received reasonable empirical support. Sargent (2004) applied a sister-group comparison approach to determine whether lineages (families or groups of families) with zygomorphic flowers tended to contain more species than their sister groups. Using the most current phylogenetic hypothesis available at the time (Soltis et al. 1999), Sargent (2004) identified 19 sister groups in which the two descendant lineages differed in the state of the trait of interest. For example, the zygomorphic family, Vochysiaceae (210 spp.), was found to contain more species than its actinomorphic sister family, Heteropyxidaceae (three spp.). Overall, in 15 out of 19 sister-group comparisons, the lineage with zygomorphic flowers contained more species than its sister group, which was significantly greater than the random expectation (see Sargent 2004).

Subsequently, two concerns were raised about the analyses, although the main results have held up in both cases. First, Sargent (2004) analyzed untransformed species diversities, which does not account for the multiplicative nature of diversification (Vamosi and Vamosi 2005b). However, a subsequent reanalysis of log-transformed species richness values (Vamosi and Vamosi, 2005b) upheld the original result. Second, a similar analysis was conducted to that of Sargent (2004) using an expanded dataset with an angiosperm supertree, which contained 22 sister-group comparisons (Kay et al. 2006). Because Kay et al. (2006) observed what they believed to be a U-shaped distribution of proportion of species in the clade with the trait of interest (*sensu* Vamosi and Vamosi 2005b), which may lead to elevated Type I error rates, they largely discounted the role of zygomorphy in elevated diversification of angiosperms. However, as the authors admitted, their analyses did actually not overturn Sargent's (2004) original finding, with the zygomorphic lineage containing more species than its sister group in 16 of 22 comparisons. Furthermore, the distribution (see Fig 17.2b in Kay et al. 2006) was actually more J-shaped than U-shaped. Taken together with the recent findings of Vamosi and Vamosi (2010), which supported a similar conclusion as Sargent (2004) despite applying a whole-tree analysis approach to an updated phylogeny, zygomorphic flowers appear to be associated with higher species richness than actinomorphic flowers.

The low explanatory power of the majority of the key innovations discussed above in determining tree imbalance has led some researchers to posit that much evolutionary success of any given lineage lies with extrinsic random events early in its history (Ricklefs 2003). One such critical extrinsic determinant of evolutionary success that could be especially important is an early rise in geographical extent i.e. the amount of space occupied by a family (Ricklefs 2003; Davies et al. 2004b; Jansson and Davies 2008), yet whether the degree to which increases in geographical extent are simply stochastic or influenced by intrinsic biological traits are not clearly understood (Phillimore et al. 2006). It has been suggested that neither geographical nor biological traits determine diversification on their own but rather certain traits (or combinations thereof) may stimulate diversification within a particular geographical context (De Queiroz 2002).

Allopatric speciation rates may increase with geographical extent as the probability of populations becoming reproductively isolated increases (Owens et al. 1999; Losos and Schluter 2000). Furthermore, there may be a tendency for certain traits to be associated with geographical extent, which may explain why a relationship between diversification and a trait has been observed. There is some evidence that phylogenetic relatedness predicts geographical extent (i.e. it exhibits a phylogenetic signal) in marine bivalves (Roy et al. 2009) and gastropods (Jablonski 1987) but not birds (Webb and Gaston 2003). Specifically, dispersal traits have emerged as most influential in determining range size in animal lineages (Bohning-Gaese et al. 2006; Roy et al. 2009) and associations between seed/fruit size and geographical extent in plants may indicate a common underlying mechanism (Morin and Chuine 2006; Kolb et al. 2007). Some indications also exist for relationships between large geographical extent and an outcrossing mating system (Lowry and Lester 2006). Thus, if zygomorphic lineages have greater ranges, any association between zygomorphy and greater species richness may be via the effect of zygomorphy on geographical extent.

Apart from effects on geographical extent, latitude is typically acknowledged to be a surrogate metric for other variables, such as increased humidity and temperature (Allen et al. 2006). Incorporating metrics of energy availability (UV, temperature, evapotranspiration) into models of latitude, geographical extent, and diversification rate, contrasts between 86 angiosperm sister-families revealed that geographical extent and latitude were inextricably linked and had greater influences on diversification rates than metrics of energy availability (Jansson and Davies 2008). Considering that latitude was retained in the minimum adequate models, it suggests that other variables, potentially correlated with latitude, may play a role in driving diversification rates, although plant–pollinator relationships were not examined. These results indicate that complex interactions between key traits, latitude, and geographical extent are likely at play. For instance, evidence suggests that zygomorphy is associated with more specialized pollination (Sargent 2004), which may be (i) more prevalent in the tropics (Olesen and Jordano 2002) and (ii) associated with small geographical extents (Stevens 1989). Such observations suggest the need for a close examination of the effects of floral symmetry on geographical extents and diversification.

11.2.2 Ecological limits versus diversification rates

Some ambiguity in the effects of traits on diversification results from the different metrics used to characterize evolutionary success. For example, recent studies have called into question the wisdom of using diversification rates (i.e. log-transformed species richness of a lineage divided by its age) because older lineages will be unduly penalized. Even when estimators of diversification rate incorporate extinction (as in Magallon and Sanderson 2001), the accumulation of species within a lineage is expected to increase with time, resulting in the null expectation that species richness will show positive linear relationships with age. Surprisingly, however, recent evidence suggests that this pattern rarely exists (Rabosky 2009a; Rabosky 2009b). In the absence of a strong relationship between lineage age and species richness (Ricklefs 2007), it has been suggested that log-transformed species richness alone may be a better measure of evolutionary success (Rabosky 2009a). Furthermore, the near-zero correlation between age and species richness is unlikely to be attributed to variation in diversification rates between lineages (Rabosky 2009b). Instead, the lack of a relationship between age and species richness is best explained by diversity-dependent diversification, where diversification is initially rapid but slows as lineages age, implying that lineages approach a "carrying capacity" set by ecological limits. Available area may be the factor that places limits on diversification (Ricklefs 2007; Rabosky 2009a), such that speciation declines as competition increases with the ever-increasing number of species within a particular clade (Phillimore and Price 2008). Extinction rates may also increase due to competition from other (i.e. unrelated groups of) species, yet this process appears to exert little influence (Rabosky 2009b).

Key innovations, such as zygomorphy, may operate to alter the evolutionary success of lineages upon this biogeographical backdrop, although the mechanism may differ from that typically envisioned. Examining these traits after the "explosive diversification" phase may instead provide evidence that certain traits influence the carrying capacity of a lineage (Rabosky 2009a). Whether traits are instrumental in altering diversification rates or carrying capacities should be evident if one examines the interaction between the presence of a trait and time. If the differences in species richness between sister clades increase with age, it implies that the trait affects net diversification rates (and not ecological limits) because the difference in species richness should increase over time.

A lack of any such interaction effect was evident in a recent analysis (Vamosi and Vamosi 2010), indicating that zygomorphy does indeed influence the carrying capacity of lineage. When different carrying capacities distinguish the species richness among clades and if carrying capacities are indeed set by available area, then key innovations may influence the carrying capacity of a lineage in two main ways. First, a trait may alter the ability of a lineage to expand its geographical extent, which then in turn increases the carrying capacity of the lineage. Of the traits that may influence geographical extent, life history (Cardillo et al. 2003) and dispersal (Roy et al. 2009) have emerged as most influential in animal lineages. In plants, associations between growth form, seed/fruit size, and geographical extent may indicate a similar pattern (Oakwood et al. 1993; Morin and Chuine 2006). Second, certain traits may appear to be key innovations by facilitating greater species packing upon a landscape (e.g. through increased specialization). Thus, ecological limits are present but the carrying capacity is set higher for some lineages over others for a given amount of space. Although not a key innovation in the same sense, an important contributor to this pattern might be if the bulk of diversification occurs within the tropics, as the tropics has long been acknowledged for supporting more species per unit area (Pianka 1966; Currie and Paquin 1987). While the latitudinal biodiversity gradient has garnered much attention, most evolutionary investigations have focused on whether diversification rates are higher in tropical lineages (through increased speciation or decreased extinction) or whether the tropics have simply had more time to accumulate more species (Mittelbach et al. 2007; Allen and Gillooly 2006). Fischer (1960) hypothesized that tropical biotas diversify more rapidly than do temperate biotas and reach a higher equilibrium number of species. In addition, if a particular key innovation has an association with latitude, the avenue by which diversification increases may be through this correlation.

The interconnected issues related to characterizing and understanding the basis of variation in diversification among lineages when controlling for phylogenetic relatedness and age was examined in a recent paper (Vamosi and Vamosi 2010). Using a phylogenetically-informed multiple regression framework, we attempted to decipher the contributions of, and plausible interactions between, four putative key traits: growth form, fruit type, sexual system, and floral symmetry, and available area (i.e. summed area of the ecoregions occupied by a lineage) to species richness of 409 angiosperm families. We also examined the effects of geographical distribution for a given available area by examining whether lineages that have a predominantly tropical distribution exert effects on species richness independently or in concert with the above traits.

In brief, it was found that only models that included available area explained meaningful amounts of the variation (~ 50 % when included, < 5 % when excluded) in species richness among families (Vamosi and Vamosi 2010). In contrast, the explanatory power of traits and trait interactions was typically rather low (Table 11.1). Additionally, age explained little of the variation in species richness, indicating diversity-dependent diversification consistent with previous studies (Rabosky 2009a). Our results indicated support, albeit rather weak, for the incorporation of phylogenetic information into the metrics (Table 11.3). We also found that most of the traits considered had moderate (fleshy fruit) or effectively no (breeding system, tropicality) phylogenetic signal, which indicates that these traits transition between families frequently upon the angiosperm phylogeny. Floral symmetry and growth form, on the other hand, exhibited significant and marked phylogenetic signals (e.g. zygomorphic (actinomorphic) families were often sister to other zygomorphic (actinomorphic) families). The results suggest a stochastic element to initial expansion of geographical extent in particular lineages, which subsequently experience higher diversification rates, with no clear signature of the prevailing effect of many of the ecological traits. However, some traits, notably zygomorphy and tropicality, exhibited consistent additive effects. The lack of a correlation between age and the strikingly high effect of area implies a diversity-dependence to diversification (Rabosky 2009b) that certainly deserves more attention.

In previous studies (Rabosky 2009b), the relationship between species richness and age has been examined through simple linear regression. The general expectation is that, if lineages simply accumulate species over time, then older lineages should be more species-rich than younger ones and logSR should exhibit a linear relationship with time. However, if traits do indeed have an effect, this linear relationship may be obscured with each group of lineages (perhaps having zygomorphic flowers) exhibiting a relationship with a different slope. The confounding influences of differing diversification rates have been examined with simulations (Rabosky 2009a), which suggest that a positive relationship between logSR and time should still be evident. Because the relationship between time and logSR

Table 11.3 Phylogenetically-informed models of species richness (N = 408 families), ranked by relative support; number of parameters (K), lambda for model (λ), Akaike's information criterion (AIC), difference between AIC of model i and best model (Δ_i), Akaike weight (w_i), R^2 values; total area of occupied ecozones (A), tropics (T), herbaceousness (H), breeding systems (B), and zygomorphy (Z). Of the three models with considerable empirical support, it can be argued that the most parsimonious one is M5 (species richness = available area + zygomorphy), given the relatively small drop in support compared to M3, which contains an extra parameter.

Model	Terms	K	λ	AIC	Δ_i	w_i	R^2
M3	A + Z + T	4	0.29	958.02	0	0.35	0.51
M5	A + Z	3	0.29	958.97	0.95	0.22	0.50
M12	A + Z + B	4	0.28	959.18	1.16	0.19	0.50
M11	A + Z + B + Z:B	5	0.29	961.00	2.98	0.08	0.51
M10	A + Z + H	4	0.29	961.91	3.89	0.05	0.50
M4	A + Z + A:Z	4	0.30	961.96	3.94	0.05	0.50
M9	A + Z + H + Z:H	5	0.31	963.04	5.02	0.03	0.50
M2	A + Z + T + A:Z + A:T + Z:T	7	0.28	963.39	5.37	0.02	0.51
M1	A + Z + T + A:Z + A:T + Z:T + A:Z:T	8	0.29	964.11	6.09	0.02	0.51
M8	Z	2	< 0.01	1215.68	257.66	0.00	0.03
M6	Z + T + Z:T	4	< 0.01	1217.41	259.39	0.00	0.04
M7	Z + T	3	< 0.01	1218.46	260.44	0.00	0.03

exhibited a very high degree of scatter (Vamosi and Vamosi 2010), we revisited the data, applying a log transformation of the x-axis, to explore whether this would provide a better fit and/or evidence for a saturating function (Fig 11.2). Although the transformation did reduce the influence of the oldest families, the effects of age on logSR remain dwarfed in comparison to the effect of factors such as available area (Fig 11.2a) and floral symmetry (Fig 11.2b) on the carrying capacities of lineages.

11.2.3 Zygomorphy and ecological limits

To summarize, with ecozone area included in the models, zygomorphy still contributed additively to increased species richness (Vamosi and Vamosi, 2010). The effects of tropicality and cosexuality also received marginal empirical support.

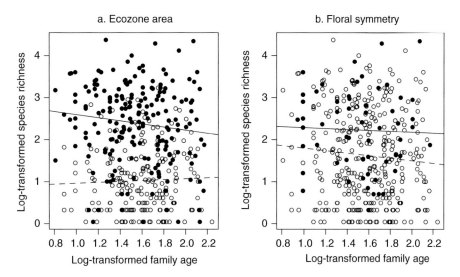

Fig 11.2 The effect of family age on species richness. While log-transforming family age improved the fit, it is evident that age does not have a strong positive effect on species richness of a family. Greater effect sizes are observed for extrinsic traits such as available area (*a*: > median with solid line and filled circles, < median with dashed line and open circles) and intrinsic traits such as zygomorphy (*b*: families with predominantly zygomorphic flowers depicted with solid line and filled circles). Figure modified from Vamosi and Vamosi (2011).

Considering that there was little evidence for interactions between traits and age, it suggests that zygomorphy conveys a higher "carrying capacity" per unit area, potentially by coevolving into ever-narrower specialist "niches" (Rabosky 2009a). This finding further suggests that the effect of key innovations on diversification rates (speciation or extinction) is minimal relative to the effect on carrying capacities within a geographical landscape.

Studies in other systems have found associations between fast life history (Cardillo et al. 2003), and dispersal (Roy et al. 2009) and geographical extent, leading to predictions that growth form (tree versus herb) and/or fruit type (fleshy versus dry) may increase the geographical extent covered within a given ecozone area, which in turn may elevate the available area (and thus the carrying capacity) of herbaceous and/or fleshy-fruited clades. Our analysis revealed that herbaceousness was indeed a strong additional predictor of geographical extent. In a post-hoc analysis, herbaceousness was also strongly correlated with greater ecozone area ($F_{1,407} = 28.79$; $P < 0.0001$), lending further support to the idea that herbs are more widely dispersed, leading to increased species richness. Fleshy fruits, on the other hand, exhibited few strong additive effects on species richness or geographical extent, even when previously reported growth form × fruit interactions (Tiffney and Mazer 1995) were included in our models, indicating that any association between species richness and fruit may be due to complex, multidimensional effects with other traits.

Other predictors of a large geographical extent within an ecozone include a tropical distribution and, marginally, cosexuality. Interestingly, a strong negative interaction between tropicality and ecozone area was present in the best models. This interaction indicates that, while tropical clades were larger in range than temperate ones for more restricted lineages (and thus could elevate the carrying capacity of a lineage), the tropical lineages occupied a smaller amount of available area than temperate clades as the ecozone area expanded. Opposite to the effect of herbaceousness, there was a nonsignificant tendency for tropical families to occupy smaller ecozone areas in general. In sum, these effects:

(1) may show some support for the tropical conservatism hypothesis (Wiens and Donoghue 2004) in that tropical clades remain constrained close to the tropical band in more widely distributed clades; and

(2) diminish the overall effect of tropicality on species richness when analyzed in this coarse manner.

Associations between zygomorphy and tropicality were not explored previously but, if present, could potentially produce the pattern of increased diversification in zygomorphic lineages in the tropics. We therefore investigate them further here (see Tables 11.2 and 11.3).

Higher geographical extent may produce higher species richness through buffering from extinction (Payne and Finnegan 2007), yet we expect this effect plays only a minor mechanistic role in most key innovations. Previous studies have had little power to tease the two (speciation and extinction) apart (Heilbuth 2000). With the analysis summarized here, we would expect such a strong effect of extinction rates to manifest itself with a trait showing significant effects in both geographical extent and species richness for a given ecoregion area and only cosexuality had equivalent dual effects (and these were relatively weak), consistent with previous studies (Vamosi and Vamosi 2005a). Thus, we venture that traits can act as key innovations through two main mechanisms: increasing speciation through increased specialization within a given area (e.g. zygomorphy), and being associated with increases in the amount of area occupied (e.g. herbaceousness), or a combination of the two (e.g. tropicality).

11.3 Zygomorphy; the role of genetics versus ecology in increasing diversification

Zygomorphy is found in c. 27 % of the angiosperm families (Neal et al. 1998) but only 14 % of families are predominantly zygomorphic (Vamosi and Vamosi 2010).

Effective pollen transfer between flowers depends on the accuracy of the physical fit between the flower and pollinator (Sargent 2004). Because zygomorphic flowers have, in contrast to actinomorphic flowers, only one plane of symmetry, the number of possible positions of the pollinator on the flower to access floral rewards and effect pollination are reduced (Berg 1960; Sargent 2004). Other factors have been proposed, but have not received as much attention, including protection from rain and increased floral constancy through pollinator recognition (Waser 1986; Neal et al. 1998). Regardless of the exact mechanism(s) of increased success, these factors have led to independent evolution of zygomorphy in a number of different lineages, yet through mutation at different transcription factor genes that influence the degree of floral symmetry. Despite the acknowledged advantages of zygomorphy, transitions to actinomorphy occur readily (Ree and Donoghue 1999), indicating that the advantages of zygomorphy are likely to be pollinator context-specific. However, phylogenetic investigations in Solanaceae of the distribution of floral characters related to pollination indicate that diversification events are often uncoupled from pollinator shifts (Knapp 2010). Dual investigations of the genomic potential to speciate within zygomorphic lineages along with the geographical distribution of pollinators therefore hold the most promise of uncovering how zygomorphy and plant diversification are linked.

11.3.1 Genetics of speciation in zygomorphic lineages

Very few studies have examined the direct effect of floral symmetry genes on the speciation process but studies performed on speciation genetics of other systems may provide insight. Some of the major genes determining floral symmetry are well characterized (Kalisz et al. 2006). Generally, the evolution of zygomorphy has involved the duplication and subsequent cooption of the duplicated transcription factors, namely *CYCLOIDEA2*-like (or CYC2-like) genes, to express later in floral development (Zhang et al. 2010). Thus, the developmental genetics of a zygomorphic flower is more complex and involves more genes than the development of an actinomorphic flower, and the different genomic architecture of zygomorphic flowers may play a role in facilitating speciation. We highlight two avenues by which the genomic architecture could play a role in changing speciation rates: the sheer number of genes and the correlations (or integration) between floral genes.

Genes involved in speciation: In other systems, traits with few loci have been observed to evolve more rapidly, which may suggest that this then hampers the rate of speciation in zygomorphic lineages (Kondrashov and Kondrashov 1999). Whether this "genic" perspective could be underlying differences in speciation rates is a novel but as yet understudied avenue in angiosperm research (Lexer and Widmer 2008). Most studies conducted thus far suggest that a few genes of large effect are all that is necessary, such that it does not seem as though speciation in actinomorphic lineages would be hampered if the geographical mosaic of pollinators imposed the necessary selection criteria, e.g. through evolving different floral colors (Hoballah et al. 2007) or floral size (Venail et al. 2010) in *Petunia*. Furthermore, other studies indicate that prezygotic mating barriers (those involved in pollinator preferences, for instance) are more important than postzygotic mating barriers in flowering plants (Lowry et al. 2008), consistent with the idea that the differing genomic architecture between zygomorphic and actinomorphic lineages is not a large contributing factor to differing speciation rates. In sum, the causal arrow tentatively appears to point towards ecological differences.

Floral integration differences: Berg (1960) predicted that floral integration (i.e. correlated variation in floral traits) should be stronger in zygomorphic species because it promotes precise pollination by restricting which pollinators can visit a flower, the direction from which they can approach, and their movement within the flower. Generally, traits of zygomorphic (bilaterally symmetrical) flowers are more strongly correlated than traits of actinomorphic (radially symmetrical) flowers (Ashman and Majetic 2006). Whether these findings conform to Berg's prediction is unclear however, because the genomic architecture of zygomorphic lineages may result in more pleiotropic interactions and this alone may result in the floral integration. Recently, a tight genetic correlation between the evolution of zygomorphy and stamen abortion was uncovered (Song et al. 2009). While some have argued that floral integration may constrain the adaptive evolution of zygomorphs (Smith and Rausher 2008), other studies find that genetic covariance generally does not constrain adaptation (Agrawal and Stinchcombe 2009) and may even facilitate it (Harder and Johnson 2009), which may in turn elevate speciation rates (Rundle and Nosil 2005). Whether the tighter correlations imply genetic constraints, or that the pollinator environment sets more severe selection pressures on zygomorphs, it does appear as though stabilizing selection can be stronger in zygomorphic lineages (van Kleunen et al. 2008). One interesting prediction of tighter stabilizing selection upon floral traits of zygomorphs would be that there would generally be little additive variance and low heritability. However, to our knowledge, no studies have compared heritabilities between zygomorphic and actinomorphic species. Thus, there are three factors that seem intertwined in zygomorphic lineages:

(1) whether genetic constraints result in greater morphological change for a given selection pressure;

(2) whether pollinators indeed exert more directional selection upon zygomorphs, via more specialized relationships; and

(3) whether selection is simply more spatially heterogeneous in zygomorphic lineages.

We address how the ecology of zygomorphy may affect the speciation genetics below.

11.3.2 Zygomorphy and plant–pollinator ecology

It is important to note that it is not the covariance between floral traits that is necessarily important in speciation but rather the covariance between traits and the associated pollinator environment that may accelerate speciation. Two main lines of inquiry should be undertaken to determine whether this occurs:

is the spatial mosaic in pollinator environments different in zygomorphic lineages different and/or

does selection on the traits imposed by their environment occur more rapidly in zygomorphic lineages?

The former question is a purely ecological approach whereas the latter also involves an investigation of genetics. If indeed floral traits are correlated, then selection on a single floral trait may allow for the hitchhiking of other floral traits to follow if they are genetically constrained. Clearly, more empirical work is required to tease apart the (relative importance of the) two possibilities.

Specialized pollination and zygomorphic diversification: Actinomorphic lineages may be quite specialized (e.g. several species of *Petunia, Aquilegia, Solanum*) in terms of their pollinators whereas many zygomorphic lineages are comparatively generalized, allowing many bees, moths, and hummingbirds, e.g. *Delphinium, Castilleja*. However, it is thought that zygomorphy provides a coarse morphological estimate that can generalize the number of pollinators (Ramírez 2003; Neal et al. 1998). What is perhaps a more important question is whether specialized pollination should theoretically be associated with higher species richness. Smith et al. (2008a), for example, found no association between specialized pollination and speciation rate in *Yucca*. Recently, some research indicates that specialists are most at risk of extinction, relying on the presence of a few key interactions (Colles et al. 2009). This would tend to indicate that zygomorphs should experience higher extinction rates and perhaps exhibit lower species richness. However, other research specifically on plant–pollinator networks indicates that plants that specialize on a certain pollinator are more likely to utilize pollinators that are extreme generalists (asymmetry in networks). If the pollinators are generalists, they themselves may be relatively immune to extinction and thus zygomorphs would experience a buffering effect through this avenue (Rezende et al. 2007).

The reasoning behind zygomorphy exhibiting higher species richness is usually thought to occur by more frequent pollinator shifts yet it is not known whether this indeed occurs (Knapp 2010). Species with very generalist pollination will have diffuse selection pressures upon them in all populations (i.e. there may be a predominant bee, fly, and hawkmoth in all communities, even if the particular species changes). Clearly, this may have different consequences for the opportunity and pace of speciation than if a single bumblebee versus a hummingbird species are the predominant pollinators in two separate populations. Therefore, not only is it necessary to have more specialized pollinators, it is important that different selection pressures are exerted on zygomorphs in different environments, which we address next.

Selection gradient for zygomorphic lineages is more heterogeneous: The pollinator-shift hypothesis developed by Grant (1949) and Stebbins (1970) is the most widely accepted explanation for how pollinators operate in speciation (Johnson 2010; Kay and Sargent 2009; Waser 1998; Smith et al. 2008b; Fenster et al. 2004). Recent tests of this prediction have produced results indicating that floral attraction traits may evolve quickly to the local pollinator environment, but mechanical-fit traits may be more constrained (Nattero et al. 2010). While it stands to reason that a smaller pool of pollinators may heighten the peaks in a selection landscape, more research is currently required to deduce whether the differences in speciation rates are due to the differences in spatial ecology that occur between zygomorphic and actinomorphic species.

In a recent meta-analysis, zygomorphy showed an association with having fewer pollinators and higher pollen limitation (Knight et al. 2005). Empirical studies are sparse but some evidence does exist that the prevalence of zygomorphy increases with species diversity (Ostler and Harper 1978). This would be consistent with the idea that plants that have higher pollen limitation inhabit more species-rich, competition-prone environments and that pollen limitation and speciation are connected (Vamosi et al. 2006). For instance, pollen limitation in species-rich environments may lead to floral specialization to ever-narrowing pollination niches through character displacement (Armbruster and Muchhala 2009). Therefore, it is further unlikely that zygomorphy is associated with higher species richness through greater efficiency of pollination leading to lower extinction rates (or they would experience lower pollen limitation), and more likely that higher speciation rates are at the heart of the association. Thus, for zygomorphic clades, they may inhabit more segregated mosaics, whereby a zygomorphic species may be under very different selection pressures from different community compositions of other plants competing for pollinators.

Finally, zygomorphy may be associated with high species richness through underlying correlations with other variables, that are themselves affecting diversification rates such as greater outcrossing (Kalisz et al. 2006; Igic et al. 2008). Zygomorphy has been associated with pollination by bees or birds, which in turn may be more prevalent in tropical environments, and herbaceous habit (Neal et al. 1998). Tropical environments may also have more heterogeneous pollinator environment because the forest canopy is more layered (Sargent and Vamosi 2008), all dominated by angiosperms as opposed to conifers of more temperate climes. Vamosi and Vamosi (2010) found little to indicate that anything other than zygomorphy was the driving trait involved in elevation of diversification rates. The one potential caveat was a weak correlation with tropicality, which we discuss in more depth next.

11.4 Biotic interactions, range and the tropical biodiversity gradient

Increased biotic interactions have been argued to influence the tropical biodiversity gradient (Schemske et al. 2009). We find little to indicate that pollinators are more specialized in the tropics, with only a weak association between zygomorphy and tropicality at the family level (likelihood-ratio test; $P = 0.05$). However, both zygomorphy and tropicality jointly provide the strongest influences beyond available area on species richness. In terms of relative proportions, zygomorphy was most prevalent in the Afrotropic ecozone, closely followed by Oceania (Fig 11.3). The idea of a greater "carrying capacity" per unit area with specialized pollination has intuitive appeal because it suggests the possibility of local adaptation (Kay and Sargent 2009; Sargent and Otto 2006; Vamosi et al. 2006). The effect of latitude on species richness per unit area was quite modest and most of the effect of latitude was through increased geographical extent (Vamosi and Vamosi 2010). Teasing apart the historical effect of area is, of course, very difficult because the area encompassed by the tropics has changed dramatically over time (Fine and Ree 2006).

In summary, we investigated whether the latitudinal gradient in angiosperm diversification relies less on the potential for more specialized relationships between plants and their pollinators in the tropics (Olesen and Jordano 2002), and more on the finding that lineages restricted to the tropical realms occupy larger areas than lineages restricted to temperate realms. In other words, zygomorphic lineages may not speciate through evolution of pollinator specialization in the tropics (Kay and Schemske 2004), but rather speciate as a result of larger ranges through their association with tropicality. One interesting previous finding was that of a strong negative interaction between tropicality and ecozone area on geographical extent. In other words, when lineages were constrained to a single ecozone, tropical lineages exhibited larger geographical extents than did temperate lineages. In more widespread clades (those that had expanded to more than one ecozone), temperate clades had larger geographical extents. Overall, there was a nonsignificant tendency for tropical families to have smaller geographical extents ($F_{1,396} = 2.38$, $P = 0.12$), and to occupy smaller available areas ($F_{1,407} = 1.43$, $P = 0.15$), than temperate families, indicating that tropicality does not spur diversification via greater geographical extents, which is consistent with previous studies (Rohde 1997; Gaston and Blackburn 2000). These observations provide some support for the tropical conservatism hypothesis in that tropical lineages do become more constrained to the tropical band even as lineages expand.

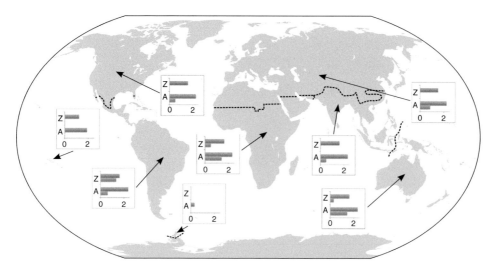

Fig 11.3 Log-transformed species richness of all (red bars) and endemic (blue bars) zygomorphic (Z) and actinomorphic (A) families in the ecozones of Udvardy (Udvardy 1975), in a clockwise direction: Nearctic, Palaearctic, Australasia, Indo–Malaya, Afrotropic, Antarctic, Neotropic, and Oceania. No ecozone contains markedly more angiosperm families than the others, zygomorphy is moderately more prevalent in tropical ecozones, and no ecozone appears to have undue influence on the proportion of zygomorphic families. With reference to endemic families, the Neotropic ecozone appears to be the one exception to the general pattern of greater numbers of actinomorphic families than zygomorphic families. Figure modified from Vamosi and Vamosi (2011).

Given that tropical families tend to occupy less of their available area in general yet still have more species per given area, this indicates that there is indeed a modest effect of tropicality on the carrying capacity of a lineage. How this operates may have something to do with greater pollinator specialization in the tropics and more varied topography in the tropics, including more island systems (Fig 11.3) and the operation of the Andes as an "island archipelago" (Hughes and Eastwood 2006). Indeed, there may have previously been a tendency to ignore the habitat diversity of the tropics, which contributes to its species richness, and instead to think of it as one large uniform rainforest (Prance 1977). Thus tropical climes may be more heterogeneous in their biotic and/or abiotic microhabitats (Thompson 2005), resulting in more varied selection pressures over the range of a species, or the evolution that occurs for a given selection pressure may simply be more rapid in the tropics (Wright et al. 2006), resulting in higher speciation rates for a given area. Regardless of the precise mechanism, a simple investigation of species richness per family reveals no effect of latitude, yet including available area as an additional factor revealed important nuanced patterns. The distinction between the two approaches implies that an explicit geographical perspective will need to be included in future investigations of speciation rates and the tropical biodiversity gradient in angiosperms.

11.5 Concluding remarks

For years, the examination of the determinants of broad-scale diversification has been studied in near isolation from the macroecology and biogeography of biodiversity, and we feel this dichotomy will soon be merged. An increasing number of recent studies have shown that diversification is diversity-dependent and the diversity-dependence is determined by spatial constraints imposed by geography, a pattern confirmed in flowering plants. Novel multi-trait analyses that can examine numerous lineages while still removing the confounding effects of time, allow for some insight into common overall patterns in large lineages such as the angiosperms. When we do this, we find that available area exerts the most important constraint upon the species richness of a lineage. There is also supporting evidence that zygomorphy and tropicality allow for more species richness per given unit area and these appear to act additively. Limited work exists on whether the genomic architecture of zygomorphy affects speciation directly, yet the existing empirical data would seem to indicate that ecology plays a more important role. The more specialized pollination systems of zygomorphic species likely make their geographical mosaic of effective pollinators and competition for pollination more heterogeneous, which produces divergent selection pressures on zygomorphic lineages. Continued work on how zygomorphy operates in concert and/or in addition to range expansion will further enlighten and expand the field of diversification research.

Acknowledgements

We thank A. Mooers for continued discussions about comparative methods, especially with regard to analyzing diversification. The project was funded through NSERC (Canada) Discovery Grants to JCV and SMV.

References

Agrawal, A. F. and Stinchcombe, J. R. (2009). How much do genetic covariances alter the rate of adaptation? *Proceedings of the Royal Society of London B*, **276**, 1183–91.

Allen, A. P. and Gillooly, J. F. (2006). Assessing latitudinal gradients in speciation rates and biodiversity at the global scale. *Ecology Letters*, **9**, 947–54.

Allen, A. P., Gillooly, J. F., Savage, V. M. and Brown J. H. (2006). Kinetic effects of temperature on rates of genetic divergence and speciation. *Proceedings of the National Academy of Sciences USA*, **103**, 9130–5.

Armbruster, W. S. and Muchala, N. (2009). Associations between floral specialization and species diversity: cause, effect, or correlation? *Evolutionary Ecology*, **23**, 159–79.

Ashman, T.-L. and Majetic, C. J. (2006). Genetic constraints on floral evolution: a review and evaluation of patterns. *Heredity*, **96**, 343–52.

Bohning-Gaese, K., Caprano, T., van Ewijk, K. and Veith, M. (2006). Range size: disentangling current traits and phylogenetic and biogeographic factors. *American Naturalist*, **167**, 555–67.

Berg, R. L. (1960). The ecological significance of correlation pleiades. *Evolution*, **14**, 171–80.

Cardillo, M., Huxtable, J. S. and Bromham, L. (2003). Geographic range size, life history, and rates of diversification in Australian mammals. *Journal of Evolutionary Biology*, **16**, 282–8.

Colles, A., Liow, L. H. and Prinzing, A. (2009). Are specialists at risk under environmental change? Neoecological, paleoecological and phylogenetic approaches. *Ecology Letters*, **8**, 849–63.

Currie, D. J. and Paquin, V. (1987). Large-scale biogeographical patterns of species richness of trees. *Nature*, **329**, 326–7.

Davies, T. J., Savolainen, V., Chase, M. W., Moat, J. and Barraclough, T. G. (2004a). Environmental energy and evolutionary rates in flowering plants. *Proceedings of the Royal Society London, Series B*, **271**, 2195–200.

Davies, T. J., Barraclough, T. G., Chase, M. W., Soltis, P. S., Soltis, D. E. and Savolainen, V. (2004b). Darwin's abominable mystery: insights from a supertree of the angiosperms. *Proceedings of the National Academy of Sciences USA*, **101**: 1904–9.

De Queiroz, A. (2002). Contingent predictability in evolution: key traits and diversification. *Systematic Biology*, **51**, 917–29.

Farrell, B. D., Dussourd, D. E. and Mitter, C. (1991). Escalation of plant defense: do latex and resin canals spur plant diversification? *American Naturalist*, **138**, 881–900.

Fenster, C. B., Armbruster, W. S., Wilson, P., Thomson, J. D. and Dudash, M. R. (2004). Pollination syndromes and floral specialization. *Annual Review of Ecology, Evolution, and Systematics*, **35**, 375–403.

Fine, P. V. A. and Ree, R. (2006). Evidence for a time-integrated species-area effect on the latitudinal gradient in tree diversity. *American Naturalist*, **168**, 796–804.

Fischer, A. G. (1960). Latitudinal variation in organic diversity. *Evolution*, **14**, 64–81.

Gaston, K. J. (1998). Species-range size distributions: products of speciation, extinction, and transformation. *Philosophical Transactions of the Royal Society, Series B*, **353**, 219–30.

Gaston, K. J. and Blackburn, T. M. (2000). *Pattern and process in macroecology.* Malden, MA: Blackwell Science.

Grant, V. (1949). Pollination systems as isolating mechanisms in angiosperms. *Evolution*, **3**, 82–97.

Harder, L. D. and Johnson, S. D. (2009). Darwin's beautiful contrivances: evolutionary and funcitonal evidence for floral adaptation. *New Phytologist*, **183**, 530–45.

Heard, S. B. and Cox, G. H. (2007). The shapes of phylogenetic trees of clades, faunas, and local assemblages: exploring spatial pattern in differential diversfication. *American Naturalist*, **169**, E107–E118.

Heilbuth, J. C. (2000). Lower species richness in dioecious clades. *American Naturalist*, **156**, 221–41.

Hoballah, M. E., Gubitz, T., Stuurman, J., Broger, L., Barone, M., Mandel, T., Dell'Olivo, A., Arnold, M. L. and Kuhlemeier C. (2007). Single gene-mediated shift in pollinator attraction in Petunia. *Plant Cell*, **19**, 779–90.

Hughes, C. and Eastwood, R. (2006). Island radiation on a continental scale: exceptional rates of plant diversfication after uplift of the Andes. *Proceedings of the National Academy of Sciences USA*, **103**, 10334–9.

Igic, B., Lande, R. and Kohn, J. R. (2008). Loss of self-incompatibility and its evolutionary consequences. *International Journal of Plant Sciences*, **169**, 93–104.

Jablonski, D. (1987). Heritability at the species level: analysis of geographic ranges of Cretaceous mollusks. *Science*, **238**, 360–3.

Jansson, R. and Davies, T. J. (2008). Global variation in diversification rates of flowering plants: energy versus climage change. *Ecology Letters*, **11**, 173–83.

Johnson, S. D. (2010). The pollination niche and its role in the diversification and maintenance of the southern African flora. *Philosophical Transactions of the Royal Society, Series B*, **12**, 499–516.

Kalisz, S., Ree, R. H. and Sargent, R. D. (2006). Linking regulatory genes to evolutionary patterns: a case for floral symmetry genes in mating system evolution. *Trends in Plant Science*, **11**, 568–73.

Kay, K. M. and Sargent, R. D. (2009). The role of animal pollination in plant speciation: integrating ecology, geography, and genetics. *Annual Review of Ecology, Evolution, and Systematics*, **40**, 637–56.

Kay, K. M. and Schemske, D. W. (2004). Geographic patterns in plant-pollinator mutualistic networks: comment. *Ecology*, **85**, 875–978.

Kay, K. M., Voelckel, C., Yang, J. Y., Hufford, K. M., Kaska, D. D. and Hodges S. A. (2006). Floral characters and species diversification. In *Ecology and Evolution of Flowers*, ed. L. Harder and S. C. H. Barrett. Oxford, UK: Oxford University Press.

van Kleunen, M., Meier, A., Saxenhofer, M. and Fischer, M. (2008). Support for the predictions of the pollinator-mediated stabilizing-selection hypothesis. *Journal of Plant Ecology*, **1**, 173–8.

Knapp, S. (2010). On "various contrivances": pollination, phylogeny and flower form in the Solanaceae. *Philosophical Transactions of the Royal Society, Series B*, **365**, 449–60.

Knight, T. M., Steets, J. A., Vamosi, J. C., Mazer, S. J., Burd, M., Campbell, D. R., Dudash, M. R., Johnston, M. O., Mitchell, R. J. and Ashman, T. L. (2005). Pollen limitation of plant reproduction: pattern and process. *Annual Review of Ecology, Evolution, and Systematics*, **36**, 467–97.

Kolb, A., Barsch, F. and Diekmann, M. (2007). Determinants of local abundance and range size in forest vascular plants. *Global Ecology and Biogeography*, **15**, 237–47.

Kondrashov, A. S. and Kondrashov, F. A. (1999). Interactions among quantitative traits in the course of symaptric speciation. *Nature*, **400**, 351–4.

Lexer, C. and Widmer, A. (2008). The genic view of plant speciation: recent progress and emerging questions. *Philosophical Transactions of the Royal Society, Series B*, **363**, 3023–36.

Losos, J. B. and Schluter, D. (2000). Analysis of an evolutionary species–area relationship. *Nature*, **408**, 847–50.

Lowry, D. B., Modliszewski, J. L., Wright, K. M., Wu, C. A. and Willis, J. H. (2008). The strength and genetic basis of reproductive isolating barriers in flowering plants. *Philosophical Transactions of the Royal Society, Series B*, **363**, 3009–21.

Lowry, E. and Lester, S. E. (2006). The biogeography of plant reproduction: potential determinants of species' range sizes. *Journal of Biogeography*, **33**, 1975–82.

Magallon, S. and Sanderson, M. J. (2001). Absolute diversification rates in angiosperm clades. *Evolution*, **55**, 1762–80.

Mittelbach, G. G., Schemske, D. W., Cornell, H. V., Allen, A. P., Brown, J. M., Bush, M. B., Harrison, S. P., Hurlbert, A. H., Knowlton, N., Lessios, H. A., McCain, C. M., McCune, A. R., McDade, L. A., McPeek, M. A., Near, T. J., Price, T. D., Ricklefs, R. E., Roy, K., Sax, D. F., Schluter, D., Sobel, J. M. and Turelli, M. (2007). Evolution and the latitudinal diversity gradient: speciation, extinction and biogeography. *Ecology Letters*, **10**, 315–31.

Morin, X. and Chuine, I. (2006). Niche breadth, competitive strength and range size of tree species: a trade-off based framework to understand species distribution. *Ecology Letters*, **9**, 185–95.

Nattero, J., Cocucci, A. and Medel, R. (2010). Pollinator-mediated selection in a specialized pollination system: matches and mismatches across populations. *Journal of Evolutionary Biology*, **23**, 1957–68.

Neal, P. R., Dafni, A. and Giurfa, M. (1998). Floral symmetry and its role in plant–pollinator systems:Terminology, distribution and hypotheses. *Annual Review of Ecology and Systematics*, **29**: 345–73.

Oakwood, M., Jurado, E., Leishman M. R. and Westoby, M. (1993). Geographical ranges of plant species in relation to dispersal morphology, growth form, and diaspore weight. *Journal of Biogeography*, **20**, 563–72.

Olesen, J. M. and Jordano, P. (2002). Geographic patterns in plant-pollinator mutualistic networks. *Ecology*, **83**, 2416–24.

Orme, C. D. L., Davies, R. G., Olson, V. A., Thomas, G. H., Ding, T. S., Rasmussen, P. C., Ridgely, R., Stattersfield, A. J., Bennett, P. M., Owens, I. P. F., Blackburn, T. M. and Gaston, K. G. (2006). Global patterns of geographic range size in birds. *PLoS Biology*, **4**, e208.

Ostler, W. K. and Harper, K. T. (1978). Floral ecology in relation to plant species diversity in the Wasatch Mountains of Utah and Idaho. *Ecology*, **59**, 848–61.

Owens, I. P. F., Bennett, P. M. and Harvey, P. H. (1999). Species richness among birds: body size, life history, sexual selection or ecology? *Proceedings of the Royal Society London, Series B*, **266**, 933–9.

Payne, J. L. and Finnegan, S. (2007). The effect of geographic range on extinction risk during background and mass extinction. *Proceedings of the National Academy of Sciences USA*, **104**, 10506–11.

Phillimore, A. B. and Price, T. D. (2008). Density-dependent cladogenesis in birds. *PLoS Biology*, **6**, e71.

Phillimore, A. B, Freckleton, R. P., Orme, C. D. L. and Owens I. P. F. (2006). Ecology predicts large-scale patterns of phylogenetic diversification in birds. *American Naturalist*, **168**, 220–9.

Pianka, E. R. (1966). Latitudinal gradients in species diversity: a review of concepts. *American Naturalist*, **100**, 33–46.

Prance, G. T. (1977). Floristic inventory of the tropics: where do we stand? *Annals of the Missouri Botanical Garden*, **64**, 659–84.

Rabosky, D. L. (2009a). Ecological limits and diversification rate: alternative paradigms to explain the variation in species richness among clades and regions. *Ecology Letters*, **12**, 735–43.

Rabosky, D. L. (2009b). Ecological limits on clade diversification in higher taxa. *American Naturalist*, **173**, 662–74.

Ramirez, N. (2003). Floral specialization and pollination: a quantitative analysis and comparison of the Leppik and the Faegri and van der Pijl classification systems. *Taxon*, **52**, 687–700.

Ree, R. H. and Donoghue, M. J. (1999). Inferring rates of change in flower symmetry in asterid angiosperms. *Systematic Biology*, **48**, 633–41.

Rezende, E. L., Lavabre, J. E., Guimaraes, P. R., Jordano, P. and Bascompte, J. (2007). Non-random coextinctions in phylogenetically structured mutualistic networks. *Nature*, **448**, 925–8.

Ricklefs, R. E. (2003). Global diversification rates of passerine birds. *Proceedings of the Royal Society of London B*, **270**, 2285–91.

Ricklefs, R. E. (2007). History and diversity: explorations at the intersection of ecology and evolution. *American Naturalist*, **170**, S56–S70.

Ricklefs, R. E. and Renner, S. S. (1994). Species richness within families of flowering plants. *Evolution*, **48**, 1619–36.

Rohde, K. (1997). The larger area of the tropics does not explain latitudinal gradients in species diversity. *Oikos*, **79**, 169–72.

Roy, K., Hunt, G., Jablonski, D., Krug, A. Z. and Valentine, J. W. (2009). A macroevolutionary perspective on species range limits. *Proceedings of the Royal Society of London B*, **276**, 1485–93.

Rundle, H. D. and Nosil, P. (2005). Ecological speciation. *Ecology Letters*, **8**, 336–52.

Sargent, R. D. (2004). Floral symmetry affects speciation rates in angiosperms. *Proceedings of the Royal Society of London B*, **271**, 603–8.

Sargent, R. D. and Otto, S. P. (2006). The role of local species abundance in the evolution of pollinator attraction in flowering plants. *American Naturalist*, **167**, 67–80.

Sargent, R. D. and Vamosi, J. C. (2008). The influence of canopy position, pollinator syndrome, and region on evolutionary transitions in pollinator guild size. *International Journal of Plant Sciences*, **169**, 39–47.

Schemske, D. W., Mittelbach, G. G., Cornell, H. V., Sobel, J. M. and Roy, K. (2009). Is there a latitudinal gradient in the importance of biotic interactions? *Annual Review of Ecology, Evolution, and Systematics*, **40**, 245–69.

Smith, C. I., Pellmyr, O., Althoff, D. M., Balcazar-Lara, M., Leebens-Mack, J. and Segraves, K. A. (2008a). Pattern and timing of diversification in Yucca (Agavaceae): specialized pollination does not escalate rates of diversification. *Proceedings of the Royal Society of London B*, **275**, 249–58.

Smith, J. F. (2001). High species diversity in fleshy-fruited tropical understory plants. *American Naturalist*, **157**, 646–53.

Smith, R. A. and Rausher, M. D. (2008). Selection for character displacement is constrained by the genetic architecture of floral traits in the ivyleaf morning glory. *Evolution*, **62**, 2829–41.

Smith, S. D., Ane, C. and Baum, D. A. (2008b). The role of pollinator shifts in the floral diversification of lochroma (Solanaceae). *Evolution*, **62**, 1-14.

Soltis, P. S., Soltis, D. E. and Chase, M. W. (1999). Angiosperm phylogeny inferred from multiple genes as a tool for comparative biology. *Nature*, **402**, 402–4.

Song, C.-F., Lin, Q.-B., Liang, R.-H. and Wang, Y.-Z. (2009). Expressions of ECE-CYC2 clade genes relating to abortion of both dorsal and ventral stamens in *Opithandra* (Gesneriaceae). *BMC Evolutionary Biology*, **9**, 244.

Stebbins, G. L. (1970). Adaptive radiation of reproductive characteristics in angiosperms I: pollination mechanisms. *Annual Review of Ecology, Evolution, and Systematics*, **1**, 307–26.

Stevens, G. C. (1989). The latitudinal gradient in geographical range-how so many species coexist in the tropics. *American Naturalist*, **133**, 240–56.

Thompson, J. N. (2005). *The Geographic Mosaic of Coevolution*. Chicago, IL Chicago University Press.

Tiffney, B. H. and Mazer, S. J. (1995). Angiosperm growth habit, dispersal and diversification reconsidered. *Evolutionary Ecology*, **9**, 93–117.

Vamosi, J. C. and Otto, S. P. (2002). When looks can kill: the evolution of sexually-dimorphic floral display and the extinction of dioecious plants. *Proceedings of the Royal Society of London B*, **269**, 1187–94.

Vamosi, J. C. and Vamosi, S. M. (2004). The role of diversification in causing the correlates of dioecy. *Evolution*, **58**, 723–31.

Vamosi, J. C. and Vamosi, S. M. (2005a). Present day risk of extinction may exacerbate the lower species richness of dioecious clades. *Diversity and Distributions*, **11**, 25–32.

Vamosi, J. C. and Vamosi, S. M. (2010). Key innovations within a geographical context: towards resolving Darwin's abominable mystery. *Ecology Letters*, **13**, 1270–79.

Vamosi, J. C. and Vamosi, S. M. (2011). Factors influencing diversification in angiosperms: at the crossroads of intrinsic and extrinsic traits. *American Journal of Botany*, **98**, 460–71.

Vamosi, J. C., Otto, S. P. and Barrett, S. C. H. (2003). Phylogenetic analysis of the ecological correlates of dioecy in angiosperms. *Journal of Evolutionary Biology*, **16**, 1006–18.

Vamosi, J. C., Knight, T., Steets, J., Mazer, S., Burd, M. and Ashman, T.-L. (2006). Pollination decays in biodiversity hotspots. *Proceedings of the National Academy of Sciences USA*, **103**, 956–61.

Vamosi, S. M. and Vamosi, J. C. (2005b). Endless tests: guidelines to analyzing non-nested sister-group comparisons. *Evolutionary Ecology Research*, **7**, 567–79.

Venail, J., Dell'Olivo, A. and Kuhlemeier, C. (2010). Speciation genes in the genus *Philosophical Transactions of the Royal Society, Series B*, **365**, 461–8.

Verdú, M. (2002). Age at maturity and diversification in woody angiosperms. *Evolution*, **56**, 1352–61.

Waser, N. M. (1986). Flower constancy: definition, cause, and measurement. *American Naturalist*, **127**, 593–603.

Waser, N. M. (1998). Pollination, angiosperm speciation, and the nature of species boundaries. *Oikos*, **82**, 198–201.

Webb, T. J. and Gaston, K. G. (2003). On the heritability of geographic range sizes. *American Naturalist*, **161**, 553–66.

Wiens, J. J. and Donoghue, M. J. (2004). Historical biogeography, ecology, and species richness. *Trends in Ecology and Evolution*, **19**, 639–44.

Wright, S., Keeling, J. and Gillman, L. (2006). The road from Santa Rosalia: a faster tempo of evolution in tropical climates. *Proceedings of the National Academy of Sciences USA*, **103**, 7718–22.

Zhang, W., Kramer, E. M. and Davis, C. C. (2010). Floral symmetry genes and the origin and maintenance of zygomorphy in a plant–pollinator mutualism. *Proceedings of the National Academy of Sciences USA*, **107**, 6388–93.

12

Ambophily and "super generalism" in *Ceratonia siliqua* (Fabaceae) pollination

Amots Dafni, Talya Marom-Levy, Andreas Jürgens,
Stefan Dötterl, Yuval Shimrat, Achik Dorchin,
H. Elizabeth Kirkpatrick and Taina Witt

12.1 Introduction

Since ancient times, the evergreen carob tree *Ceratonia siliqua* L. (Fabaceae: Caesalpinoideae) has been grown in most countries of the Mediterranean basin for its edible seed pots, which are an important crop (von Hasselberg 2000). It has been used historically as feed for domesticated animals (sometimes referred to as "locust beans"), and the current cultivars of the carob tree were probably selected by the Arabs (Ramón-Laca and Mabberley 2004). Carob seed pots were also used to supplement the human diet (e.g. known as "St. John's bread") and its products are used even nowadays in many ways (e.g. as thickening agents). Carob trees were traditionally interplanted with olives, grapes, almonds, and barley in low-intensity farming systems in most carob-producing countries (Battle and Tous 1997). The carob tree is a large, sclerophyllous tree of the Mediterranean evergreen maquis (von Hasselberg 2000; Zohary and Orshan 1959). The tree is usually dioecious (hermaphrodites occur rarely: Zohary 1972: 32; Tucker 1992) and produces many-flowered catkin-like inflorescences (von Hasselberg 2000; Battle and Tous 1997; Feinbrun-Dothan and Danin 1998: 294)

Evolution of Plant–Pollinator Relationships, ed S. Patiny. Published by Cambridge University Press. © The Systematics Association 2012.

with strongly reduced flowers. The pentamerous flowers of both sexes are 6–12 mm long, yellowish–green, apetalous and consist merely of sexual organs (von Haselberg et al. 2004). Male flowers have five stamens and an abortive pistil, whereas female flowers have abortive staminodia and a fully developed pistil formed of a single carpel (Tucker 1992). The oval, two-lobed stigma is about 2.5 × 2.3 mm in size (von Hasselberg 2000), peltate, wet, and covered by verrucate papillae (Tucker 1992). The floral nectar produced is exposed (Battle and Tous 1997) on the broad hypogynous disk (Polhill et al. 1981) and therefore easily accessible for flower-visiting insects. The strongly scented inflorescences (Custodio et al. 2004, 2006), usually bearing 20–50 single flowers, arise as short lateral racemes mainly on branches (cauliflorous flowering) and on the trunk (ramiflorous flowering) (von Haselberg et al. 2004). The prolonged flowering season is mainly from September to December, which is regarded as a harsh pollination environment due to climatic conditions, a low diversity of potential pollinators, and a low number of individuals (Dafni 1986).

The literature concerning the pollination of the carob is scant and equivocal. The strong odor emitted from the flowers (Ortiz et al. 1996; Custódio et al. 2004, 2006), the secretion of nectar, and the high frequency of insect floral visitors suggest that the carob is predominantly an entomophilous species (Ortiz et al. 1996). Several authors mentioned bees as main pollinators and Diptera as secondary (Ortiz et al. 1996; Arista et al. 1999) or vice versa (Retana et al. 1990). Arista et al. (1999) found Vespidae as main visitors rather than bees and flies. Retana et al. (1990) mentioned moths as night visitors but also stated that they did not carry any *Ceratonia* pollen.

The role of wind pollination in the carob is under debate. Retana et al. (1990) found no wind pollination even though other authors have suspected wind pollination based on the floral structure (Hillcoat et al. 1980; Passos de Caravalho 1988; Tous and Battle 1990), and Goor's (1965) conclusion that "pollination is carried out especially by wind." The observation that 27 % of all the airborne pollen found in October near the Mediterranean in Israel was of *C. siliqua* (Waisel et al. 1997) offers circumstantial evidence for pollen transport by wind. The objectives of our study were to answer the following three questions:

(1) How do dioecious *C. siliqua* trees achieve pollination in the harsh pollination environment of their flowering season?

(2) What are the relative contributions of day, night, and wind pollination, and who or what are the effective pollinating agents?

(3) Can floral attractants (scents) and rewards be interpreted as adaptations to attract and guide effective pollinating insect agents?

12.2 Choice of the study sites

The observations and investigations were carried out during 2005–2009 at various sites, all of which are under typical Mediterranean conditions. Several localities are within Haifa city boundaries:

(1) Nahal Lotem 246°4′N:148°2′E, 160 a.s.l.

(2) Khalissa 244°8′N:150°9′E, 160 a.s.l.

(3) Tchernichovsky St. 246°8′N:147°4′E, 200 a.s.l.

(4) Marcus St. 247°4′N:147°3′E, 220 a.s.l.

(5) Ovadia St. 246°5′N:148°3′E, 180 a.s.l.

(6) Stella Maris 248°3′N:147°3′E, 100 a.s.l.

(7) Mt. Carmel at Haifa University 245°6′N:152°5′E, 440 a.s.l.

(8) Rom Carmel 240°3′N:152°3′E, 450 a.s.l.

(9) Wadi Oren 235°5′N:147°5′E, 40 a.s.l.

(10) Coastal plain at Caesarea, 214°3′N:147°3′E, 20 a.s.l.

The nearest climatological stations (between 20 and 450 a.s.l.) report an average annual precipitation of 550 mm at the coastline in Haifa, 591.7 mm at Atlit (nearest station to Caesarea), and 715.6 mm at Haifa University (nearest station to Rom Carmel). In Haifa, average maximal temperatures range from 17.1 to 31.4 °C, and minimal temperatures from 8.7 to 23.6 °C, respectively (data from http://www.ims.gov.il/ims/climate).

12.3 Observations of day-active flower visitors

To indentify the day-active floral visitor composition and their frequencies, day observations were conducted mostly at the Rom Carmel site and partly at the Wadi Oren site on four male and six female trees on 11 days between October 3 and November 26, 2005. Observation times were chosen to represent various weather conditions and times of the day. In order to reduce biases introduced by different abiotic conditions (temperature, humidity, light) which may have concealed any gender-derived patterns of flower visits to the trees, most observations were carried out simultaneously by two observers on male and female trees growing in close proximity (20–100 m). Flowering branches of approximately equal height and under similar light conditions were chosen. On the observed branches, the number of inflorescences with open flowers that offer a reward were counted. Each observational unit typically covered an area containing 8–60 such inflorescences and lasted 15–30 minutes. Any new flower visitor entering the observation area was counted as a separate individual. Flower visitors were identified on the wing to lowest taxonomic level possible. The number of visiting individuals as well as visit frequency were quantified for all repetitions for each nominated hour interval during the whole observation period between sunrise and sunset (5.30 a.m. and 5.30 p.m.), resulting in 14 observation hours for each sex. For later graphical presentation, data from 35 observation units from each sex were pooled, and day visits per inflorescence for each plant sex and hourly day time interval were averaged. Hourly averages represent the data from at least two (up to six) observation units.

12.4 Observations of night-active flower visitors

Direct consistent observations of the behavior of night-active flower-visiting moths on trees were not possible because light sources used to enable the observer to actually see the visitors' behavior may themselves attract moths. As preliminary observations showed that moths exemplify the same behavior under red and white light, we used short-term, white-light torch flashes to check moth activity at various times after sunset, between 7.00 p.m. and 10.00 p.m. Moths drinking nectar while sitting and crawling on the inflorescences were caught directly. It was, however, practically impossible to collect quantitative data on flower-visit frequency and stigma contacts because moths were found to stay at the same flower for longer periods (up to one hour) and moved only very slowly between flowers and inflorescences. Furthermore, it was very difficult to confirm whether moths made contact with the stigmas because their wings covered the flowers. For a representative sample of the various taxa, we spent in total 60 hours (ten nights) at various localities in 2005 observing and collecting nocturnal flower-visiting moths. Moth species were identified by Dr. V. Kravchenko (Tel Aviv University, Israel).

Occasional sightings of adult flower-visting lacewings (Chrysopidae) during moth observations prompted us to pay more attention to them and their activity on *C. siliqua* trees. It is known that lacewings complement their prey diet with nectar and pollen (Principi and Canard 1984; Villenave et al. 2006). However, we were unable to find any literature about them as pollinators. As observations are quite difficult with respect to their low abundance and night-time conditions, we aimed our observations specifically at lacewings found on female *C. siliqua* flowers. In order to confirm their putative role as pollinators, we checked the following parameters: presence on female *C. siliqua* flowers and stigma contacts during visits, and *C. siliqua* pollen loads on the bodies of specimens captured from female flowers.

Lacewing observation was conducted in 2007 on six nights at five different localities (total observation time 21 hours) during their main activity period (6.00–10.00 p.m.), as determined by preliminary observations. We screened whole female trees for the presence of lacewings on flowers and then observed the occupied flowers from a distance of 50 cm. We checked whether the lacewings contacted the floral stigmas during their flower visits. At the end of each observation period we collected individual lacewings in separate vials for later examination of pollen presence and distribution on their bodies. The lacewings were then identified by Dr. Peter Duelli of the Swiss Federal Research Institute (Zurich, Switzerland).

Due to the low visitation rate, high flower density, dense structure of the individual inflorescences, and the inabilty to check at night if the flowers are receptive, determination of nocturnal visit frequencies per flower or inflorescence was not possible for moths or lacewings.

12.5 Pollen loads on nocturnal flower visitors

In order to examine the possible role of nocturnal visitors as pollinators in *C. siliqua*, we quantified the pollen load on insects visiting female flowers at night. Various individual night-active flower visitors (50 lacewings, 100 moths) were caught after or while visiting flowers of female trees in several localities (Nahal Lotem, Tchenchivsky, Stella Maris, Khalisa, Ovadia, and Marcus) between October 20 and November 20, 2007. Caught insects were kept individually in a clean vial for further examination in the lab. Each flower visitor was screened under a stereomicroscope (60x magnification) to determine the number of *Ceratonia* pollen grains that it carried. In addition, the distribution of pollen grains on the various body parts was recorded. Identifying the pollen was relatively easy because less than 0.1 % of the pollen grains found were from other plant species. We also analyzed whether the type (lacewings versus moths) or size (large moths > 2 cm versus small moths < 2 cm) was correlated with the pollen grain numbers found on the night-active insects.

12.6 Wind pollination

For the assessment of the contribution of wind pollination to total pollination success, we studied seven trees, each at four different localities (Nahal Lotem, Haifa University, Wadi Oren, Caesarea) during October and November, 2009. Before flowering commenced, we marked each tree with one untreated control sample of inflorescences and covered another corresponding sample of inflorescences with a mesh bag. In preceding experiments, we had ensured that glycerin-covered microscope slides placed outside and within fine mesh bags (mesh size ca. 5 mm) in trees received similar amounts and types of airborne pollen. As thrips were not observed on *C. siliqua* flowers in any of our study sites, 5 mm mesh size was sufficient to exclude any potential flower visitor. Thus, these nets used on flowering trees prevented pollination by insects but allowed wind pollination.

Three weeks after the end of the flowering of the marked and enclosed inflorescences, we counted the swollen ovaries as an indication for successful pollination. The ovary assessment was done early so as not to lose fruits due to fruit abortion, which is quite common in *C. siliqua* (von Hasselberg 2000), and which would potentially mask pollination success.

12.7 Day and night pollination versus wind pollination

To assess the contribution of each pollination mode to total pollination success, we subjected inflorescences of eight trees to four different treatments in 2009 at Nahal Lotem:

(1) Open (free) pollination as a control

(2) Permanent enclosure in 5 mm mesh size bags to allow only wind pollination

(3) Cover with bags during night and exposure to visitors daily between 6.00 a.m. and 6.00 p.m.

(4) Cover with bags during day and exposure to visitors nightly from 6.00 p.m. to 6.00 a.m.

On each tree, all four treatments were applied to inflorescences in close proximity, ca 50–70 cm apart to minimize microsite effects. Bags were fitted before commencement of main flowering, and all flowers that were open before the bagging experiments started were removed. The experiments were laid out for two weeks and all experimental inflorescences were covered with pollen-proof paper bags to prevent any additional uncontrolled pollination. All flowers that opened after that were removed. Swollen ovules indicating pollination were counted three weeks after the experimental period ended.

It should be noted that individual flowers may be receptive for at least 11 days (Dafni personal observation). Therefore, the late flowers in the experiment had much shorter exposure times than flowers that had started flowering at the beginning of the two-week period. In conclusion, pollination rates under the specific experimental bagging conditions can be compared among our four different treatments, but they do not reflect pollination success under natural pollination regimes, when all flowers are exposed for their whole lifespan.

12.8 Nectar measurements

Standing crop nectar volume and concentration in male and female trees of *Ceratonia siliqua* were measured in hourly intervals in relation to air temperature and relative humidity at Nahal Lotem from October 24–25, 2009. To minimise the effect of uncontrolled variables (such as exposure to sun and position in the canopy), we used the following procedure: two trees, one male and one female of the same size, 10 m from each other and with the same exposure to wind and sun. All the nectar samples (n = 9 for each time point) were taken at the same part of each tree (at the western side of the canopy, height 2 m from the ground and about 40–50 cm within the canopy to avoid direct sunlight.

Nectar was drawn with microcapillaries of 1 or 5 µl according to nectar availability from a single flower. Nectar concentration as percentages of sucrose equivalents was measured by using a Bellingham and Stanley refractometer adapted for small quantities (down to 0.2 µl). When nectar was too scarce for measuring, several flowers were pooled to enable measurement, and the result adjusted to the number of flowers used. All the flowers were chosen to be at the same developmental stage (stage 4 according to Retana et al. 1990) when nectar secretion is maximal. Humidity and temperature data were collected using a digital Thermo–Hygrometer (MICRONTA™) as close as possible to the sampled trees.

Furthermore, during all nocturnal lacewing observations in 2007, the nectar of 30 random flowers on the observed tree was collected, and nectar volume per flower and nectar concentration were measured.

12.9 Volatile sampling

Floral volatile samples were taken at Rom Carmel in October 2005 from different plant individuals during day and night time. For volatile sampling we used miniature trapping vials (microvials) that were loaded into a modified GC injector for direct thermal desorption (Gordin and Amirav 2000). The microvials were prepared from quartz sample vials (15 × 1.9 mm internal diameter; Varian, Inc., Palo Alto, CA, USA) that were open on both sides. These microvials were filled with tenax and carbotrap (3–6 mm), and glass wool was added on both sides of the adsorption material to keep it in place. The microvials were cleaned with an acetone wash and heated for 30 minutes at 250 °C. Inflorescences were enclosed in polyvinylacetate oven bags (Toppits[®]; Melitta, Germany) for 20 minutes to concentrate volatiles before sampling. Volatile-containing air was then sucked through the microvials (flow rate 200 ml minutes^{-1}) with a battery-operated membrane pump (G12/01 EB, Rietschle Thomas, Memmingen, Germany) for 20 minutes.

12.10 Gas chromatography–mass spectrometry

The volatiles trapped in the microvials were analyzed on a Varian Saturn 2000 mass spectrometer coupled to a Varian Saturn 3800 gas chromatrograph using a 1079 injector that had been fitted with the ChromatoProbe device (see Dötterl et al. 2005). This device allows the thermal desorption of small amounts of solids or liquids contained in quartz microvials (Amirav and Dagan 1997). A microvial was loaded into the probe, which was then inserted into the modified GC injector. A ZB-5 column (5 % phenyl polysiloxane) was used for the analyses (60 m long, inner diameter 0.25 mm, film thickness 0.25 μm; Phenomenex, Torrance, CA, USA). Electronic flow control was used to maintain a constant helium carrier gas flow of 1.8 ml minutes^{-1}. The GC oven temperature was held at 40 °C for 4.6 minutes, then increased by 6 °C minutes^{-1} to 260 °C and held for 1 minute. The MS interface was 260 °C and the ion trap worked at 175 °C. The mass spectra were taken at 70 eV (in EI mode) with a scanning speed of one scan per second from m/z 30–350. The GC–MS data were processed using the Saturn Software package 5.2.1. Component identification was carried out using the NIST 2005 and Mass Finder 2.1 mass spectral databases and confirmed by comparison of Kovats' relative retention times with the MassFinder 2.1 software. Identification of individual components was confirmed by comparison of both mass spectrum and GC retention data with those of authentic standards. For description of volatile spectra, relative amounts of volatile compounds (total volatile spectrum equaling 100 %) were calculated.

12.11 Results

12.11.1 Diurnal flower visitors

In both male and female trees of *C. siliqua*, Hymenoptera (comprising Apidae, Vespidae and Halictidae), dominated the flower visitor spectra (Table 12.1). In male trees, Hymenoptera contributed 77.3 % of the observed diurnal insect visitors to inflorescences and 78.6 % of the total number of observed visits. In female trees, the data were 65 % and 75.4 %, respectively.

Visit frequency to male trees was generally higher than to female trees (Fig 12.1). In particular, visit frequency to inflorescences on male trees peaked around late morning (10.00 a.m.) with about five visits per inflorescence and hour observed on males compared to only about 0.5 on females.

Table 12.1 Day-active flower visitors on *Ceratonia siliqua* at Rom Carmel, Israel. Number of individual visitors and number of flower visits (shown in parentheses) during observations in 2005.

Order	Family	Genus/Species	Individual visitors (visits)	
			Male trees	**Female trees**
Hymenoptera	Apidae	*Apis mellifera* L.	175 (522)	22 (44)
	Apidae	*Amegilla* sp.	7 (16)	0 (0)
	Vespidae	*Vespula germanica* F.	4 (6)	0 (0)
	Vespidae	*Vespa orientalis* F.	8 (16)	2 (2)
	Halictidae	*Halictus sp.*	4 (5)	1 (1)
	Unidentified	1 species	6 (9)	1 (2)
Diptera	Syrphidae	*Eristalinus taenipos* Wiedemann	32 (87)	6 (8)
	Muscidae	*Musca domestica* L.	11 (24)	2 (2)
	Calliphoridae	*Lucilia sericata* Mg.	4 (7)	2 (2)
	Unidentified	1 species	13 (38)	4 (4)

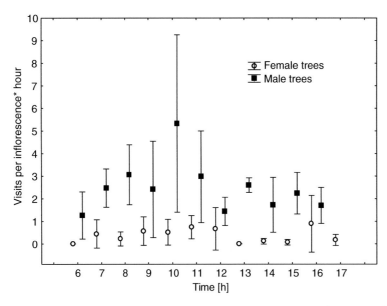

Fig 12.1 Average visit frequency per hour and inflorescence (mean ± SD) by Hymenoptera and Diptera to male and female inflorescences (curves slightly shifted against each other for better resolution) of *Ceratonia siliqua* in 2005. For each sex data from 35 observation units covering 14 hours were pooled and analyzed in reference to day time. Each data point represents between two and six observation units.

12.11.2 Nocturnal flower visitors

Nocturnal visitors included several species of settling moths, mainly Noctuidae (19 species, see Table 12.2) and rarely Geometridae (*Xanthorboe* sp.), Pyralidae (2 spp.), and Eponomeutidae (1 sp.) were found. Furthermore, five species of Chrysopidae (Neuroptera) were identified as visitors of *Ceratonia siliqua* female trees (Table 12.2).

At least one *Ceratonia siliqua* pollen grain was found on 73 % of the 100 moths investigated (on 68 % of large moths and on 78 % of small moths), and on 92 % of all 50 Chrysopid individuals examined. Chrysopidae carried a median number of 40 pollen grains compared to small and large moths which carried only 20 and 10 pollen grains, respectively (Fig 13.2). Variation was quite high with minimum and maximum values ranging between 0 and 1000 (lacewings), 600 (small moths), and 400 (large moths), respectively. Differences between visitor types were significant according to Kruskal–Wallis test (H (2, N= 150) = 20.35, p < 0.001). Multiple comparisons of mean ranks showed that large moths differed significantly from lacewings (p < 0.001) and small moths (p < 0.05), while differences between lacewings and small moths were not significant.

On the moths, the pollen was highly dispersed over their bodies and most of it seemed unlikely to come into contact with stigmas during flower visits to female trees. Moreover, most of the pollen was hidden between hairs, which may impede pollen deposition on the stigma. Most of the lacewings carried pollen on the wings (79 %), followed by abdomen (61 %), and thorax (40 %), while individuals with pollen on the legs (19 %) and head (6 %) were less frequent. More than a quarter (Table 12.3) of the lacewing individuals observed during female flower visits touched the stigma, mainly with their wings and abdomens, and those parts were more likely to carry pollen than any other body part.

12.11.3 Pollination success

The rate of wind pollination in the four investigated populations was between 6.5 % and 20.8 % (Fig 12.3). Wind pollination resulted in up to 48 % of the total open pollination (Table 12.4). The more-detailed pollination experiments conducted at Nahal Lotem showed that flowers under open pollination (i.e. the control) had a natural pollination rate of 61.8 ± 18.7 % (mean ± SD; Fig 12.3). In day-exposed flowers, the pollination rate was 33.0 ± 14.1 %; in night-exposed flowers, pollination was 22.7 ± 13.9 %; and wind pollination resulted in only 6.5 ± 5.0 % of the flowers developing fruits. Differences between all treatments were significant (LSD test: p < 0.001, except for day versus night with p < 0.05).

12.11.4 Nectar standing crop

Nectar standing crop patterns and amounts of male and female flowers were very similar and therefore data for both sexes were pooled (Fig 12.4). Nectar standing

Table 12.2 Night-active flower visitors on *Ceratonia siliqua* at Mt. Carmel, Israel.

Lepidoptera	Neuroptera
Eponomeutidae	**Chrysopidae**
1sp.	*Chrysopa viridana* (Schneider 1845)
	Chrysoperla agilis (Henry et al. 2003)
Geometridae	*Chrysoperla lucasina* (Lacroix 1912)
Xanthorboe sp.	*Dichochrysa flavifrons* (Brauer 1850)
	Dichocrysa zelleri (Schneider 1851)
Noctuidae	
Agrotis bigramma (Esper 1790)	
Agrotis ipsilon (Hufnagel 1766)	
Agrotis puta (Hübner 1803)	
Agrotis trux (Hübner 1824)	
Aporophyla australis (Boisduval 1829)	
Aporophyla canescens (Duponchel 1826)	
Caradrina (Paradrina) amseli (Boursin 1936)	
Caradrina (Paradrina) atriluna (Guenée 1852)	
Dryobotodes eremita (Fabricius 1775)	
Helicoverpa armigera (Hübner 1808)	
Leucania (Acantholeucania) loreyi (Duponchel 1827)	
Leucania (Leucania) punctosa (Treitschke 1825)	
Noctua comes (Hübner 1813)	
Noctua pronuba (Linnaeus 1758)	
Olivenebula subsericata (Herrich-Schäffer 1861)	
Polymixis trisignata (Ménétriés 1847)	
Spodoptera cilium (Guenée 1852)	
Spodoptera littoralis (Boisduval 1833)	
Xestia xanthographa (Denis and Schiffermüller 1775)	
Pyralidae	
2 spp.	

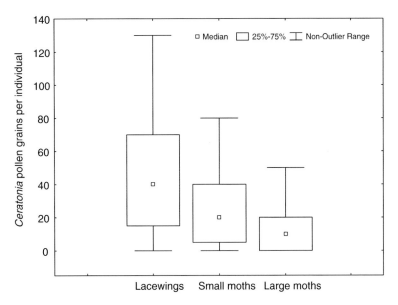

Fig 12.2 Median number of pollen grains found on the different night-active flower visitors (lacewings, small moths < 2 cm, and large moths > 2 cm; n = 50 for each group) of *Ceratonia siliqua* in 2007. Differences significant (Kruskal–Wallis test: H (2, N= 150) = 20.35, p < 0.001) between large moths and lacewings (p < 0.001), and large and small moths (p < 0.05).

crop seemed to be correlated with air temperature (negatively) and relative humidity (positively). Nectar volumes and relative humidity peaked during the night when nectar concentration was lowest. Nectar concentration was highest during midday and the early afternoon hours when temperature was highest and nectar volume was lowest.

Although conclusions can only be drawn with care, as data on nectar standing crop and diurnal visitor frequencies were not collected at the same time, it seems that the increase in day-visitor activity towards midday coincides with the decrease in nectar volume and increase in sugar concentration. In our investigation, carried out on two consecutive days in 2009 (Fig 12.4), nectar volume reached a low-point after midday (around 2.00 p.m., less than 0.5 μl in flowers of both sexes) while nectar concentration was highest at the same time (71-76 % in both sexes).

In contrast to diurnal visitors, the main activity period of nocturnal visitors seems to be parallel to increasing nectar volume and decreasing nectar sugar concentration in carob flowers. During nectar standing crop investigations, flowers of both sexes reached a volume high (5-6 μl) of low-concentrated nectar (10-18 %) per flower at 2.00 a.m. (Fig 12.4). Moth activity is usually highest between

Table 12.3 Records of lacewings visiting female trees of *Ceratonia siliqua* at six different localities in 2007 and numbers of lacewings contacting a stigma, as well as time of observation, range of relative humidity and temperature, and average nectar volume per flower and nectar concentration (mean ± SD) recorded during that time.

Location	Date	Time	Rel. Hum.	Temp.	Nectar Vol.	Nectar Conc.	Lace–wings	Individuals contacting stigmas
	dd.mm	[h]	[%]	[°C]	[μl]	[%]	n	[%]
Khalissa	16.10	19–22	50–60	25–27	2.8 ± 2.7	17.5 ± 4.5	6	16.6
Tchernichovsky	25.10	18–22	70–81	24–27	8.7 ± 5.6	12.9 ± 3.4	13	38.4
Stella–maris	26.10	19–22	65–78	21–25	9.3 ± 3.4	13.9 ± 3.4	18	44.4
Marcus	27.10	18–22	69–88	23–25	15.2 ± 2.6	10.3 ± 2.2	27	25.9
Ovadia	28.10	18–22	67–75	24–26	6.3 ± 2.5	14.0 ± 4.1	7	28.5
Marcus	1.11	18–21	67–75	24–26	7.2 ± 3.1	13.4 ± 2.6	7	28.5
Sum	6 d	22 h	–	–	–	–	78	–
Overall mean	–	–	–	–	8.3 ± 4.1	13.7 ± 2.3	–	28.0 ± 11.6

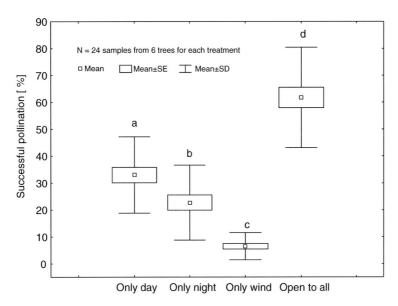

Fig 12.3 Pollination rate by different pollination modes (only day-pollination, only night-pollination, only wind-pollination) compared to pod set by open pollination in *Ceratonia siliqua* in 2009. Different superscript letters indicate significant differences (LSD test:p < 0.001 for all except for day versus night with p < 0.05).

sunset and midnight; moth activity generally drops after 10.00 p.m. before nectar concentration reaches its lowest.

The main activity time of lacewings on the female trees was from 6.00–10.00 p.m. (similiar to moths), at the same time when copious amounts (on average between 2.8 and 15.2 μl) of relatively low-concentrated nectar (between 10.3 % and 17.5 %) was available in the flowers of the observed trees (Table 12.3).

12.11.5 Floral volatiles

The floral volatiles emitted by *C. siliqua* inflorescences comprise a mixture of compounds from five different compound classes, dominated by monoterpenoids, hydrocarbon esters, and benzenoids (Table 12.5, Fig 12.5). Thirty-eight compounds were detected, out of which 37 were identified to compound class. The highest number of compounds (36) was found in day samples of male inflorescences; the lowest number of compounds (31) was found in day samples of female inflorescences. The major compounds in the floral volatiles of *C. siliqua* inflorescences of both sexes were the monoterpenoids linalool and (*E*)-linalool oxide (furanoid); followed by (*Z*)-linalool oxide (furanoid); 2,6,6-trimethyl-2-vinyl-5-ketotetrahydropyran; (*E*)-linalool oxide (pyranoid); and the fatty acid derivative methyl-2-hydroxy-3-methylbutanoate. The benzenoid 3,5-dimethylbenzaldehyde and the

Table 12.4 Wind pollination success (mean ± SD) in comparison to open pollination success of *Ceratonia siliqua* at different sites during 2009. Different superscript letters within a column indicate significant differences among sites (LSD test: p < 0.05).

	Only wind pollination	Open pollination	% wind from open
Nahal Lotem	6.5[b] ± 5.0	61.8 [a] ± 18.7	11.6 [a] ± 9.2
Caesarea	18.4[a] ± 5.6	44.3 [b] ± 11.6	43.2 [b] ± 15.2
Beit Oren	20.8 [a] ± 8.7	43.7 [b] ± 10.3	48.2 [b] ± 19.2
Haifa University	7.1 [b] ± 4.8	33.3[c] ± 10.0	23.3 [c] ± 17.4
Over all sites	12.5 ± 8.8	47.6 ± 17.4	29.4 ± 21.3

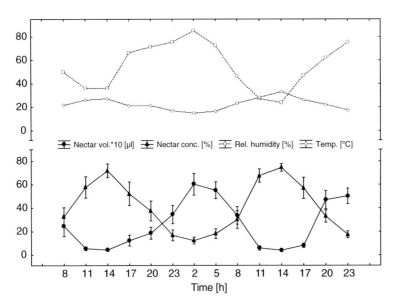

Fig 12.4 Standing crop nectar volume per flower and nectar concentration (mean ± SD) in hourly intervals in relation to air temperature and relative humidity in male and female trees of *Ceratonia siliqua* at Nahal Lotem from October 24–25, 2009.

fatty acid derivatives butyl acetate, and especially hexyl acetate, were found in considerable proportions in day samples from female flowers only. All linalool derivatives (linalool and its furanoids and pyranoids and 2,6,6-trimethyl-2-vinyl-5-ketotetrahydropyran) taken together, contributed 84.8 % in day samples from males and a similar 81.8 % in night samples from males, but only 47.3 % in females at day compared with 76.9 % in females at night (Table 12.5). In other words, with respect to monoterpenoids (Fig 12.5), and especially linalool and its derivatives,

Table 12.5 Mean relative amounts [%] of floral volatiles in day and night samples of male and female inflorescences of *Ceratonia siliqua*. The compounds are listed according to class and relative retention time (RRT), tr = trace amounts (< 0.01 %). Unknowns were included when present with more than 1.0 % of the total amount in any sample.

	RRT	Male			Female		
		All	Day	Night	All	Day	Night
Number of samples collected		**17**	**9**	**8**	**17**	**11**	**6**
Total number of compounds		36	36	34	35	31	32
FATTY ACID DERIVATIVES							
Nonanal	1453	0.2	0.3	0.1	0.5	tr	1.4
Butyl acetate	928	0.2	0.3	—	4.5	7.0	—
Methyl 2-hydroxy-3-methylbutanoate	1084	5.4	4.8	6.1	6.6	6.6	6.0
Methyl 2-hydroxy-4-methylpentanoate	1252	0.1	0.1	0.1	—	—	—
(Z)-3-Hexenyl acetate	1281	0.4	0.2	0.5	0.6	0.6	0.5
Hexyl acetate	1297	0.3	0.5	tr	11.8	18.3	—
(Z)-3-Hexenyl butyrate	1579	0.3	0.2	0.5	0.3	0.3	0.3
2,4-Hexadiene	618	1.2	1.3	1.1	0.2	0.1	0.3
BENZENOIDS							
2-Phenylethyl alcohol	1479	0.4	0.2	0.6	0.4	0.4	0.2
Benzyl acetate	1553	0.9	0.7	1.2	0.6	0.3	1.3
Methyl salicylate	1604	0.7	0.7	0.8	0.1	tr	0.3
3,5-Dimethylbenzaldehyde	1644	0.4	0.4	0.4	4.6	7.0	0.2
(E)-Cinnamaldehyde	1722	tr	0.1	—	Tr	tr	0.1
Methyl anisate	1805	1.2	1.0	1.4	1.4	1.4	1.6
3-Hydroxy-4-phenyl-2-butanone	1819	0.5	0.2	0.8	1.4	1.7	0.8

Table 12.5 (cont.)

	RRT	Male			Female		
		All	Day	Night	All	Day	Night
Number of samples collected		**17**	**9**	**8**	**17**	**11**	**6**
Total number of compounds		**36**	**36**	**34**	**35**	**31**	**32**
Methyleugenol	1890	0.4	0.1	0.7	0.1	—	0.4
MONOTERPENOIDS							
Unidentified monoterpene	1235	0.2	0.2	0.1	0.9	0.8	1.2
(Z)-Ocimene	1342	0.9	1.6	0.2	1.1	0.5	2.1
(E)-Ocimene	1364	0.3	0.3	0.3	0.7	0.3	1.4
(Z)-linalool oxide (furanoid)	1411	10.6	9.9	11.5	6.5	5.3	8.7
(E)-linalool oxide (furanoid)	1436	24.8	22.3	27.7	19.1	15.5	25.8
Linalool	1449	32.3	38.0	25.8	15.4	11.4	22.8
Hotrienol	1453	0.2	0.3	0.1	0.5	tr	1.4
2,6,6-Trimethyl-2-vinyl-5-ketotetrahydropyran	1468	7.3	7.1	7.4	9.8	8.5	12.2
2,6-Dimethyl-2,4,6-octatriene	1489	0.1	tr	0.2	0.1	0.1	0.1
(Z)-linalool oxide (pyranoid)	1566	2.3	1.8	2.8	2.3	2.6	1.9
(E)-linalool oxide (pyranoid)	1571	6.0	5.6	6.6	4.5	4.0	5.5
SESQUITERPENOIDS							
α-Copaene	1875	0.6	0.6	0.6	2.3	3.0	0.9
Unidentified sesquiterpene	1918	tr	tr	0.1	0.1	0.1	—

Compound	RI						
(E)-Caryophyllene	1940	0.1	tr	0.1	0.1	tr	0.1
Unidentified sesquiterpene	1988	—	—	—	0.1	—	0.3
Unidentified sesquiterpene	1997	—	—	—	0.1	—	0.2
(E,E)-α-farnesene	2022	0.1	0.2	tr	tr	—	Tr
Unidentified sesquiterpene	2032	tr	tr	tr	—	—	—
δ-Cadinene	2053	tr	tr	tr	—	—	—
Unidentified sesquiterpene	2060	0.3	0.4	0.3	2.2	2.9	1.0
NITROGEN-CONTAINING COMPOUNDS							
Methyl nicotinate	1510	0.8	0.6	1.0	0.7	0.6	0.9
UNKNOWNS							
Unidentified	1623	0.7	0.3	1.1	0.9	0.7	1.2
TOTALS BY GROUP:							
FATTY ACID DERIVATIVES		8.1	7.7	8.4	24.5	32.9	8.5
BENZENOIDS		4.5	3.4	5.9	8.6	10.8	4.9
MONOTERPENOIDS		85.0	87.1	82.6	61.0	48.9	83.1
SESQUITERPENOIDS		0.9	0.9	0.8	2.5	3.1	1.4
NITROGEN CONTAINING COMPOUNDS		0.8	0.6	1.0	0.7	0.6	0.9
UNKNOWNS		0.7	0.3	1.1	0.9	0.7	1.2

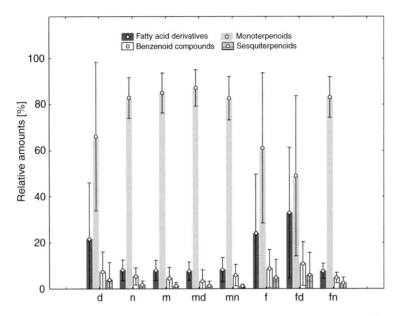

Fig 12.5 Relative amounts (mean ± SD) of volatile compound classes (fatty acid derivatives, benzenoids, monoterpenoids, and sesquiterpenoids) emitted from day (d) and night (n) samples of male (m) and female (f) *Ceratonia siliqua* flowers.

floral scent of male and female flowers is more similar at night than during the day. During the day, the main difference between males and females is the lower percentage of linalool derivatives and particularly high proportion of the fatty acid derivatves hexyl acetate, butyl acetate, and the benzenoid 3,5-dimethylbenzalde-hyde in females compared to males.

12.12 Discussion

12.12.1 Diurnal flower visitors

In the Carmel region of Israel, the main flower visitors observed during daylight hours belonged to the orders Hymenoptera and Diptera; this finding agrees with the results of studies conducted in Southern Spain (Arista et al. 1999; Ortiz et al. 1996) and in Portugal (Linskens and Scholten 1980; Passos de Carvalho 1988). The most frequent visitor to flowers during the day, for both male and female trees, was the honeybee (*Apis mellifera*), which made about 70 % of the total visits to trees of both sexes. Other studies conducted in Southern Spain and Portugal also found *Apis mellifera* to be a very frequent visitor to the carob tree (Ortiz et al. 1996; Linskens and Scholten 1980; Passos de Carvalho 1988). One study (Retana et al. 1990), conducted in Northern Spain, found that during the day carob flowers are visited mainly by flies, and this was explained by Retana et al. (1994) as being a result of the scarcity of bees during the flowering season of the carob. The fact that the time and length of the flowering period depends on local climatic conditions (Battle and Tous 1997) may explain the contradiction with the results of Retana et al. (1990) and the general variation in flower visitors to the carob tree recorded by different authors. A slightly later start to the flowering season in Northern Spain could bring it in or out of phase with the activity season of specific flower visitors. The presence or absence of honeybees in the reported flower visitor spectra (compare Retana et al. 1990, and this study) may be an artefact, introduced by the presence or absence of bee hives in the study area; this does not necessarily represent the natural ecological or evolutionary context in which the floral traits of this species have evolved.

The male and the female trees offered similar amounts of nectar while the males have also a lot of exposed pollen. This asymetric reward structure may explain the differences in the diurnal flower visitor frequencies observed on the two genders. Bees and flies may seek both pollen and nectar, and the male trees can attract visitors from both the nectar and pollen reward-seeking fraction. In addition, male trees have also far more inflorescences than female trees (von Hasselberg 2000) and, in our semi-randomly chosen observation units, male units had on average more rewarding inflorescences than units on female trees (min–max / mean ± SD: males 9–60 / 20 ± 12; females: 8–40 / 14 ± 7), which may enhance their attractivity further and attract more visitors. The day-active flower visitors, both Hymenoptera and Diptera, changed their foraging activities on the male trees from mainly pollen collecting between 5.30–9.30 a.m. to both pollen collecting and nectar drinking between 9.30–10.30 a.m., and then mostly to nectar drinking until the end of the day (data not shown). This is likely caused by the daily nectar standing crop patterns and the preference of bees for higher concentrated nectar (Heinrich 1975; Corbet 1978).

12.12.2 Nocturnal flower visitors

Settling moths are common at the Mediterranean during the flowering time of *C. siliqua* (Kravchenko et al. 2006; Yela and Herrera 1993). Settling moths could easily be attracted to the exposed nectariferous flowers of *C. siliqua* simply by the production of olfactory attractants as linalool and its derivates, which is regarded as common attractant in moth-pollinated flowers (Raguso and Pichersky 1999).

In total, 23 different species of settling moths were found visiting carob flowers and about 73 % of the individuals caught on the female trees were loaded with *C. siliqua* pollen. The large abundance of moths on *C. siliqua* female trees may indicate that moths contribute to pollination. However, pollen placement on the moths' bodies and their movements on the inflorescences suggest that the grains have a low chance of reaching a carob flower's stigma. Furthermore, in contrast to our findings, moths caught on male and female trees in Spain were not loaded with pollen (Retana et al. 1990).

In this study, we found at least five species of Chrysopidae as pollinators of *C. siliqua*. Adults feed on honeydew, nectar, and pollen (Principi and Canard 1984: 85; Villenave et al. 2006), and the females, in particular, need pollen for ovule maturation (Principi and Canard 1984: 90). These charcteristics, and the fact that lacewings caught on female trees were carrying a fair amount of pollen, support the assumption that they might be involved in the pollination of *C. siliqua*. Because *C. siliqua* pollen is viable, under field conditions for at least ten days (Dafni unpublished data), pollen deposition on female trees would not have to be accomplished the same night that the pollen was acquired. As far as the authors are aware, this is the first record of the possible involvement of Chrysopidae as active pollinators, though they are well-known for carrying pollen (Silberbauer et al. 2004; Villenave et al. 2006). Clinebell et al. (2004) were the first to show the involvement of Neuroptera in pollination. They found that the antlion *Scotoleon minusculus* (Myrmeleontidae) was a major pollen carrier of *Gaura villosa* ssp. *villosa* while another antlion species (*Vella fallax*) was observed in contact with stamens and stigma, suggesting that antlions may be effective pollinators.

12.12.3 Floral scent and pollination

Custodio et al. (2004, 2006) analyzed the floral scent composition of *C. siliqua* flowers and identified more than 25 compounds. Similar to our findings, linalool and its derivatives ((*Z*)-linalool oxide (furanoid); (*E*)-linalool oxide (furanoid); 2,2,6-trimethyl-3-keto-6-vinyltetrahydropyran; (*Z*)-linalool oxide (pyranoid)), were the dominant volatiles in the emitted scent. However, there are also differences in the odor composition found by these authors and our data. For example, Custodio et al. (2004, 2006) found high relative amounts of α-pinene that was not present in our samples. They also found two carotenoids, theaspirane A and theaspirane B, that we did not find in our samples. On the other hand, our samples contained several hydrocarbon esters (butyl acetate; methyl 2-hydroxy-3-methylbutanoate; methyl-2-hydroxy-4-methylpentanoate; (*Z*)-3-hexenyl acetate; hexyl acetate; (*Z*)-3-hexenyl butyrate) that they did not report. These differences in the floral scent composition can be due to the different methods used – SPME versus direct thermodesoption using microvials – although both methods are powerful solventless techniques (Dötterl et al. 2005; Raguso and Pellmyr 1998).

Floral scent is known to play a role as an important modulator of pollinator behavior not only in intensely scented, deceptive, or specialized pollinator systems but also in species visited by generalist pollinators (Ashman et al. 2005), such as the carob, which bears small flowers and is pollinated by various generalist insects (this study; Retana et al. 1990; Ortiz et al. 1996; Arista et al. 1999). The scent of carob flowers is most likely an olfactory attractant to the observed spectrum of potential day and night pollinators. Its scent is mainly attributable to a high abundance of monoterpenoids (this study; Custódio et al. 2004). The most abundant monoterpenoids in the present study were linalool, (E)-linalool oxide, and (Z)-linalool oxide. At night, floral scent of males and females consisted of about 80 % linalool and its oxides, and male and female inflorescences were relatively similar regarding their scent composition. This makes it more difficult for night-active flower visitors to differentiate between male and female inflorescences by scent only. However, at least for nectar-seeking moths, this is not a disadvantage as male and female flowers offer a similar nectar standing crop. Linalool and linalool oxides (furanoids and pyranoids) are typical and widespread floral scent compounds (see Knudsen et al. 2006) and are likely attractants for the observed moths, as they have been shown to be electrophysiologically active compounds in Lepidoptera and are known as moth attractants (Raguso et al. 1996; Raguso and Pichersky 1999; Andersson 2003; Piechulla and Pott 2003). However, linalool is not restricted to moth-pollinated flowers and occurs also in many diurnal flowers pollinated by a broad spectrum of pollinators, including bees (Raguso and Pichersky 1999; Knudsen et al. 2006) for which it is known to be attractive (Dötterl and Vereecken 2010). It seems that carobs produce a mixture of compounds that have a potential for attracting both bees (and possibly flies) during daylight hours, and moths and lacewings at night. Interestingly, in males, linalool and its derivatives were present during the day in similar relative amounts as during the night, whereas in females the proportion dropped during the day to about 40 %. In contrast, hexyl acetate, butyal acetate, and 3,5-dimethylbenzaldehyde were produced in considerable amounts by females during the day only (Table 12.5). Hexyl acetate is known as a green-leaf volatile and herbivore-induced plant volatile (HIPV) (e.g. in another legume *Vicia faba;* Frati et al. 2009) and floral-scent compound (e.g. in Lepidoptera-visited *Buddleja davidii;* Guédot et al. 2008), as well as a constituent of many insect pheromones, e.g. the alarm pheromone of bees recruiting other bees from the colony (Wager and Breed 2000). In *Vicia faba*, green-leaf volatiles emitted from whole plants were found in greater amounts during the day than at night (Webster et al. 2010). According to Hatanaka (1993), green-leaf volatiles require lipoxygenase for their biosynthesis, the activity of which is drastically reduced in dark compared to light conditions. This may explain the lack of hexyl acetate in the night samples from male inflorescences in carob but not its lack in the day samples. Butyl acetate, a typical fruit odor, was also found in bee alarm pheromones, enhancing bee recruitment from colonies and localization of moving targets (Wager and Breed 2000).

Some of the (mostly minor) floral volatiles that were identified in *C. siliqua* flowers are already known as HIPVs, which are usually induced in response to herbivore attack. HIPVs signal the presence of prey to other predators and provide an indirect defense via tritrophic interactions (Heil 2008). Some of these compounds, which are present in the floral emission of *C. siliqua*, have previously been experimentally found to be attractants of several species of lacewings: (Z)-3-hexenyl acetate (*Chrysoperla carnea* in Reddy 2002; *Chrysopa sinica* in Han and Chen 2002); methyl salicylate (*Chrysoperla carnea* in Molleman et al. 1997; *Chrysopa nigricornis* in James 2003; James et al. 2005; *Chrysopa occulata* in James 2003, 2006); methyleugenol (*Chrysoperla caernea* in Molleman et al. 1997; *Chrysoperla basalis* in Umeya and Hirao 1975; *Chrysoperla* sp. in Suda and Cunningham 1970); and nonanal (*Chyrysopa oculata, Ch. nigricornis, Ch. coloradensis* in Zhang et al. 2006). It might be speculated that lacewings, especially when looking for oviposition sites, are attracted by floral-scent compounds of *C. siliqua* that are normally part of the herbivore-induced communication channel that attracts the predators of leaf herbivores. Lacewing larvae are strictly carnivorous and many of them feed on aphids, which are known to induce emission of compounds, such as (Z)-3-hexenyl acetate (Du et al. 1998), a typical green-leaf volatile. Therefore, female lacewings may use such induced compounds to find appropriate oviposition sites. However, the HIPV compounds occur only in small relative amounts in flower samples of carob. Moreover, many of the compounds are widespread, occur in plants not pollinated by lacewings, and are emitted from almost all plant parts, making it unlikely that their emission by carob flowers targets the attraction of potential lacewing pollinators in particular. Typical green-leaf volatiles such as (Z)-3-hexenyl acetate are likely to be emitted in higher amounts from leaves (but see Brodmann et al. 2008). Non-HIPV compounds may play a much more important role for attracting adult lacewings because all adult Chrysopidae (with the exception of *Chrysopa* species) are not merely carnivorous, but feed on pollen, nectar, and honeydew (Principi and Canard 1984). Therefore, it is likely that the non-*Chrysopa* visitors found in the present study are attracted by common flower volatiles such as linalool and its derivatives, which dominate carob flower scent.

12.12.4 Ambophily and the shift to night pollination

Members of the former family of Caesalpiniaceae (now recognized as subfamily Caesalpinioideae) are typically entomophylous with showy flowers (Endress 1994) and to our knowledge only one case of secondary wind pollination in Caesalpiniaceae has been reported so far (Shi-Jing et al. 2000). The flowers of *C. siliqua* show a mixture of traits attributed to wind and to insect pollination (Culley et al. 2002) and some authors indicate the possibility of wind pollination in *C. siliqua*. According to Goor (1965), "pollination is mainly by wind" while Hillcoat et al. (1980) mentioned that "the floral structure suggests wind pollination while the odor suggests fly pollination." In the present work, it was found that wind pollination may achieve from 6–21 % of the total pollination when insects were excluded. The lowest rate of wind pollination was found at a closed Mediterranean maqui and the maximum at an open habitat in which the trees were highly dispersed. *C. siliqua*, thus, shows a combination of wind and insect pollination, termed ambophily. Ambophily has indeed shown to be more common than previously thought (Meeuse 1978; Meeuse et al. 1990; Sacchi and Price 1988; Vroege and Stelleman 1990; Karrenberg et al. 2002; Culley et al. 2002), although it is still controversial whether it is a stable system or a transient stage towards absolute anemophily or entomophily (Culley et al. 2002; Friedmann and Barrett 2009). Insects may play an important role in the reproduction of species with typically anemophilous flowers, i.e. unisexual, small in size, rather inconspicuous, and with a low reward for pollinators (e.g. Gulías et al. 2004 in Bandera and Traveset 2006). Likewise, typically entomophilous flowers have shown to be pollinated also by wind (Dafni and Dukas 1986; Scariot et al. 1991; Bullock 1994). The evolution of wind pollination in animal-pollinated lineages is thought to occur when physical and biological conditions render biotic pollination less reliable (Whitehead 1968; Regal 1982; Cox 1991; Culley et al. 2002; Dafni and Dukas 1986; Duan et al. 2009; Bandera and Traveset 2006; Friedmann and Barrett 2009), when delivered pollen is of poor quality (Weller et al. 1998; Goodwillie 1999), and when plants colonize areas with low insect abundance (Berry and Calvo 1989; Gómez et al. 1996; Totland and Sottocornola 2001; Shi-Jing et al. 2000). Under such circumstances a shift to wind pollination may be regarded as reproductive assurance (Anderson and Overal 1988; Douglas 1997; Mahy et al. 1998; Peeters and Totland 1999; Bandera and Traveset 2006). For the genus *Salix*, for instance, Karrenberg et al. (2002) have recently argued that ambophily is the ancestral stage. The importance of either pollinator vector may vary spatially, and even temporally, and thus both systems may be maintained through time without any strong selection against either one (Vroege and Stelleman 1990; Gómez et al. 1996; Totland and Sottocornola 2001).

We may conclude that the shift of *C. siliqua* from typical showy entomophilous flowers into night as well as wind pollination, in addition to general diurnal pollination by bees and flies, is a kind of "super generalism." Retana et al. (1994) have mentioned that the extended flowering season of *C. siliqua* compensates for the unstable weather at this time of the year and ensures that at least some flowers will be pollinated in "a period of good weather and insect activity." All this can be interpreted as maximizing pollination chances under the harsh pollination environment, in terms of high temperature and aridity, which causes low diversity and abundance of potential pollinators in the Mediterranean during the flowering priod of *C. siliqua*.

Acknowledgements

Thanks to Dr. Vasily Kravchenko at Tel Aviv University, Israel and Dr. Peter Duelli, Swiss Federal Research Institute, Zurich for identifying the insects. We thank The Dorothy and Henk Schussheim Foundation for Ecological Research on Mt. Carmel for their generous support. Andreas Jürgens is supported by the NRF (South African National Research Foundation).

References

Amirav, A. and Dagan, S. (1997). A direct sample introduction device for mass spectrometry studies and GC–MS analysis. *European Journal of Mass Spectrometry*, **3**, 105–11.

Anderson, A. B. and Overal, W. L. (1988). Pollination ecology of forest-dominant palm (*Orbignya phalerat* Mart.) in Northern Brazil. *Biotropica*, **20**, 192–205.

Andersson, S. (2003). Antennal responses to floral scents in the butterflies *Inachis io, Aglais urticae* (Nymphalidae), and *Gonepteryx rhamni* (Pieridae). *Chemoecology*, **13**, 13–20.

Arista, M., Ortiz, P. and Talavera, S. (1999). Apical pattern of fruit production in the racemes of *Ceratonia siliqua* (Leguminosae:Caesalpinioideae): role of pollinators. *American Journal of Botany*, **86**, 1708–16.

Ashman, T.-L., Cole, D. H., Bradburn, M., Blaney, B. and Raguso, R. A. (2005). Scent of a male: the role of floral volatiles in pollination of a gender dimorphic plant. *Ecology*, **86**, 2099–105.

Bandera, M. C. dela and Traveset, A. (2006). Breeding system and spatial variation in the pollination ecology of the heterocarpic *Thymelaea velutina*

(Thymelaeaceae). *Plant Systematics and Evolution*, **257**, 9–23.

Battle, I. and Tous, J. (1997). *Carob tree: Ceratonia siliqua L. Promoting the Conservation and Use of Under-Utilized and Neglected Crops.* Rome: Plant Genetic Resource Institute, Gatersleben/International.

Berry, P. E. and Calvo, R. N. (1989). Wind pollination, self-incoptability, and altitunal shift in pollination system in the high Andean genus *Espeletia* (Asteraceae). *Americam Journal of Botany*, **76**, 1602–14.

Brodmann, J., Twele, R., Francke, W., Holzler, G., Zhang, Q. H., and Ayasse, M. (2008). Orchids mimic green-leaf volatiles to attract prey-hunting wasps for pollination. *Current Biology*, **18**, 740–4.

Bullock, S. H. (1994). Wind pollination of neotropical dioecious trees. *Biotropica*, **26**, 172–9.

Clinebell, R. R, Crowe, A., Gregory, D. P. and Hoch, P. C. (2004). Pollination ecology of *Gaura* and *Calylophus* (Onagraceae, Tribe Onagreae) in western Texas, USA. *Annals of the Missouri Botanical Garden*, **91**, 369–400.

Corbet, S. A. (1978). A bee's view of nectar. *Bee World*, **59**, 25–32.

Cox, P. A. (1991). Abiotic pollination: an evolutionary escape for animal-pollinated angiosperms. *Philosophical Transactions of the Royal Society Series B*, **333**, 217–224.

Culley, T. M., Weller, S. G. and Sakai, A. K. (2002). The evolution of wind pollination in angiosperms. *Trends in Ecology and Evolution*, **17**, 361–9.

Custódio, L., Nogueira, J. M. F. and Romano, A. (2004). Sex and developmental stage of carob flowers affects composition of volatiles. *Journal of Horticultural Science and Biotechnology*, **75**, 689–92.

Custódio, L., Serra, H., Nogueira, J. M. F., Gonçalves, S. and Romano, A. (2006). Analysis of the volatiles emitted by whole flowers and isolated flower organs of the carob tree using HS–SPME–GC/MS. *Journal of Chemical Ecology*, **32**, 929–42.

Dafni, A. (1986). Autumnal and winter pollination adaptations under Mediterranean conditions. *Bocconea*, **5**, 171–81.

Dafni, A. and Dukas, R. (1986). Wind and insect pollination in *Urginea maritima* (L.) Bak. (Liliaceae). *Plant Systematics and Evolution*, **152**, 1–10.

Dötterl, S. and Vereecken, N. (2010). The chemical ecology and evolution of bee–flower interactions: a review and perspectives. *Canadian Journal of Zoology*, **88**, 668–97.

Dötterl, S., Wolfe, L. M. and Jürgens, A. (2005). Qualitative and quantitative analyses of flower scent in *Silene latifolia*. *Phytochemistry*, **66**, 203–13.

Douglas, D. A. (1997). Pollination, capsule damage, and production of seeds in *Salix setchelliana* (Salicaceae), an Alaskan glacial river gravel bar willow. *Canadian Journal of Botany*, **75**, 1182–7.

Du, Y-J., Poppy, G. M., Powell, W., Pickett, J. A., Wadhams, L. J. and Woodcock, C. M. (1998). Identification of semiochemicals released during aphid feeding that attract parasitoid *Aphidius ervi*. *Journal of Chemical Ecology*, **24**, 1355–68.

Duan, Y. W., Zhang, T. F., He, Y. P. and Liu, J. Q. (2009). Insect and wind pollinartion of an alpine biennial *Aconitum gymnandrum* (Ranunculaceae). *Plant Biology*, **11**, 796–802.

Endress, P. K. (1994). *Diversity and Evolutionary Biology of Tropical flowers*. Cambridge, UK: Cambridge University Press.

Feinbrun-Dothan, N. and Danin, A. (1998). *Analytical Flora of the Land of Israel*. Jerusalem, Israel: CANA Publishing House Ltd.

Frati, F., Chamberlain, K., Birkett, M., Dufour, S., Mayon, P., Woodcock, C., Wadhams, L., Pickett, J., Salerno, G., Conti, E. and Bin, F. (2009). *Vicia faba-Lygus rugulipennis* interactions: induced plant volatiles and sex pheromone enhancement. *Journal of Chemical Ecology*, **35**, 201–8.

Friedmann, J. and Barrett, S. C. H. (2009). A phylogenetic analysis of the evolution of wind pollination in the angiosperms. *International Journal of Plant Sciences*, **169**, 49–58.

Gómez, J. M., Zamora, R., Hódaz, J. D. and Carcía, D. (1996). Experimental study of pollination by ants in Mediterranean high mountain and arid habitats. *Oecologia*, **105**, 236–42.

Goodwillie, C. (1999). Wind pollination and reproductive assurance in *Linanthus parviflorus* (Polemoniaceae), a self-incompatible annual. *American Journal of Botany*, **86**, 948–54.

Goor, A. (1965). *Ceratonia siliqua*. In *Encyclopaedia Hebraica Vol. 17.* Tel Aviv, Israel: Massaada, pp 985–8, (in Hebrew).

Gordin, A. and Amirav, A. (2000). SnifProbe: new method and device for vapor and gas sampling. *Journal of Chromatography A*, **903**: 155–72.

Guédot, C., Landolt, P. J. and Smithhisler, C. L. (2008). Odorants of the flowers of butterfly bush, *Buddleja davidii*, as possible attractants of pest species of moths. *Florida Entomologist*, **91**, 576–82.

Gulías, J., Traveset, A., Mus, M. and Riera, N. (2004). Critical stages in the recruiment process of *Rhamnus alaternus*. *Annals of Botany*, **93**, 723–31.

Han, B. Y. and Chen, Z. M. (2002). Composition of the volatiles from intact and mechanically pierced tea aphid-tea shoot complexes and their attraction to natural enemies of the tea aphid. *Agricultural and Food Chemistry*, **50**, 2571–5.

von Haselberg, C. D. (2000). Vegetative growth and fruit development in Carob trees (*Ceratonia siliqua* L.) with special emphasis on environmental conditions at Marginal production sites in South Portugal. Thesis, Humboldt-Universität, Berlin.

von Haselberg, C. D., Lüdders, P. and Stösser, R. (2004). Pollen tube growth, fertilization and ovule longevity in the carob tree (*Ceratonia siliqua* L.). *Journal of Applied Botany-Angewandte Botanik*, **78**, 32–40.

Hatanaka, A. (1993). The biogeneration of green odor by green leaves. *Phytochemistry*, **34**, 1201–18.

Heil, M. (2008). Insect defence via tritrophic interactions. *New Phytologist*, **178**, 41–61.

Heinrich, B. (1975). Energetics of pollination. *Annual Review of Ecology and Systematics*, **6**, 139–70.

Hillcoat, D., Lewis, G. and Verdcourt, B. (1980). A new species of *Ceratonia* (Leguminosae, Caesalpiniodieae) from Arabia and the Somali Republic. *Kew Bulletin*, **35**, 261–71.

James, D. G. (2003). Field evaluation of herbivore-induced plant volatiles as attractants for beneficial insects: methyl salicylate and the green lacewing *Chrysopa nigricornis*. *Journal of Chemical Ecology*, **29**, 1601–9.

James, D. G. (2006) Methyl salicylate is a field attractant for the goldeneyed lacewing, *Chrysopa oculata*. *Biocontrol Science and Technology*, **16**, 107–10.

James, D. G., Castle, S. C., Grassiwitz, T. and Reyna, V. (2005) Using synthetic herbivore-induced plant volatiles to enhance conservation biological control: field experiments in hops and grapes. Second International Symposium on Biological Control of Arthropods. Davos, Switzerland, 12–16.9.2005. USDA Forest Publication FHTET-2005–08, pp 192–205.

Jürgens A, and Dötterl, S. (2004). Chemical composition of pollen volatiles in Ranunculaceae: genera-specific profiles in *Anemone, Aquilegia, Caltha, Pulsatilla, Ranunculus*, and *Trollius* species. *American Journal of Botany*, **91**, 1969–80.

Karrenberg, S., Kollman, J. and Edwards, P. J. (2002). Pollen vectors and inflorescence morphology in four species of *Salix*. *Plant Systematics and Evolution*, **235**, 181–8.

Knudsen, J. T., Eriksson, R., Gershenzon, J. and Ståhl, B. (2006). Diversity and distribution of floral scent. *The Botanical Review*, **72**, 1–120.

Kravchenko, V. D., Figiger, M., Mosser, J. and Muller, G. C. (2006). The Noctuinae of Israel (Lepidoptera:Noctuidae). *SHILP Revista de Lepideptorologia*, **34**, 353–70.

Linskens H. F. and Scholten W. (1980). The flower of carob. *Portugaliae Acta Biologica*, **16**, 95–102.

Mahy G., de Sloover, J. and Jacquemart, A. L. (1998). The generalist pollination system and reproductive success of *Calluna vulgaris* in the Upper Ardenne. *Canadian Journal of Botany*, **76**, 1843–51.

Meeuse, A. D. J. (1978). Entomophily in *Salix*: theoretical considerations. In *The Pollination of Flowers by Insects*, ed. A. J. Richards. London, UK: Academic Press, pp. 47–50.

Meeuse A. D. J., de Meijer, A. H., Mohr, O. W. P. and Wellinga, S. M. (1990). Entomophily in the diecious gymnosperm *Ephedra aphylla* Frusk (=*E. alte* C.A. Mey), with some notes on *Ephedra campylopoda* C. A. Mey. III. Further anthecological studies and relative importance of entomophyli. *Israel Journal of Botany*, **39**, 113–29.

Molleman, F., Drukker, B. and Blommers, L. (1997) A trap for monitoring pear psylla predators using dispenser with the synomone methyl salicylate. *Proceedings in Experimental and Applied Entomology* N.E.V. Amsterdam, **8**, 177–82.

Ortiz, P., Arista, M. and Traveset, S. (1996). Distance-independent fruit: set pattern in dioecious population of *Ceratonia siliqua* L. (Casalpiniaceae). *Flora*, **194**, 277–80.

Passos De Carvalho, J. (1988). Carob pollination aspects. *Actas del II Symposium Internacional Sobre la Garrofa*, 255–64.

Peeters, L. and Totland, Ø. (1999). Wind to insect pollination ratio and floral traits in five alpine *Salix* species. *Canadian Journal of Botany*, **77**, 556–63.

Piechulla, B. and Pott, M. B. (2003). Plant scents: mediators of inter- and intraorganismic communication. *Planta*, **217**, 687–9.

Polhill, R. M., Raven, P. H. and Stirton, C. H. (1981). *Evolution and systematics of the Leguminosae*. In *Advances in Legume Systematics, Vol. 1*, ed. C. H. Stirton. Kew, UK: Royal Botanic Gardens, pp. 1–26.

Principi, M. M. and Canard, C. (1984). *Feeding habits*. In *Biology of Chrysopidae. Series Entomologica 27*, ed. M. Y. Canard and T. R. Séméria. The Hague, The Netherlands: Dr. W. Junk Publishers, pp. 76–92.

Ramon-Lacca, L. and Mabberley, D. J. (2004). The ecological status of carob-tree (*Ceratonia siliqua*, Legumonosae) in the Mediterranean. *Botanical Journal of the Linnean Society*, **144**, 431–6.

Raguso, R. A. and Pellmyr, O. (1998). Dynamic headspace analysis of floral volatiles: a comparison of methods. *Oikos*, **81**, 238–54.

Raguso, R. A. and Pichersky, E. (1999). New perspectives in pollination biology: floral fragrances. A day in the life of a linalool molecule: chemical communication in a plant–pollinator system. Part 1: linalool biosynthesis in the flowering plants. *Plant Species Biology*, **14**, 95–120.

Raguso, R. A., Light, D. M. and Pichersky, E. (1996). Electroantennogram responses of *Hyles lineata* (Sphingidae:Lepidoptera) to volatile compounds from *Clarkia breweri* (Onagraceae) and other moth-pollinated flowers. *Journal of Chemical Ecology*, **22**, 1735–66.

Regal, P. J. (1982). Pollinationby wind and animals: ecology of geographic patters. *Annual Review of Ecology*, **13**, 497–524.

Reddy, G. V. P. (2002). Plant volatiles mediate orientaion and plant preference of *Chrysoperla carnea* Stephens (Neuroptera: Chrysopidae). *Biological Control*, **25**, 49–55.

Retana, J., Ramoneda, J. and Garcia del Pino, F. (1990). Importancia del los insectos en la polinización del algarrobo. *Boletín de Sanidad Vegetal Plagas*, **16**, 143–150.

Retana J., Ramoneda, J., Garcia del Pino, F. and Bosch, J. (1994). Flowering phenology of carob, *Ceratonia siliqua* L. (Cesalpinaceae). *Journal of Horticultural Science*, **69**, 97–103.

Sacchi, C. F. and Price, P. M. (1988). Pollination of the arroyo willow, *Salix lasiolepis*: role of insects and wind. *American Journal of Botany*, **75**, 1387–93.

Scariot, A. O., Lleras, A. and Hay, J. D. (1991). Reproductive biology of the palm *Acrocomia aculeata* in Central Brazil. *Biotropica*, **23**, 12–22.

Shi-Jing, L., Zhang, D. X., Lin, L. and Chen, Z. Y. (2000). Pollination ecology of *Caesalpinia crista* (Leguminosae:Ca esalpinioideaea). *Acta Botanica*, **46**, 271–8.

Silberbauer, L., Yee, M., del Soccoro, A., Wratten, S., Griegg, S. and Bowie, M. (2004). Pollen grains as markers to track the movements of generalist predatory insects in agroecosystems. *International Journal of Pest Management*, **50**, 165–71.

Suda, D. Y. and Cunningham, R. T. (1970) *Chrysopa basalis* captured in plastic traps containing methyl eugenol. *Journal of Economic Entomology*, **63**, 1706–9.

Totland, Ø. and Sottocornola, M. (2001). Pollen limitation of reproductive success in two sympatric alpine willows (Salicaceae) with contrasting pollination strategies *American Journal of Botany*, **88**, 1011–15.

Tous, J. and Battle, I. (1990). *El algarrobo*. Madrid, Spain: Ediciones Mundi-Pres.

Tucker S. C. (1992). The development basis for sexual expression in Ceratonia siliqua (Leguminosae: Cassieae) *American Journal of Botany*, **78**, 318–327.

Tucker, S. C. (1992). The developmental basis of sexual expression in *Ceratonia siliqua* (Leguminosae:Cassieae). *American Journal of Botany*, **79**, 318–27.

Umeya, K. and Hirao, J. (1975). Attraction of the Jackfruit Fly, *Dacus umbrosus* F. (Diptera: Tephritidae) and lacewing, *Chrysopa sp.* (Neuroptera: Chrysopidae) by lure traps baited with methyl eugenol and cue-lure in the Philippines. *Applied Entomology and Zoology*, **10**, 60–2.

Villenave, J., Duetsch, B., Lode, T. and Rat-Morris, E. (2006). Pollen preference of the *Chrysoperla* species (Neuroptra:Chrysopidae) occuring in the crop environment in western France. *European Journal of Entomology*, **103**, 771–7.

Vroege, P. W. and Stelleman, P. (1990). Insect and wind pollination in *Salix repens* L. and *Salix carpea* L. *Israel Journal of Botany*, **39**, 125–32.

Wager, B. R. and Breed, M. D. (2000). Does honeybee sting alarm pheromone give orientation information to defensive bees? *Annals of the Entomological Society of America*, **93**, 1329–32.

Webster, B., Gezan, S., Bruce, T., Hardie, J. and Pickett, J. (2010). Between plant and diurnal variation in quantities and ratios of volatile compounds emitted by *Vicia faba* plants. *Phytochemistry*, **71**, 81–9.

Waisel, Y., Ganor, E., Glikman, M., Epstein, V. and Brenner, S. (1997). Seasonal distribution of airborne pollen in the coastal main of Israel. *Aerobiologia*, **13**, 127–34.

Weller, S. G., Sakai, A. K., Rankin, A. E., Golonka, A., Kutcher, B. and Ashby, K. E. (1998). Dioecy and evolution of pollnation systems in *Schiedea* and *Alsinidendron* (Caryophyllaceae:Alsinoideae) in the Hawaiian Islands. *American Journal of Botany*, **85**, 1377–88.

Whitehead, D. R. (1968). Wind pollination in angiosperms: evolution and environment considerations. *Evolution*, **23**, 28–35.

Yela, J. L. and Herrera, C. M. (1993). Scasonality and lifecycles of woody plant-feeding noctuid moths Lepidoptera:Noctuidae) in Mediterranean habitats. *Ecological Entomology*, **18**, 259–69.

Zhang, Q.-H., Schneidmiller, R. G., Hoover, D. R., Young., K., Welshons, D. O., Margaryan, A., Aldrich, J. F. and Chauhan, K. R. (2006). Male-produced pheromones in the green lacewing *Chrysopa nigricornis*. *Journal of Chemical Ecology*, **32**, 2163–76.

Zohary, M. (1972). *Flora Palaestina Vol. 2.* Jerusalem, Israel: The Israel Academy of Science and Humanities.

Zohary, M. and Orshan, G. (1959). The maquis of *Ceratonia siliqua* in Israel. *Palestine Journal of Botany*, **8**, 385–97.

13

Structure and dynamics of pollination networks: the past, present, and future

JENS M. OLESEN, YOKO L. DUPONT, MELANIE HAGEN,
CLAUS RASMUSSEN AND KRISTIAN TRØJELSGAARD

13.1 Introduction

By far, most studies in ecology are about single species and their interactions with the surroundings, and this is also true in pollination ecology. However, species are members of communities of interacting species, i.e. networks. According to our definition, a network only includes species, whose linkage is spatially unconstrained, i.e. species may potentially meet in nature.

During the last decade, this gap between 1-species ecological research and nature's overwhelming complexity has rapidly been bridged by a new generation of studies taking place at the network level. These studies offer fascinating new insight into a kind of natural history, which we term link ecology. In the first section, we trace the roots of pollination network ecology. Then we describe what this discipline is doing today, and finally we attempt to predict what is coming up in the near future.

13.2 The past

13.2.1 The Canadian beginning

Community studies in pollination ecology began in arctic Canada. The study by Mosquin and Martin (1967) was made on the 42 000 km²-large, high-arctic Melville

Evolution of Plant–Pollinator Relationships, ed S. Patiny. Published by Cambridge University Press. © The Systematics Association 2012.

Island at 75°N, and the studies of Hocking (1968) and his Ph.D student Kevan (1970) on the even larger Ellesmere Island (196 000 km²) further north at 82°N. In Mosquin and Martin's (1967) study, plant–pollinator links are for the first time presented in a matrix form. Each entry in the matrix tells whether a link, i.e. visits, has been observed between a pollinator and a plant species or not, and eventually how many visits were observed. However, most beginnings are modest and Mosquin and Martin (1967), during only three days, recorded 38 links between only 18 flower-visitor species and 11 plants. This quartet of Canadian researchers also examined floral nectar, odor and colors at the level of the community. The first three did not relate their research to earlier community ecology and only referred to a few other studies in pollination ecology and arctic entomology (but see Kevan, 1970). Hocking and Mosquin had "ecological bends" and were interested in natural history, whereas Martin focused more upon diversity and systematics (P. G. Kevan, personal communication, 2001).

These Canadian studies, however, were preceded by extensive published lists of regional flower visitors compiled from USA and Europe, e.g. the studies by Müller (1883) of European flowers and their visitors, Clements and Long (1923) at the Carnegie Institution's Alpine Laboratory on Pike's Peak at an altitude of 2600 m in Colorado, and Robertson (1929), a study made around Carlinville in Illinois. However, these large studies and other similar ones are not community studies according to our current definition because a proportion of the species are spatially uncoupled (but see Medan et al. 2007; Memmott et al. 2007).

13.2.2 Integration of pollination community ecology into general community and food web ecology

The next step in our story is Percival (1974), who made a study of flower-visitation interactions between 36 animals and 61 plant species from a shrubland in Jamaica. She stated that the study is "an attempt to elucidate the floral biology of a community as a whole and to show that it is a biotic factor equal in importance to the classical habitat factors," which is an interesting goal, but again the study did not refer to other community studies.

This was certainly not the case for another paper published simultaneously by Heithaus (1974). He may be the first who formally introduced pollination ecology and community ecology to each other. His study might be inspired by a larger "macroecological" research program initiated by P. H. Raven and coworkers, intended to compare flower-visitor communities in California and Chile (Cowling and Campbell 1980). The title of the paper by Heithaus (1974), "The role of plant–pollinator interactions in determining community structure," sounds very modern. Heithaus' (1974) work was inspired by system ecology, which he criticized for not taking reproduction into account when it constructed its flow diagrams of energy and matter, because, as he said, reproduction is a prerequisite to biomass production. Heithaus (1974) took the most recent community ecology, developed by Levin

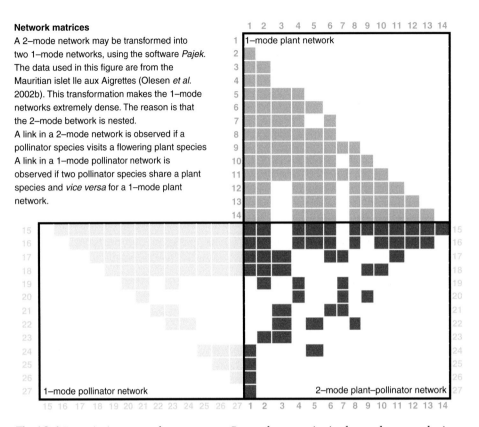

Network matrices

A 2–mode network may be transformed into two 1–mode networks, using the software *Pajek*. The data used in this figure are from the Mauritian islet Ile aux Aigrettes (Olesen *et al.* 2002b). This transformation makes the 1–mode networks extremely dense. The reason is that the 2–mode betwork is nested.

A link in a 2–mode network is observed if a pollinator species visits a flowering plant species

A link in a 1–mode pollinator network is observed if two pollinator species share a plant species and *vice versa* for a 1–mode plant network.

Fig 13.1 Descriptive network parameters: P, no. plant species in the study network; A, no. pollinator species in the study network. AP, matrix size or total no. potential links in the 2-mode network. $0.5P(P-1)$, total no. potential links in the 1-mode pollinator network. I, total no. observed links in the network. C, 2-mode network connectance $= I/(AP)$ or 1-mode network connectance $= I/(0.5A(A-1))$ L_1, linkage level of species $1 = 1$ link, i.e. no. links from species 1 to other species. $<L_A>$, average linkage level for all pollinator species in either 1- or 2-mode networks. The following two parameters are only calculated for 1-mode networks. l_{1-2}, shortest path length between species 1 and 2, i.e. no. links between the two species. $<l>$, average shortest path length among all species. c_3, clustering coefficient of species 3, i.e. link density among neighbors of species 3. Species 3 interacts with three species. Max. no. links among these species is 3. Observed no. links among these three species is 1 (between species 2 and 4). $c =$ observed no. links among neighbors/potential no. links among neighbors. NB. If a species has only one link, it has no clustering coefficient. The software *Pajek* calculates these parameters. See plate section for color version.

and Anderson (1970), MacArthur (1972), and May (1973) in particular, and analyzed a series of Costa Rican flower-visitor communities, but he also linked his work to that of Hocking (1968). Among other results, he showed that niche width declined with increasing species richness. Stiles (e.g. 1975) studied subcommunities of

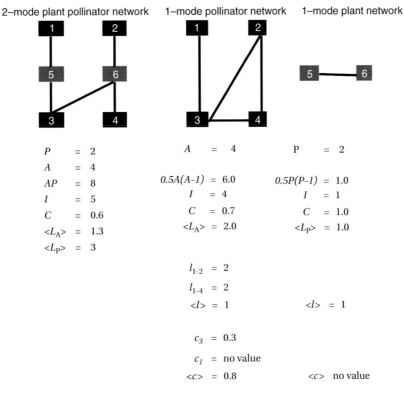

2-mode plant pollinator network   1-mode pollinator network   1-mode plant network

$$P = 2$$
$$A = 4$$
$$AP = 8$$
$$I = 5$$
$$C = 0.6$$
$$\langle L_A \rangle = 1.3$$
$$\langle L_P \rangle = 3$$

$$A = 4$$
$$0.5A(A-1) = 6.0$$
$$I = 4$$
$$C = 0.7$$
$$\langle L_A \rangle = 2.0$$

$$P = 2$$
$$0.5P(P-1) = 1.0$$
$$I = 1$$
$$C = 1.0$$
$$\langle L_P \rangle = 1.0$$

$$l_{1-2} = 2$$
$$l_{1-4} = 2$$
$$\langle l \rangle = 1$$

$$\langle l \rangle = 1$$

$$c_3 = 0.3$$
$$c_1 = \text{no value}$$
$$\langle c \rangle = 0.8$$

$$\langle c \rangle \quad \text{no value}$$

Fig 13.1 (*cont.*)

hummingbirds and their flowers, asking similar questions. Within the framework of the Chilean–Californian research program, A. R. Moldenke, R. M. Primack, M. T. K. Arroyo and others published papers about pollination community ecology in the southwest USA, Chile, and New Zealand, notably Moldenke and Lincoln (1979), Arroyo et al. (1982) and Primack (1983). These studies focused particularly upon the composition of the flower-visitor communities and level of specialization. The method used was observation of visitation to all flowering plant species within a defined area or along a transect and sampling specimens of all visitor species ("the perfect observant vacuum cleaner approach," sensu Moldenke 1979).

In 1978, Schemske et al. published a study of the spring flora and its flower visitors in a forest in Illinois (see also Schemske et al. 1978, Schemske 1983). They also addressed community questions, such as "Is there competition among simultaneously blooming species for pollinators?" – a question originating from Levin and Anderson (1970) and achieving considerable research interest in the 1970s. Schemske et al. (1978) refers to Charles Robertson and his visitor lists, and to competition and coevolution in communities of interacting species (Levin and Anderson 1970; Macior 1971; Mosquin 1971; Straw 1972; Heithaus 1974; Reader 1975).

In 1987, P. Jordano (1987) published a seminal paper, in which he compares mutualistic networks. His database consisted of 36 pollination and 19 seed-dispersal networks. Here, for the first time, several network metrics were calculated, such as connectance (Fig 13.1), link strength and dependence asymmetry. His study is a macroecological study, including mainly partial networks ("mutualistic modules"), i.e. networks between subsets of communities or guilds, especially bumblebees and their flowers (e.g. Inouye 1978; Bauer 1983), but also total pollination networks, i.e. networks including all interacting species within an area, e.g. Herrera (1988). The latter analyzed a pollination network from the large *Parque Nacional de Doñana* in Southern Spain. In fact, in 1988 this network was the largest available, including 205 species and 412 links. The work of Herrera (1988) was fully embedded in earlier pollination community ecology, but less so in general community ecology. Jordano (1987) showed that connectance decreased with increasing network size and that the pattern was somewhat similar to that found for food webs, suggesting an overall invariant link structure among interacting species irrespective of the nature of the links. In addition, he found that frequency distributions of species dependencies were strongly skewed and dependency values between interacting species pairs were asymmetrical, i.e. if species i was strongly dependent upon species j, then j was often weakly dependent upon i. This told him the improbability of pair-wise coevolution. He also suggested that the link structure observed in mutualistic networks might facilitate the persistence of rare species. In the discussion, he writes, "Since most of our knowledge of mutualism comes from studies of mutualistic modules, we can only speculate about the evolutionary dynamics of complete systems. For example, if modules act as attraction domains, on evolutionary time, to canalize coadaptations of new species to module-specific partners, this would explain why C decreases as the number of species S increases. Note that secondary (weak) relations, which interconnect species from different modules [actually termed connectors, Olesen et al. 2007], can act as potential evolutionary bridges to canalize new adaptations." Back in 1987, this may have sounded almost metaphysical, but it turned out to predict some of the most recent developments in network analysis of mutualistic networks (e.g. Thompson 2005; Olesen et al. 2007; Jordano 2010). The analysis of Jordano (1987) was fully integrated into food web analysis, e.g. the work of J. E. Cohen, F. Briand, R. Margalef, R. M. May, J. H. Lawton, P. Yodzis and G. Sugihara (see Jordano 1987 for references).

13.2.3 The largest pollination networks

In the 1990s, we get the largest pollination networks published, viz. a series of Japanese studies, especially Kato et al. (1990), Inoue et al. (1990), and Kakutani et al. (1990), and a Greek study of Petanidou (1991). Inoue et al. (1990) included 952 species and 1876 links, and Petanidou (1991) 797 species and an astonishing 2933 links.

These four networks were compiled over 3–4 years, and linkage among species was constrained by phenological uncoupling, i.e. proportions of the species in the network had phenophases that did not overlap (Jordano 1987; Olesen et al. 2010b).

13.2.4 The generalization–specialisation debate

We bridge the past and the present by three important papers (Herrera 1996; Waser et al. 1996; Kearns et al. 1998). These papers were a showdown with the ruling idea about specialization and species–species coevolution ("We must abandon the perspective that to lose one plant species is to lose one or more animal species via linked extinction, and vice versa," Kearns et al. 1998). Waser et al. (1996) conclude their analysis by stating that generalization "appears to be the rule rather than the exception," and urge research in the field to produce "community-wide or taxon-wide studies of pollination" and "more effort in combing the existing literature." Level of generalization is also discussed in Johnson and Steiner (2000) and Olesen (2000). The first argues that generalization-specialization is a continuum and that some floras, e.g. the South African, indeed have a high proportion of specialists, especially if one looks, not just at the linkage level L, i.e. number of links of a species (Fig 13.1), but more detailed and accurately at female and male fitness components. Olesen (2000) is an analysis of the influence of the local pollinator fauna on L of individual species. He showed that in order to call a species a generalist or specialist one needs to know the composition of the local visitor fauna, e.g. a plant species being visited by five bee species may be regarded as more specialized than one that receives visits from one fly species if the plant grows in a fly-dominated habitat. An important reason for the deviating conclusions in Herrera (1996), Waser et al. (1996), Johnson and Steiner (2000), and Olesen (2000) is the extent of pooling visit-link data, i.e. are data pooled regionally or analyzed at the level of the population? In Olesen (2000), each of the 21 network study sites included was analyzed separately, and as many as 64 % of all visitor species and 32 % of all plant species were very specialized, i.e. they had an L of only 1 or 2. However, despite some species displaying specialization, Olesen et al. (2007) and Joppa et al. (2009) concluded that reciprocal specialization is very rare in both mutualistic and parasitic networks. During the following decade, this generalization–specialization debate became an inspiring part of the research agenda.

13.3 The present

13.3.1 The growth of pollination network research

In pollination networks, links represent exchanges of ecological services, and evolutionarily, they are icons of selection factors. Thus, fundamental to any analysis of ecology and evolution of plants and their pollinators is knowledge of their network context and its spatio–temporal variation. Networks are ecological-evolutionary roadmaps showing the flows of energy, matter and services, and the selective factors among interacting species.

Since 1999, 90 studies of pollination network analysis have been published. About half of these are single-network studies (including those with replicates) (55 % of all publications, e.g. Dicks et al. 2002 (two sites), González-Álvaro, 2004 (one site), Gibson et al. 2006 (five sites); for more references see Olesen et al. (2007). The remaining ones are macroecological comparisons of a set of networks (21 %, e.g. Ollerton and Cranmer 2002; Bascompte et al. 2003, 2006; Jordano et al. 2003; Olesen et al. 2006, 2007; Rezende et al. 2007; Saavedra et al. 2008), reviews (Jordano et al. 2006; Montoya et al. 2006; Bascompte and Jordano 2006, 2007; Bascompte 2009a, 2009b; Ings et al. 2009), and analyses of specific network variables, e.g. body size (Woodward et al. 2005), asymmetry (Vázquez and Aizen 2004), and functional specialization (Dalsgaard et al. 2008, 2009).

This most recent development of pollination network analysis began with the study of Memmott (1999). She made an analog to food webs and suggested in concordance with Jordano (1987) to use the food-web approach as a protocol in the study of pollination networks. Memmott (1999) quantified link strength (visitation rate) and stressed the importance of including all flower visitors into the network. Memmott et al. (2004) suggested to simulate species extinction by removing species from the network and analyze its properties before and after (see also Kaiser-Bunbury et al. 2009). The study network of Memmott (1999) is illustrated with pollinators and plants in two rows interconnected by their links, a representation that has been widely used since (see also Memmott et al. 1994). Potts et al. (2003) defined factors shaping a pollinator community and pointed toward the floral diversity and abundance along with available nectar resources, and Stang et al. (2006, 2007) assessed the relative roles of abundance and morphology in shaping multispecific patterns in pollination networks.

13.3.2 Network analysis

Food-web research has been far ahead, but studies of mutualism networks are now catching up. This is most important in formulating a general theory about ecological networks. After a couple of decades of efforts from many researchers, we now feel we know mutualism networks; in particular pollination networks. Let us, as a start, ignore spatio–temporal dynamics and describe a network as a static structure (Fig 13.1).

Networks are described at three levels:

(1)   Macroscopic or global, i.e. properties of the entire network, perhaps emergent properties that could not be deduced from information about single species, e.g. connectance C and nestedness N (Bascompte et al. 2003).

(2) Microscopic or local, i.e. properties of the individual species or nodes in the network, e.g. linkage level L.

(3) Mesoscopic or modular, i.e. properties of the modules – small groups of highly linked species or nodes, e.g. proportion of connector species in a module (Olesen et al. 2007).

Thus (3) becomes a level intermediate between (1) and (2). As a first step, the values of a set of simple basic descriptors are estimated (Fig 13.1, see Olesen et al. 2006). Historically, these descriptors either come from food-web theory, social network analysis, or mathematics.

A network study is delimited in space and time by decisions made by the researcher (Bundgaard 2003). Most often these decisions are based on habitat borders as perceived by the researcher and knowledge about the extent of the flowering season. The method most often used is to choose a study plot of a type of vegetation and then score interactions between all flowering plant and flower-visitor species through, most often, a season. Links may be sampled by single-plant observations (Elberling and Olesen 1999) or transects (e.g. Memmott 1999). The first method corrects for variation in plant-species abundance, whereas the latter does not. No study corrects for variation in phenophase length within the study period, i.e. species may cease their phenophase just after the beginning of the study period or beginning shortly before the end of the study. Sampling should be continued until a robust estimate of the actual interaction richness is reached (Nielsen and Bascompte 2007; Jordano et al. 2009; Olesen et al. 2010b). A single-animal observation protocol is also possible. Kanstrup and Olesen (2000), Bosch et al. (2009), Alarcón (2010), and Olesen et al. (2010b) sampled flower visitors and analyzed their pollen load. However, changing focus does not seem to affect $A{:}P$ (number of animal species: number of plant species) ratio and L very much, but the use of both a plant- and an animal-focused approach seems valuable because more links are detected faster (Philipp et al. 2006; Olesen et al. 2008).

Prior to data analysis, species and their links may be organized in two ways: (i) as a matrix of size AP, i.e. of A rows and P columns (Fig 13.1), or (ii) as a network visualization, showing species as nodes and links as lines between nodes (Lewinsohn et al. 2006). In a qualitative matrix, a cell element a_{ij} between pollinator species i and plant species j is 1, if a link is present, and 0 if absent. In a quantitative matrix, a_{ij} gives the value of the link strength and may take any value dependent upon choice of link currency, e.g. number of visiting individuals of pollinator species i to one flower of plant species j during one hour. This matrix is used to produce input files to different kinds of analytical network software, e.g. *Pajek* (Batagelj and Mrvar 2009), *Foodweb3D* (Williams et al. 2002), *Aninhado* (Guimarães and Guimarães 2006), *R* ('bipartite') (Dormann et al. 2008, 2009), and *SA* (Guimerà and Amaral 2005).

13.3.3 The pollination network database

At present, our database includes 53 total pollination networks, including $S = A + P =$ 10 016 species and $I = 20\,368$ links. The locations of the study networks span in latitude from 43°S (New Zealand: Primack 1983) to 81°N (Canada: Hocking 1968; Kevan 1970) and in altitude from sea level (Canada: Mosquin and Martin 1967; Hocking 1968; Jamaica: Percival 1974; Spain: Herrera 1988; Azores and Mauritius: Olesen et al. 2002; Denmark: Bundgaard 2003; Montero 2005; Galápagos: Philipp et al. 2006) to 3200–3600 m (Chile: Arroyo et al. 1982; Ramos-Jiliberto et al. 2010; Argentina: Medan et al. 2002). Number of species S $(= A + P)$ in a network ranges from 16 (Galápagos: Philipp et al. 2006) to 952 (Japan: Inoue et al. 1990). The $A{:}P$ ratio is 3.5. Total number of interactions I ranges from 16 (Galápagos: Philipp et al. 2006) to 2933 (Greece: Petanidou 1991), connectance C ranges from 1.7 % (Japan: Kato 2000) to 39.5 % (Denmark: Bundgaard 2003). C decreases with increasing S (Olesen and Jordano 2002). Network specialization or C differs between different kinds of subnetwork. Jordano (1987) observes a decrease in C from plant–hummingbird to plant–bumblebee and to plant–bee subnetworks/modules. Thus pollinator groups differ in their linkage pattern. In an English meadow, both plants and pollinators are relatively generalized, i.e. high L (Memmott 1999). Memmott (1999) introduces the idea of coining the term keystone species to the most generalized species in the network, which in the case of her network is the umbellifer wild carrot (*Daucus carota*), linking to 61 % of all visitor species (see also Memmott et al. 2004; Pauw 2007). The frequency distribution of L (the degree distribution) has received a lot of attention, and its analysis has followed a tradition laid down by networkers outside biology (e.g. Albert et al. 1999). The actual distribution has been fitted to a power law, exponential or truncated power law distribution (Jordano et al. 2006). Underlying causes of these patterns have been discussed widely, and Jordano (1987) suggested two: (1) skewed species abundance distribution in most communities, and (2) number of links increases with abundance according to a saturated exponential (see also Vazquez and Aizen 2003). This suggests that the frequency distribution of L also becomes skewed.

More recently, however, both preferential attachment (Jordano et al. 2003) and differences in species richness of plants and animals (Guimarães et al. 2007) have been added to the list of factors contributing to the exponential truncation often observed in mutualistic networks. Preferential attachment means that new species arriving to a network preferentially link to well-linked old species (Olesen et al. 2008).

All networks are nested, except a few small ones (Bascompte et al. 2003). A nested structure is believed to confer robustness against disintegration of the network, especially if weakly linked species have a higher probability of extinction (Burgos et al. 2007). Nielsen and Bascompte (2007) concluded that nestedness is relatively insensitive towards sampling effort and it is therefore likely that the nested structure of many ecological networks is a true inherent property and not simply a sampling artifact.

Without using the term "nestedness," Jordano (1987) describes the link pattern as "The interaction of most ... of the species pairs ... lie in the upper left side of the table ... The remaining pairs lie along both the top and left side of the table. The relationships are therefore very asymmetrical."

In his analysis, Jordano (1987) also looked at network modularity. He used the concept of module (Paine 1980) and clique (Yodzis 1980) and finds that module number increases with S. A module is defined as, for example, a set of fruit-eating birds sharing an abundant, "major" fruit-plant species, which is not eaten by birds outside the module. Thus, modules are not completely isolated from each other. "Minors" (or in later papers "connectors") link modules together into a coherent network. There are several ways to identify modules within networks (Dicks et al. 2002; Lewinsohn et al. 2006). Olesen et al. (2007) used an algorithm by Guimerà and Amaral (2005), and so have several other recent studies (Saavedra et al. 2008; Dupont and Olesen 2009) (for details about modularity analysis see Olesen et al. 2007). All networks with $S > 150$, some with $50 < S < 150$, and none with $S < 50$ were modular (Olesen et al. 2007; Saavedra et al. 2008). A positive correlation between nestedness and modularity has been demonstrated for networks with low C, i.e. most real networks, while high C makes the correlation negative (Fortuna et al. 2010). A high C of small networks destroys modularity. Large networks with modules containing more than 70 species even become hierarchical, i.e. with submodules within modules (Olesen et al. unpublished data). Individual modules may be stable over larger geographic distances but also show a distinct seasonal development (Dupont and Olesen 2009). Furthermore, modules within a network can themselves be nested (Lewinsohn et al. 2006). Pollination networks also show link asymmetry (Vazquez and Aizen 2004; Bascompte et al. 2006; Stang et al. 2007). All networks between plants and pollinators can be transformed to two one-mode networks (Fig 13.1), which are their one-mode projections. Such networks were analyzed in Olesen et al. (2006). Most have a very short path length and high clustering coefficient, i.e. they are small worlds (Lundgren and Olesen 2005; Olesen et al. 2006). That means that any disturbance quickly spreads to the entire network.

13.3.4 Linkage constraints

Network structure is commonly constrained by phenological uncoupling, abundance, body size, and population structure (e.g. Hegland et al. 2009). Some potential links are never observed, i.e. they are "forbidden" (fully constrained) or "missing" (a sampling effect), and their absence can be just as ecologically informative as their presence (Olesen et al. 2010b). Jordano (1987) asked the question: why are so many null observations made in nature? He suggested that "temporal noncoincidence" ("temporal matching") or weak coincidence or weak coupling might explain a large fraction of these null observations.

13.3.5 Network dynamics

During the last decade, some studies of networks have moved from static to dynamic analyses, and have attained a deeper insight into their internal structure, heterogeneity, and temporal and spatial resolution (e.g. Lundgren and Olesen 2005; Basilio et al. 2006; Olesen et al. 2008; Petanidou et al. 2008). A common pattern has appeared: that is we see a stable core of species surrounded by a group of dynamic peripherals, which enter the web via preferential linkage to the most generalist species.

However, many aspects of modern pollination network analysis have to be left out here because of page constraints, in particular the importance of invasive species (e.g. Olesen et al. 2002; Traveset and Richardson 2006; Aizen et al. 2008; Padrón et al. 2009).

13.4 … and the future

During the next decade we predict that research will focus upon temporal dynamic, space-for-time substitutions, linkage constraints, habitat borders, colonization/extinction dynamics, network modularity, individual-based networks, species invasion and extinction in networks, biological diversification in networks, network genetics and evolution, and super-networks, integrating different kinds of network (Ings et al. 2009; Bascompte 2009a,b; Scheffer et al. 2009; Sugihara and Ye 2009; Olesen et al. 2010a).

To some extent temporal and spatial scales are interchangeable, and future studies will explore how space-for-time substitutions can be used in the study of networks or in the study of most biological objects. Spatial habitat borders can add heterogeneity to network structure, but their importance has rarely been studied. Many networks are hierarchically structured, with modules forming the basic building blocks, which may even result in self-similarity. Scaling down from networks of species reveals another, finer-grained level of individual-based organization, the ecological consequences of which have yet to be fully explored (Araújo et al. 2008; Fortuna et al. 2008; Dupont et al. 2010). In individual-based networks, individuals are the focal nodes. However, more studies are required to link the individual and species levels in networks. Invasions by alien species can be tracked by following the topological "career" of the invader as it establishes itself within a network, with implications for conservation biology (Memmott and Waser 2002; Morales and Aizen 2006; Aizen et al. 2008; Bartomeus et al. 2008; Tylianakis 2008; Padrón et al. 2009). By scaling up to a higher level of organization, it is possible to combine different network types (e.g. food webs and mutualistic networks) to form super-networks (Melián et al. 2008), but this approach has yet to be integrated into mainstream ecological research. Finally, network analysis will become a very important tool in our attempts to formulate more general theories about ecocomplexity or bio-complexity (e.g. McCann 2007; Butts 2009; Vespignani 2009) and in many applied projects in ecology in general, including restoration ecology and network management (Memmott 2009).

Acknowledgements

We are most grateful to P. G. Kevan for personal communication and to P. Jordano and Sébastien Patiny for their critique and many comments to the chapter. The study was supported by the Danish National Research Council (JMO, YLD), the Carlsberg Foundation (CR), and the Aarhus University Research Foundation (KT).

References

Aizen, M. A., Morales, C. L. and Morales, J. M. (2008). Invasive mutualists erode native pollination webs. *PLoS* Biology, **6**, 396–403.

Alarcón, R. (2010). Congruence between visitation and pollen-transport networks in a California plant–pollinator community. *Oikos*, **119**, 35–44.

Albert, R., Jeong, H. and Barabási, A.-L. (1999). Diameter of the world-wide web. *Nature*, **401**, 130–1.

Araújo, M. S., Guimarães, P. R. J., Svanbäck, R., Pinheiro, A., Guimarães, P., Dos Reis, S. F. and Bolnick, D. I. (2008). Network analysis reveals contrasting effects of intraspecific competition on individuals versus population diets. *Ecology*, **89**, 1981–93.

Arroyo, M. T. K., Primack, R. and Armesto, J. (1982). Community studies in pollination ecology in the high temperate Andes of Central Chile. I. Pollination mechanisms and altitudinal variation. *American Journal of Botany*, **69**, 82–97.

Bartomeus, I., Vilà, M. and Santamaría, L. (2008). Contrasting effects of invasive plants in plant–pollinator networks. *Oecologia*, **155**, 761–770.

Bascompte, J. (2009a). Disentangling the web of life. *Science*, **325**, 416–419.

Bascompte, J. (2009b). Mutualistic networks. *Frontiers in Ecology and the Environment*, **7**, 429–36.

Bascompte, J. and Jordano, P. (2006). The structure of plant–animal mutualistic networks. In *Ecological Networks: Linking Structure to Dynamics in Food Webs*, ed. M. Pascual and J. A. Dunne. Oxford, UK: Oxford University Press, pp. 143–59.

Bascompte, J. and Jordano, P. (2007). Plant–animal mutualistic networks: the architecture of biodiversity. *Annual Review of Ecology, Evolution and Systematics*, **38**, 567–93.

Bascompte, J., Jordano, P., Melián, C. J. and Olesen, J. M. (2003). The nested assembly of plant–animal mutualistic networks. *Proceedings of the National Academy of Sciences USA*, **100**, 9383–7.

Bascompte, J., Jordano, P. and Olesen, J. M. (2006). Asymmetric coevolutionary networks facilitate biodiversity maintenance. *Science*, **312**, 431–3.

Basilio, A. M., Medan, D., Torretta, J. P. and Bartoloni, N. J. (2006). A year-long plant–pollinator network. *Austral Ecology*, **31**, 975–83.

Batagelj, V. and Mrvar, A. (2009). *Pajek* – Program for Large Network Analysis. Home page: http://vlado.fmf.uni-lj.si/pub/networks/pajek/

Bauer, P. J. (1983). Bumblebee pollination relationships on the Beartooth plateau tundra of southern Montana. *American Journal of Botany*, **70**, 134–44.

Bosch, J., Martin, A. G., Anselm, R. and Navarro, D. (2009). Plant–pollinator networks: adding the pollinator's perspective. *Ecology Letters*, **12**, 409-19.

Bundgaard, M. (2003). *Tidslig og rumlig variation i et plante-bestøvernetværk.* MSc Thesis, Aarhus University, Denmark.

Burgos, E., Ceva, H., Perazzo, R. P. J., Devoto, M., Medan, D., Zimmermann, M. and Delbue, A. M. (2007). Why nestedness in mutualistic networks? *Journal of Theortical Biology*, **249**, 307-13.

Butts, C. T. (2009). Revisiting the foundations of network analysis. *Science*, **325**, 414-6.

Clements, R. E. and Long, F. L. (1923). *Experimental pollination. An outline of the ecology of flowers and insects.* Carnegie Institute of Washington, Washington DC.

Cowling, R. M. and Campbell, B. M. (1980). Convergence in the vegetation structure in the Mediterranean communities of California, Chile, and South Africa. *Plant Ecology*, **43**, 191-7.

Dalsgaard, B., Martín, A. M. G., Olesen, J. M., Ollerton, J., Timmermann, A., Andersen, L. H. and Tossas, A. G. (2009). Plant-hummingbird interactions in the West Indies: floral specialisation gradients associated with environment and hummingbird size. *Oecologia*, **159**, 757-66.

Dalsgaard, B., Martín, A. M. G., Olesen, J. M., Timmermann, A., Andersen, L. H. and Ollerton, J. (2008). Pollination networks and functional specialization: a test using Lesser Antillean plant-hummingbird assemblages. *Oikos*, **117**, 789-93.

Dicks, L. V., Corbet, S. A. and Pywell, R. F. (2002). Compartmentalization in plant-insect flower webs. *Journal of Animal Ecology*, **71**, 32-43.

Dormann, C. F., Fründ, J., Blüthgen, N. and Gruber, B. (2009). Indices, graphs and null models: analyzing bipartite ecological networks. *Open Ecology Journal*, **2**, 7-24.

Dormann, C. F., Gruber, B. and Fründ, J. (2008). Introducing the bipartite package: analysing ecological networks. *R News*, **8**, 8-11.

Dupont, Y. L. and Olesen, J. M. (2009). Modules and roles of species in heathland pollination networks. *Journal of Animal Ecology*, **78**, 346-53.

Dupont, Y. L., Trøjelsgaard, K. and Olesen, J. M. (2010, online) Scaling down from species to individuals: a flower-visitation network between individual honeybees and thistle plants. *Oikos*.

Elberling, H. and Olesen, J. M. (1999). The structure of a high latitude plant-pollinator system: the dominance of flies. *Ecography*, **22**, 314-23.

Fortuna, M. A., García, C., Guimarães, P. R. J. and Bascompte, J. (2008). Spatial mating networks in insect-pollinated plants. *Ecology Letters*, **11**, 490-8.

Fortuna, M. A., Stouffer, D. B., Olesen, J. M., Jordano, P., Mouillot, D., Krasnov, B. R., Poulin, R. and Bascompte, J. (2010). Nestedness versus modularity in ecological networks: two sides of the same coin? *Journal of Animal Ecology*, **79**, 811-7.

Gibson, R. H., Nelson, I. L., Hopkins, G. W., Hamlett, B. J. and Memmott, J. (2006). Pollinator webs, plant communities and the conservation of rare plants: arable weeds as a case study. *Journal of Applied Ecology*, **43**, 246-57.

González-Álvaro, V. (2004). *Estimación de la riqueza y abundancia de polinizadores potenciales basada y parámetros morfológicos florales de plantas ruderales.* PhD Thesis, University of Castilla-La Mancha, Toledo, Spain.

Guimarães, P. R. and Guimarães, P. (2006). Improving the analyses of nestedness for large sets of matrices.

Environmental Modelling and Software, **21**, 15112–13.

Guimarães, P. R., Machado, G., Aguiar, M. A. M. d., Jordano, P., Bascompte, J., Pinheiro, A. and Furtado dos Reis, S. (2007). Build-up mechanisms determining the topology of mutualistic networks. *Journal of Theoretical Biology*, **249**, 181–9.

Guimerà, R. and Amaral, L. A. N. (2005). Functional cartography of complex metabolic networks. *Nature*, **433**, 895–900.

Hegland, S. J., Nielsen, A., Lázaro, A., Bjerknes, A.-L. and Totland, Ø. (2009). How does climate warming affect plant–pollinator interactions? *Ecology Letters*, **12**, 184–95.

Heithaus, E. R. (1974). The role of plant-pollinator interactions in determining community structure. *Annals Missouri Botanical Garden*, **61**, 675–91.

Herrera, C. M. (1996). Floral traits and plant adaptation to insect pollinators: a devil's advocate approach. In *Floral Biology*, ed. D. G. Lloyd and S. C. H. Barrett. London, UK: Chapman and Hall, pp. 65–87.

Herrera, J. (1988). Pollination relationships in Southern Spanish Mediterranean shrublands. *Journal of Ecology*, **76**, 274–87.

Hocking, B. (1968). Insect–flower association in the high Arctic with special reference to nectar. *Oikos*, **19**, 359–88.

Ings, T. C., Bascompte, J., Blüthgen, N., Brown, L., Dormann, C. F., Edwards, F., Figueroa, D., Jacob, U., Jones, J. I., Lauridsen, R. B., Ledger, M. E., Lewis, H. M., Montoya, J. M., Neutel, A.-M., Olesen, J. M., Veen, F. J. v., Warren, P. H. and Woodward, G. (2009). Ecological networks: food webs and beyond. *Journal of Animal Ecology*, **78**, 253–69.

Inoue, T., Kato, M., Kakutani, T., Suka, T. and Itino, T. (1990). Insect–flower

relationship in the temperate deciduous forest of Kibune, Kyoto: an overview of the flowering phenology and the seasonal pattern of insect visits. *Contributions from the Biological Laboratory, Kyoto University*, **27**, 377–463.

Inouye, D. W. (1978). Resource partitioning in bumblebees: experimental studies of foraging behavior. *Ecology*, **59**, 672–8.

Johnson, S. D. and Steiner, K. E. (2000). Generalization versus specialization in plant pollination systems. *Trends in Ecology and Evolution*, **15**, 140–3.

Joppa, L. N., Bascompte, J., Montoya, J. M., Solé, R., Sanderson, J. and Pimm, S. L. (2009). Reciprocal specialization in ecological networks. *Ecology Letters*, **12**, 961–9.

Jordano, P. (1987). Patterns of mutualistic interactions in pollination and seed dispersal: connectance, dependence, and coevolution. *American Naturalist*, **129**, 657–77.

Jordano, P. (2010). Coevolution in multispecific interactions among free-living species. *Evolution, Education, and Outreach*, **3**, 40–6.

Jordano, P., Bascompte, J. and Olesen, J. M. (2003). Invariant properties in coevolutionary networks of plant-animal interactions. *Ecology Letters*, **6**, 69–81.

Jordano, P., Bascompte, J. and Olesen, J. M. (2006). The ecological consequences of complex topology and nested structure in pollination webs. In: *Plant-Pollinator Interactions: From Specialization to Generalization*, ed. N. M. Waser and J. Ollerton. Chicago, IL: Chicago University Press, pp. 173–99.

Jordano, P., Vázquez, D. P. and Bascompte, J. (2009). Redes complejas de interacciones mutualistas planta-animal. In: *Ecología y Evolución de Interacciones Planta–Animal: Conceptos y Applicaciones*, ed. R. Medel, M. A. Aizen and R. Zamora. Santiago de Chile, Chile: Editorial Universitaria, pp. 17–41.

Kaiser-Bunbury, C. N., Memmott, J. and Müller, C. B. (2009). Community structure of pollination webs of Mauritian heathland habitats. *Perspectives in Plant Ecology, Evolution and Systematics*, **11**, 241–54.

Kakutani, T., Inoue, T., Kato, M. and Ichihashi, H. (1990). Insect–flower relationship in the campus of Kyoto University, Kyoto: an overview of the flowering phenology and the seasonal pattern of insect visits. *Contributions from the Biological Laboratory, Kyoto University*, **27**, 465–521.

Kanstrup, J. and Olesen, J. M. (2000). Plant-flower visitor interactions in a neotropical rain forest canopy: community structure and generalisation level. In *The Scandinavian Association for Pollination Ecology honours Knut Fægri*, ed. Ø. Totland, W. S. Armbruster, C. Fenster, U. Molau, L. A. Nilsson, J. M. Olesen, J. Ollerton, M. Philipp and J. Aagren. Oslo, Norway: The Norwegian Academy of Science and Letters, pp. 33–41.

Kato, M. (2000). Anthophilous insect community and plant-pollinator interactions on Amami Islands in the Ryukyu Archipelago, Japan. *Contributions from the Biological Laboratory, Kyoto University*, **29**, 157–252.

Kato, M., Kakutani, T., Inoue, T. and Itino, T. (1990). Insect–flower relationship in the primary beech forest of Ashu, Kyoto: an overview of the flowering phenology and the seasonal pattern of insect visits. *Contributions from the*

Biological laboratory, Kyoto University, **27**, 309–75.

Kearns, C. A., Inouye, D. W. and Waser, N. M. (1998). Endangered mutualisms: the conservation of plant–pollinator interactions. *Annual Review of Ecology and Systematics*, **29**, 83–112.

Kevan, P. G. (1970). *High Arctic Insect–Flower Relations: the Interrelationships of Arthropods and Flowers at Lake Hazen, Ellsmere Island, Northwest Territories, Canada.* PhD Thesis, University of Alberta, Edmonton, Canada.

Levin, D. A. and Anderson, W. W. (1970). Competition for pollinators between simultaneously flowering species. *American Naturalist*, **104**, 455–67.

Lewinsohn, T., Prado, P., Jordano, P., Bascompte, J. and Olesen, J. M. (2006). Structure in plant-animal interaction assemblages. *Oikos*, **113**, 174–84.

Lundgren, R. and Olesen, J. M. (2005). The dense and highly connected world of Greenland's plants and their pollinators. *Arctic, Antarctic, and Alpine Research*, **37**, 514–20.

MacArthur, R. H. (1972). *Geographical Ecology.* New York, NY: Harper and Row.

Macior, L. W. (1971). Coevolution of plants and animals. Systematic insights from plant-insect interactions. *Taxon*, **20**, 17–28.

Martín, A. M. G., Dalsgaard, B., Ollerton, J., Timmermann, A., Olesen, J. M., Andersen, L. H. and Tossas, A. G. (2009). Effects of climate on pollination networks in the West Indies. *Journal of Tropical Ecology*, **25**, 493–506.

May, R. M. (1973). *Stability and Complexity in Model Ecosystems.* Princeton, NJ: Princeton University Press.

McCann, K. (2007). Protecting biostructure. *Nature*, **446**, 29.

Medan, D., Montaldo, N. H., Devoto, M., Mantese, A., Vasellati, V. and Bartoloni, N. H. (2002). Plant–pollinator relationships at two altitudes in the Andes of Mendoza, Argentina. *Arctic, Antarctic, and Alpine Research*, **34**, 233–41.

Medan, D., Perazzo, R. P. J., Devoto, M., Burgos, E., Zimmermann, M. G., Ceva, H. and Delbue, A. M. (2007). Analysis and assembling of network structure in mutualistic systems. *Journal of Theoretical Biology*, **246**, 510–21.

Melián, C. J., Bascompte, J., Jordano, P. and Krivan, V. (2008). Diversity in a complex ecological network with two interaction types. *Oikos*, **118**, 122–30.

Memmott, J. (1999). The structure of a plant–pollinator food web. *Ecology Letters*, **2**, 276–80.

Memmott, J. (2009). Food webs: a ladder for picking strawberries or a practical tool for practical problems. *Philosophical Transactions from the Royal Society in London, B Biological Sciences*, **364**, 1693–9.

Memmott, J., Craze, P. G., Waser, N. W. and Price, M. V. (2007). Global warming and the disruption of plant–pollinator interactions. *Ecology Letters*, **10**, 710–7.

Memmott, J., Godfray, H. C. J. and Gauld, I. D. (1994). The structure of a tropical host-parasitoid community. *Journal of Animal Ecology*, **63**, 521–40.

Memmott, J. and Waser, N. M. (2002). Integration of alien plants into a native flower–pollinator visitation web. *Proceedings of the Royal Society, London B*, **269**, 2395–9.

Memmott, J., Waser, N. M. and Price, M. V. (2004). Tolerance of pollination networks to species extinctions. *Proceedings of the Royal Society, London B*, **271**, 2605–11.

Moldenke, A. R. (1979). Pollination ecology within the Sierra Nevada. *Phytologia*, **42**, 223–82.

Moldenke, A. R. and Lincoln, P. G. (1979). Pollination ecology in montane Colorado: a community analysis. *Phytologia*, **42**, 349–79.

Montero, A. C. (2005). *Pollination Generalization Level and Habitat Width of Temperate Forest Insects*. MSc Thesis, Aarhus University, Denmark.

Montoya, J. M., Pimm, S. L. and Solé, R. V. (2006). Ecological networks and their fragility. *Nature*, **442**, 259–64.

Morales, C. L. and Aizen, M. A. (2006). Invasive mutualisms and the structure of plant–pollinator interactions in the temperate forests of north-west Patagonia, Argentina. *Journal of Ecology*, **94**, 171–80.

Mosquin, T. (1971). Competition of pollinators as a stimulus for evolution of flowering time. *Oikos*, **22**, 398–402.

Mosquin, T. and Martin, J. E. (1967). Observations on the pollination biology of plants on Melville Island, NWT, Canada. *Canadian Field-Naturalist*, **81**, 201–5.

Müller, H. (1883). *The Fertilization of Flowers*. London, UK MacMillan.

Nielsen, A. and Bascompte, J. (2007). Ecological networks, nestedness and sampling effort. *Journal of Ecology*, **95**, 1134–41.

Olesen, J. M. (2000). Exactly how generalised are pollination interactions? In: *The Scandinavian Association for Pollination Ecology honours Knut Fægri*, ed. Ø. Totland, W. S. Armbruster, C. Fenster, U. Molau, L. A. Nilsson, J. M. Olesen, J. Ollerton, M. Philipp and J. Aagren. Oslo, Norway: Norwegian Academy of Science and Letters, pp. 161–178.

Olesen, J. M. and Jordano, P. (2002). Geographic patterns in plant/pollinator mutualistic networks. *Ecology*, **83**, 2416–24.

Olesen, J. M., Bascompte, J., Dupont, Y. L. and Jordano, P. (2006). The smallest of all worlds: pollination networks. *Journal of Theoretical Biology*, **240**, 270–6.

Olesen, J. M., Bascompte, J., Dupont, Y. L. and Jordano, P. (2007). The modularity of pollination networks. *Proceedings of the National Academy of Sciences USA*, **104**, 19891–6.

Olesen, J. M., Bascompte, J., Elberling, H. and Jordano, P. (2008). Temporal dynamics of a pollination network. *Ecology*, **89**, 1573–82.

Olesen, J. M., Dupont, Y. L., Bascompte, J., Elberling, H. and Jordano, P. (2010a online). Missing and forbidden links in mutualistic networks. *Proceedings of the Royal Society, London B*.

Olesen, J. M., Dupont, Y. L., O'Gorman, E., Ings, T. C., Layer, K., Melían, C. J., Trøjelsgaard, K., Pichler, D. E., Rasmussen, C. and Woodward, G. (2010b). From Broadstone to Zackenberg: space, time and hierarchies in ecological networks. *Advances in Ecological Research*, **42**, 1–69.

Olesen, J. M., Eskildsen, L. I. and Venkatasamy, S. (2002). Invasion of pollination networks on oceanic islands: importance of invader complexes and endemic super generalists. *Diversity and Distribution*, **8**, 181–92.

Ollerton, J. and Cranmer, L. (2002). Latitudinal trends in plant-pollinator interactions: are tropical plants more specialized? *Oikos*, **98**, 340–50.

Padrón, B., Traveset, A., Biedenweg, T., Diáz, D., Nogales, M. and Olesen, J. M. (2009). Impact of alien plant invaders on pollination networks from oceanic and continental islands. *PLoS ONE*, **4**, e6275.

Paine, R. T. 1980. Food webs: linkage, interaction strength and community infrastructure. *Journal of Animal Ecology*, **49**, 667–85.

Pauw, A. (2007). Collapse of a pollination web in small conservation areas. *Ecology*, **88**, 1759–69.

Percival, M. (1974). Floral ecology of coastal scrub in southeast Jamaica. *Biotropica*, **6**, 104–29.

Petanidou, T. (1991). *Pollination Ecology in a Phryganic Ecosystem*. PhD Thesis, Aristotelian University, Thessaloniki, Greece.

Petanidou, T., Kallimanis, A. S., Tzanopoulos, J., Sgardelis, S. P. and Pantis, J. D. (2008). Long-term observation of a pollination network: fluctuation in species and interactions, relative invariance of network structure and implications for estimates of specialization. *Ecology Letters*, **11**, 564–75.

Philipp, M., Böcher, J., Siegismund, H. R. and Nielsen, L. R. (2006). Structure of a plant–pollinator network on a pahoehoe lava desert of the Galápagos Islands. *Ecography*, **29**, 531–40.

Potts, S. G., Vulliamy, B., Dafni, A., Ne'eman, G. and Willmer, P. G. (2003). Linking bees and flowers: how do floral communities structure pollinator communities? *Ecology*, **84**, 2628–2642.

Primack, R. B. (1983). Insect pollination in the New Zealand mountain flora. *New Zealand Journal of Botany*, **21**, 317–33.

Ramos-Jiliberto, R., Domínguez, D., Espinoza, C., López, G., Valdovinos, F. S., Bustamante, R. O. and Medel, R. (2010). Topological change of Andean plant-pollinator networks along an altitudinal gradient. *Ecological Complexity*, **7**, 86–90.

Reader, R. J. (1975). Competitive relationships of some bog ericads for major insect pollinators. *Canadian Journal of Botany*, **53**, 1300–5.

Rezende, E. L., Lavabre, J. E., Guimarães, P., Jordano, P. and Bascompte, J. (2007). Non-random coextinctions in phylogenetically structured mutualistic networks. *Nature*, **448**, 925–9.

Robertson, C. (1929). *Flowers and Insects: List of Visitors of Four Hundred and Fifty-Three Flowers*. Carlinville, IL: Science Press.

Saavedra, S., Reed-Tsochas, F. and Uzzi, B. (2008). A simple model of bipartite cooperation for ecological and organizational networks. *Nature*, **457**, 463–6.

Scheffer, M., Bascompte, J., Brock, W. A., Brovkin, V., Carpenter, S. R., Dakos, V., Held, H., Nes, E. H. v., Rietkerk, M. and Sugihara, G. (2009). Early-warming signals for critical transitions. *Nature*, **461**, 53–9.

Schemske, D. (1983). Limits to specialization and coevolution in plant-animal mutualisms. In *Coevolution*, ed. M. Nitecki. Chicago, IL: Chicago University Press, pp. 67–110.

Schemske, D., Willson, M., Melampy, M., Miller, L., Verner, L., Schemske, K. and Best, L. (1978). Flowering ecology of some spring woodland herbs. *Ecology*, **59**, 351–66.

Stang, M., Klinkhamer, P. G. L. and Meijden, E. v. d. (2006). Size constraints and flower abundance determine the number of interactions in a plant–flower visitor web. *Oikos*, **112**, 111–21.

Stang, M., Klinkhamer, P. G. L. and Meijden, E. v. d. (2007). Asymmetric specialization and extinction risk in plant-flower visitor webs: a matter of morphology or abundance? *Oecologia*, **151**, 442–53.

Stiles, F. G. (1975). Ecology, flowering phenology and hummingbird pollination of some Costa Rican *Heliconia* species. *Ecology*, **56**, 285–301.

Straw, R. M. (1972). A Markov model for pollinator constancy and competition. *American Naturalist*, **106**, 597–620.

Sugihara, G. and Ye, H. (2009). Cooperative network dynamics. *Nature*, **458**, 979–80.

Thompson, J. N. (2005). *The Geographic Mosaic of Coevolution*. Chicago, IL: Chicago University Press.

Traveset, A. and Richardson, D. M. (2006). Biological Invasions as disruptors of plant reproductive mutualisms. *Trends in Ecology and Evolution*, **21**, 208–16.

Tylianakis, J. M. (2008). Understanding the web of life: the birds, the bees, and sex with aliens. *PLoS Biology*, **6**, 224–7.

Vázquez, D. P. and Aizen, M. A. (2003). Null model analyses of specialization in plant–pollinator interactions. *Ecology*, **84**, 2493–501.

Vázquez, D. P. and Aizen, M. A. (2004). Asymmetric specialization: a pervasive feature of plant–pollinator interactions. *Ecology*, **85**, 1251–7.

Vespignani, A. (2009). Predicting the behavior of techno-social systems. *Science*, **325**, 425–8.

Waser, N. M., Chittka, L., Price, M. V., Williams, N. and Ollerton, J. (1996). Generalization in pollination systems, and why it matters. *Ecology*, **77**, 1043–60.

Williams, R. J., Berlow, E. L., Dunne, J. A., Barabási, A.-L. and Martinez, N. D. (2002). Two degrees of separation in complex food webs. *Proceedings from the National Academy of Sciences USA*, **99**, 12913–6.

Woodward, G., Ebenman, B., Emmerson, M., Montoya, J. M., Olesen, J. M., Valido, A. and Warren, P. H. (2005). Body size in ecological networks. *Trends in Ecology and Evolution*, **20**, 402–9.

Yodzis, P. (1980). The connectance of real ecosystems. *Nature*, **284**, 544–5.

14

Pollinators as drivers of plant distribution and assemblage into communities

Loïc Pellissier, Nadir Alvarez
and Antoine Guisan

14.1 Introduction

Understanding the factors that mold species distributions and communities has a long tradition in ecology and biogeography (Wallace 1876; Clements 1916; Phillips 1931). Recently, this topic has greatly benefited from technical and statistical developments, notably those that allow the prediction of the nature and distribution of species assemblages under different environmental conditions (Ferrier and Guisan 2006). Given the current perspective of climate change, this matter is critical for yielding realistic forecasts of the responses of species and communities to global change scenarios (Adler and HilleRisLambers 2008). However, whereas a large number of studies have focused on abiotic drivers, such as climatic (Guisan and Zimmermann 2000) or edaphic (Alvarez et al. 2009) factors, it is widely recognized that biotic factors can additionally strongly influence the distribution and assemblage of species (Pulliam 2000; Lortie et al. 2004). For example several studies have emphasized the importance of competitors and facilitators (Leathwick and Austin 2001; Heikkinen et al. 2007, Pellissier et al. 2010a) in delimiting species ranges. Pollination is among the main biotic factors that control the ecology, distribution, and assemblage of vascular plants. Whereas the pollen of gymnosperm species is predominantly dispersed by wind,

Evolution of Plant–Pollinator Relationships, ed S. Patiny. Published by Cambridge University Press. © The Systematics Association 2012.

the majority of angiosperms are dispersed by animal vectors (Barth 1991). Despite the recognition of pollination as a major facet of plant ecology, the importance of pollinators for predicting plant distribution has not been thoroughly investigated in recent decades. There is a strong need to characterize plant–pollinator interactions at large spatial scales and especially with respect to dynamic communities, whose compositions and patterns of relative species abundance vary in time and space.

Recent research on pollinators, while considering spatial variation, has mostly focused on how coevolution with pollinators can generate within-species geographic variation in the morphology of plant species, leading to plant speciation. In contrast, studies on the ecological links between plant and pollinator species have generally focused on a limited number of taxa and sites, with temporal replicates instead of observations across geographic space. Yet, the intensity and nature of interactions between plants and their pollinators often varies spatially, possibly because the ranges of interacting species do not completely overlap (Thompson 1988) and can cause plant fitness variation across its range. For instance, Espíndola et al. (2011) showed, by examining the entire distribution range of the lure-and-trap *Arum maculatum*, that pollination was not accomplished by a single specialized fly species as previously thought; instead, pollination was achieved by two fly species that showed distinct regional relative densities, despite being sympatric over a large portion of the plant's distribution. Segregation of sites with either one or two pollinators followed the cline of environmental gradients related to precipitation. But even for less specialized species, changes in pollinators' density along environmental gradients can, in turn, affect the pollination of the species and cause pollen limitation (Gómez et al. 2010). In our opinion, the lack of spatial replicates in biotic pollination studies can be mostly explained by the large sampling effort required to properly describe the plant–pollinator network throughout the entire distribution range of a plant. Another frequent bias in such studies is the absence of an accurate examination of pollen transfer, which is associated with plant fitness (Alarcón 2010). Observing the biotic vectors that visit flowers does not necessarily provide information on the nature and efficiency of the pollination process (e.g. the visitors could steal nectar without pollinating the plant, or local change in pollinator behavior could modify the intensity of pollination). Although a plant may attract a wide range of flower visitors, only a few groups can act as efficient pollinators (Bawa 1990). Consequently, drawing conclusions about the dependency of plants on particular biotic vectors may be misleading if pollination efficiency is not examined for each floral visitor (Reynolds and Fenster 2008; Kay and Sargent 2009).

As an alternative to direct observations, other approaches have been developed in plant functional ecology, which can provide a deeper ecological insight into the role of biotic interactions in shaping plant distributions and assemblages. One approach is to focus on traits associated with important functions. For instance, examining plant characters associated with growth form and leaf structure has permitted a better understanding of the effect of facilitative and competitive interactions (Lavorel et al. 1997; Kraft et al. 2008; Pellissier et al. 2010a).

Traditionally, in comparative functional biology, researchers have assumed that several pollination syndromes reflect convergent evolution via pollinator-mediated selection. This leads to phenotypes composed of specific scents and colors or floral morphologies that attract or restrict pollinator access to reward (Grant and Grant 1965; Faegri and van der Pijl 1979; Fenster et al. 2004). Olesen et al. (2011) argued that the reward, as well as the size and morphology of the flower, could impose constraints on the pollinator, preventing mutualistic interaction. In this view, a plant is considered as phenotypically specialized (sensu Ollerton et al. 2007) if it displays specific traits permitting its flowers to be successfully pollinated only by a subset of functionally similar, but not necessarily phylogenetically related, pollinators (e.g. long- and short-tongued insects, long-tongued bees, moths, birds, and bats; Fenster et al. 2004), for instance by exhibiting specific flower size, color, and shape (Lázaro et al. 2008, Campbell et al. 2010). An extreme case is when a plant is visited only by one sole lineage of pollinators – either one single species or a group of sister species (i.e. in such a case, it can be considered specialized both functionally and phylogenetically). Such a case can be encountered when, for example, the chemical composition, amount, and accessibility of the reward are highly specific (Fenster et al. 2004). As an illustration, the use of oil instead of nectar in *Lysimachia vulgaris* is strongly associated with pollination by oil-collecting bees *Macropis europaea* and *M. fulvipes* (Michez and Patiny 2005), two species showing overlapping but distinct ecological niches (Bassin et al. 2011). In contrast, a large number of plant species display flower structures, rendering pollination possible by many different pollinators (i.e. they are phenotypically generalized).

The evolution from generalization to specialization has been explained by the "most effective pollination principle" (Stebbins 1970), which states that natural selection is expected to modify the plant phenotype to optimize the frequency of interaction with the pollinator providing the most efficient pollen transfer. This evolutionary specialization leads to pollination by fewer functional groups or fewer pollinators compared to the pollinators of the ancestral state (Fenster et al. 2004). However, the advantage provided by focusing on a limited number of efficient pollinators to ensure reproduction, only persists as long as the pollinator or group of pollinators remain available. Specialization in floral displays should thus be selected for under undisturbed environment conditions, while the opposite, that is, generalisation, should be selected for when the availability of the most effective pollinator(s) is unpredictable in space and time (Waser et al. 1996). Still, the role of coevolution between plants and pollinators in shaping the large diversity of floral morphologies is controversial, especially considering the asymmetry in specialization between plant and pollinators (Vasquez and Aizen 2004): whereas many plants have flowers specialized for a single functional group of pollinators, pollinators are likely to forage any available flower. It is likely that other evolutionary processes alongside coevolution can drive the variation in plant traits under selection (Johnson and Anderson 2010).

Even if pollinator–mediated selection is an ongoing process in the evolution of lineages and species (Johnson and Anderson 2010), displaying one trait rather than another may strongly influence the current ecology of a given plant species. Because floral traits are directly related with the ecology, physiology, and behavior of pollinator agents, they are likely to have a strong influence on the current distribution of the plant species, from local to global scales, and on co-occurring species within communities (Sargent and Ackerly 2008). However, studies investigating such aspects are scarce.

In the following sections, we first focus on how pollination may influence the distribution of plants, anchoring our view into the theoretical concept of the Hutchinsonian niche (Hutchinson 1957). Second, we address the way pollination should influence how single plant species assemble into communities, following the ideas proposed by Sargent and Ackerly (2008). Here, we also demonstrate that our understanding of the way pollination influences plant assembly may be improved upon by using an approach based on functional traits. Finally, we illustrate these concepts with empirical data showing among-community variation in flowers' morphological traits, along an elevation gradient in the Western Swiss Alps (Pellissier et al. 2010c).

14.2 Abiotic versus biotic niches and species distribution

Because plant species distributions are driven by abiotic and biotic factors, their occurrences can be represented in a multivariate environmental space whose dimensions are those factors, or combinations of them, according to the framework introduced by Hutchinson (1957, Pulliam 2000). Variations in mutualistic and antagonistic interactions can thus potentially modify the population dynamics and the occupied range of a plant species. For instance, pollinator abundance is recognized as a primary driver of pollen limitation in many plant species and therefore of population fitness over the landscape (Cosacov et al. 2008). More severely, the presence or absence of an appropriate pollen vector can dictate whether or not a particular plant species can survive in a given area (Baker and Hurd 1968).

Hutchinson defined the *fundamental environmental niche* as the n-dimensional hypervolume composed of n environmental axes where a species, in the absence of competitive exclusion, is able to persist indefinitely. In contrast, the *realised environmental niche* is the part of the *fundamental niche* where the species is restricted due to negative biotic interactions with other species (e.g. competitors, predators, pathogens) excluding it from the other parts of the *fundamental niche* (e.g. through competitive exclusion). However, in its original definition, Hutchinson mostly considered constraints through competition and it still remains unclear how and where in the definition of niche spaces Hutchinson considered positive biotic interactions (e.g. facilitation). One potential solution is that they could be considered as implicitly comprised within the axes defining the fundamental niche (Araújo and Guisan 2006). To better illustrate the possible effect of pollinators on plant species distribution, we propose here to further distinguish the environmental abiotic niche from the biotic niche, at least for those positive biotic interaction that are strictly necessary for plant (Fig 14.1). The *fundamental abiotic niche* can be defined by resources and physiological limitations only, in absence of any biotic constraints, and can include aspects of climate, physical environment, or edaphic conditions that impose physiological limits on the ability of populations of a species to persist in an area (Guisan and Thuiller 2005). The *fundamental biotic niche*, in contrast, can be defined by the range of positive biotic interactions that are necessary for the plant (e.g. one of the dimensions of this space is defined by suitable conditions for associated pollinators) or, in other words, by biotic constraints in the absence of any other direct abiotic limitations to the plant. The fundamental niche *sensu stricto* can finally be seen as the overlap between the fundamental abiotic and biotic niches. Note that this should not be confounded with the pollination niche as defined by Elton (1927), which quantifies the impact of the species in its environment. Rather, the environmental niche (sensu Hutchinson 1957) describes the species response – in our case, the plant – to abiotic or biotic resources.

Alternatively, when positive biotic interactions do not come into play, the fundamental biotic niche is null. For example, plant species that possess alternative reproductive mechanisms such as selfing and clonality can fill their entire fundamental abiotic niche without any need for a pollinator. However, in such situations, the presence of pollinators may not only improve individual reproductive success, but also drive changes in the relative fitness of the progeny. This may happen if, for example, cross-fertilization induces genetic recombination that eventually enhances tolerance to a wider range of environmental conditions via an increased heterozygosity level, and contributes in modifying boundaries of the fundamental niche. Positioning the place of positive biotic interactions in the niche framework may not be as straightforward as illustrated here and can depend on the studied organism perspective. Here, for a better illustration of the

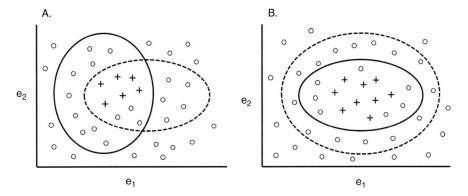

Fig 14.1 Illustration of the environmental fundamental abiotic (solid line) and biotic (dashed line) niches with presence (+) and absences (o) of the plant. The two axes (e1, e2) represent environmental resources or limiting factors (e.g. temperature, humidity). (A) In the case of a specialized pollination system, we would expect the species to be able to establish only in conditions corresponding to the overlap between the fundamental abiotic and biotic niches while in (B), in the case of generalized pollination, the plant could find pollinators all over the range of its fundamental abiotic niche. This figure was inspired by Pulliam et al. (2000).

case drawn by pollinators, we will consider the dichotomous view of fundamental biotic and abiotic niches.

14.2.1 Partial overlap between biotic and abiotic fundamental niches

Because pollinators differ in their distributions along ecological abiotic gradients (Cruden 1972; Arroyo et al. 1982; Warren et al. 1988; Kearns 1992; Devoto et al. 2005; Dalsgaard et al. 2009; González et al. 2009), their abiotic fundamental and realised niches (i.e. defined when taking into account, or not, negative biotic factors such as competition with other species) may be different from that of their associated plant species. As a consequence of this potential mismatch, the overlap between the biotic fundamental niche of the plant, for example, defined by the range of the pollinators, and its abiotic fundamental niche can be only partial. While plants with generalized pollination systems may find suitable pollen vectors in any environment for which they are physiologically adapted, plants with pollination systems specialized in a few pollinator functional groups (or even in a few pollinator species) may be limited to areas suitable for those pollinators. For example, it is recognized that specialized plants will be more prone to pollen limitation because they are less likely to interact with any available pollinator (Ashman et al. 2004, Knight et al. 2005). As a result, the plant will only be able to maintain dynamically in the environment corresponding to the overlap between fundamental abiotic and biotic niches, the latter being defined by the environment suitable for its pollinators (Fig 14.1).

For example, plants displaying a hawkmoth-pollination syndrome are relatively common in low- and mid-elevation ecosystems, but are absent or rare at high elevations (Cruden et al. 1976). Because temperature decreases with increasing elevation and moth activity decreases with decreasing temperature (Harling 1968), there is a negative relationship between elevation and fecundity in such plants (Cruden et al. 1976). The reduced hawkmoth activity, and the resulting decreased fecundity, as well as the small number of populations of hawkmoth-pollinated plants in high-elevation ecosystems, all support a limiting effect of the pollinator on plant distribution (Baker and Hurd 1968). In such cases, it is also likely that the plant species would be maladapted to abiotic high-altitude conditions, because the lack of pollinator prevents individuals from establishing at the abiotic range margin. In this situation, alleles conferring tolerance to colder conditions would never be selected for. While more extreme conditions may be unfavorable for plants with one type of pollination system, they may promote the establishment of others. Several studies have documented a higher proportion of plants adapted to bird pollination in wetter (Aizen 2003, Dalsgaard et al. 2009) and colder (Cruden 1972) conditions, corresponding to a higher proportion of hummingbirds as flower visitors. Dalsgaard et al. (2009) argued that hummingbirds, which have high energetic demands, feed on nectar even during the rain, while other pollinators such as bees are unable to, favoring hummingbird-pollinated plants in rainy areas. In another study, Johnson and Bond (1992) compared the pollination success of the orchid *Disa uniflora* in two contrasted habitats and found that plants occurring in a rocky gorge had a higher pollination success than plants occurring in an adjacent open valley because the exclusive obligate pollinator of the orchid, the butterfly *Meneris tulbaghia*, showed preferences for rocky, sheltered habitats. Another example at a larger spatial scale is the blooming of hummingbird flowers in California, which coincides in time and space with bird migration patterns (Grant and Grant 1967). Partial overlap of biotic and abiotic fundamental niches is, however, not restricted to mutualisms and can also arise in antagonistic pollination systems. Pellissier et al. (2010d) found that the relative frequency of food-deceptive orchids decreased with increasing elevation, suggesting that deception may be less profitable at high compared to low elevations, perhaps because of reduced pollinator activity in the more stressful environmental conditions.

The previous examples focus on plants displaying particular floral characteristics, whose spatial pattern could be investigated to yield deeper understanding of the effect of pollinators on the distribution of such specialized plants. Because plant responses to the environment can be considered as functions of their morphological and physiological traits, such an approach may allow investigating the drivers of the plant species niches.

14.2.2 Cases of biotic fundamental niche nested within the abiotic fundamental niche

As discussed above, the fundamental niches of plants showing generalized pollination syndromes are mostly constrained by their abiotic component. In extreme cases, the biotic fundamental niche may even be fully nested into the abiotic fundamental niche, when direct abiotic environmental pressure is more limiting than the absence of efficient pollinators (Fig 14.1). For example, some plants typically display radially symmetrical flowers with open access to the reward. In such plants, the identity of pollinators changes through time and space, and is largely determined by the relative abundance of each potential pollinator species (Ollerton et al. 2007). But because pollen transfer occurs irrespective of the composition of pollinator communities (i.e. such plants are phenotypically generalized), this floral character allows the plant to thrive in any environment compatible with its abiotic fundamental niche. We can thus assume that this trait should be particularly common among invasive species. Indeed, introduced animal-pollinated plants can become invasive only if they manage to reproduce in the new colonized range, where they may suffer from low population densities and pollinator-limitation decreasing the possibility of pollen transfer. Baker's law (Baker 1955, Stebbins 1957) suggests that guarantees for reproductive success after long-distance dispersal are self-compatibility in the case of mate limitation, and autonomous seed production and vegetative reproduction in the case of pollinator limitation. When expanding Baker's law in the context of pollination, not only may invasive species be prone to selfing and cloning, but they are expected to be associated with abiotic pollen vectors or to display a more generalized pollination syndrome to expand into new areas. In such cases, plant distribution is mostly driven by abiotic factors, while biotic pollination has only a minor role or no role at all in driving plant species distribution.

14.2.3 Cases of interaction between abiotic and biotic fundamental niches

Interactions between fundamental abiotic and biotic niches can potentially decrease the size of the fundamental niche sensu stricto, and accordingly restrict the species distribution (if these conditions are themselves geographically restricted). Even if the suitable pollinator is present, direct physiological conditions may limit the possibility of interaction, notably through limitations in rewards production. For example, the most important adaptation for attracting birds as pollinators is the production of abundant nectar (Stiles 1977). Most bird flowers secrete higher nectar volumes than bee flowers (Stiles 1977, Opler 1978). Bird pollination, therefore, requires a much higher energy investment per flower than insect pollination. Ecological factors (e.g. temperature, humidity, soil moisture) and the physiological condition of the plant can both limit production of large nectar volumes in various ways (Shuel 1967). The decline in sugar concentration of bird–flower nectars with increasing elevation (Hainsworth and Wolf 1972) can therefore lead to a lower attraction of pollinators, which in turn can reduce plant fecundity. As a result, the decline in nectar production becomes one of the main factors reducing the frequency of bird-pollinated flowers at higher elevations (Stiles 1977).

14.3 From single species to communities

The nature and quantity of pollinators do not solely influence individual plant species distribution but also the way species assemble into communities. It has been shown experimentally that persistence of a given plant community can be affected by a reduction in the diversity of pollinators' functional groups, which is directly related to the diversity of plant floral traits (Fontaine et al. 2006). Changes in the proportion of pollinator functional groups is thus likely to trigger changes in the structure and composition of natural plant communities. Incorporation of trait data to investigate the drivers of community assemblage can help unravelling the mechanisms of species coexistence (Sargent and Ackerly 2008). Investigations in community ecology have recognized three main situations in which the structure of species assemblages can be detected from plant traits: habitat filtering, competition, and facilitation.

14.3.1 Habitat filtering

Habitat filtering (also called environmental filtering) predicts that the environmental conditions affecting the local community act as a sieve that filters which species can establish and persist. Consequently, biotic factors such as the local pollinators' assemblage can operate as an environmental filter in the plant community. For example, the absence of a particular pollinator functional group can prevent the sustainable establishment of plant species with a specialized floral display. According to Sargent and Ackerly (2008), such filtering can be either direct, when a plant species cannot reproduce (e.g. when there are no suitable pollinators), or indirect, when the physical environment itself influences the interaction between a plant and its pollinators. This filtering determines which pollination systems can persist (see the hummingbird case previously discussed). As a consequence of habitat filtering, we should observe under-dispersion (or clustering) of traits related to pollination within communities (Sargent and Ackerly 2008). For instance, a given flower morphology (Pellissier et al. 2010c) or color (Arnold et al. 2009) should be filtered locally along environmental gradients. This situation can be considered an extension of the effect of pollinator(s) on plants distributions by considering that several species, not just one, may have their distribution limited by similar environmental constraints.

14.3.2 Competition and limiting similarity

Because resources are generally limited, the species that most efficiently exploits them should exclude the others. The term *limiting similarity* was introduced by MacArthur and Levins (1967) to describe the maximum similarity in resource-use patterns that is consistent with the coexistence of two or more competing species. Consequently, species assemblages within communities are arranged in relation to the availability of resources; therefore, species can more readily coexist if they show diverse functional traits and thus differ in resource use (i.e. competition between them is reduced). Synchronously flowering (hereafter coflowering) plant species frequently share pollinators, leading to competition for pollination that can often be detrimental to one or more of these species (reviewed in Mitchell et al. 2009). Such competition can be direct when related to pollinator preference or indirect through interspecific pollen transfer (Sargent and Ackerly 2008). Because decreased pollinator availability can induce pollen limitation and reduced fitness (Cosacov et al. 2008), the fitness of a given plant could be enhanced by attracting a different group of pollinators from neighboring species. For example, a plant may use distinct floral displays; in such a case, a pattern of character displacement, translated into an over-dispersion of floral display traits (when compared to random), is expected. McEwen and Vamosi (2010) proposed that displaying a rare color in the community may be advantageous and could improve the pollination success of the species. Also, Gumbert et al. (1999) showed that rare plants show a significant tendency to diverge from coflowering plants, with regard to color, compared to frequent plants, and proposed that rare species may fare best if they possess a strongly divergent signal.

Competition for resources can even indirectly limit interactions with some functional groups of pollinators and therefore restrict the floral display found in the community. For example, hummingbird-pollinated plants tend to produce much higher quantities of nectar than insect-pollinated species (Stiles 1977). As harsher climatic conditions limit nectar production, hummingbird-pollinated plants are less common in the understory. This is likely because competition for light limits the rate of photosynthesis and does not allow such species to thrive in forests (Sargent and Vamosi 2008).

14.3.3 Facilitation

In the case of *facilitation*, the occurrence of a plant species may enhance the establishment of other species within the community. Such enhancement can thus also be produced through facilitation of pollinator visitation (Waser and Real 1979; Thompson 1988; Moeller 2004). Particularly attractive plant species may indirectly improve rates of pollinator visitation in a neighboring species by increasing visitation of shared pollinators. For example, Johnson et al. (2003)

showed that the food-deceptive orchid species *Anacamptis morio* could benefit from visitors attracted to a nearby nectar-producing species (*Geum rivale, Allium schoenoprasum*). Similarly, Ghazoul (2006) found that the number of flower visitors to *Raphanus raphanistrum* increased when the plant was in the same patch as other species (*Circium arvense, Hypericum perforatum*, and *Solidago canadensis*) compared to when the species occurred alone. Hegland et al. (2009) revealed that such positive (or facilitative) interactions for pollinator attraction were far more frequent than negative (or competitive) ones.

The facilitative effect is more likely to occur if plants share floral traits with coflowering species, rendering insects less able to distinguish between them (Juillet et al. 2007). De Jager et al. (2011) found a significant clustering of flower colors in South African *Oxalis* communities, suggesting pollinator-mediated facilitation for attracting honeybees. Also, contrasting with the findings of Gumbert et al. (1999), Feldman et al. (2004) proposed that rare species show little divergence in floral color with their neighbors, as they likely share pollinators with at least one attractive common species. Common species may act as magnets for pollinators of rare species, resulting in stabilizing forces maintaining diversity within communities (McEwen and Vamosi 2010). Hence, facilitation is expected to yield communities composed of plants with similar floral display. As a consequence, the effect of pollinators in driving plant community assemblages could be better understood by recording and analysing trait similarities and divergences in coflowering species (e.g. amount of nectar reward, flower color, pollen placement). One should keep in mind that facilitative effects other than those involving pollinator activity could also be effective and that their relative impact would influence the extent to which pollinators drive community assembly.

14.3.4 The benefit of phylogeny

Identifying whether phylogenetically related species resemble or differ from each other for particular traits may enable the formulation of hypotheses regarding the evolutionary ecology of plant community assembly (Cavender-Bares et al. 2009). Indeed, interactions with pollinators can influence the phylogenetic community structure in either direction mentioned previously (i.e. clustering or over-dispersion), depending on the nature of the interactions (Sargent and Ackerly 2008). For instance, facilitation predicts that pollination should increase phylogenetic clustering when increased benefits are provided to congeners through shared pollinators (Moeller 2005; Sargent and Ackerly 2008). Similarly, habitat filtering should also promote phylogenetic clustering in situations where spatially aggregated pollinators filter closely related plant species sharing similar floral display (i.e. specialized on the same pollinator functional group) (Cavender-Bares et al. 2009). Finally, competition is predicted to generate communities of species less closely related than communities formed by random assembly, again when similarity in pollination strategies is associated with phylogenetic proximity among plant species. Correlating phylogenetic patterns to the distribution of functional traits (e.g. related to floral display) may thus allow a better understanding of how plant species assemble into communities. However, when testing such hypotheses, one should account for putative phylogenetic constrains that may act as confounding factors. For instance, closely related plant species (i.e. phylogenetically clustered) are likely to have similar floral display and occur in a similar environment.

Floral differences that arose through pollinator-mediated selection or pollination ecotypes could evolve in response to a geographical mosaic of pollinators, creating extensive geographic variation in floral displays in association with contrasting pollination environments. This may contribute to species diversification (Harder and Johnson 2009). At a wider evolutionary scale, phylogenetic approaches may allow a better understanding of how evolution shaped floral displays when traits are not conserved along phylogenetic trees (i.e. when phylogenetic constraints are weak for a given character). Such approaches, which use a larger time scale, can thus also inform us about how community structures evolved (Harder and Johnson 2009). Pollinators do not only affect current plant distributions and assemblages, but they represent available niche dimensions for long-term ecological diversification in associated plants (Johnson 2010). Plant community assemblages are thus the net result of processes taking place at different time scales (Kembel 2009). For example, McEwen and Vamosi (2010) found that even if coflowering species within communities tended to be more divergent in floral color than expected by chance, they were not phylogenetically dispersed. Floral color is therefore a labile trait displaying a symplesiomorphic phylogenetic component (i.e. it evolved and converged multiple times independently), which can reduce pollen transfer between two close relatives increasing the probability of coexistence within communities. Hence, selection may favor divergence in floral color among close relatives to attract different pollinators and avoid outbreeding (McEwen and Vamosi 2010). Also, Armbruster et al. (1994) showed that the distribution and overlap of traits associated with pollen placement in several species of *Stylidium* in Western Australia was consistent with post-assembly character displacement, and more recent studies have documented similar patterns both for bat (Muchhala and Potts 2007) and bird-pollinated plants (Botes et al. 2008).

14.4. Pollinator availability drives floral traits along an altitudinal gradient

Heithaus (1974) proposed that the availability of pollinators may limit floral diversity within a given pollination syndrome: pollinators may filter which plants can persist, as a function of their floral display (Sargent and Ackerly 2008). The filtering level may thus vary according to floral morphology, and it is widely accepted that this filter ranges from generalization to specialization as a function of flower openness (Olesen et al. 2007). The openness can be characterized, for instance, by the blossom classes described by Faegri and van der Pijl (1979) (i.e. these classes can be considered as discrete cases along the continuum between specialization and generalization). While actinomorphic blossoms such as open disks (see Faegri and van der Pijl 1979) are accessible to any pollinators, even those with short mouthparts, flowers displaying bilabiate or tubular morphologies restrict access to rewards by allowing only specific functional pollinator groups (e.g. mostly bees in the case of bilabiate; see Müller 1881; Ramírez 2003; Westerkamp 2007). Displaying one of such particular blossom types may influence plant distributions because the corresponding pollinators may not be available under all environmental conditions. As mentioned above, we can expect that plants that depend on a limited number of pollinators can be particularly constrained by the distribution of their pollinators, while plants with relatively generalized pollination systems are resilient to the loss of some pollinators and thus able to thrive in more harsh and stressful conditions.

To assess the role of pollinators on plant distribution in an alpine landscape, Pellissier et al. (2010a) investigated the spatial structure and patterns of functional diversity – in particular, following habitat filtering – of mountain plant communities for seven blossom types (wind, disk, funnel, tube, bilabiate, head, or brush, see Faegri and van der Pijl 1979; Ramírez 2003; Olesen et al. 2007) in 870 vegetation plots of an alpine landscape in the western Swiss Alps. For each plot, the deviation from random assembly of plant species according to traits was evaluated using null-models. The proportions of the seven blossom types were weighted on the basis of the relative abundance of each species in the 870 plots. These proportions were related to the first axis of an environmental PCA, representing strong drivers of insect distribution (Dillon et al. 2006) highly correlated to elevation.

Evidence for habitat filtering with regard to blossom types was found in 12 % of the plots, which were located at significantly higher elevation than plots not differing from random assembly. This pattern was related to the change of abundance in the blossom type along elevation. At higher elevations, environmental conditions are incredibly stressful for insects, and both pollinator diversity and activity tend to decrease (Warren et al. 1988). Cold-adapted (i.e. arctic and alpine) insect communities are notably characterized by the large dominance of flies, a group that shows a good ability to tolerate cold temperatures (Kevan 1972; Arroyo et al. 1982; Elberling and Olesen 1999). Under such climate pressures and pollinator limitations, we may expect that cold-adapted plants shift reproduction strategies towards possible alternatives to heterogamous sexual reproduction, for example clonality and selfing. However, it is recognized that a large proportion of alpine plants are entomophilous (Körner 2003), even if mixed strategies between sexual and selfing are also common (Pellissier et al. 2010b). Pollen exchange via insects may favor genetic exchanges, allowing faster adaptation in response to changing conditions. While this is also true for wind pollination, using wind as a pollen vector is usually not as effective as insect pollination when plant interindividual distances are large, which might be the case at higher elevations (Regal 1982).

Disk blossoms display a simple structure where the reward in the form of pollen and nectar is not restricted by appendices within the flower, but is available even to small insects with a short proboscis, such as flies that are dominant at high elevations (Müller 1881). This open floral structure also permits access to other guilds of pollinators occurring at higher elevations, such as butterflies and some bumblebee species (Müller 1881; Mani 1962). Under conditions of pollinator limitation at high elevation (i.e. an habitat showing high temporal heterogeneity in biotic and abiotic factors), plants could benefit from interacting with any pollinator available, leading to more randomly organised plant–pollinator networks in more severe abiotic conditions (Ramos-Jiliberto et al. 2010).

In contrast, plants with more specialized morphologies may be limited to milder conditions. In our survey, bilabiate blossoms were proportionally more important at lower elevation where such milder conditions occur. This is mainly explained by bilabiate blossoms being predominately pollinated by bees (Müller 1881; Ramírez 2003; Westerkamp 2007), which are more frequent in the warmer and drier conditions characteristic of lower elevations (Müller 1881; Arroyo et al. 1982; Warren et al. 1988; Devoto et al. 2005). Similar observations were previously made by Müller (1881) and Mani (1962), who found that bilabiate blossoms were more rarely represented in high elevation plant communities in the Alps and Himalayas. In addition, Collins et al. (1983) noticed an overrepresentation of flowers with restricted rewards in plant communities at lower elevations in Utah, while open-access flowers were overrepresented at higher elevations. At lower elevations, where bees are abundant, displaying a more complex flower structure, which is attractive to this species group, may enhance reproductive success, whereas at higher elevations, dominated by flies, disk blossoms easily accessible to all insects – even those with a short proboscis – should have an advantage (Fig 14.2).

Our approach does not provide absolute evidence about the existence of filter-impeding plants that display more specialized blossoms (e.g. bilabiate) to survive in high-elevation conditions. While we most often assume this filtering to be related to pollinator-mediated traits, it could also be partly due to physiological characteristics of the plants. However, even if pollination systems only partially drive current plant species distributions at high elevations, particular morphologies should have evolved in parallel with physiological traits at a wider evolutionary scale and should therefore have participated in shaping current species' response to the environment. By combining an investigation of (1) convergence between traits related to pollination and (2) knowledge about the environmental

Fig 14.2 Proportion of bilabiate (top) and disk blossoms (bottom) along the first axis of an environmental PCA corresponding to elevation. To better visualize the pattern, a boxplot of the vegetation plots belonging to each interval with a length of 1 ranging from –4 to 4 in the first PCA axis were used instead of drawing points. The dashed line represents the GLMs applied independently on the proportion of the two blossom types with the first axis of the environmental PCA as explaining factor (see Pellissier et al. 2010c, pictures: L. Pellissier). See plate section for color version.

niche of species within a phylogenetic framework as described previously, we understand better the extent to which pollination can drive plant species distribution in short- (i.e. ecological) and long-term (i.e. evolutionary) perspectives. This area also requires further field studies to investigate the components of the plant and insect environmental niches.

14.5 Conclusion

In this chapter, we illustrate how investigating plant traits should advance understanding of how pollinators affect plant distribution. While it is likely that pollinators play a major role in molding plant distributions, only a few case studies so far have addressed these effects. There is a large potential for future research. In particular, plants characterized by similar flower traits (e.g. flower color or shape, amount or depth of reward) are likely to be visited by similar functional groups of pollinators and thus to show similar distributions. By comparing niches of plants with similar floral displays and linking them to the niches of functional groups of pollinators, a better understanding of how pollinators drive species distributions can be achieved. Alternatively, contrasting the environmental requirements of plants with pollination systems that range from generalized to specialized could provide insight into whether some traits restrict plant species distributions. The increase in available data on the distribution range of multiple species (e.g. Global Biodiversity Information Facility, Atlas Flora Europaea) coupled with the enhancement of plant traits databases (e.g. Biolflor, Klotz et al. 2002; LEDA, Kleyer et al. 2008) provides promising possibilities for better understanding the effect of pollinators on plant distributions as mediated by plant traits.

Plant functional traits have been increasingly used to improve our understanding of how plant communities assemble, but this approach has rarely been applied to floral traits at the community level to understand how pollination shapes plant communities. Understanding the assemblage of plant communities can thus greatly benefit from investigating patterns of traits dispersion within and between communities, especially when combined with phylogenetic approaches. Trait-based approaches are promising as a shortcut to increasing our understanding of such patterns without needing actual pollinator observations. We encourage the development of more trait databases on pollination for both plants and insects.

Acknowledgements

We are grateful to Luc Gigord for comments on a previous version of this manuscript. The redaction of this book chapter was supported by the Centre de conservation de la faune et de la nature (Canton de Vaud), the European Commission

(ECOCHANGE project, contract no. FP6 2006 GOCE 036866), and the Swiss National Science Foundation grant no. 31003A-125145 (BIOASSEMBLE project). Nadir Alvarez was funded by an AMBIZIONE grant from the Swiss National Science Foundation (PZ00P3-126624).

References

Adler, P. B. and HilleRisLambers, J. (2008). The influence of climate and species composition on the population dynamics of ten prairie forbs. *Ecology*, **89**, 3049–60.

Aizen, M. A. (2003). Down-facing flowers, hummingbirds and rain. *Taxon*, **52**, 675–80.

Alarcón R. (2010). Congruence between visitation and pollen-transport networks in a California plant-pollinator community. *Oikos*, **119**, 35–44.

Alvarez, N., Thiel-Egenter, C., Tribsch, A., Holderegger, R., Manel, S., Schönswetter, P., Taberlet, P., Brodbeck, S., Gaudeul M., Gielly, L., Küpfer, P., Mansion, G., Negrini, R., Paun, O., Pellecchia, M., Rioux, D., Schüpfer, F., Van Loo, M., Winkler, M. and Gugerli, F. (2009). History or ecology? Substrate type as a major driver of spatial genetic structure in Alpine plants. *Ecology Letters*, **12**, 632–40.

Araújo, M. B. and Guisan, A. (2006). Five (or so) challenges for species distribution modelling. *Journal of Biogeography*, **33**, 1677–88.

Armbruster, W. S., Edwards, M. E. and Debevec, E. M. (1994). Floral character displacement generates assemblage structure of Western Australian Triggerplants (Stylidium). *Ecology*, **75**, 315–29.

Arnold, S. E. J., Savolainen, V. and Chittka, L. (2009). Flower colours along an alpine altitude gradient, seen through the eyes of fly and bee pollinators. *Arthropod–Plant Interaction*, **3**, 27–43.

Arroyo, M. T. K., Armesto, J. J. and Primack, R. (1982). Community studies in pollination ecology in the high temperate Andes of central Chile. I. Pollination mechanisms and altitudinal variation. *American Journal of Botany*, **69**, 82–97.

Ashman, T. L., Knight, T. M., Steets, J. A. Amarasekare, P., Burd, M., Campbell, D. R., Dudash, M. R., Johnston, M. O., Mazer, S. J., Mitchell, R. J., Morgan, M. T. and Wilson, W. G. (2004) Pollen limitation of plant reproduction: ecological and evolutionary causes and consequences. *Ecology*, **85**, 2408–21.

Baker, H. G. (1955). Self-compatibility and establishment after long-distance dispersal. *Evolution*, **9**, 347–9.

Baker, H. G., and Hurd, P. D. JR. (1968). Intrafloral ecology. *Annual Review of Entomology*, **13**, 385–414.

Barth, F. G. (1991). *Insects and Flowers: the Biology of a Partnership*. Princeton, NJ: Princeton University Press.

Bassin, L., Alvarez, N., Pellissier, L. and Triponez, Y. (2011). Ecological niche overlap in sister species: how do oil-collecting bees Macropis europaea and M. fulvipes (Hymenoptera: Melittidae) avoid competition and hybridization? *Apidologie*, **42**.

Bawa, K. S. (1990). Plant–pollinator interaction in tropical rain forests.

Annual Review of Ecology and Systematics, **21**, 399–422.

Botes, C., Johnson, S. D., Cowling and R. M. (2008). Coexistence of succulent tree aloes: partitioning of bird pollinators by floral traits and flowering phenology. *Oikos*, **117**, 875–82.

Campbell, D., Bischoff, M., Lord, J. and Robertson, A. (2010). Flower color influences insect visitation in alpine New Zealand. *Ecology*, **91**, 2638–49.

Cavender-Bares, J., Kozak, K. H., Fine P. V. A. and Kembel S. W. (2009). The merging of community ecology and phylogenetic biology. *Ecology Letters*, **12**, 693–715.

Clements F. E. (1916). *Plant Succession: an Analysis of the Development of Vegetation*. Washington, DC: Carnegie Institution of Washington.

Collins, P. D., Harper, K. T. and Pendelton B. K. (1983). Comparative life history and floral characteristics of desert and mountain floras in Utah. *Great Basin Naturalist*, **43**, 385–93.

Cosacov, A., Naretto, J. and Cocucci, A. A. (2008) Variation of pollinator assemblages and pollen limitation in a locally specialized system: the oil-producing Nierembergia linariifolia (Solanaceae). *Annals of Botany*, **102**, 723–34.

Cruden, R. W. (1972). Pollinators in high-elevation ecosystems: relative effectiveness of birds and bees. *Science*, **176**: 1439–40.

Cruden R. W. (1976). Fecundity as a function of nectar production and pollen–ovule ratios. In: *Variation, Breeding and Conservation of Tropical Trees*, ed. J. Burley and B. T. Styles. Academic Press, pp. 171–178.

Cruden, R. W., Kinsman, S., Stockhouse II R. E. and Linhart Y. B. (1976). Fecundity, and the distribution of moth-flowered plants. *Biotropica*, **8**, 204–10.

Dalsgaard, B., González, A. M. M., Olesen J. M., Ollerton, J., Timmermann, A., Andersen, L.H. and Tossas, A.G. (2009). Plant–hummingbird interactions in the West Indies: floral specialisation gradients associated with environment and hummingbird size. *Oecologia*, **159**, 757–76.

De Jager, M. L., Dreyer, L. L. and Ellis, A. G. (2011). Do pollinators influence the assembly of flower colours within plant communities? *Oecologia*, **166**, 543–53.

Devoto, M., Medan, D. and Montaldo N. H. (2005). Patterns of interaction between plants and pollinators along an environmental gradient. *Oikos*, **109**, 461–72.

Dillon, M. E., Frazier, M. R. and Dudley, R. (2006). Into thin air: physiology and evolution of alpine insects. *Integrative and Comparative Biology*, **46**, 49–61.

Elberling, H. and Olesen, J. M. (1999). The structure of a high latitude plant-flower visitor system: the dominance of flies. *Ecography*, **22**, 314–23.

Elton, C. (1927). *Animal Ecology*. London, UK: Sidgwick and Jackson.

Espíndola A., Pellissier L. and Alvarez, N. (2011). Variation in the proportion of flower visitors of A. maculatum along its distributional range in relation with community-based climatic niche analysis. *Oikos*, **120**, 728–34.

Faegri, K. and van der Pijl, L. (1979). *The Principles of Pollination Ecology*. Oxford, UK: Pergamon Press.

Feldman, T. S., Morris, W. F. and Wilson, W. G. (2004). When can two plant species facilitate each other's pollination? *Oikos*, **105**, 197–207.

Fenster, C. B., Armbruster, W. S., Wilson, P., Dudash, M. R. and Thompson J. D. (2004). Pollination syndromes and floral specializations. *Annual Review of Ecology, Evolution and Systematic*, **35**, 375–403.

Ferrier, S. and Guisan, A. (2006) Spatial modelling of biodiversity at the community level. *Journal of Applied Ecology*, **43**, 393–404.

Fontaine, C., Dajoz, I., Meriguet, J. and Loreau, M. (2006). Functional diversity of plant–pollinator interaction webs enhances the persistence of plant communities. *PLoS Biology*, **4**, 129–35.

Ghazoul, J. (2006). Floral diversity and the facilitation of pollination. *Journal of Ecology*, **94**, 295–304.

Gómez, J. M., Abdelaziz, M., Lorite, J., Muñoz-Pajares, A. J. and Perfectti, F. (2010) Change in pollinator fauna cause spatial variation in pollen limitation. *Journal of Ecology*, **98**, 1243–52.

González, A. M. M., Dalsgaard, B., Ollerton, J., Timmermann, A., Olesen, J. M., Andersen, L. and Tossas, A. G. (2009). Effects of climate on pollination networks in the West Indies. *Journal of Tropical Ecology*, **25**, 493–506.

Grant, V. and Grant, K. A. (1965). *Flower Pollination in the Phlox Family*. New York: Columbia University Press, p. 224.

Grant, V. and Grant, K. A. (1967). Effects of hummingbird migration on plant speciation in the California flora. *Evolution*, **21**, 457–65.

Guisan, A. and Zimmermann, N. E. (2000). Predictive habitat distribution models in ecology. *Ecological Modelling*, **135**, 147–86.

Guisan, A. and Thuiller, W. (2005). Predicting species distribution: offering more than simple habitat models. *Ecology Letters*, **8**, 993–1009.

Gumbert, A., Kunze, J. and Chittka, L. (1999). Floral colour diversity in plant communities, bee colour space and a null model. *Proceedings of the Royal Society of London, Series B, Biological Sciences*, **266**, 1711–6.

Hainsworth, F. R. and Wolf, L. L. (1972). Power for hovering flight in relation to body size in hummingbirds. *The American Naturalist*, **106**, 589–96.

Harder, L. D. and Johnson S. D. (2009). Darwin's beautiful contrivances: evolutionary and functional evidence for floral adaptation. *New Phytologist*, **183**, 530–45.

Harling, J. (1968). Meterological factors affecting the activity of night flying macrolepidoptera. *Entomologist*, **101**, 83–93.

Hegland, S. J., Grytnes, J. A. and Totland, Ø. (2009) The relative importance of positive and negative interactions for pollinator attraction in a plant community. *Ecological Research*, **24**, 929–36.

Heikkinen, R. K., Luoto, M., Virkkala, R., Pearson, R. G. and Körber J. H. (2007). Biotic interactions improve prediction of boreal bird distributions at macroscales. *Global Ecology and Biogeography*, **16**, 754–63.

Heithaus, R. E. (1974). The role of plant–pollinator interactions in determining community structure. *Annals of the Missouri Botanical Garden*, **61**, 675–91.

Hutchinson, G. E. (1957) Concluding remarks. *Cold Spring Harbor Symposia on Quantitative Biology*, **22**, 145–59.

Johnson, S. D. and Bond, W. J. (1992). Habitat dependent pollination success in a Cape orchid. *Oecologia*, **91**, 455–6.

Johnson, S. D., Peter, C. I., Nilsson, L. A. and Ågren, J. (2003). Pollination success in a deceptive orchid is enhanced by co-occurring rewarding magnet plants. *Ecology*, **84**, 2919–27.

Johnson, S. D. (2010). The pollination niche and its role in the diversification and maintenance of the southern African flora. *Proceedings of the Royal Society of London, Series B, Biological Sciences*, **365**, 499–516.

Johnson, S. D. and Anderson, B. (2010). Coevolution between food-rewarding flowers and their pollinators. *Evolution, Education and Outreach*, **3**, 32–9.

Juillet N., Gonzalez, M. A., Page, P. A. and Gigord, L. D. B. (2007). Pollination of the European food-deceptive *Traunsteinera globosa* (Orchidaceae): the importance of nectar-producing neighbouring plants. *Plant Systematics and Evolution*, **265**, 123–29.

Kay, K. M. and Sargent, R. D. (2009). The role of animal pollination in plant speciation: integrating ecology, geography, and genetics. *Annual Review of Ecology, Evolution, and Systematics*, **40**, 637–56.

Kearns, C. A. (1992). Anthophilous fly distribution across an elevation gradient. *American Midland Naturalist*, **127**, 172–82.

Kembel, S. W. (2009). Disentangling niche and neutral influences on community assembly: assessing the performance of community phylogenetic structure tests. *Ecology Letters*, **12**, 949–60.

Kevan, P. G. (1972). Insect pollination of high arctic flowers. *Journal of Ecology*, **60**, 831–47.

Kleyer, M., Bekker, R. M., Knevel, I. C. Bakker, J. P., Thompson, K., Sonnenschein, M., Poschold, P., Van Groenendael, J. M., Klimes, L., Klimesova, J., Klotz, S., Rusch, G. M., Hermy, M., Adriaens, D., Boedeltje, G., Bossuyt, B., Dannemann, A., Endels, P., Götzenberger, L., Hodgson, J. G., Jackel, A.-K., Kühn, I., Kunzmann, D., Ozinga, W. A., Römermann, C., Stadler, M., Schlegelmilch, J., Steendam, H. J., Tackenberg, O., Wilmann, B., Cornelissen, J. H. C., Eriksson, O., Garnier, E. and Peco, B. (2008). The LEDA Traitbase: a database of life-history traits of Northwest European flora. *Journal of Ecology*, **96**, 1266–74.

Klotz, S., Kühn, I. and Durka, W. (2002). *BIOLFLOR – eine Datenbank zu biologisch-ökologischen Merkmalen der Gefäßpflanzen in Deutschland. Schriftenreihe für Vegetationskunde*, **38**. Bonn, Germany: Bundesamt für Naturschutz.

Knight, T. M., Steets, J. A., Vamosi, J. C., Mazer, S. J., Burd, M., Campbell, D. R., Dudash, M. R., Johnston, M., Mitchell, R. J. and Ashman, T. L. (2005) Pollen limitation of plant reproduction: ecological and evolutionary causes and consequences. *Annual Review of Ecology, Evolution and Systematics*, **36**, 467–97.

Körner, C. (2003). *Alpine Plant Life*, 2nd edn. Berlin, Germany: Springer.

Kraft, N. J. B, Renato, V. and David, D. (2008). Functional traits and niche-based tree community assembly in an Amazonian forest. *Science*, **322**, 580–2.

Lavorel, S., McIntyre, S., Landsberg, J. and Forbes, T. D. A. (1997). Plant functional classifications: from general groups to specific groups based on response to disturbance. *Trends in Ecology and Evolution*, **12**, 474–8.

Lázaro, A. Hegland, S. J., Totland, Ø. (2008). The relationships between floral traits and specificity of pollination systems in three Scandinavian plant communities. *Oecologia*, **157**, 249–57.

Leathwick, J. R. and Austin, M. P. (2001). Competitive interactions between tree species in New Zealand's old-growth indigenous forests. *Ecology*, **82**, 2560–73.

Lortie, C. J., Brooker, R. W., Choler, P., Kikvidze, Z., Michalet, R., Pugnaire, F. I. and Callaway, R. M. (2004). Rethinking plant community theory. *Oikos*, **107**, 433–8.

MacArthur, R. H. and Levins, R. (1967). The limiting similarity, convergence, and divergences of coexisting species. *The American Naturalist*, **101**, 377–85.

Mani, M. S. (1962). *Introduction to High Altitude Entomology: Insect Life Above Timberline in the North-Western Himalayas*. London, UK: Methuen.

McEwen, J. R. and Vamosi, J. C. (2010). Floral colour versus phylogeny in structuring subalpine flowering communities. *Proceedings of the Royal Society of London, Series B, Biological Sciences*, **277**, 2957–65.

McGill, B. J. Enquist, B. J., Weiher, E. and Westoby, M. (2006). Rebuilding community ecology from functional traits. *Trends in Ecology and Evolution*, **21**, 178–85.

Moeller, D. A. (2004). Facilitative interactions among plants via shared pollinators. *Ecology*, **85**, 3289–1.

Moeller, D. A. (2005) Pollinator community structure and sources of spatial variation in plant–pollinator interactions in Clarkia xantiana ssp xantiana. *Oecologia*, **142**, 28–37.

Michez, D. and Patiny, S. (2005). World revision of the oil-collecting bee genus *Macropis* Panzer 1809 (Hymenoptera: Apoidea:Melittidae) with a description of a new species from Laos. *Annales de la Société entomologique de France*, **41**, 15–28.

Mitchell, R. J., Flanagan, R. J., Brown, B. J., Waser, N. M., and Karron, J. D. (2009). New frontiers in competition for pollination. *Annals of Botany*, **103**, 1403–13.

Müller, H. (1881). *Die Alpenblumen, ihre Befruchtung durch Insecten und ihre Anpassungen an dieselben*. Leipzig, Germany: W. Englemann.

Muchhala, N. and Potts M. D. (2007). Character displacement among bat-pollinated flowers of the genus *Burmeistera*: analysis of mechanism, process, and pattern. *Proceedings of the Royal Society B*, **274**, 2731–7.

Olesen, J. M., Dupont, Y. L., Ehlers, B. K. and Hansen, D. M. (2007). The openness of a flower and its number of flower-visitor species. *Taxon*, **56**, 729–36.

Olesen, J. M., Bascompte, J., Dupont, Y. L., Elberling, H., Rasmussen C. and Jordano, P. (2011). Missing and forbidden links in mutualistic networks. *Proceedings of the Royal Society B*, **278**, 725–32.

Ollerton, J., Killick, A., Lamborn, E., Watts, S. and Whiston, M. (2007). Multiple meanings and modes: on the many ways to be a generalist flower. *Taxon*, **56**, 717–28.

Opler, P. A. (1978). Nectar production in a tropical ecosystem. In *The Biology of Nectaries*, ed. T. S. Elias and B. Bentley. New York, NY: Columbia University Press.

Pellissier, L., Bråthen, K. A., Pottier, J., Randin, C. F., Vittoz, P., Dubuis, A., Yoccoz, N., G., Alm, T., Zimmerman, N., E. and Guisan, A. (2010a). Species distribution models reveal apparent competitive and facilitative effects of a dominant species on the distribution of tundra plants. *Ecography*, **33**, 1004–14.

Pellissier, L., Fournier, B., Guisan, A. and Vittoz, P. (2010b). Plant traits covary with altitude in grasslands and forests in the European Alps. *Plant Ecology*, **211**, 351–65.

Pellissier, L., Pottier, J., Vittoz, P., Dubuis, A. and Guisan A. (2010c). Spatial pattern of floral morphology: a possible insight into the effects of pollinators on plant distribution. *Oikos*, **119**, 1805–13.

Pellissier, L., Vittoz, P., Internicola, N. I. and Gigord, L. D. B, (2010d). Generalized food-deceptive orchid species flower earlier and occur at lower altitudes than rewarding ones. *Journal of Plant Ecology*, **3**, 243–50.

Phillips, J. (1931). The biotic community. *Journal of Ecology*, **19**, 1–24.

Pulliam, H. R. (2000). On the relationship between niche and distribution. *Ecology Letters*, **3**, 349–61.

Ramírez, N. (2003). Floral specialization and pollination: a quantitative analysis and comparison of the Leppik and the Faegri and van der Pijl classification systems. *Taxon*, **52**, 687–700.

Ramos-Jiliberto, R., Domínguez, D., Espinoza, C. Lópeza, G., Valdovinosa, F. S., Bustamantea, R. O. and Medela, R. (2010). Topological change of Andean plant-pollinator networks along an altitudinal gradient. *Ecological Complexity*, **7**, 86–90.

Regal, P. J. (1982). Pollination by wind and animals: ecology of geographic patterns. *Annual Review of Ecology, Evolution and Systematics*, **13**, 497–524.

Reynolds, R. J. and Fenster, C. B. (2008). Point and interval estimation of pollinator importance: a study using pollination data of *Silene caroliniana*. *Oecologia*, **156**, 325–32.

Sargent, R. D. and Ackerly, D. D. (2008). Plant-pollinator interactions and the assembly of plant communities. *Trends in Ecology and Evolution*, **23**, 123–30.

Sargent R. D. and Vamosi J. C. (2008). The influence of canopy position, pollinator syndrome, and region on evolutionary transitions in pollinator guild size. *International Journal of Plant Sciences*, **169**, 39–47.

Shuel, R. W. (1967). The influence of external factors on nectar production. *American Bee Journal*, **107**, 54–56.

Stebbins G. L. (1957). Self-fertilization and population variability in the higher plants. *The American Naturalist*, **91**, 337–54.

Stebbins, G. L. (1970). Adaptive radiation of reproductive characteristics in angiosperms. I: pollination mechanisms. *Annual Review of Ecology and Systematics*, **1**, 307–26.

Stiles, F. G. (1977). Coadapted competitors: flowering seasons of hummingbird-pollinated plants in a tropical forest. *Science*, **198**, 1177–8.

Thompson, J. N. (1988). Coevolution and alternative hypotheses on insect/plant interactions. *Ecology*, **69**, 893–5.

Vázquez, D. P. and Aizen, M. A. (2004). Asymmetric specialization: a pervasive feature of plant–pollinator interactions. *Ecology*, **85**, 1251–7.

Wallace, A. R. (1876). *The Geographical Distribution of Animals*. New York, NY: Harper and Brothers.

Warren, S. D., Harper, K. T. and Booth G. M. (1988). Elevational distribution of insect pollinators. *American Midland Naturalist*, **120**, 325–30.

Waser N. M. and Real L. A. (1979). Effective mutualism between sequentially flowering plant species. *Nature*, **281**, 670–2.

Waser N. M., Chittka, L., Price, M. V., Williams, N. M. and Ollerton, J. (1996). Generalization in pollination systems, and why it matters. *Ecology*, **77**, 1043–60.

Westerkamp, C. (2007). Bilabiate flowers: the ultimate response to bees? *Annals of Botany*, **100**, 361–74.

15

Effects of alien species on plant–pollinator interactions: how can native plants adapt to changing pollination regimes?

Gideon Pisanty and Yael Mandelik

15.1 Introduction

Invasive alien species are a major concern in the management and conservation of habitats and species worldwide (Crooks 2002; Bax et al. 2003; Levine et al. 2003; Vilà et al. 2010). The direct effects of these species may further cascade in the ecosystem and affect inter- and intraspecific ecological interactions. The introduction of alien plants and animals can have severe consequences, not only for individual native plant and pollinator species, but also for their ecological interactions through plant–pollinator networks (Morales and Traveset 2009; Dohzono and Yokoyama 2010; Schweiger et al. 2010). Integration of alien plant and pollinator species into pollination networks inevitably creates new interactions and may also affect the strength and quality of existing ones. These changes are open niches for novel evolutionary adaptations of both alien and native species (Mooney and Cleland 2001). However, research in this topic is very limited, and has focused mostly on adaptations of alien plant species to pollinator-independent reproduction modes (Barrett et al. 2008). We know of no study investigating adaptations of native plant and pollinator species to invaders, and the ecological and possibly evolutionary consequences of these adaptations in the

Evolution of Plant–Pollinator Relationships, ed S. Patiny. Published by Cambridge University Press. © The Systematics Association 2012.

context of plant–pollinator networks. Such adaptations might have far-reaching ecological and evolutionary implications, as has been shown in plant–herbivore and predator–prey interactions (Cox 2004). Here we outline the main effects of species invasions on plant–pollinator interactions, and deduce the main adaptive mechanisms that native plant species can exhibit in response to changes in their pollination regime. Finally, we explore the characteristics of plant populations that are likely to affect their probability of exhibiting such adaptations and their conservation implications.

15.2 Effects of alien plant and animal species on native plant pollination

Several groups of alien organisms have been shown to affect native plant pollination. Most research has focused on alien plants (Morales and Traveset 2009) and flower visitors (Lach 2003; Dohzono and Yokoyama 2010); however, other groups, such as alien herbivores and predators, can also be influential (Traveset and Richardson 2006). In the following, we explore the possible effects of different groups of alien organisms on pollination of native plants.

15.2.1 Effects of alien plants

Alien plant species can exert both positive and negative effects on native plant pollination. Especially important in this regard are alien plant species that are highly attractive for pollinators. Such species often display conspicuous advertisements such as large, showy flowers, offer high rewards for their visitors, and/or employ a super-generalist pollination strategy (Morales and Traveset 2009). The attraction of pollinators to these invasive plants can have major effects on native plant species in the invaded community (Bjerknes et al. 2007; Morales and Traveset 2009). The frequency of visits to native plants can either decrease, if pollinators visit alien plants instead of natives (pollinator usurpation; Chittka and Schürkens 2001; Brown et al. 2002) or increase, if more pollinators are attracted to native plants that grow near highly attractive aliens (pollinator facilitation; Moragues and Traveset 2005; Nielsen et al. 2008). In addition, the composition of the pollinator fauna that visits native species can be changed, possibly affecting also the quality of individual visits (Ghazoul 2002; Muñoz and Cavierez 2008). The movement of pollinators between alien and native plants may increase heterospecific pollen deposition on native plant stigmas (Grabas and Laverty 1999; Ghazoul 2002), as well as loss of native plant's pollen (Larson et al. 2006; Flanagan et al. 2009); both of these processes may impede plant reproduction. These effects can be changed and even reversed when tested across varying plant densities or spatial scales, if different interaction mechanisms (e.g. pollinator usurpation versus facilitation) operate at different plant densities or geographical distances (Muñoz and Cavieres 2008; Jakobsson et al. 2009). When larger spatio–temporal scales are considered, alien plant invasion may change the overall carrying capacity of pollinators in the ecosystem, which can also affect native plant pollination (Bjerknes et al. 2007; Tepedino et al. 2008; see below).

It is not yet fully understood why in certain situations, alien animal-pollinated plants facilitate the pollination of natives, whereas in others, pollinators are usurped. Multiple factors are involved in such interactions, and the final outcome will depend on the relative characteristics of the native versus alien plant species, such as flower density, morphology, and attractiveness to pollinators, as well as on the unique pollinator species involved. Theoretically, the larger the niche over-lap between alien and native plant species, the higher the chances that pollina-tors' visits to the native plant will be affected (Goodell 2008). In particular, plants sharing similar floral traits and pollination syndromes have increased chances of sharing also their pollinator guild and therefore having interspecific pollinator transitions (Schemske 1981; Internicola et al. 2007). There is evidence that when alien and native plants share flower shape and/or color, the probable outcome for the native species will be pollinator usurpation rather than facilitation, and ultim-ately, decreased reproductive success (Morales and Traveset 2009). However, we hypothesize that both pollinator usurpation and facilitation will be more prob-able when floral advertizement traits are similar, because some pollinators that are attracted to the more showy or rewarding alien, may mistake a nearby grow-ing native plant for the alien, thus facilitating visits to the native (Dafni and Ivri 1981a, 1981b; Johnson et al. 2003). The unique outcome under such circumstances will depend on the extent of similarity in visual and/or olfactory signals and in rewards between the two plant species, as well as on the pollinator's sensory and learning capabilities. A possible scenario is one of mixed effects – the alien plant will usurp the more skilled and loyal pollinators that distinguish between the two plants, but will facilitate visits of other, more naive pollinator species to the native. If the alien plant offers a higher reward than the native, usurpation of pollinators may also increase gradually during the flowering season, as naive newly emerged pollinators will learn over time to discriminate between the two plant species (Dafni 1984).

15.2.2 Effects of alien flower visitors

Like alien plants, alien species of pollinators and flower visitors may also either hamper or facilitate native plant pollination and seed set. The pollination ser-vices delivered by alien visitor species may differ markedly from those provided by native visitors, due to behavioral and/or morphological differences (Dafni and Shmida 1996; Lach 2003; Dohzono and Yokoyama 2010). Moreover, alien visitors often usurp native plants of their native visitors, by depletion of rewards

(Dafni and Shmida 1996; Hingston and McQuillan 1999), damage to floral tissues (Dohzono et al. 2008), or physical deterrence (Gross and McKay 1998; Hansen and Muller 2009). Theoretically, however, it is possible that deterrence by alien visitors will enhance native pollinators' efficiency by forcing native pollinators to move more frequently among flowers, thus increasing their visit frequencies (Lach 2007; see also Greenleaf and Kremen 2006). In addition, deterred pollinators may fly greater distances between consecutive visits, possibly enhancing outcrossing. Both of these processes might benefit plant reproduction. Furthermore, native pollinators that are deterred from visiting native focal plants may switch to foraging on other native plant species, affecting these latter species' reproduction as well (Roubik and Villanueva-Gutiérrez 2009). The effects of alien flower visitors on native plant pollination have been studied mostly in alien species of social bees (reviewed in Vergara 2008; Dohzono and Yokoyama 2010) and ants (Lach 2003, 2007, 2008a, 2008b; Roberts and McGlynn 2004; Blancafort and Gómez 2005; Hansen and Müller 2009). However, other groups of alien flower visitors may also be influential, for instance solitary bees (Cane 2003; Pemberton and Liu 2008), birds (Cox 1983; Kelly et al. 2006), and wasps (Morales and Aizen 2002). There are many examples of native plant species that suffer reduced pollination services due to the effects of alien visitors (e.g. Dafni and Shmida 1996; do Carmo et al. 2004; Hansen and Müller 2009). However, in many other cases, alien visitors have no effect on native plant seed set (e.g. Dupont et al. 2004; Lach 2007), and sometimes even positive influences have been documented (Chamberlain and Schlising 2008), particularly when the native plant was dependent upon a native pollinator species that had gone extinct locally or globally (Traveset and Richardson 2006; Cox 1983; Lord 1991). Hence, empirical work to date does not point to any general trend regarding the impacts of alien flower visitors on native plant species.

The integration of an alien pollinator into a native plant–pollinator network can result in significant breakage of pollination syndromes. For example, several cases have been documented of honeybees visiting native plants adapted for bird pollination, especially in Australia. In some of these cases, honeybees were the main visitor; some plant species were efficiently pollinated by honeybees, whereas others only poorly or not at all (Paton 2000; Fumero-Cabán and Meléndez-Ackerman 2007).

15.2.3 Indirect effects

Alien species can also influence native plant pollination indirectly, by affecting native pollinator populations through diverse ecological interactions and mechanisms, including competition, predation, herbivory, parasitism, and habitat modification. Highly attractive alien plants can increase native pollinators' carrying capacities by providing increased forage resources (Bjerknes et al. 2007;

Tepedino et al. 2008). Some alien plants can also provide feeding substrates for herbivorous pollinator larvae such as butterfly caterpillars (Graves and Shapiro 2003), or nesting substrates for bees (Hurd 1978). An opposite effect may be induced by unattractive invading plant species that spread vigorously and create dense monospecific stands, thereby outcompeting native flowering plants that provide forage resources, and transforming nesting habitats such as bare ground (Johnson 2008; Moroń et al. 2009). Alien animal species can also affect pollinator abundance and diversity. For instance, alien flower visitor species may compete with native pollinators for forage resources (Thomson 2004; Paini and Roberts 2005) or nesting substrates (Inoue et al. 2008); alien predators may prey heavily on pollinators, and even cause their extinction (Fritts and Rodda 1998; Abe et al. 2010); and alien herbivores may consume important forage plants or trample them (Traveset and Richardson 2006). However, secondary species interactions may also induce positive effects, e.g. an alien predator that preys on a native herbivore. Alien species of parasites and pathogens, often introduced with alien animal species, can also have disastrous consequences for native pollinator faunas (Cox and Elmqvist 2000). Generally speaking, the effects of alien species on native pollinator populations are still poorly understood and need to be further explored.

Table 15.1 summarizes the different impacts exerted by each group of alien organisms on native plant pollination and seed set. As can be seen, alien species from diverse functional groups can impose drastic positive or negative effects; there seems to be a greater focus on the negative aspects in the literature, although this may represent a methodological bias. From a conservational point of view, negative effects seem to be the most important, given the frequent evidence of species decline and extinction due to alien species invasions (Coblentz 1990; Mooney and Cleland 2001). Therefore, in the following sections, we will focus mainly on the negative effects of alien species on native plant pollination and reproduction, and the potential of evolutionary adaptations to overcome them.

15.3 Possible adaptive mechanisms of native plants in response to alien plant and pollinator invasions

Native plant species experiencing changes in pollination and/or reproductive success due to the processes described above, may adapt to their altered environments in several, not necessarily mutually exclusive ways (Bjerknes et al. 2007; Harder and Aizen 2010). These can be broadly classified into two categories:

(1) Alteration of flower traits and/or blooming characteristics to attract the highest number of efficient pollinators.

Table 15.1 Summary of the effects of alien species on the abundance and behavior of native pollinators, and the consequences for native plant pollination and seed set.

Invading organism	Effects on native pollinators	References	Effects on native plants	References
Alien plants	Pollen source	Moragues & Traveset 2005; Stubbs et al. 2007; Tepedino et al. 2008	Reduced visit frequencies	Chittka & Schürkens 2001; Brown et al. 2002; Moragues & Traveset 2005; Larson et al. 2006
	Nectar source	Chittka & Schürkens 2001; Moragues & Traveset 2005; Tepedino et al. 2008; Kandori et al. 2009	Increased visit frequencies	Moragues & Traveset 2005; Nielsen et al. 2008; Jakobsson et al. 2009
	Caterpillar forage host	Graves & Shapiro 2003	Increased visit lengths	Muñoz & Cavieres 2008
	Bee nesting substrate	Hurd 1978	Change in composition of visitor fauna	Ghazoul 2002; Muñoz & Cavieres 2008; Nienhuis et al. 2009; Tscheulin et al. 2009
	High flower attractiveness	Chittka & Schürkens 2001; Moragues & Traveset 2005; Vanparys et al. 2008; Kandori et al. 2009	Low flower constancy	Brown et al. 2002; Flanagan et al. 2009; Kandori et al. 2009
	Increased pollinator abundance / diversity	Stubbs et al. 2007	Reduced conspecific pollen deposition	Larson et al. 2006; Flanagan et al. 2009
	Reduced pollinator abundance / diversity	Johnson 2008; Moroń et al. 2009	Heterospecific pollen deposition	Grabas & Laverty 1999; Ghazoul 2002; Nielsen et al. 2008; Montgomery 2009
	Change in composition of pollinator fauna	Nienhuis et al. 2009	Increased seed set	Muñoz & Cavieres 2008
			Reduced seed set	Grabas & Laverty 1999; Chittka & Schürkens 2001; Brown et al. 2002; Muñoz & Cavieres 2008

Table 15.1 (*cont.*)

Invading organism	Effects on native pollinators	References	Effects on native plants	References
Alien flower visitors	Reduced abundance / diversity of native pollinators	Roubik 1978	Reduced abundance / diversity of native / legitimate visitors	Dafni & Shmida 1996; Lach 2007, 2008a,b
	Physically deterring native pollinators from forage plants	Dafni & Shmida 1996; Gross & Mackay 1998; Hansen & Müller 2009	Increased frequencies of total legitimate visits	Osorio-Beristain et al. 1997; Madjidian et al. 2008
	Native pollinators shift their foraging to other plant species	Roubik & Villanueva-Gutiérrez 2009	Reduced visit frequencies of native / legitimate visitors	Dafni & Shmida 1996; Blancafort & Gómez 2005; Dohzono et al. 2008; Hansen & Müller 2009
	Competition for forage resources	Thomson 2004; Paini & Roberts 2005	Shortened visit lengths of native / legitimate visitors	Hingston & McQuillan 1999; Lach 2008a
	Competition for nest sites	Inoue et al. 2008	Reduced average efficiency per visit (illegitimate visits not included)	Osorio-Beristain et al. 1997; Hansen et al. 2002; do Carmo et al. 2004; Madjidian et al. 2008
	Competition through interspecific mating	Kanbe et al. 2008; Kondo et al. 2009	Nectar robbery	Dafni & Shmida 1996; Roberts & McGlynn 2004; Lach 2005; Dohzono et al. 2008
			Pollen robbery	Gross & Mackay 1998; do Carmo et al. 2004
			Nectar depletion	Dafni & Shmida 1996; Hingston & McQuillan 1999; Dupont et al. 2004; Roberts & McGlynn 2004

Invading organism	Effects on native pollinators	References	Effects on native plants	References
Alien flower visitors (contd.)			Pollen depletion	do Carmo et al. 2004
			Damage to floral tissues	Kenta et al. 2007; Dohzono et al. 2008
			Reduced seed set	Gross & Mackay 1998; Roberts & McGlynn 2004; Dohzono et al. 2008; Hansen & Müller 2009
			Reduced pollinator movement among plants	Dupont et al. 2004
			Increased pollinator movement among plants	Celebrezze & Paton 2004
			Reduced outcrossing	England et al. 2001; Dick et al. 2003
			Increased gene flow distances	Dick et al. 2003
			Compensation for rare or extinct pollinators	Cox 1983; Lord 1991; Kelly et al. 2006

Table **15.1** (cont.)

Invading organism	Effects on native pollinators	References	Effects on native plants	References
Alien herbivores	Reduced pollinator abundance / diversity	Spurr & Anderson 2004		
Alien predators	Reduced pollinator abundance / diversity	Kelly et al. 2006; Abe et al. 2010		
	Local or global pollinator extinction	Nogales & Medina 1996; Fritts & Rodda 1998		
Alien parasites & pathogens	Local or global pollinator extinction	Cox & Elmqvist 2000		

(2) Development of reproductive modes that are not animal-mediated, or increased reliance on such mechanisms that already exist.

Next, we explore both of these adaptive paths and their evolutionary consequences.

15.3.1 Optimization of biotic pollination

Flower morphology. Major changes in the composition of pollinator species visiting a plant species can induce morphological changes in flowers that will allow a better fit to the behavioral and/or morphological characteristics of the new visitors, especially to those species that are the most common and/or efficient pollinators (Bernardello et al. 2001; Johnson 2006). For example, a shift to pollinators with larger bodies and shorter tongues will select for wider and shorter corollas, respectively, and vice versa (e.g. Dohzono et al. 2008; but see Harder and Aizen 2010). The more generalist and attractive the plant, the higher the chances that spatio–temporal changes in the relative abundances of different pollinator species will eliminate any adaptive effect that a particular pollinator exerts on the flowers (Johnson and Steiner 2000; Gomez and Zamora 2006). If, however, a plant is pollinated exclusively by a single species or a narrow suit of closely related species in its invaded environment, these pollinators will select for flower morphologies that fit them best, potentially initiating a process of specialization. Conversely, if the plant's main pollinator becomes rare, and other visitors are also uncommon or inefficient, a process of generalization will initiate (Harder and Aizen 2010).

Floral mimicry. Selection can favor floral advertising cues that mimic those of a highly attractive invasive plant growing nearby, so that some visitors will move between the two species indiscriminately, leading to pollinator facilitation (Mullerian mimicry, Dafni 1984) (Dafni and Ivri 1981a, 1981b; Johnson et al. 2003). The extent of the similarity in advertisement depends on relative flower sizes, colors, shapes and scents. This mechanism may be problematic, however, if the frequent movement of visitors between species negatively affects the native plant due to heterospecific pollen deposition or major losses of conspecific pollen. The problem of interspecific pollen transfer may be reduced by a shift in the sexual organs' point of contact with the pollinator in the native plant, such that different areas on the pollinator's body come into contact with different plant species' sexual organs (Caruso 2000).

Flower attractiveness. Changes in the quantity or quality of pollinator visits can affect the attractiveness of flowers. Plants often respond to decreased/increased pollination services by respectively increasing/decreasing various parameters of reward and/or advertisement (Ashman and Morgan 2004). Such parameters include: nectar sugar content, nectar volume, corolla size, scent, flower longevity, and blooming synchronization. Different pollinator species are attracted by different floral cues and rewards, and thus changes in the visiting fauna can also induce changes in flower attractiveness. Blooming synchronization among flowers on different plants and/or on an individual plant, can also be modified as a means of minimizing negative effects of interspecific pollen transfer by disloyal visitors on the one hand, and excess geitonogamy (i.e. the transfer of pollen among flowers of the same individual plant) by more constant visitors on the other (Harder and Aizen 2010). Plants that receive increased pollination services and can invest more resources in reproduction, can further increase their fitness by producing more flowers per plant.

Adaptations to illegitimate visitors. A high incidence of pollen or nectar robbing by alien visitor species may select for morphologies that better conceal these rewards, and that allow access only to legitimate pollinators, or (in the case of pollen robbery) for no visitors whatsoever. For example, small bees that rob nectar from a large flower may select for concealment of the nectar deeper inside the corolla; corolla piercing by alien bumblebees might be prevented by selecting for a thicker and/or longer calyx (Maloof and Inouye 2000); and nectar robbing by ants may select for hairy stems (Howarth 1985). If, however, robbing cannot be prevented, plants may adapt by producing more rewards, to compensate for the amount robbed (Maloof and Inouye 2000).

Spatio–temporal adaptations. Plants adapt to changes in pollination regimes also by shifting their spatio–temporal flowering niches (Waser 1978; Ghazoul 2002). In the case of usurpation of pollinators by alien plant species, competitive exclusion can trigger adaptation to a new habitat, by favoring plants or populations that grow relatively far from the invader (Waser 1978). Similarly, a shift in the blooming period will reduce temporal overlap with the alien (Waser 1978; Ghazoul 2002). Facilitation will favor opposite trends. These trends can also operate on smaller scales, for instance height of flowers on plants, and daily timing of flower opening and closing. Different pollinators can also prefer flowers located on different parts of plants (e.g. inner versus outer branches, lower versus upper branches), and thus a change in the pollinator fauna can induce a change in the position of blossoms. Furthermore, if different species of pollinators consistently visit the same different parts of the plant, the flowers in each part can develop somewhat different morphologies or offer different amounts of rewards (Colwell et al. 1974; Willmer and Corbet 1981; Maloof and Inouye 2000). However, some alien plant species (Ghazoul 2002), and most alien species of ants and social bees (Vergara 2008), have relatively wide and flexible spatial and/or temporal niches, and it is therefore less likely that a shift in blooming time or location will help minimize any negative effects they may have.

15.3.2 Reproductive modes which are not animal-mediated

Three major reproductive modes that are independent of animal visitation are known in terrestrial plants: autonomous self-pollination, wind pollination, and asexual reproduction. Each of these modes can be utilized as either a complementary strategy to biotic pollination, providing reproductive assurance, or a sole, obligate strategy. Species that rely on two or more reproductive strategies may shift between them according to their environmental conditions and developmental stages. Thus, for a given species, the relative importance of each reproductive mode may change among different populations, individuals, flowers, and seasons, and even during the lifetime of a single flower. The facultative reliance on these reproductive modes as reproductive assurance mechanisms may enable some highly pollinator-specialist plant species to persist for prolonged periods with little or no biotic pollination (Bond 1994).

Autonomous self-pollination. Plants that experience a reduction in quantity and/ or efficiency of visits, leading to pollen limitation, often compensate by increasing their rates of autonomous self-pollination (Fishman and Wyatt 1999; Barrett et al. 2009; Eckert et al. 2009). Adaptation for increased rates of autonomous selfing usually involves a relaxation of spatio–temporal and genetic mechanisms designed to minimize selfing, i.e. dichogamy, herkogamy, and self-incompatibility. Thus, the distance between anthers and stigmas is often reduced, the overlap in functional male and female periods is increased, and self-incompatibility is broken (Harder and Aizen 2010). By analogy, dioecious species become andro/gynodioecious, and heterostylous species become monostylous (Barrett et al. 2009). There is, however, a genetic limitation to successful autonomous selfing. Increased reliance on selfing may have severe consequences on population dynamics if the population has high levels of inbreeding depression, as most of the selfed progeny will not reach maturity. Autonomous selfing may operate at different stages of the flower's lifespan. Very low visit frequencies, or a major negative effect of pollen robbing or heterospecific pollen deposition, are expected to select for prior selfing. This often occurs already in the unopened bud (cleistogamy), before any visitors have contacted the flower. On the other hand, high spatio–temporal variation in pollinator activity is more likely to select for competing or delayed selfing, a compromise which maintains reproductive assurance without completely losing the advantages of outcrossing (Eckert et al. 2009).

Asexual reproduction. Asexual modes of reproduction, such as vegetative growth and apomixis (asexual seed production), can also compensate for reduced pollination services, although they rarely appear as a sole reproductive strategy (Eckert 2002; Bicknell and Koltunow 2004). Resources freed up by reduced flower production could potentially allow increased asexual reproduction (Fischer and van Kleunen 2002; Eckert 2002). The evolution of clonal plants from non-clonal ancestors has appeared frequently among the angiosperms; however, the adaptive evolution of vegetative reproduction has been poorly studied (Fischer and van Kleunen 2002). Apomixis is not as common as vegetative reproduction or autonomous selfing, probably because it requires two or three mutations, each of which is disadvantageous when appearing in isolation (Marshall and Brown 1981). Hence, apomictic mutants are rare in plant populations, and are often polyploids derived from hybridization between reproductively incompatible progenitors, such as interspecific hybrids. However, once such a mutant is formed, it has an automatic selection advantage, and thus spreads rapidly (Holsinger 2000).

Wind pollination. An alternative pollination mode that is independent of animal visitors yet does achieve significant levels of outcrossing is wind pollination. Wind pollination seems to be especially common in plants inhabiting oceanic islands, which seems to be related to the limited pollinator faunas in these ecosystems (Harder and Aizen 2010). Efficient wind pollination is dependent upon several factors, including exposed stigmas and anthers, relatively open habitats or deciduous vegetation, and relatively high conspecific densities (Culley et al. 2002; Friedman and Barrett 2009). Shifts to wind pollination are more likely to evolve in plants with floral morphologies that favor pollen dispersal and capture by wind, such as small flowers, exerted stamens, and short or absent corollas. Such species usually have inconspicuous floral advertisements and attract generalist pollinators (Friedman and Barrett 2009). Plants that receive reduced biotic pollination services due to effects such as pollinator usurpation by alien plants or illegitimate flower visitors, or pollinator predation by alien animals, can compensate for reduced visits by gradually shifting their pollination strategy to wind pollination. However, wind pollination may not be an effective strategy if pollen is robbed by alien flower visitors such as honeybees, which can efficiently locate pollen sources even in the absence of floral advertising structures, as is common in wind-pollinated plants. There is evidence that wind-pollinated species can be more resistant than animal-pollinated species to negative processes associated with species invasions such as habitat fragmentation; this could be related to increased levels of long-distance pollen dispersal in wind-pollinated species (Friedman and Barrett 2009).

15.4 Which species and populations of native plants are most likely to undergo adaptation, and in what direction?

Although plants have numerous different strategies to adapt to changing pollination regimes, not all plant populations are equally likely to undergo adaptive selection in response to species invasion. Several conditions have to be met to allow adaptation. Moreover, the unique adaptive path taken may change among different species and possibly even among different populations of the same species, depending on several environmental, demographic, genetic, and phylogenetic factors. In the following, we discuss the conditions that allow for adaptive selection to occur and the factors that determine its direction.

15.4.1 Conditions required for adaptive selection

Plant life cycle. For adaptation to occur, pollination and/or seed production must be a major limiting factor in the plant's lifecycle. In many cases, even a significant change in seed set has no effect on overall plant fitness (Ashman et al. 2004; Gomez and Zamora 2006). The life strategies of many flowering plants are characterized by the production of a vast amount of offspring per individual, the majority of which do not survive to maturity. The processes governing seed, seedling, and juvenile plant mortalities often surpass any effect the amount of seed sired might have on plant fitness, with any modest change in the pollination regime proving completely irrelevant for the demographics of the population (Ashman et al. 2004; Gomez and Zamora 2006). Furthermore, a trait that is advantageous for pollinator attraction or seed production may be disadvantageous for other life stages, such as seedling survival. For example, increased seed set is often associated with a decreased maternal investment per individual seed, potentially leading to decreased seedling survival (Gomez and Zamora 2006). Density-dependent processes, such as seed or seedling predation and intraspecific competition, can also eliminate any positive effect of increased pollination services (Ashman et al. 2004).

Pollen limitation. Most studies stress the importance of pollen limitation as a precondition for environmental effects on plant reproduction. In general, pollen-limited plants are expected to be more sensitive to changes in the pollination regime, since any change in pollinator type, abundance or behavior may influence the amount and/or quality of seeds sired. Furthermore, maternal selection for traits that reduce pollen limitation have been shown to vary positively with the intensity of the phenomenon (Harder and Aizen 2010). However, differences in the quality of seeds may also appear in plants that are seed- or resource-limited. For example, a plant that compensates for low visitation rates by delayed selfing can still achieve full seed set and remain pollen-unlimited, but if the species has a high rate of late-acting inbreeding depression, most of the progeny will not reach maturity, and fitness will decline (Harder and Aizen 2010). Furthermore, pollen-unlimited plants that experience increased visitation frequencies (e.g. because of facilitation by an invasive plant) can also increase their fitness by reducing their investment in advertisement and/or reward or by reducing flower longevity, and reallocating resources to other physiological processes (Harder and Aizen 2010).

Magnitude of alien species' invasion. Plant populations that are likely to adapt to species invasions are those that experience, on both spatial and temporal scales, significant and prolonged negative or positive effects due to these invasion events, such as pollen or resource limitation, and pollinator usurpation or facilitation (Harder and Aizen 2010). Thus, the entire plant population should be affected by the invasive species acting as a selective agent, and there should be relatively little gene flow from adjacent populations that are not under such influence. Therefore, adaptation is most likely where species invasion follows a uniform pattern across large areas, rather than a patchy pattern, as well as in well-isolated stands of the native species upon which selection can act.

Population size. The size and density of the plant population are of the utmost importance in determining its fate in the event of invasion. Small plant populations occupying anthropogenically transformed habitats, where alien species often predominate, are particularly prone to decline. Small populations are subject to Allee effects, which may also manifest in pollination, by reduced pollinator attraction, increased interspecific pollen movement, decreased mating opportunities, and increased inbreeding (Ashman et al. 2004). Furthermore, the smaller genetic reservoir available for small populations reduces the chances of successfully adapting to the changing environment. Thus, adaptive evolution is less likely to salvage small populations that are on the brink of extinction from the detrimental effects of species invasions.

15.4.2 Factors influencing the direction of adaptation

Among the different available adaptive mechanisms that optimize biotic pollination, adaptations that increase plant attractiveness to pollinators or compensate for illegitimate reward consumption often require increased allocation of resources to the floral tissues. However, the availability of such resources can be severely limited in habitats invaded by some alien species, especially plants (Levine et al. 2003). Furthermore, many showy invasive plants maintain an unusually high attractiveness to a wide range of pollinator species, which often far outcompetes that of native plant species (Morales and Traveset 2009); in some cases, rates of nectar or sugar production differ by an order of magnitude between natives and aliens (e.g. Chittka and Schürkens 2001). Under such harsh competitive conditions, it is unlikely that any modest increase in advertisement or reward in the native species will prevent pollinator usurpation by the alien. Rather, usurpation of pollinators by a highly attractive alien plant is more likely to select for an increasing investment in animal-independent reproductive modes, possibly coupled with a reduction in the amount of advertisement and reward.

Among the animal-independent reproductive adaptations, evolution of autonomous selfing is one of the most common transitions during angiosperm history (Harder and Aizen 2010), and is one of the chief paths that species, subjected to pollen or resource limitation, are likely to take. Several explanations can be given for why this adaptation is so common. To name only two: selfing strategies have a two-fold advantage over outcrossing strategies in the rates of genetic transmission; and autonomous selfing is a "safe bet" in most environments, since it does not depend on any outside vector for efficient pollination. Therefore, the genetic sequences that code for autonomous selfing may remain largely conserved within lineages, even after prolonged periods when they are not in use. Evolution of autonomous selfing is especially likely in invaded habitats, if both adequate pollen vectors and potential mates are in short supply (Eckert et al. 2009). A recent study estimated that increased selfing may evolve about three to four times more often than increased outcrossing in response to anthropogenic disturbance, especially in short-lived herbs (Harder and Aizen 2010).

Which type of selfing mechanism is most likely to evolve? According to some recent models, prior selfing is more likely to evolve than delayed selfing under pollen limitation, especially in annual species, and even with strong inbreeding depression (Harder and Aizen 2010). However, in species that have already acquired delayed selfing, when pollination services are improved and the opportunities for outcrossing increase, the presence of delayed selfing may slow down selection towards the optimal mating system, which should then rely more upon outcrossing (Harder and Aizen 2010).

Although a widespread mechanism, there are situations in which autonomous self-pollination is less likely to develop. The presence of strong inbreeding depression can prevent selection towards autonomous selfing (Harder and Aizen 2010). Some strategies of sex segregation are also unlikely to revert to allow selfing, especially dioecy and dicliny (flower unisexuality) (Culley et al. 2002; Friedman and Barrett 2009). Indeed, other modes of animal-independent reproduction are often associated with increased sex segregation. Thus, apomictic reproduction is especially important in self-incompatible, dioecious, and heterostylous taxa (Bicknell and Koltunow 2004; Barrett et al. 2008); and dioecious and diclinous lineages usually shift to wind pollination instead of autonomous selfing (Culley et al. 2002; Friedman and Barrett 2009). These adaptations may be viewed as alternative modes of reproductive assurance (Friedman and Barrett 2009).

The various adaptive paths followed by different plant species are also determined to a certain degree by the evolutionary history of the clade (Harder and Aizen 2010). Species are more likely to shift toward pollination modes that are common among their closely related taxa. However, phylogenetic evidence suggests that some transitions between reproductive modes tend to be irreversible. In general, shifts from partial or obligate animal pollination to obligate uniparental reproduction or wind pollination are rarely reversed (Harder and Aizen 2010; Culley et al. 2002). Moreover, being an absorbing state with reduced genetic diversity and accumulation of deleterious mutations, obligate selfing lineages, like obligate asexual lineages, are often short-lived and prone to frequent extinctions (Holsinger 2000; Harder and Aizen 2010). The shift from short-tongued to long-tongued pollination is also often irreversible (Harder and Aizen 2010). Thus, plants adapted for bird pollination that are visited by alien honeybees, for instance, are likely to resist radical shifts of their pollination syndrome, and be more susceptible to chronic pollen limitation. However, some degree of adaptation to the alien visitor may still be possible in such instances.

15.5 Conclusions

We have delineated different paths of adaptation that are available for native plant populations. Which of these alternative paths will be "chosen" by a given plant species suffering decreased reproductive output due to the invasion of aliens? As a rule of thumb, we suggest that plant species that are highly dependent on pollinator visits to achieve significant seed set, exhibit strong sexual segregation and/or inbreeding depression, and enjoy abundant abiotic resources and potential mating individuals, will maintain a reproductive mechanism that is dependent upon external vectors such as animals or wind, and will undergo selection to achieve optimal biotic or wind pollination; of these, wind pollination will prevail in dioecious and diclinous species with exposed sexual organs. On the other hand, species that are capable, to some extent, of reproducing without the aid of external pollen vectors (or have close relatives that do so), and occupy habitats that are severely resource-limited and with a low density of potential mates, will tend to increasingly rely on uniparental reproductive strategies, such as autonomous self-pollination, vegetative growth, and apomixis (see also Eckert et al. 2009).

Not all native plant species will survive the environmental changes induced by species invasions. Adaptive evolution can allow some plant species to meet their biotic pollination needs, thus protecting them from decline. Other species, especially those suffering high levels of competition or herbivory, will not be able to allocate enough resources to attract sufficient visitors in light of increased competition for pollinators. In some scenarios, wind pollination can provide a suitable alternative for these species. In other cases, adaptation will favor modes of reproduction that do not involve outcrossing, with consequent reductions in effective population size. Small, isolated populations, species with heavy inbreeding depression and those with strong self-incompatibility mechanisms, are expected to suffer the severest declines, with some populations and species reaching extinction. The reduction in gene flow among individuals in these instances will further limit the capacity to adapt to invaded environments. In the minority of cases, populations will increase due to facilitative effects of species invasions.

Pollination constitutes only one step in the lifecycle of plants. The more limiting the pollination step on the species' reproduction and survival, the more significant will be any change in the pollination regime induced by alien species. In addition to changing pollination regimes, alien species exert many other direct and indirect effects on native plants, such as interspecific competition, herbivory, and habitat modification (Levine et al. 2003; Crooks 2002). Many native plant populations have been affected by such processes, some reaching the brink of extinction (Coblentz 1990; Mooney and Cleland 2001). Efforts to overcome negative impacts by evolutionary adaptation are more likely to occur in populations that have maintained some genetic variability, and not in the weakest, most threatened ones. However, the potential for the occurrence of adaptive processes across different native taxa and invaded ecosystems has not been studied yet. Assessing which native plant species and communities have the capacity to adapt to species invasions, and which adaptation mechanisms are most likely to occur under different circumstances, should be a major goal for future research. Research in this field will improve our ability to manage pivotal pollination services and maintain functioning ecosystems.

Acknowledgements

We are grateful to Amots Dafni, Yuval Sapir, and an anonymous reviewer for their helpful comments on the manuscript. GP is supported by the Robert H. Smith Fellowship and the Whole Organism Fellowship at the Hebrew University of Jerusalem.

References

Abe, T., Makino, S. and Okochi, I. (2010). Why have endemic pollinators declined on the Ogasawara Islands? In *Restoring the Oceanic Island Ecosystem: Impact and Management of Invasive Alien Species in the Bonin Islands*, ed. K. Kawakami and I. Okochi. Tokyo, Japan: Springer, pp. 75–83.

Ashman, T. and Morgan, M. T. (2004). Explaining phenotypic selection on plant attractive characters: male function, gender balance or ecological context? *Proceedings of the Royal Society of London B*, **271**, 553–9.

Ashman, T., Knight, T. M., Steets, J. A., Amarasekare, P., Burd, M., Campbell, D. R., Dudash, M. R., Johnston, M. O., Mazer, S. J., Mitchell, R. J., Morgan, M. T. and Wilson, W. G. (2004). Pollen limitation of plant reproduction: ecological and evolutionary causes and consequences. *Ecology*, **85**, 2408–21.

Barrett, S. C. H., Colautti, R. I. and Eckert, C. G. (2008). Plant reproductive systems and evolution during biological invasion. *Molecular Ecology*, **17**, 373–83.

Barrett, S. C. H., Ness, R. W. and Vallejo-Marín, M. (2009). Evolutionary pathways to self-fertilization in a tristylous plant species. *New Phytologist*, **183**, 546–56.

Bax, N., Williamson, A., Aguero, M., Gonzalea, E. and Geeves, W. (2003). Marine invasive alien species: a threat to global biodiversity. *Marine Policy*, **27**, 313–23.

Bernardello, G., Andreson, G. J., Stuessy, T. F. and Crawford, D. J. (2001). A survey of floral traits, breeding systems, floral visitors, and pollination systems of the angiosperms of the Juan Fernández Islands (Chile). *The Botanical Review*, **67**, 255–308.

Bicknell, R. A. and Koltunow, A. M. (2004). Understanding apomixis: recent advances and remaining conundrums. *The Plant Cell*, **16**, S228–S245.

Bjerknes, A., Totland, Ø., Hegland, S. J. and Nielsen, A. (2007). Do alien plant invasions really affect pollination success in native plant species? *Biological Conservation*, **138**, 1–12.

Blancafort, X. and Gómez, C. (2005). Consequences of the Argentine ant, *Linepithema humile* (Mayr), invasion on pollination of *Euphorbia characias* (L.) (Euphorbiaceae). *Acta Oecologica*, **28**, 49–55.

Bond, W. J. (1994). Do mutualisms matter? Assessing the impact of pollinator and disperser disruption on plant extinction. *Philosophical Transactions of the Royal Society of London B*, **344**, 83–90.

Brown, B. J., Mitchell, R. J. and Graham, S. A. (2002). Competition for pollination between an invasive species (purple loosestrife) and a native congener. *Ecology*, **83**, 2328–36.

Cane, J. H. (2003). Exotic nonsocial bees (Hymenoptera:Apiformes) in North America: ecological implications. In *For Nonnative Crops, Whence Pollinators of the Future?* ed. K. Strickler and J. H. Cane. Lanham, MD: Entomological Society of America, pp. 113–26.

do Carmo, R. M., Franceschinelli, E. D. and Silveira, F. A. (2004). Introduced honeybees (*Apis mellifera*) reduce pollination success without affecting the floral resource taken by native pollinators. *Biotropica*, **36**, 371–6.

Caruso, C. M. (2000). Competition for pollination influences selection on floral traits of *Ipomopsis aggregata*. *Evolution*, **54**, 1546–57.

Celebrezze, T. and Paton, D. C. (2004). Do introduced honeybees (*Apis mellifera*, Hymenoptera) provide full pollination service to bird-adapted Australian plants with small flowers? An experimental study of *Brachyloma ericoides* (Epacridaceae). *Austral Ecology*, **29**, 129–36.

Chamberlain, S. A. and Schlising, R. A. (2008). Role of honey bees (Hymenoptera:Apidae) in the pollination biology of a California native plant, *Triteleia laxa* (Asparagales:Themidaceae). *Environmental Entomology*, **37**, 808–16.

Chittka, L. and Schürkens, S. (2001). Successful invasion of a floral market: an exotic Asian plant has moved in on Europe's riverbanks by bribing pollinators. *Nature*, **411**, 653.

Coblentz, B. E. (1990). Exotic organisms: a dilemma for conservation biology. *Conservation Biology*, **4**, 261–5.

Colwell, R. K., Betts, B. J., Bunnell, P., Carpenter, F. L. and Feinsinger, P. (1974). Competition for the nectar of *Centropogon valerii* by the hummingbird *Colibri thalassinus* and the flower-piercer *Diglossa plumbea*, and its evolutionary implications. *The Condor*, **76**, 447–52.

Cox, G. W. (2004). *Alien Species and Evolution: The Evolutionary Ecology of Exotic Plants, Animals, Microbes, and Interacting Native Species*. Washington, DC: Island Press.

Cox, P. A. (1983). Extinction of the Hawaiian avifauna resulted in a change of pollinators for the ieie, *Freycinetia arborea*. *Oikos*, **41**, 195–9.

Cox, P. A. and Elmqvist, T. (2000). Pollinator extinction in the Pacific Islands. *Conservation Biology*, **14**, 1237–9.

Crooks, J. A. (2002). Characterizing ecosystem-level consequences of biological invasions: the role of ecosystem engineers. *Oikos*, **97**, 153–66.

Culley, T. M., Weller, S. G. and Sakai, A. K. (2002). The evolution of wind pollination in angiosperms. *Trends in Ecology and Evolution*, **17**, 361-9.

Dafni, A. (1984). Mimicry and deception in pollination. *Annual Review of Ecology and Systematics*, **15**, 259-78.

Dafni, A. and Ivri, Y. (1981a). Floral mimicry between *Orchis israelitica* Baumann and Dafni (Orchidaceae) and *Bellevalia flexuosa* Boiss. (Liliaceae). *Oecologia*, **49**, 229-32.

Dafni, A. and Ivri, Y. (1981b). The flower biology of *Cephalanthera longifolia* (Orchidaceae) – pollen limitation and facultative floral mimicry. *Plant Systematics and Evolution*, **137**, 229-240.

Dafni, A. and Shmida, A. (1996). The possible ecological implications of the invasion of *Bombus terrestris* (L.) (Apidae) at Mt. Carmel, Israel. In *The Conservation of Bees,* ed. A. Matheson. London, UK: The Linnean Society of London and the International Bee Research Association, pp. 183-200.

Dick, C. W., Etchelecu, G. and Austerlitz, F. (2003). Pollen dispersal of tropical trees (*Dinizia excelsa*: Fabaceae) by native insects and African honeybees in pristine and fragmented Amazonian rainforest. *Molecular Ecology*, **12**, 753-64.

Dohzono, I. and Yokoyama, J. (2010). Impacts of alien bees on native plant–pollinator relationships: a review with special emphasis on plant reproduction. *Applied Entomology and Zoology*, **45**, 37-47.

Dohzono, I., Kunitake, Y. K., Yokoyama, J. and Goka, K. (2008). Alien bumblebee affects native plant reproduction through interactions with native bumblebees. *Ecology*, **89**, 2082-3092.

Dupont, Y. L., Hansen, D. M., Valido, A. and Olesen, J. M. (2004). Impact of introduced honey bees on native pollination interactions of the endemic *Echium wildpretii* (Boraginaceae) on Tenerife, Canary Islands. *Biological Conservation*, **118**, 301-11.

Eckert, C. G. (2002). The loss of sex in clonal plants. *Evolutionary Ecology*, **15**, 501-20.

Eckert, C. G., Kalisz, S., Geber, M. A., Sargent, R., Elle, E., Cheptou, P.-O., Goodwillie, C., Johnston, M. O. Kelly, J. K., Moeller, D. A., Porcher, E., Ree, R. H., Vallejo-Marin, M. and Winn, A. A. (2009). Plant mating systems in a changing world. *Trends in Ecology and Evolution*, **25**, 35-43.

England, P. R., Beynon, F., Ayre, D. J. and Whelan, R. J. (2001). A molecular genetic assessment of mating-system variation in a naturally bird-pollinated shrub: contributions from birds and introduced honeybees. *Conservation Biology*, **15**, 1645-55.

Fischer, M. and van Kleunen, M. (2002). On the evolution of clonal plant life histories. *Evolutionary Ecology*, **15**, 565-82.

Fishman, L. and Wyatt, R. (1999). Pollinator-mediated competition, reproductive character displacement, and the evolution of selfing in *Arenaria uniflora* (Caryophyllaceae). *Evolution*, **53**, 1723-33.

Flanagan, R. J., Mitchell, R. J., Knutowski, D. and Karron, J. D. (2009). Interspecific pollinator movements reduce pollen deposition and seed production in *Mimulus ringens* (Phrymaceae). *American Journal of Botany*, **96**, 809-15.

Friedman, J. and Barrett, S. C. H. (2009). Wind of change: new insights on the ecology and evolution of pollination and mating in wind-pollinated plants. *Annals of Botany*, **103**, 1515-27.

Fritts, T. H. and Rodda, G. H. (1998). The role of introduced species in the degradation of island ecosystems: a case history of Guam. *Annual Review of Ecology and Systematics*, **29**, 113–40.

Fumero-Cabán, J. J. and Meléndez-Ackerman, E. (2007). Relative pollination effectiveness of floral visitors of *Pitcairnia angustifolia* (Bromeliaceae). *American Journal of Botany*, **94**, 419–24.

Ghazoul, J. (2002). Flowers at the front line of invasion? *Ecological Entomology*, **27**, 638–40.

Gomez, J. M. and Zamora, R. (2006). Ecological factors that promote the evolution of generalization in pollination systems. In *Plant–Pollinator Interactions – from Specialization to Generalization*, ed. N. M. Waser and J. Ollerton. Chicago, IL: The University of Chicago Press, pp. 145–66.

Goodell, K. (2008). Invasive exotic plant-bee interactions. In *Bee Pollination in Agricultural Ecosystems*, ed. R. R. James and T. L. Pitts-Singer. New York, NY: Oxford University Press.

Grabas, G. P. and Laverty, T. M. (1999). The effect of purple loosestrife (*Lythrum salicaria* L.; Lythraceae) on the pollination and reproductive success of sympatric coflowering wetland plants. *Ecoscience*, **6**, 230–42.

Graves, S. D. and Shapiro, A. M., (2003). Exotics as host plants of the California butterfly fauna. *Biological Conservation*, **110**, 413–33.

Greenleaf, S. S. and Kremen, C. (2006). Wild bees enhance honeybees' pollination of hybrid sunflower. *Proceedings of the National Academy of Sciences USA*, **103**, 13890–5.

Gross, C. L. and Mackay, D. (1998). Honeybees reduce fitness in the pioneer shrub *Melastoma affine* (Melastomataceae). *Biological Conservation*, **86**, 169–78.

Hansen, D. M. and Müller, C. B. (2009). Invasive ants disrupt gecko pollination and seed dispersal of the endangered plant *Roussea simplex* in Mauritius. *Biotropica*, **41**, 202–8.

Hansen, D. M., Olesen, J. M. and Jones, C. G. (2002). Trees, birds and bees in Mauritius: exploitative competition between introduced honeybees and endemic nectarivorous birds? *Journal of Biogeography*, **29**, 721–34.

Harder, L. D. and Aizen, M. A. (2010). Floral adaptation and diversification under pollen limitation. *Philosophical Transactions of the Royal Society B*, **365**, 529–43.

Hingston, A. B. and McQuillan, P. B. (1999). Displacement of Tasmanian native megachilid bees by the recently introduced bumblebee *Bombus terrestris* (Linnaeus, 1758) (Hymenoptera:Apidae). *Australian Journal of Zoology*, **47**, 59–65.

Holsinger, K. E. (2000). Reproductive systems and evolution in vascular plants. *Proceedings of the National Academy of Sciences USA*, **97**, 7037–42.

Howarth, F. G. (1985). The impacts of alien land arthropods and mollusks on native plants and animals. In *Hawaii's Terrestrial Ecosystems: Protection and Management*, ed. C. P. Stone and J. M. Scott. Honolulu, HI: University of Hawaii Press.

Hurd, P. D. (1978). Bamboo-nesting carpenter bees (Genus *Xylocopa* Latreille) of the subgenus *Stenoxylocopa* Hurd and Moure (Hymenoptera:Anthophoridae). *Journal of the Kansas Entomological Society*, **51**, 746–64.

Inoue, M. N., Yokoyama, J. and Washitani, I. (2008). Displacement of Japanese native bumblebees by the recently introduced *Bombus terrestris* (L.)

(Hymenoptera:Apidae). *Journal of Insect Conservation*, **12**, 135–46.

Internicola, A. I., Page, P. A., Bernasconi, G. and Gigord, L. D. B. (2007). Competition for pollinator visitation between deceptive and rewarding artificial inflorescences: an experimental test of the effects of floral colour similarity and spatial mingling. *Functional Ecology*, **21**, 864–72.

Jakobsson, A., Padrón, B. and Traveset, A. (2009). Competition for pollinators between invasive and native plants: effects of spatial scale of investigation (note). *Ecoscience* **16**, 138–41.

Johnson, R. L. (2008). *Impacts of Habitat Alterations and Predispersal Seed Predation on the Reproductive Success of Great Basin Forbs*. PhD Diss., Brigham Young University.

Johnson, S. D. (2006). Pollinator-driven speciation in plants. In *Ecology and Evolution of Flowers,* ed. L. D. Harder and S. C. H. Barrett. Oxford, UK: Oxford University Press.

Johnson, S. D. and Steiner, K. E. (2000). Generalization versus specialization in plant pollination systems. *Tree*, **15**, 140–3.

Johnson, S. D., Peter, C. I., Nilsson, L. A. and Aegren, J. (2003). Pollination success in a deceptive orchid is enhanced by co-occurring rewarding magnet plants. *Ecology*, **84**, 2919–27.

Kanbe, Y., Okada, I., Yoneda, M., Goka, K. and Tsuchida, K. (2008). Interspecific mating of the introduced bumblebee *Bombus terrestris* and the native Japanese bumblebee *Bombus hypocrita sapporoensis* results in inviable hybrids. *Naturwissenschaften*, **95**, 1003–8.

Kandori, I., Hirao, T., Matsunaga, S. and Kurosaki, T. (2009). An invasive dandelion unilaterally reduces the reproduction of a native congener through competition for pollination. *Oecologia*, **159**, 559–69.

Kelly, D., Robertson, A. W., Ladley, J. J., Andreson, S. H. and McKenzie, R. J. (2006). Relative (un)importance of introduced animals as pollinators and disperses of native plants. In *Biological Invasions in New Zealand*, ed. R. B. Allen and W. G. Lee. Berlin, Germany: Springer, pp. 227–45.

Kenta, T. Inari, N., Nagamitsu, T., Goka, K. and Hiura, T. (2007). Commercialized European bumblebee can cause pollination disturbance: an experiment on seven native plant species in Japan. *Biological Conservation*, **134**, 298–309.

Kondo, N. I., Yamanaka, D., Kanbe, Y., Kunitake, Y. K., Yoneda, M., Tsuchida, K. and Goka, K. (2009). Reproductive disturbance of Japanese bumblebees by the introduced European bumblebee *Bombus terrestris*. *Naturwissenschaften*, **96**, 467–75.

Lach, L. (2003). Invasive ants: unwanted partners in ant–plant interactions? *Annals of the Missouri Botanical Garden*, **90**, 91–108.

Lach, L. (2005). Interference and exploitation competition of three nectar-thieving invasive ant species. *Insectes Sociaux*, **52**, 257–62.

Lach, L. (2007). A mutualism with a native membracid facilitates pollinator displacement by Argentine ants. *Ecology*, **88**, 1994–2004.

Lach, L. (2008a). Argentine ants displace floral arthropods in a biodiversity hotspot. *Diversity and Distributions*, **14**, 281–90.

Lach, L. (2008b). Floral visitation patterns of two invasive ant species and their effects on other hymenopteran visitors. *Ecological Entomology*, **33**, 155–60.

Larson, D. L., Royer, R. A. and Royer, M. R. (2006). Insect visitation and pollen deposition in an invaded prairie plant community. *Biological Conservation*, **130**, 148–59.

Levine, J. M., Vilà, M., D'Antonio, C. M. et al. (2003). Mechanisms underlying the impacts of exotic plant invasions. *Proceedings of the Royal Society of London B*, **270**, 775–81.

Lord, J. M. (1991). Pollination and seed dispersal in *Freycinetia baueriana*, a dioecious liane that has lost its bat pollinator. *New Zealand Journal of Botany*, **29**, 83–6.

Madjidian, J. A., Morales, C. L. and Smith, H. G. (2008). Displacement of a native by an alien bumblebee: lower pollinator efficiency overcome by overwhelmingly higher visitation frequency. *Oecologia*, **156**, 835–45.

Maloof, J. E. and Inouye, D. W. (2000). Are nectar robbers cheaters or mutualists? *Ecology*, **81**, 2651–61.

Marshall, D. R. and Brown, A. H. D. (1981). The evolution of apomixis. *Heredity*, **47**, 1–15.

Montgomery, B. R. (2009). Effect of introduced *Euphorbia esula* on the on the pollination of *Viola pedatifida*. *Botany*, **87**, 283–92.

Mooney, H. A. and Cleland, E. E. (2001). The evolutionary impact of invasive species. *Proceedings of the National Academy of Sciences USA*, **98**, 5446–51.

Moragues, E. and Traveset, A. (2005). Effect of *Carpobrotus* spp. on the pollination success of native plant species of the Balearic Islands. *Biological Conservation* **122**, 611–9.

Morales, C. A. and Aizen, M. A. (2002). Does invasion of exotic plants promote invasion of exotic flower visitors? A case study from the temperate forests of the southern Andes. *Biological Invasions*, **4**, 87–100.

Morales, C. A. and Traveset, A. (2009). A meta-analysis of impacts of alien versus native plants on pollinator visitation and reproductive success of coflowering native plants. *Ecology Letters*, **12**, 716–28.

Moroń, D., Lenda, M., Skórka, P., Szentgyorgyi, H., Settele, J. and Woyciechowski, M. (2009). Wild pollinator communities are negatively affected by invasion of alien goldenrods in grassland landscapes. *Biological Conservation*, **142**, 1322–32.

Muñoz, A. A. and Cavieres, L. A. (2008). The presence of a showy invasive plant disrupts pollinator service and reproductive output in native alpine species only at high densities. *Journal of Ecology*, **96**, 459–67.

Nielsen, C., Heimes, C. and Kollmann, J. (2008). Little evidence for negative effects of an invasive alien plant on pollinator services. *Biological Invasions*, **10**, 1353–63.

Nienhuis, C. M., Dietzsch, A. C. and Stout, J. C. (2009). The impacts of an invasive alien plant and its removal on native bees. *Apidologie*, **40**, 450–63.

Nogales, M. and Medina, F. M. (1996). A review of the diet of feral domestic cats (*Felis silvestris* f. *catus*) on the Canary Islands, with new data from the laurel forest of La Gomera. *Z. Saugertierkunde*, **61**, 1–6.

Osorio-Beristain, M., Domínguez, C. A., Eguiarte, L. E. and Benrey, B. (1997). Pollination efficiency of native and invading Africanized bees in the tropical dry forest annual plant, *Kallstroemia grandiflora* Torr ex Gray. *Apidologie*, **28**, 11–6.

Paini, D. R. and Roberts, J. D. (2005). Commercial honey bees (*Apis mellifera*) reduce the fecundity of an Australian native bee (*Hylaeus alcyoneus*). *Biological Conservation*, **123**, 103–12.

Paton, D. C. (2000). Disruption of bird–plant pollination systems in Southern Australia. *Conservation Biology*, **14**, 1232–4.

Pemberton, R. W. and Liu, H. (2008). Naturalization of the oil collecting bee *Centris nitida* (Hymenoptera, Apidae, Centrini), a potential pollinator of selected native, ornamental, and invasive plants in Florida. *Florida Entomologist*, **91**, 101–9.

Roberts, D. L. and McGlynn, T. P. (2004). *Tetramorium insolens* Smith (Hymenoptera:Formicidae): a new record for Mauritius, Indian Ocean. *African Entomology* **12**, 265–7.

Roubik, D. W. (1978). Competitive interactions between Neotropical pollinators and Africanized honey bees. *Science*, **201**, 1030–2.

Roubik, D. W. and Villanueva-Gutiérrez, R. (2009). Invasive Africanized honey bee impact on native solitary bees: a pollen resource and trap nest analysis. *Biological Journal of the Linnean Society*, **98**, 152–60.

Schemske, D. W. (1981). Floral convergence and pollinator sharing in two bee-pollinated tropical herbs. *Ecology*, **62**, 946–54.

Schweiger, O., Biesmeijer, J. C., Bommarco, R., Hickler, T., Hulme, P. E., Klotz, S., Kühn, I., Moora, M., Nielsen, A., Ohlemüller, R., Petanidou, T., Potts, S. G., Pyšek, P., Stout, J. C., Sykes, M. T., Tscheulin, T., Vilà, M., Walther, G. R., Westphal, C., Winter, M., Zobel, M. and Settele J. (2010). Multiple stressors on biotic interactions: how climate change and alien species interact to affect

pollination. *Biological Reviews*, DOI:10.1111/j.1469-185X.2010.00125.x

Spurr, E. B. and Anderson, S. H. (2004). Bird species diversity and abundance before and after eradication of possums and wallabies on Rangitoto Island, Hauraki Gulf. *New Zealand Journal of Ecology*, **28**, 143–9.

Stubbs, C. S., Drummond, F. and Ginsberg, H. (2007). *Effects of invasive plant species on pollinator service and reproduction in native plants at Acadia National Park*. Technical Report NPS/NER/NRTR – 2007/096, National Park Service, US Department of the Interior.

Tepedino, V. J., Bradley, B. A. and Griswold, T. L. (2008). Might flowers of invasive plants increase native bee carrying capacity? Intimations from Capitol Reef National Park, Utah. *Natural Areas Journal*, **28**, 44–50.

Thomson, D. (2004). Competitive interactions between the invasive European honeybee and native bumblebees. *Ecology*, **85**, 458–470.

Traveset, A. and Richardson, D. M. (2006). Biological invasions as disruptors of plant reproductive mutualisms. *Trends in Ecology and Evolution*, **21**, 208–16.

Tscheulin, T., Petanidou, T., Potts, S. G. and Settele, J. (2009). The impact of *Solanum elaeagnifolium*, an invasive plant in the Mediterranean, on the flower visitation and seed set of the native coflowering species *Glaucium flavum*. *Plant Ecology*, **205**, 77–85.

Vanparys, V., Meerts, P. and Jacquemart, A. (2008). Plant–pollinator interactions: comparison between an invasive and a native congeneric species. *Acta Oecologica*, **34**, 361–9.

Vergara, C. H. (2008). Environmental impact of exotic bees introduced for crop pollination. In *Bee Pollination in Agricultural Ecosystems*, ed. R. R. James and T. L. Pitts-Singer. New York, NY: Oxford University Press.

Vilà, M., Basnou, C., Pyšek, P., Josefsson, M., Genovesi, P., Gollasch, S., Nentwig, W., Olenin, S., Roques, A., Roy, D., Hulme, P. E. and DAISIE partners. (2010). How well do we understand the impacts of alien species on ecosystem services? A pan-European cross-taxa assessment. *Frontiers in Ecology and the Environment*, **8**, 135–44.

Waser, N. M. (1978). Interspecific pollen transfer and competition between co-occurring plant species. *Oecologia*, **36**, 223–36.

Willmer, P. G. and Corbet, S. A. (1981). Temporal and microclimatic partitioning of the floral resources of *Justicia aurea* amongst a concourse of pollen vectors and nectar robbers. *Oecologia*, **51**, 67–78.

16

Pollen resources of non-*Apis* bees in southern Africa

MICHAEL KUHLMANN AND CONNAL D. EARDLEY

16.1 Introduction

Southern Africa, which is the region south of the Rivers Cunene and Zambezi, is one of the world's bee diversity hotspots (Kuhlmann 2009). As bees are the most important pollinators of flowering plants, including crops, they are ecological and economic keystone species (Corbet et al. 1991; Allen-Wardell et al. 1998; Klein et al. 2007). Pollinators are believed to have played an important role in plant speciation in southern Africa, especially in the Cape Floral Kingdom (Kreft and Jetz 2007; van der Niet and Johnson 2008; Waterman et al. 2008). Notwithstanding the great economic, ecological, and evolutionary significance of wild bees, knowledge of this important group of pollinators and their floral relationships in this region is poor.

Struck (1990, 1994a, 1994b, 1995) was the first to extensively study the relationships between flowers and solitary bees in southern Africa. In his pioneering work in Namaqualand, he used pollen analyses for recording flower visitation to investigate flower specialization of bees and other flower visiting insects. But due to taxonomic uncertainties in many of the bee genera in those days, the published data is of limited usefulness without the re-examination of the specimens, and thus it is not considered here.

Since that time, bee taxonomy has progressed considerably (summarized in Kuhlmann 2009) and substantial amounts of new flower-related data have become available. The most important and extensive data resource on flower visitation by southern African bees are the detailed observations of Sarah and Friedrich Gess over the past 40 years. This data was recorded in an electronic database that is continuously updated, and much of it has been published (Gess and Gess 2003, 2004a, 2006 and references therein). All the observations are completely documented by voucher specimens that are housed in the Albany Museum, Grahamstown, making it a most valuable source of information. Additional records of flower visitation, mostly without details about the sex of the bees or further circumstances, are widely scattered in the bee taxonomic literature and summarized in Eardley and Urban (2010).

The purpose of this paper is to give an overview of the data (pollen analyses and flower visiting records) that are available for the bee genera of southern Africa. It will also summarize current knowledge on bee–flower specialization. The geographical and taxonomic gaps in our knowledge are identified. And lastly, the informative value of pollen analyses versus flower visitation records, using data from southern Africa, is discussed.

Detailed information about pollen collecting and flower visitation preferences of single bee species will be published elsewhere.

16.2 Taxonomic framework

Bee classification on family and genus levels follows Michener (2007). The classification and nomenclature of flowering plant families follows that of the Angiosperm Phylogeny Group (APG III 2009). Information about species numbers and occurrence of bee genera in southern Africa have been extracted from Eardley and Urban (2010), Kuhlmann (2009), Pauly (2008), and Timmermann and Kuhlmann (2009). For this paper, only pollen-collecting bees are considered and cleptoparasitic species have been omitted.

Terminology and classification of flower specialization of pollen collection follows Müller and Kuhlmann (2008) and Murray et al. (2009). The term "eclectic oligolecty" is not used here because in a number of cases pollen could not be reliably identified to plant genus level. Bees potentially belonging to this category are summarized as mesolectic instead. The classification of flower specialization is purely based on pollen samples and not on flower visitation records. Bees are only classified when at least ten pollen samples were available or, in case of less than ten samples, when the samples originated from at least five different localities. For comparison of the informative value of pollen analyses versus flower visitation records, bee species have been selected with both > 15 pollen samples and > 15 flower visitation records.

16.3 Origin of flower visitation records

Flower visitation records of bees have been extracted from two sources: first, from the database "A catalogue of flower visiting records for aculeate wasps and bees in the semi-arid to arid areas of southern Africa" (version May 3, 2010) compiled by F. W. and S. K. Gess (Department of Entomology and Arachnology, Albany Museum in Grahamstown). Origin and collection of data as well as the structure of the database are described in detail by Gess and Gess (2003, 2004a, 2006). Here only data on female bees, identified to species, have been used.

Second, data has been taken from Eardley and Urban (2010), who exhaustively compiled bee–flower visitation records from a broad range of sources in their catalogue of Afrotropical bees. However, the visitation records are neither related to the sex of the bee, locality, or date nor do they specify how frequently observations have been made (e.g. how many specimens were involved). This information, if at all, is only available in the original sources cited by Eardley and Urban (2010), limiting its informative value and, thus has not been analyzed in detail. Also, because no geographical origin of data is given for widespread bee species, flower visitation records from Central and East Africa, with its distinct flora, are included and significantly broadening the flora-host records beyond the southern African flora.

16.4 Data on pollen collecting

Data about the pollen composition of female scopal pollen loads are mainly based on Timmermann (2005) and subsequent unpublished pollen analyses that have been partly published (Pauly et al. 2008; Timmermann and Kuhlmann 2008; Kuhlmann and Timmermann 2009). Additional sources of information are available for the bee genera *Capicola, Meganomia, Haplomelitta* (Michez et al. 2008, 2010 and unpublished data), and *Colletes* (Kuhlmann 2006 and unpublished data).

16.5 Geographical, phenological and taxonomic data coverage for southern Africa

In southern Africa, 89 genera of non-*Apis* bees (1416 species) are found, of which 19 genera represent cleptoparasitic cuckoo-bees (159 described species). They have been omitted, leaving a total of 1259 described pollen-collecting bee species (Table 16.1) in southern Africa. Flower visitation records are available for 80 bee

Table 16.1 List of pollen-collecting bee genera known to occur in southern Africa including number of described species (Kuhlmann, 2009; Eardley and Urban, 2010), flower visitation (FV) recorded in the database of Sarah and Friedrich Gess (G) and from other sources (O) summarized in Eardley & Urban (2010) for N bee species (N FV / for N species), pollen analyses (PA) of female scopal loads (N PA / for N species), and host-plant preferences. Bees of the genus *Hylaeus* transport pollen internally in their crop, making pollen analysis impossible without dissection (n/a). Flower visitation records of bees identified to genus level only are not included in the table. In a number of bee genera, flower visitation records and pollen analyses are available for undescribed species. Except for *Colletes* (comprehensive taxonomic data on new species available) they have been excluded when the proportion (%) of available data in relation to the number of species was calculated. Definition of host-plant preferences are according to the terminology used by Müller & Kuhlmann (2008) and Murray et al. (2009) and are only based on pollen analyses. For abbreviations of plant family names see Table 16.4. mono: monolecty, noligo: narrow oligolecty, boligo: broad oligolecty, meso: mesolecty, polyp: polylecty with strong preference, polys: polylecty sensu stricto.

Genus	N Species	FV(G + O) / N	%	PA / N	%	Mono	Noligo	Boligo	Meso	Polyp	Polys
Andrenidae	**12**										
Pollen collecting	**12**	**(257 + 24) / 12**	**100**	**0.0**	**0.0**	**0**	**0**	**0**	**0**	**0**	**0**
Andrena	1	(0 + 9) / 1	100	–	–	–	–	–	–	–	–
Melitturga	3	(12 + 2) / 3	100	–	–	–	–	–	–	–	–
Meliturgula	7	(239 + 12) / 7	100	–	–	–	–	–	–	–	–
Mermiglossa	1	(6 + 1) / 1	100	–	–	–	–	–	–	–	–
Apidae	**280**										
Pollen collecting	**213**	**(1484 + 804) / 114**	**54**	**57 / 12**	**5.6**	**0**	**0**	**1 Ast**	**0**	**0**	**1**
Allodape	17	(33 + 6) / 7	41	3 / 1	6	–	–	–	–	–	–
Allodapula	13	(25 + 0) / 2	15	–	–	–	–	–	–	–	–
Amegilla	31	(364 + 170) / 20	65	5 / 2	7	–	–	–	–	–	–
Anthophora	31	(171 + 71) / 19	61	22 / 3	10	–	–	–	–	–	1
Braunsapis	17	(144 + 88) / 8	47	–	–	–	–	–	–	–	–

Ceratina	37	(233 + 93) / 17	46	7 / 3	8	–	–	–	–	–
Compsomelissa	1	(4 + 0) / 1	100	–	–	–	–	–	–	–
Ctenoplectra	4	(0 + 2) / 2	50	–	–	–	–	–	–	–
Hypotrigona	3	–	–	–	–	–	–	–	–	–
Liotrigona	2	(0 + 48) / 2	100	–	–	–	–	–	–	–
Macrogalea	1	(0 + 3) / 1	100	–	–	–	–	–	–	–
Meliponula	3	(21 + 84) / 3	100	–	–	–	–	–	–	–
Pachymelus	3	(62 + 5) / 2	67	–	–	–	–	–	–	–
Plebeina	1	(0 + 2) / 1	100	–	–	–	–	–	–	–
Tetralonia	6	(10 + 4) / 2	33	–	–	–	–	–	–	–
Tetraloniella	19	(49 + 28) / 9	47	18 / 1	5	–	–	1 Ast	–	–
Xylocopa	24	(368 + 197) / 18	75	2 / 2	8	–	–	–	–	–
Colletidae	202									
Pollen collecting	202	(352 + 116) / 65	32	383 / 51	33.3[1]	0	1 *Oxalis*	4 Ast	1	1 Aiz, 2 4 Ast
Colletes[2]	101	(88 + 67) / 29	29	261 / 38	38	–	–	3 Ast	–	1 Aiz, 1 4 Ast
Hylaeus	54	(1 + 12) / 8	15	n / a	n/a	–	–	–	–	–
Scrapter	43	(263 + 37) / 28	65	122 / 13	16	–	1 *Oxalis*	1 Ast	1	1 Ast
Halictidae	384									
Pollen collecting	361	(618 + 556) / 93	26	548 / 31	7.5	0	0	1 Ast	5	1 Aiz, 1 2 Ast

Table **16.1** (cont.)

| Genus | N Species | FV(G + O) / N | % | PA / N | % | Host plant preferences | | | | | |
						Mono	Noligo	Boligo	Meso	Polyp	Polys
Cellariella	4	(0 + 4) / 2	50	–	–	–	–	–	–	–	–
Ceylalictus	3	(0 + 42) / 2	67	1 / 1	33	–	–	–	–	–	–
Halictus	11	(0 + 48) / 3	27	73 / 5	46	–	–	–	1	1 Ast	1
Lasioglossum	133	(0 + 102) / 18	14	114 / 10	5	–	–	–	2	–	–
Lipotriches	68	(0 + 164) / 22	32	–	–	–	–	–	–	–	–
Nomia	20	(167 + 90) / 14	70	–	–	–	–	–	–	–	–
Nomioides	1	(0 + 3) / 1	100	2 / 1	100	–	–	–	–	–	–
Patellapis	95	(240 + 21) / 19	20	351 / 13	13	–	–	1 Ast	2	1 Aiz	1
Pseudapis	13	(211 + 69) / 8	62	7 / 1	8	–	–	–	–	–	–
Spatunomia	1	–	–	–	–	–	–	–	–	–	–
Steganomus	1	(0 + 1) / 1	100	–	–	–	–	–	–	–	–
Systropha	5	(0 + 1) / 1	20	–	–	–	–	–	–	–	–
Thrinchostoma	5	(0 + 11) / 2	40	–	–	–	–	–	–	–	–
Megachilidae	**484**										
Pollen collecting	**417**	**(1154 + 463) / 141**	**34**	**213 / 40**	**4.8**	**0**	**0**	**5 Ast**	**2**	**1 Fab**	**0**
Afranthidium	41	(126 + 20) / 17	41	33 / 11	10	–	–	1 Ast	–	–	–
Afroheriades	6	(3 + 11) / 4	67	7 / 3	33	–	–	–	–	–	–
Anthidiellum	13	–	–	–	–	–	–	–	–	–	–

Genus											
Anthidioma	2	–	–	–	–	–	–	–	–	–	–
Anthidium	13	(2 + 2) / 2	15	–	–	–	–	–	–	–	–
Aspidosmia	2	(23 + 3) / 2	100	1 / 1	50	–	–	–	–	–	–
Cyphanthidium	2	(3 + 0) / 1	50	–	–	–	–	–	–	–	–
Eoanthidium	3	(7 + 3) / 1	33	–	–	–	–	–	–	–	–
Fidelia	11	(159 + 51) / 10	91	–	–	–	–	–	–	–	–
Gnathanthidium	1	–	–	–	–	–	–	–	–	–	–
Heriades	59	(11 + 21) / 7	12	13 / 3	0	–	1 Ast	–	–	–	–
Hoplitis	19	(6 + 6) / 3	16	13 / 7	5	–	1 Ast	–	–	–	–
Lithurgus	3	(24 + 2) / 2	67	–	–	–	–	–	–	–	–
Megachile	192	(636 + 267) / 69	36	66 / 7	2	–	1	–	–	–	–
Noteriades	8	(0 + 1) / 1	13	–	–	–	–	–	–	–	–
Ochreriades	1	–	–	–	–	–	–	–	–	–	–
Othinosmia	11	(35 + 23) / 7	64	26 / 3	27	–	2 Ast	–	–	–	–
Pachyanthidium	6	(3 + 3) / 2	33	–	–	–	–	–	–	–	–
Plesianthidium	7	(69 + 36) / 7	100	50 / 4	57	–	1	1 Fab	–	–	–
Pseudoanthidium	4	(6 + 9) / 3	75	–	–	–	–	–	–	–	–
Pseudoheriades	2	(9 + 0) / 1	50	–	–	–	–	–	–	–	–
Serapista	2	(32 + 5) / 2	100	4 / 1	50	–	–	–	–	–	–
Stenoheriades	5	–	–	–	–	–	–	–	–	–	–
Trachusa	1	–	–	–	–	–	–	–	–	–	–

Table 16.1 (cont.)

Genus	N Species	FV(G + O) / N	%	PA / N	%	Host plant preferences					
						Mono	Noligo	Boligo	Meso	Polyp	Polys
Wainia	3	–	–	–	–	–	–	–	–	–	–
Melittidae	**56**										
Pollen collecting	**56**	**(335 + 373) / 49**	**88**	**183 / 24**	**37.5**	**1 Crassula dichotoma**	**0**	**1 Aiz, 2 Fab**	**0**	**1 Cam, 1 Scr, 1 Zyg**	**1**
Afrodasypoda	1	(13 + 0) / 1	100	–	–	–	–	–	–	–	–
Capicola	13	(136 + 32) / 11	85	68 / 7	54	–	–	1 Aiz, 1 Fab	–	–	–
Ceratomonia	1	(16 + 0) / 1	100	–	–	–	–	–	–	–	–
Haplomelitta	8	(84 + 10) / 4	50	50 / 8	63	1 Crassula	–	1 Fab	–	1 Cam	–
Meganomia	2	(43 + 2) / 2	100	26 / 2	100	–	–	–	–	–	1
Melitta	6	(26 + 19) / 5	83	16 / 2	33	–	–	–	–	1 Zyg	–
Rediviva	24	(15 + 310) / 24	100	22 / 4	17	–	–	–	–	1 Scr	–
Redivivoides	1	(2 + 0) / 1	100	1 / 1	0	–	–	–	–	–	–
Total	**1416**										
Pollen collecting	**1259**	**(4200 + 2336) / 474**	**37.6**	**1384 / 158**	**9.9**	**1**	**1**	**14**	**8**	**9**	**8**

[1] Without *Hylaeus*, which are collecting pollen internally in their crop (pollen unavailable for analysis).
[2] Including undescribed species (Kuhlmann, unpublished data).

genera (89.9 %) and pollen samples from 29 genera (32.6 %). Data for 1384 pollen samples from female scopal loads of 158 bee species (9.9 % of the nonparasitic bee species) and 6536 flower visitation records for 474 species (37.6 % of the pollen-collecting bee fauna) are available for southern Africa with 4200 records (64.3 %) for 239 species (50.4 %) from the Albany Museum database alone (Table 16.1). From the latter, 59.1 % of all observations are from South Africa, 40.4 % from Namibia and 0.5 % from Lesotho, with the vast majority of records originating from the arid to semiarid west of South Africa and Namibia. For the pollen data, 96.6 % are from South Africa and 3.4 % from Namibia with 75.5 % of all samples from the Nieuwoudtville vicinity in the west of South Africa.

Among bee families, data quality and quantity shows a lot of variation (Table 16.1). Recording of flower visitation and pollen sampling is often highly biased to a small number of species, so the taxonomic coverage is variable between families (Table 16.1). Regarding pollen, Melittidae (37.5 % of all species) and Colletidae (33.3 % of all species) are best covered, while there are no data for Andrenidae.

For South Africa, the bulk of the flower visitation records (77.0 %) are from spring and early summer (September to December), during the main blooming season in the winter rainfall area, and only 0.4 % are from winter (June to August). In Namibia, the observation focus was during March through April (90.5 % of all records). In contrast, the majority of pollen samples (98.4 %) are from winter and early spring (July to October) with few records taken in autumn.

16.6 Flower visitation patterns

Flowers of 40 families of flowering plants have been recorded as receiving visits from bees, and pollen of 30 plant families has been identified from pollen in the scopae of female bees. Amaryllidaceae, Caryophyllaceae, Liliaceae, Loranthaceae, Montiniaceae, Oxalidaceae, Papaveraceae, Tecophileaeceae, and Verbenaceae are only represented in the latter sample. The 15 most important plant families based on pollen samples are shown in Table 16.2, and for flower visitation records in Table 16.3. The four most important plant families for bees in both the pollen analyses and flower visitation records are Asteraceae, Aizoaceae, Fabaceae, and Zygophyllaceae. Pollen of these families has been found in female scopae of 146 species (92.4 %) and flower visitations on them have been recorded for 203 species (84.9 %). Also 21 (84 %) of the identified pollen-specialist bees are dependent on one of these plant families. Of the remaining "top 15" of the most frequently recorded plant families, only three are shared between both lists: Campanulaceae, Molluginaceae, and Scrophulariaceae.

Table 16.2 List of the 15 most important plant families based on pollen collected by female bees, including the number of host-plant specialists. A bee species is regarded as a specialist when it belongs to one of the following categories (Müller and Kuhlmann, 2008, Murray et al. 2009): monolectic, all forms of oligolecty, and polylectic with strong preference.

Plant family	N bee species collecting pollen	%	N specialist bees
Asteraceae	102	64.6	13
Zygophyllaceae	53	33.5	2
Aizoaceae	47	29.8	3
Fabaceae	26	16.4	3
Scrophulariaceae	19	12.0	1
Oxalidaceae	13	8.2	1
Xanthorrhoeaceae	11	7.0	–
Iridaceae	10	6.3	–
Asparagaceae	9	5.7	–
Campanulaceae	9	5.7	1
Euphorbiaceae	8	5.1	–
Molluginaceae	6	3.8	–
Brassicaceae	5	3.2	–
Crassulaceae	5	3.2	1
Proteaceae	4	2.5	–
Other 15 families / unidentified	70	44.3	–
Total	**158**		**35**

16.7 Flower specialisation

For 41 bee species (29.9 % of the species with pollen data), sufficient pollen samples were available to assess their specialisation status. Of these 25 species (61 %) are regarded as specialists to various degree (monolectic, all forms of oligolecty, and polylectic with strong preference) (Table 16.2). Asteraceae are the most important host plants for specialist bees with 13 species (52 % of the specialists) collecting pollen exclusively on them, followed by Aizoaceae and Fabaceae (3 specialists (12 %)

Table 16.3 List of the 15 most important plant families based on flower visitation records of female bees (Albany Museum bee database only).

Plant family	N bee species visiting flowers	%
Fabaceae	107	44.8
Asteraceae	85	35.6
Aizoaceae	63	26.4
Zygophyllaceae	43	18.0
Malvaceae	39	16.3
Acanthaceae	38	15.9
Lamiaceae	30	12.6
Scrophulariaceae	30	12.6
Boraginaceae	27	11.3
Capparaceae	27	11.3
Campanulaceae	17	7.1
Amaranthaceae	10	4.2
Apocynaceae	10	4.2
Molluginaceae	10	4.2
Solanaceae	10	4.2
Other 25 families	95	39.5
Total	**239**	

each), Zygophyllaceae (2 specialists (8 %) each) and Campanulaceae, Crassulaceae, Oxalidaceae and Scrophulariaceae each hosting a single specialized bee species.

Pollen specialisation is unevenly distributed among bee families (Table 16.1). When estimated using the proportion of specialization (monolectic, all forms of oligolecty and polylectic with strong preference), in all species with pollen data available, the Apidae (8 %) and Halictidae (10 %) have the smallest percentage of bees collecting pollen from certain plants, followed by Megachilidae (15 %), which generally seem to have a preference for Fabaceae, and Colletidae (16 %). Most specialist bees are found in the Melittidae (29 %) while there is no data available for Andrenidae.

The highest degree of specialization has been found in two undescribed species with *Scrapter* spec. nov. being a narrow oligolege on *Oxalis*, and *Haplomelitta spinosa* monolectic on *Crassula dichotoma*. For the iconic South African endemic oil-collecting *Rediviva* bees, the situation regarding flower specialization is more complex. The females collect floral oil from a limited range of oil-producing host-plants, some of which have long twin spurs. Beside their specialized oil-collecting habits, *Rediviva* bees are fairly generalized when it comes to pollen collecting, but more data is needed to clarify the situation.

For 53 bee species (4.2 % of the fauna), both pollen analyses and flower visitation records were available, but for only nine of them the number of records and pollen samples was sufficiently high to permit a comparison of both datasets (Table 16.4). They represent three different levels of specialization (broad oligolecty, polylecty with strong preference, polylecty s.str.) each with three species. For the broadly oligolectic bee species, data on flower visitation records and pollen analyses generally coincide, reflecting their status of specialisation. However, when only looking at flower visitation records for *Capicola micheneri*, some uncertainty is left because it has been observed visiting Molluginaceae several times instead of its regular Fabaceae host plants. For two of the polylectic species, data on both methods come to the same conclusion about their status. In contrast, *Meganomia binghami* can be seen as polylectic with strong preference for Fabaceae when only flower visitation records are considered. For all polylectic species, records of plant families used differ significantly between the pollen and the flower visitation record. The bees classified as polylectic with a strong preference for a certain plant family show most discrepancies between the results of pollen analyses and flower visitation records, leading to potentially misleading interpretations. Data for *Haplomelitta ogilviei* are almost identical. But the flower visitation records for *Scrapter niger* suggest a specialist for Asteraceae, and pollen analyses confirm that other plant families are also visited for pollen. The differences between both methods for *Colletes capensis* are striking and indicate a strong preference (80 % of all pollen) for Aizoaceae in the pollen record, while this species has been observed on flowers of this family only in 20 % of all recorded cases. Generally, flower visitation records when used cautiously can be used to identify potentially specialized bee species that should then be studied in more detail using pollen analysis to confirm their status.

16.8 Discussion

Data of both flower visitation records and especially pollen samples are strongly geographically biased towards the arid to semiarid western parts of South Africa and Namibia. This region certainly has the richest bee fauna with many endemic elements and basal lineages important for understanding bee evolution, but the totally different bee fauna of the moister and, several times larger, eastern parts of the subcontinent has largely been ignored (Kuhlmann 2009). Within the dry west of southern Africa, observations and collection of pollen samples have been mostly restricted

Table 16.4 Comparisons of flower visitation records and results of pollen analyses of female scopal loads for nine bee species of different levels of specialization. Aiz: Aizoaceae, Apo: Apocynaceae, Asp: Asparagaceae, Ast: Asteraceae, Bra: Brassicaceae, Bor: Boraginaceae, Cam: Campanulaceae, Cap: Capparaceae, Cra: Crassulaceae, Cuc: Cucurbitaceae, Eup: Euphorbiaceae, Fab: Fabaceae, Ger: Geraniaceae, Gis: Gisekiaceae, Iri: Iridiaceae, Mal: Malvaceae, Mol: Molluginaceae, Nyc: Nyctaginaceae, Oxa: Oxalidaceae, Pro: Proteaceae, Scr: Scrophulariaceae, Ver: Verbenaceae, Xan: Xanthorrhoeaceae, Zyg: Zygophyllaceae, ???: unidentified pollen.

	Flower visitation records			Femal scopal pollen loads			
	N Records	N Localities	Flower visitation (%)	N Records	N Localities	Pollen load (%)	Host plant preference
Capicola micheneri Michez	19	6	Fab 90, Mol 10	16	4	Fab 100	Broad oligolecty (Fab)
Patellapis communis (Smith)	46	2	Ast 100	39	5	Ast 96, Xan 2, (Aiz, Mol, Iri, Pro) 2	Broad oligolecty (Ast)
Scrapter nitidus (Friese)	27	6	Ast 96, Zyg 4	34	2	Ast 100	Broad oligolecty (Ast)
Colletes capensis Cameron	15	5	Apo 67, Aiz 20, Ast 7, Iri 6	27	16	Aiz 80, Eup 11, Ast 8, (Cam, Scr, ???) 1	Polylecty with strong preference (Aiz)
Haplomelitta ogilviei (Cockerell)	67	5	Cam 79, Ast 18, Aiz 3	17	5	Cam 81, Ast 16, ??? 3	Polylecty with strong preference (Cam)
Scrapter niger Lepeletier & Serville	16	3	Ast 100	24	2	Ast 90, Bra 5, Mol 3, Cra 2	Polylecty with strong preference (Ast)
Anthophora diversipes Friese	15	6	Aiz 47, Mal 33, Bor 13, Zyg 7	20	3	Xan 38, ??? 27, Asp 9, Aiz 7, Mal 7, Iri 6, Bra 3, Ast 1, Scr 1, Zyg 1	Polylecty s. str.
Colletes rufotibialis Friese	42	6	Fab 52, Aiz 43, Zyg 5	16	6	Fab 57, Mal/Ver 31, Aiz 6, Zyg 6	Polylecty s. str.
Meganomia binghami (Cockerell)	28	10	Fab 71, Zyg 11, ?Nyc 7, Cuc 4, Gis 4, Mol 3	23	4	Zyg 48, Cap 40, ??? 6, Ger 3, Aiz 3	Polylecty s. str.

to the main flowering season in spring and early summer when weather conditions are more favorable for insects, while data for the winter and especially autumn are scarce and only available from the Nieuwoudtville vicinity (pollen samples). Records for the dry and hot summer are virtually absent, except for some parts of Namibia.

This pronounced geographical and seasonal bias has to be taken into account for data analysis and is here exemplified for the plant genus *Oxalis*. With 210 described species (Dreyer and Makgakga 2003), *Oxalis* has a centre of diversity in the Greater Cape Floristic Region (Oberlander et al. 2002), and it is the seventh largest genus within the Cape Flora (Goldblatt and Manning 2000). The vast majority of *Oxalis* species are insect pollinated, flowering in autumn and winter, and are partly producing spectacular floral mass displays (Dreyer et al. 2006). Despite the significance of *Oxalis* as a floral element investigated in the dry west of South Africa, no flower visitation data for bees has been recorded (Gess and Gess 2003, 2004a, 2006). Own observations (Timmermann and Kuhlmann unpublished data) show that some *Oxalis* species are frequently and abundantly visited by a range of bee species with at least 13 of them collecting its pollen (Table 16.2). This includes one pollen specialists of the genus *Scrapter* (Table 16.1) and makes it the sixth most important pollen source for bees in the Nieuwoudtville area (Table 16.2).

As demonstrated, the overall taxonomic coverage of the available flower visitation records and pollen analyses is unevenly distributed, with flower visitation records lacking for 62.4 % and pollen data missing for 90.1 % of all bee species. This lack of information is especially common in some species, often abundant, potentially important pollinators of the halictid genera *Lasioglossum*, *Patellapis*, and *Lipotriches*, as well as the large megachilid genus *Megachile*.

Records of flower visitation can give valuable information about flower relationships, but even when carefully done (discrimination between sexes, nectaring versus pollen collecting), they only represent a snap-shot of a bees' life. Analysis of scopal pollen instead, documents the flower visitation behavior over a much longer period of time. In the case of pollen from a bees' nest, even the complete lifespan of an individual can be covered, providing much more reliable data. But there are also potential pitfalls for the interpretation of pollen data. As Timmermann and Kuhlmann (2008) demonstrated for *Patellapis doleritica* (as *Patellapis* spec.) pollen collecting by females is highly dependent on floral display and availability of food plants that can vary dramatically between seasons and years. This is especially true for the dry west of southern Africa where the amount, time, and distribution of rainfalls varies enormously between years (Cowling et al. 1999), which influences floral display (Struck 1990, Johnson 1993). Thus, observations of flower visitation behavior or pollen collecting from a single year, or even part of the season, can be misleading (Struck 1990) if flower availability at a site is not considered, as demonstrated for the Brazilian bee *Ptilothrix plumata* (Schlindwein et al. 2009).

Clearly a much more extensive and geographically, as well as seasonally, balanced dataset of flower visitation records are needed. This is especially true for pollen samples, which are needed for a better understanding of the spatial and temporal structure of bee–flower relationships in southern Africa, as shown for pollen wasps (Vespidae: Masarinae) (Gess and Gess 2004b). Undoubtedly a large number of further specialist bees exist in southern Africa, like the Fideliinae whose flower visitation is well-documented (Whitehead 1984). However, quantitative pollen analyses of female scopal loads is needed for confirmation, and reliable determination of the degree of specialization to avoid potential misinterpretation, as in *Haplomelitta ogilviei* and its preferred host-plant *Monopsis debilis* (Campanulaceae). Although data of flower visitation records and pollen analyses are almost identical (Table 16.4), both have been interpreted in different ways. Based on observational data, Rozen (1974) and Gess and Gess (1994) cautiously concluded that *H. ogilviei* is a specialist and potentially even monolectic, while pollen analysis revealed that pollen collecting on Asteraceae is more frequent than previously thought, classifying it as polylectic with a strong preference.

The substantial gaps exist in our knowledge and missing reliable data about bee–flower relationships over large geographical areas and many seasons of the year is troublesome. However, the complete lack of information for the majority of the bee and plant species is particularly worrying, given the enormous ecological and economic significance of bees as the most important group of pollinators, and pollination in general as a vital ecosystem service. However, despite the gaps in our knowledge of bee–flower relationships in southern Africa it seems to be clear that, at least in the arid and semiarid west of the subcontinent, more than half of the bee species are specialists to various degrees, ranging from monolecty, all forms of oligolecty to polylecty with strong preference for a plant taxon. The most important host plants that are visited to a variable extend by the vast majority of all bee species are Aizoaceae, Asteraceae, Fabaceae, and Zygophyllaceae. The proportion of specialist bee species vary between families, with the Apidae and Halictidae having relatively few specialists, while the number in Colletidae, Megachilidae, and especially Melittidae is considerably higher. The latter agrees with the assumption of Michez et al. (2010) that melittids, as the most basal bee clade, should have the highest degree of flower specialization.

Our findings are generally in accordance with the observations of Struck (1990), who reported a strong preference of Namaqualand bees for Aizoaceae and Asteraceae. He also noted the differences in flower specialization between bee families and some genera, but underestimated the overall degree of specialization of the fauna (30 %).

However, levels of flower specialization found in arid to semiarid regions in Chile and California (51–61 %) (Moldenke 1976) correspond to the 61 % specialists found among the 41 bee species from similar habitats in western South Africa. Also, as the number of records is too low for making general statements, it is supporting the view that flower specialization of bees is most common in xeric ecosystems (Michener 1979). In temperate regions, generally fewer specialists exist, representing between 30–42 % of all species in western North America (Moldenke 1976) and 23 % in Germany (Westrich 1989).

Despite the fact that Western Australia (WA) in its southwestern corner is climatically and ecologically similar to the Cape Floral Kingdom (Hopper and Gioia 2004; Hopper 2009), its bee faunas and flower preferences are quite different. Unlike the rest of the continent (Batley and Hogendoorn 2009), catalogs of flower visitation records for pollinators (Brown et al. 1997), especially bees (Houston 2000), are available for WA. This data was recently analyzed by Phillips et al. (2010), and although no general conclusions can be drawn about the overall degree of bee specialization, the most species-rich bee family, Colletidae, and the Australian endemic, Stenotritidae, tend to have the highest degree of flower specialization, including some true monolectic species. For the Colletidae, this is in accordance with observations from southern Africa where one of the highest proportions of specialist bees is found in this family, exceeded only by Megachilidae, which are severely understudied in WA, and the Melittidae, which are missing in Australia (Batley and Hogendoorn 2009). In most cases, specialist bees in WA visit the abundant and ubiquitous Myrtaceae and Proteaceae (Phillips et al. 2010). In contrast, the few southern African species of Myrtaceae are virtually inexistent in the flower visitation record and even Proteaceae, although much more abundant and partly dominant in the Cape Floral Kingdom, seem to play only a minor role as a pollen and nectar source for bees.

Acknowledgements

We are very much indebted to Sarah and Friedrich Gess for making their database of flower visitation records available for this study. It is based on many years of field work in southern Africa. MK is very grateful for the analysis of pollen samples performed by Sue de Villiers, formerly Cape Town (South Africa), and Kim Timmermann, formerly Münster (Germany). Denis Michez, Mons (Belgium), kindly made pollen data for *Capicola*, *Samba*, and *Meganomia* available to us. The quality of the manuscript has been improved by insightful comments of the reviewers.

References

Allen-Wardell G., Bernhardt, P., Bitner, R., Burquez, A., Buchmann, S., Cane, J., Cox, P., Dalton, V., Feinsinger, P., Ingram, M., Inouye, D., Jones, C., Kennedy, K., Kevan, P., Koopowitz, H., Medellin, R., Medellin-Morales, S., Nabhan, G., Pavlik, B., Tepedino, V., Torchio, P. and Walker, S. (1998). The potential consequences of pollinator declines on the conservation of biodiversity and stability of food crop yields. *Conservation Biology*, **12**, 8–17.

APG III. (2009). An update of the Angiosperm Phylogeny Group classification for the orders and families of flowering plants: APG III. *Botanical Journal of the Linnean Society*, **161**, 105–21.

Batley, M. and Hogendoorn, K. (2009). Diversity and conservation status of native Australian bees. *Apidologie*, **40**, 347–54.

Brown, E. M., Burbidge, A.H., Dell, J., Edinger, D., Hopper, S. D. and Wills, R. T. (1997). *Pollination in Western Australia: a Database of Animals Visiting Flowers*. Perth, WA: Handbook No. 15. WA Naturalists' Club.

Corbet, S. A., Williams, I. H. and Osborne, J. L. (1991). Bees and the pollination of crops and wild flowers in the European Community. *Bee World*, **72**, 47–59.

Cowling, R. M., Esler, K. J. and Rundel, P. W. (1999). Namaqualand, South Africa – an overview of a unique winter-rainfall desert ecosystem. *Plant Ecology*, **142**, 3–21.

Dreyer, L. L. and Makgakga, M. C. (2003). Oxalidaceae. In *Plants of Southern Africa: An Annotated Checklist*. *Strelitzia*, **14**, 174–94.

Dreyer, L. L., Esler, K. J. and Zietsman, J. (2006). Flowering phenology of South African *Oxalis*: possible indicator of climate change? *South African Journal of Botany*, 72, 150–6.

Eardley, C. and Urban, R. (2010). Catalogue of Afrotropical bees (Hymenoptera: Apoidea:Apiformes). *Zootaxa*, **2455**, 1–548.

Gess, S. K. and Gess, F. W. (1994). Love among the flowers. The small annual, *Monopsis debilis*, and its pollinator, the bee, *Haplomelitta ogilviei*. *Veld & Flora*, **80**, 18–9.

Gess, S. K. and Gess, F. W. (2003). *A Catalogue of Flower Visiting Records for Aculeate Wasps and Bees in the Semi-Arid to Arid Areas of Southern Africa*. Grahamstown, South Africa: Albany Museum.

Gess, S. K. and Gess, F. W. (2004a). A comparative overview of flower visiting by non-*Apis* bees in the semi-arid to arid areas of southern Africa. *Journal of the Kansas Entomological Society*, **77**, 602–18.

Gess, S. K. and Gess, F. W. (2004b). Distributions of flower associations of pollen wasps (Vespidae: Masarinae) in southern Africa. *Journal of Arid Environments*, **57**, 17–44.

Gess, S. K. and Gess, F. W. (2006). Survey of flower visiting by aculeate wasps and bees in the semi-arid to arid areas of southern Africa. *Annals of the Eastern Cape Museums*, **5**, 1–51.

Goldblatt, P. and Manning, J. (2000). Cape plants: a conspectus of the Cape Flora of South Africa. *Strelitzia*, **9**, 12–3.

Hopper, S. D. (2009). OCBIL theory: towards an integrated understanding of the evolution, ecology and conservation of biodiversity on old, climatically buffered, infertile landscapes. *Plant and Soil*, **322**, 49–86.

Hopper, S. D. and Gioia, P. (2004). The Southwest Australian Floristic Region: evolution and conservation of a global hot spot of biodiversity. *Annual Review of Ecology, Evolution and Systematics*, **35**, 623–50.

Houston, T. F. (2000). *Native Bees on Wildflowers in Western Australia*. Perth, WA: Special Publication No. 2 of the Western Australian Insect Study Society.

Johnson, S. D. (1993). Climatic and phylogenetic determinants of flowering seasonality in the Cape Flora. *Journal of Ecology*, **81**, 567–72.

Klein, A.-M., Vaissière, B. E., Cane, J. H., Steffan-Dewenter, I., Cunningham, S. A., Kremen, C. and Tscharntke, T. (2007). Importance of pollinators in changing landscapes for world crops. *Proceedings of the Royal Society B*, **274**, 303–13.

Kreft, H. and Jetz, W. (2007). Global patterns and determinants of vascular plant diversity. *Proceedings of the National Academy of Sciences USA*, **104**, 5925–30.

Kuhlmann, M. (2006). Scopa reduction and pollen collecting of bees of the *Colletes fasciatus*-group in the winter rainfall area of South Africa (Hymenoptera:Colletidae). *Journal of the Kansas Entomological Society*, **79**, 165–75.

Kuhlmann, M. (2009). Patterns of diversity, endemism and distribution of bees (Insecta:Hymenoptera:Anthophila) in southern Africa. *South African Journal of Botany*, **75**, 726–38.

Kuhlmann, M. and Timmermann, K. (2009). Nest architecture and floral hosts of the South African endemic solitary bee *Othinosmia* (*Megaloheriades*) *schultzei* (Hymenoptera:Megachilidae). *Entomologia Generalis*, **32**, 1–9.

Michener, C. D. (1979). Biogeography of the bees. *Annals of the Missouri Botanical Garden*, **66**, 277–347.

Michener, C. D. (2007). *The Bees of the World*, 2nd edn. Baltimore, MD: Johns Hopkins University Press.

Michez, D., Eardley, C. D., Kuhlmann, M., Timmermann, K. and Patiny, S. (2010). The bee genera *Haplomelitta* and *Samba* (Hymenoptera, Anthophila, Melittidae): phylogeny, biogeography and host-plants. *Invertebrate Systematics*, **24(4)**, 327–347.

Michez, D., Patiny, S., Rasmont, P., Timmermann, K. and Vereecken, N. J. (2008). Phylogeny and host-plant evolution in Melittidae s.l. (Hymenoptera:Apoidea). *Apidologie* **39**, 146–62.

Moldenke, A. R. (1976). Evolutionary history and diversity of the bee faunas of Chile and Pacific North America. *Wasmann Journal of Biology*, **34**, 147–78.

Müller, A. (1996). Host-plant specialization in western palearctic anthidiine bees (Hymenoptera:Apoidea:Megachilidae). *Ecological Monographs*, **66**, 235–57.

Müller, A. and Kuhlmann, M. (2008). Pollen hosts of western palaearctic bees of the genus *Colletes* (Hymenoptera:Colletidae): the Asteraceae paradox. *Biological Journal of the Linnean Society*, **95**, 719–33.

Murray, T. E., Kuhlmann, M. and Potts, S. G. (2009). Conservation ecology of bees: populations, species and communities. *Apidologie*, **40**, 211–36.

van der Niet, T. and Johnson, S. D. (2008). Patterns of plant speciation in the Cape floristic region. *Molecular Phylogenetics and Evolution*, **51**, 85–93.

Oberlander, K. C., Dreyer, L. L. and Esler, K. J. (2002). Biogeography of *Oxalis* (Oxalidaceae) in South Africa: a preliminary study. *Bothalia*, **32**, 97–100.

Pauly, A. (2008). Catalogue of the sub-Saharan species of the genus *Seladonia* Robertson, 1918, with description of two new species (Hymenoptera:Apoidea:Halictidae). *Zoologische Mededelingen*, **82**, 391–400.

Pauly, A., Timmermann, K. and Kuhlmann, M. (2008). Description of a new interesting species from South Africa, *Evylaeus* (*Sellalictus*) *fynbosensis* n.sp. (Hymenoptera, Apoidea, Halictidae). *Journal of Afrotropical Zoology*, **4**, 85–92.

Phillips, R. D., Hopper, S. D. and Dixon, K. W. (2010). Pollination ecology and the possible impacts of environmental change in the Southwest Australian Biodiversity Hotspot. *Philosophical Transactions of the Royal Society B*, **365**, 517-28.

Rozen, J. G. (1974). The biology of two African melittid bees (Hymenoptera, Apoidea). *Journal of the New York Entomological Society*, **82**, 6-13.

Schlindwein, C., Pick, R. A. and Martins, C. F. (2009). Evaluation of oligolecty in the Brazilian bee *Ptilothrix plumata* (Hymenoptera, Apidae, Emphorini). *Apidologie*, **40**, 106-16.

Struck, M. (1990). *Vergleichende phänologische und blütenbiologische Studien an Daueruntersuchungsflächen im ariden Winterregengebiet des südlichen Afrika*. Hamburg, Germany: PhD thesis, University of Hamburg.

Struck, M. (1994a). Flowers and their visitors in the arid winter rainfall region of southern Africa: observations on permanent plots. Composition of the anthophilous insect fauna. *Journal of Arid Environments*, **28**, 45-50.

Struck, M. (1994b). A checklist of flower visiting insects and their host plants of the Goegap Nature Reserve, northwestern Cape, South Africa. *Bontebok*, **9**, 11-21.

Struck, M. (1995). Land of blooming pebbles: flowers and their pollinators in the Knersvlakte. *Aloe*, **32**, 56-64.

Timmermann, K. (2005). *Bee communities (Hymenoptera:Apoidea) of Different Vegetation Types in the Semiarid Part of Western South Africa: Structure and Pollen Specialisation* (in German). Münster, Germany: Diploma thesis, University of Münster.

Timmermann, K. and Kuhlmann, M. (2008). The biology of a *Patellapis* (s. str.) species (Hymenoptera:Apoidea:Halictidae): sociality described for the first time in this bee genus. *Apidologie*, **39**, 189-97.

Timmermann, K. and Kuhlmann, M. (2009). Taxonomic revision of the African bee subgenera *Patellapis*, *Chaetalictus* and *Lomatalictus* (Hymenoptera:Halictidae, genus *Patellapis* Friese 1909). *Zootaxa*, **2099**, 1-188.

Waterman, R. J., Pauw, A., Barraclough, T. G. and Savolainen, V. (2008). Pollinators underestimated: a molecular phylogeny reveals widespread floral convergence in oil-secreting orchids (sub-tribe Coryciinae) of the Cape of South Africa. *Molecular Phylogenetics and Evolution*, **51**, 100-10.

Westrich, P. (1989). *Die Wildbienen Baden-Württembergs*. Stuttgart, Germany: Ulmer.

Whitehead, V. B. (1984). Distribution, biology and flower relationships of fideliid bees of southern Africa (Hymenoptera, Apoidea, Fideliidae). *South African Journal of Zoology*, **19**, 87-90.

17

Advances in the study of the evolution of plant–pollinator relationships

Sébastien Patiny

17.1 Introduction

The scheme behind the present book was to draw the big picture of pollination, gathering contributions from the different domains within biology. Throughout, authors have examined many different types of data to address questions of an evolutionary flavor. Authors have not only considered the evolution of the inter-acting plants and pollinators, and the complexity of the adaptations shown by one and the other, but they also have discussed the complexity of the relation-ships between these forms of life, as well as what can be ascertained of the evo-lution of these relationships through time. The chapters were organized into a series of main topics, according to which the present conclusions are also structured.

Phylogenies of plants and pollinators are the natural backbone upon which sev-eral of the chapters were written, and evolutionary theory defines the framework in which various types of data on pollination relationships were presented and discussed. Phylogenetics and evolutionary theory were used or at least conceptu-alized by a number of contributing authors. Chapters such as the ones by Michez et al. (Chapter 5) and Hu et al. (Chapter 6) for example, take their complete sense from studies using their results, parallel with phylogenetic approaches, to shed light on pollinator evolution. Opening the present volume, Paul Wilson dedicated

Evolution of Plant–Pollinator Relationships, ed S. Patiny. Published by Cambridge University Press. © The Systematics Association 2012.

his remarkable chapter (Chapter 1) to how fundamental components of the evolutionary process should be applied hierarchically to micro- and macroscales, where the macroscale is a phylogenetic one. A series of other contributions focus on other aspects of evolution in plant–pollinator systems.

The mapping of host breadths or pollinator breadths over phylogenies leads to an important first-take on the evolution of relationships between plants and pollinators. Such approaches have been conducted for many groups in varied conditions leading to two key observations:

(1) Coevolution or cocladogenesis do not provide a suitable framework in which to consider the evolution of pollination (e.g. Cruaud et al. Chapter 4).

(2) Pollination evolves by changes in specialization or generalization, often depending to the scale being considered.

For example, in bees, the most important group of pollinators, there has been a tendency toward generalization as seen in both phylogenies and in ecological networks (Danforth et al. 2006a, 2006b; Michez et al. 2008; Olesen et al. Chapter 13).

During the last decade or so, notable progress, key for the study of the relationships between plant and pollinator, has been made in two important domains within biology:

(1) Research on the fine mechanisms of communication between flowers and pollinators (Armburster Chapter 3; Dafni et al. Chapter 12; Leonard et al. Chapter 9)

(2) Use of network analysis for describing and modeling pollination systems.

There is a long precedent for studying pollination relationships at the community level (see Olesen et al.'s introduction, Chapter 13), and the concept of a pollination network dates back at least to the 1970s. However, advances in computation, as well as the accumulation of raw data, have recently allowed this type of ecological analysis to explode (Olesen et al. Chapter 13). A pollination network is now regarded as a primary outlook upon which to discuss the nature of pollination relationships. It yields a very integrative picture. By itself the network approach could be conceived as phenomenological, that is, it shows a picture of the functioning of plant-pollinator relationships. So far, however, the phenomenon has not been studied in terms of how it changes through an evolutionary process over large timescales.

The raw material feeding network studies is typically derived from more local data sets. Within an ecological community at a moment in time, it is indeed important to answer queries about the relationships between a plant and its pollinators or about a pollinator and the plants it visits. These local studies have documented intuitive perceptions, yielded unexpected results, and revealed surprisingly important players (Dafni et al. Chapter 12). In addition to its intrinsic interest, the network approach is also valuable because (i) it broadens the general field of research, and (ii) it builds the components from which more comparative studies can be derived.

At the same time as there has been fast growth in the application of network analysis to pollination, there has been recent rapid progress in the study of communication mechanisms between insects and their host-plants using varied visual and chemical cues and rewards. Etho-physiological theories and technology have developed and cast the background for a great amplification of these approaches in recent years, as referred to in the chapters by Armburster (Chapter 3), Dafni et al. (Chapter 12), and Leonard et al. (Chapter 9) (see also Sedivy et al. 2011 for a puzzling example of competition through rewards).

17.2 Phylogenetic framework of the evolution of pollination systems

Paul Wilson's chapter (Chapter 1) outlined how adaptations in plant reproduction feed into a macroevolutionary dynamic. He cited Gould's (2002) "grand analogy" between macro- and microevolution, applying it to the specific case of our "abominable mystery" (Darwin in Darwin and Seward (1903). The grand analogy points out a similarity in process between clade selection shaping a biota and individual selection shaping the traits of organisms. In varied examples, an increased diversification of the clades expressing certain characters is observed and can be interpreted as due to positive selection among clades. Wilson showed how this model of evolution applies to pollination strategies constituting a flora. It could also be applied to the pollinators of a faunistic region.

If we consider pollination in terms of selection, observations suggest that, for pollinators or plants diversifying in the contest of one another, local counterparts are generally favored by selection. On the short term, such a strategy means that those who survive are those that use trustworthy resources supported by other approximately substitutable species. On the longer term, the favored macroevolutionary strategies are those in which diversification has proceeded by the clade having, in effect, taken out a kind of life insurance that protects it as it faces the eventuality of ecological changes such as occur from time to time.

Another component of the grand analogy is between conservatism, as clades undergo cladogenesis, and the heritability of traits among related individuals. Conservatism addresses the question of the maintenance of the mutualistic relationships by the ecological web itself, for example, how the substitutability of pollinators preserves the diversity of generalist flowers. We can understand the argument of increased stability of characters as favored by the relationships with numerous participants in the pollination web. Within species and at the local level, heritability, or in other words the genetic basis for trait variation, constitutes the seeds of future local adaptive differentiation and thus cladogenesis.

As Wilson concluded, the "grand analogy" is far from perfect. Still, it seems almost inevitable that progress in the methods used for inferring evolutionary history will improve the way in which we conceptualize the evolutionary process. Our conceptualization of the evolution of pollination systems at this point ought to be hierarchical.

Testing evolutionary hypotheses regarding pollination will require larger trees or large sets of small trees, and new methods for tree comparison and combination, etc. Advances in phylogenetics are thus of primary importance in making our studies of macroevolutionary and adaptive aspects of pollination empirical and rigorous. The methods for inference of phylogenies and for gathering raw data into large datasets are quickly expanding. The use of molecular data has constituted a big step forward. Sequencing technology is advancing exponentially. The increased use of molecular data has demanded an improvement in alignment algorithms and software. Some are now performing well with very large datasets, including many taxa and long sequences (Liu et al. 2010). For phylogenetic inference itself, new programs are providing enormous improvements. For example, maximum likelihood was once too slow and computationally too demanding for large phylogenies based on multiple gene regions. Programs like RAxML now permit extensive use of maximum likelihood methods (Stamatakis 2006). Along with other innovations in the use of maximum likelihood, this approach to phylogenetic analysis is yielding much-improved topologies and better assessment of confidence. At the same time, programs such as BEAST (Drummond and Rambaut, 2007), based on Bayesian statistics, have facilitated access to the computation of molecular clocks. This allows us to put phylogenetic events (like the radiation of bees) into historical time frames (how far angiosperms had progressed in their diversification), if not with an absolute date, at least with a narrative of what happened first and what happened later. This molecular phylogenetic revolution is far from complete, but progress is breathtaking. Second generation sequencing, now well underway, is likely to provide amounts of data that are orders of magnitude greater than was possible a few years ago. The paper by Emerson et al. (2010) exemplifies what high throughput sequencing can do for addressing pending questions. Larger and better trees will be the underpinning of many evolutionary studies on pollination biology.

Studying mutualisms, we are not only interested in inferring single plant and pollinator phylogenies, but also in determining and testing the hypothesis of possible coadaptation or cocladogenesis. The software mentioned above are for the reconstruction of phylogenies and of no help in addressing questions specific to coevolution. A cartoon of the evolution of plant–pollinator relationships follows from Farenholz's rule (Farenholz, 1913), which postulates a natural matching of the phylogenies of the groups involved in mutualistic relationships. We can reasonably agree with Page et al. (1996) in pointing out the simple-mindedness of this rule. As exposed by Cruaud et al. (Chapter 4), and numerous other authors, the relationships between plants and their pollinators are rather more complex than suggested by a strict coevolutionary presumption.

There is a parallel to note between the methods developed for inference in pollination biology and in biogeography for the confrontation of partners' evolutionary relationships. In both cases, the basic methods of inference were derived from the study of parasites and their hosts. However in both cases (i.e. pollination and biogeography), more appropriate models have been needed to better understand how evolution operates in more relaxed contexts than parasites and their hosts. As with biogeography (Ree and Sanmartín 2009), more comprehensive models would account for the likelihood of a shift (e.g. from one general type of plant to another) conditioned by an array of parameters (e.g. related to plants in the community, niches filled by other animals, and abiotic variables). Several groups are currently focusing on the development of methods of interest, for example V. Berry and his team (Montpellier, France). The research led by de Vienne et al. (2009 and other publications) also constitutes an important contribution. The further step needed is captured in a sentence of Bascompte and Jordano (2007), "The role of past evolutionary history in explaining network patterns highlights the limitations of explanations based exclusively on ultimate ecological factors." The evolution of pollination networks in deep time will be an important topic to address in the near future.

17.3 Plant strategies

The figure 2.3 is an introduction to strategies that plants use in competing for access to pollinators. The graph pictures the (female) fitness of plants increasing as the pollination rate rises. Depending on pollination efficiency (in the broadest sense), the fitness curve is more or less accelerating. Competition between plant species for pollinators, for example, impacts this curve. Selection and evolution, therefore, are consequences of the various ways in which plant–pollinator relationships affect the shape of such fitness gain curves.

In plants, competition for access to pollinators as a resource has notably resulted in adaptations of display and reward to the plant's pollinator. Armburster (Chapter 3) reviewed the evolution of flower reward in the context of relationships between plants and insects. The patterns characterized by Armburster are particularly interesting and call upon such concepts as homoplasy lability and exaptation to explain what one once called diffuse coevolution (Michez et al. 2008). The hypotheses put forward by Armburster are of particular interest because they highlight the link between this chapter and the previous one by Wilson, providing some concrete sense to theoretical hypotheses. Exaptation in evolution of the rewards is one example of drawing ingredients from deep history. Complementing Armbruster's discussion of rewards, Leonard et al. (Chapter 9) reviewed the functioning and evolution of floral signal complexity and the kind of message send by the pollinated plant to its pollinator through characters associated with pollination syndromes. Several hypotheses, namely reinforcement of the message efficacy and the sending of elaborated messages, can be put forward to explain the evolutionary advantage and ecological importance of multicomponent signaling. The addition of some complexity to the delivered signal can optimize the message, in terms of specificity and efficacy. A link with evolutionary ideas can be made in this case also.

Flower signaling in many cases incorporates chemical communication. It is nowadays well-known that chemical communication is central in the orientation of many animals in their environment. In the case of pollination by insects (but not only insects), it has been shown that volatiles emitted by flowers help guide pollinators to find the plant's sex organs (e.g. Leonard et al. 2011). The case of the carob tree developed by Dafni et al. is a good illustration of the importance of plant volatiles in pollination (see terminal part of Chapter 12).

Questions of signaling and its importance in the relationships between plants and pollinators could even be addressed one step further, at the molecular scale. In the few last years, gene repertoires involved in the animal chemoperception have been identified in bees, flies, and lepidopterans, etc. (De Bruyne and Baker 2008). This is a huge new field of investigation that is opening and may well cast light on previously obscure aspects of pollination biology. It is now possible to investigate the relationships between subgenomes, transcriptomes, specific behaviors, and ecological preferences. This will allow questions about variations among these genes, and how they impact coadaptation and reproductive isolation barriers, to be addressed. The expected outcomes should not only lead to reductionistic explanations of plant–pollinator relationships at a molecular level, but should also illuminate mechanisms of evolution through time.

The evolution of displays and rewards is often a consequence of the competition between plant species for access to pollinator resources. Parallel to that interspecific competition for the pollinator resources, plants also compete between the sexes, as exposed in de Jong's Chapter (Chapter 2). Still focusing on the chart in Fig 2.3, the fitness gain curve of males need not follow the females' curve. This brings up another dimension of plant–pollinator relationships.

Geographical specialization among populations according to the local niche has been documented for a number of pollination systems. In an evolutionary sense, competition for the access to resources, or at least selection based on differential function, also exists among populations and clades (see Chapter 1). There is a metapopulation dynamic based on plant–pollinator specialization. In addition to this selection among populations or clades, the competition that opposes the sexes within populations or species is more insidious and impacts directly upon the plant sex systems evolution. Those systems are key in evolution of the pollination relationships. This latter feature of the pollination dynamic illustrates also, in part, the reasons why the reconstruction of species to species relationships can be that difficult to infer.

17.4 Pollination systems and ecological networks

Evolutionary relationships between pollinators and host-plants are commonly depicted as an example of coevolution. However, this is a crude characterization because, in most cases, pollinators and plants are connected via unspecialized mutualisms. *Species* of pollinator are not evolving reciprocally with *species* of plant. The exact identities of the actors are constantly changing. Simple models of coadaptation and cocladogenesis have even been challenged as they apply to very specialized partnerships, such as that of fig-wasps and figs (see for instance, references in Chapter 4). A more sophisticated depiction of the evolution of relationships has yet to crystallise.

In a snapshot in time, the description of pollination systems using the methods of network analysis constitutes a big step forward (Olesen et al. Chapter 13). The great benefit of network statistics is that they account for both the vertical trophic relationships (between pollinators and host-plants) and the horizontal interactions (among pollinators, among hosts). The biology underlying the trophic relationships is provided by floral display and floral rewards in conjunction with pollinator perception and learning skills (see Chapter 9). The horizontal currency is usurpation of rewards among pollinators and distraction of pollen carriers among plants.

How does one comprehend the evolution of the network through time? From one year to the next (and for relatively small faunal entities) some research groups have showed that assessment of a previously studied network structure provides insights about the perturbations following from the collapse of some nodes or the apparitions of certain new species (Olesen et al. 2008; Díaz-Castelazo et al. 2010). By contrast, larger time scales remains difficult to follow. One possibility would be to compare networks in space as though locations were separated by time. Another possibility would be to trace phylogenetic relationships of the component organisms that make up a network. Renner (2007) used the work of Rezende et al. (2007) to show that phylogenetic relationships are not related to, and thus do not determine, interactions in local webs. The usual methods in evolutionary biology will likely be of very little help in redrawing the evolution of the ecological relationships between interdependent species.

There is, however, an opportunity to study the evolution of the characters underlying the pollinator–plant match. Armburster's chapter showed the role and impact of reward evolution and Leonard and coauthors' chapter traced display evolution. Regarding the progress made in transcriptomics, epigenetics, and ecology at the level of populations or individuals, we can hope that outcomes from such an approach will provide insights for viewing phylogenies and webs within one unique evolutionary theory. Progress in all the domains of biology provide a vague but illuminated sense of how pollination networks evolve.

17.5 The impact of ecological change: invasive species

Another aspect of pollination biology in which the network approach is of interest is its application to relatively rapid ecological change, notably the invasion of ecosystems by alien species. Invasive species constitute a fast-growing topic of study (Chapter 15). Even if we only consider bees, a group with generally low invasiveness, there are several examples of species that have become naturalized in a new biota and are expanding their ranges (Rasplus et al. 2010). One interesting example is the case of the dwarf honeybee. It is expanding in the Near East, probably from one introduced colony (Haddad et al. 2008; Moritz et al. 2010). Others cases can be found in Megachilidae, and in *Andrena*, North American populations of *Andrena wilkella* are likely alien and invasive even if they are not causing problems.

A pollinator introduction may not sound like a dire threat to local pollination networks, but, besides providing new pollination services, they also compete with native pollinators. An alien pollinator will not share exactly the host-breadth of the natives that it competes with. This inexact substitution might be a threat to native plants and pollinators. Whether competition goes to completion or only depletes native pollinator populations, some plants are likely to be deprived of pollination services and have their population ecology altered. In the light of this theoretical scenario, the local pollination network can be highly destabilized by punctuational introductions of foreign pollinators. By the interplay of the relationships asymmetries, both plant and pollinator native populations might be disrupted, or maybe not. At this point, we do not know what aspects of a pollination network are stable in the face of alien additions. We might expect that aliens would be more problematic in island ecosystems or in biogeographic provinces where the biota had recently undergone rapid diversification.

Pollinators are, of course, not the only ecosystem component threatened by alien invasions. Invasive plants can usurp a large part of the pollination service provided by the native pollinators (Morales and Traveset 2009). Presumably the introduction of alien grasses that displaced the former dominance of wildflowers, as in the conversion of California forb fields to grasslands, must have had an enormous effect on the economy of the insect community. Likewise, indirect effects, like the impact of invasive predators of native pollinators or of alien herbivores that consume the local flora, must be taken into account (Traveset and Richardson 2006; Abe et al. 2010). Pisanti and Mandelik' chapter (Chapter 15) reviewed the problems caused by aliens and the role played by networks in the severity of invasions.

17.6 What's new in pollination? Perspectives

Throughout the present book, authors have explained the complexity of pollination following from (i) the diversity of plants and pollinators; (ii) the contingent nature of their relationships; and (iii) the existence of distinct evolutionary strategies in the hierarchy of clade, species, population, and even gender within populations.

With every issue of a dozen relevant journals, obvious and significant progress is made in the many domains of pollination biology. Pollination biology also progresses circuitously by contributions from other areas of science, like phylogenetics, behavioral sciences, genomics, etc. that allow for subsequent studies addressing the topic of pollination. Taxonomic work on pollinators helps, as does a report of an unusual behavior in a wasp that visits flowers, and a new way of measuring the cost of nectar secretion, and a new statistic that weights edges in networks, etc. The unending challenge taken on by the authors of this book was to apply these advances to the study of the evolution of plant–pollinator relationships through space and time. Evolution along large timescales of pollination relationships remains a subject in which almost all understanding must come from inference. It is not like there are fossils of the many behaviors of pollinators. The values of rewards to those animals, and the amounts of pollen transferred among the plants involved is poorly documented by the known fossil archives.

The progress made in chemical ecology, neuroethology, and the related genomic and transcriptomic areas promises to aid in the reconstruction of evolutionary history. Surely, improving these aspects of the knowledge and applying these advances in a comparative way will help. Just as evo–devo has proceeded by combining phylogenetics with developmental biology, we are on the brink of an outpouring of research combining phylogenetics with pollination biology. Knowing the molecular pathways used by plants and animals to communicate with each other, it is likely to become possible to compute the ancestral states

ADVANCES IN PLANT–POLLINATOR RELATIONSHIPS **467**

of these systems and acquire insights into the evolutionary processes that have transformed relationships of one sort into those of another.

References

Abe, T., Makino, S. and Okochi, I. (2010). Why have endemic pollinators declined on the Ogasawara Islands? In *Restoring the Oceanic Island Ecosystem: Impact and Management of Invasive Alien Species in the Bonin Islands*, ed. K. Kawakami and I. Okochi. Tokyo, Japan: Springer.

Bascompte, J. and Jordano, P. (2007). Plant–animal mutualistic networks: the architecture *Annual Review of Ecological and Evolutionary Systematic*, **38**, 567–93.

de Bruyne, M. and Baker, T. (2008). Odor detection in insects: volatile codes. *Journal of Chemical Ecology*, **34**, 882–97.

Danforth, B. N., Fang, J. and Sipes, S. D. (2006a). Analysis of family-level relationships in bees (Hymenoptera:Apiformes) using 28S and two previously unexplored nuclear genes: CAD and RNA polymerase II. *Molecular Phylogenetics and Evolution*, **39**, 358–72.

Danforth, B. N., Sipes, S. D., Fang, J. and Brady, S.G. (2006b). The history of early bee diversification based on give genes plus morphology. *Proceedings of the National Academy of Sciences USA*, **103**, 15118–23.

Darwin, F., and Seward, A. C. (eds.) (1903). *More Letters of Charles Darwin*. London, UK: John Murray.

Díaz-Castelazo, C., Guimarães, P. R., Jordano, P., Thompson, J. N., Marquis, R. J. and Rico-Gray, V. (2010). Changes of a mutualistic network over time: reanalysis over a 10-year period. *Ecology*, **91**, 793–801.

Drummond, A. J. and Rambaut, A. (2007). BEAST: Bayesian evolutionary analysis by sampling trees. *BMC Evolutionary Biology*, **7**, 214.

Emerson, K. J., Merz, C. R., Catchen, J. M., Hohenlohe, P. A., Cresko, W. A., Bradshaw, W. E. and Holzapfel, C. M. (2010). Resolving postglacial phylogeography using high-throughput sequencing. *Proceedings of the National Academy of Sciences USA*, **107**, 16196–200.

Farenholz, H. (1913) Ectoparasiten und Abstammungslehre. *Zoologische Anzeiger*, **41**, 371–4.

Gould, S. J. (2002). *The Structure of Evolutionary Theory*. Cambridge, MA: The Belknap Press of Harvard University Press.

Haddad, N. J., De Miranda, J. and Bataaneh, A. (2008) The discovery of *Apis florea* in Jordan. *Journal of Apiculural Research*, **47**, 172–3.

Leonard, A. S., Dornhaus, A. and Papaj, D. R. (2011). Forget-me-not: complex floral displays, intersignal interactions, and pollinator cognition. *Current Zoology*. Online first.

Liu, K., Linder, C. R., and Warnow, T. (2010). Multiple sequence alignment: a major challenge to large-scale phylogenetics. *PLoS Currents: Tree of Life*, **2**.

Michez, D., Patiny, S., Rasmont, P., Timmermann, K. and Vereecken, N. J. (2008). Phylogeny and host-plant evolution in Melittidae s.l. (Hymenoptera, Apoidea). *Apidologie*, **39**, 146–62.

Morales, C. A. and Traveset, A. (2009). A meta-analysis of impacts of alien versus native plants on pollinator visitation and reproductive success of coflowering native plants. *Ecology Letters*, **12**, 716–28.

Moritz, R., Haddad, N., Bataieneh, A., Shalmon, B. and Hefetz, A. (2010). Invasion of the dwarf honeybee Apis florea into the near East. *Biological Invasions*, **12**, 1093–9.

Olesen, J. M., Bascompte, J., Elberling, H. and Jordano P. (2008). Temporal dynamics in a pollination network. *Ecology*, **89**, 1573–82.

Page, R. D. M., Clayton, D. H. and Paterson, A. M. (1996) Lice and cospeciation: a response to Barker. *International Journal for Parasitology*, **26**, 213–8.

Rasplus, J. Y., Villemant, C., Paiva, M. R., Delvare, G. and Roques, A. (2010). Hymenoptera Chapter 12. *BioRisk*, **4(2)**, 669–776.

Ree, R. H. and Sanmartín I. 2009. Prospects and challenges for parametric models in historical biogeographical inference. *Journal of Biogeography*, **36**, 1211–20.

Renner, S. S. (2007). Structure in mutalistic networks. *Nature*, **448**, 877–8.

Rezende, E. L., Lavabre, J. E., Guimaraes, P. R., Jordano, Jr. P., and Bascompte, J. (2007). Non-random coextinctions in phylogenetically structured mutualistic networks. *Nature*, **448**, 925–8.

Sedivy, C., Müller, A. and Dorn, S. (2011). Closely related pollen generalist bees differ in their ability to develop on the same pollen diet: evidence for physiological adaptations to digest pollen. *Functional Ecology*, online early.

Stamatakis, A. (2006). RAxML-VI-HPC: maximum likelihood-based phylogenetic analyses with thousands of taxa and mixed models. *Bioinformatics*, **22**, 2688–90.

Traveset, A. and Richardson, D. M., (2006). Biological invasions as disruptors of plant reproductive mutualisms. *Trends in Ecology and Evolution*, **21**, 208–16.

de Vienne, D. M., Hood, M. E. and Giraud, T. (2009). Phylogenetic determinants of potential host shifts in fungal pathogens. *Journal of Evolutionary Biology*, **22**, 2532–41.

Index

resource, 463
reward, 32, 44, 45, 58
scarcity, 40
sensory bias, 293
service, 6
sharing, 270
shift, 237, 238, 239, 253, 254, 331, 333, 334, 423
  specialized, 334, 364
substitutability, 460
switch, 250, 265
system, 459
trait, 239, 242, 253
transition, 239
usurpation, 415, 416, 424, 426, 427, 428
visit, 39
  decline, 40
  low-quality, 36
Polylectic, 129, 139, 141, 143, 146, 448, 449, 450, 453
Polylecty, 127, 136, 150, 450, 453
Polyphyletic, 87, 88, 92
Population, 2, 4, 6, 7, 8, 9, 10, 11, 12, 17, 18, 20, 29, 30, 35, 37, 39, 40, 142, 238, 239, 242, 243, 245, 246, 247, 251, 252, 253, 264, 265, 266, 333, 353, 383, 395, 396, 398, 399, 417, 418, 424, 425, 426, 427, 428, 430, 463, 465, 466
  allopatric, 245
  decline, 41
  management, 94
  sympatric, 245
Preaptation, 55, 57
Principle
  Bateman, 35, 36, 37
  Bell, 37, 38, 39
  most effective pollination, 394
Protein, 32, 46, 94, 134, 137
Punctuated equilibrium, 2, 12

Ratio
  animal/plant, 381
  blend, 292
  chemical, 281
  likelihood, 335
  male/female, 33
  pollen/ovule, 20
RAxML, 95, 461
Reconstruction, 464, 466
  least costly, 76
  parsimony, 169
  phylogenetic, 301, 303, 461
Relaxation, 5, 425
Reproductive

adaptation, 428
association, 237
assurance, 29, 32, 41, 367, 425, 426
barrier, 3, 5
behavior, 267, 270
biology, 19
isolation, 19, 50, 264, 266, 463
organ, 104
structure, 133
success, 34, 35, 68, 286, 290, 396, 399, 405, 416, 418
system, 130
Resin, 51, 52, 53, 58, 59, 127, 129, 132, 141, 143, 145, 146, 147, 247
  triterpene, 57
Reversal, 8, 11, 50, 54, 55, 306
Reward, 16, 19, 29, 39, 44, 45, 57, 59, 60, 104, 130, 133, 137, 146, 247, 264, 284, 288, 289, 290, 293, 294, 304, 331, 345, 346, 363, 367, 394, 399, 404, 405, 415, 416, 423, 424, 427, 428, 460, 462, 463, 464, 465, 466
  brood-site, 47
  depletion, 416
  fragrance, 47, 55, 57
  nectar, 268, 402
  non food, 268
  pollen, 249
  resin, 47, 51, 57
  scent, 270
  specialized, 44, 47, 54, 58, 59
Rewardless, 245, 284

Seed, 7, 9, 13, 20, 22, 29, 30, 33, 34, 36, 37, 38, 39, 40, 44, 47, 54, 57, 68, 132, 165, 251, 324, 326, 399, 416, 425, 427, 461
  bank, 4
  dispersal, 19, 54
  dormancy, 4, 5
  fossil, 142
  production, 41, 245, 251, 427
  set, 29, 36, 37, 38, 251
Selection, 2, 5, 7, 8, 10, 11, 12, 14, 15, 18, 19, 39, 40, 46, 47, 104, 136, 237, 238, 239, 241, 242, 243, 245, 246, 247, 249, 250, 251, 252, 253, 265, 266, 280, 282, 331, 332, 333, 334, 336, 367, 379, 394, 395, 403, 423, 426, 427, 428, 429, 430, 460, 462
  clade, 2, 5, 6, 7, 10, 11, 12, 17, 18, 22
  directional, 15
  natural, 1, 18, 36, 394
  pattern, 238
  reciprocal, 244
  sexual, 12, 19, 50, 279

Systematics Association Publications

1. Bibliography of Key Works for the Identification of the British Fauna and Flora, 3rd edition (1967)[†]
   Edited by G. J. Kerrich, R. D. Meikie and N. Tebble

2. The Species Concept in Palaeontology (1956)[†]
   Edited by P. C. Sylvester-Bradle

3. Function and Taxonomic Importance (1959)[†]
   Edited by A. J. Cain

4. Taxonomy and Geography (1962)[†]
   Edited by D. Nichols

5. Speciation in the Sea (1963)[†]
   Edited by J. P. Harding and N. Tebble

6. Phenetic and Phylogenetic Classification (1964)[†]
   Edited by V. H. Heywood and J. McNeill

7. Aspects of Tethyan Biogeography (1967)[†]
   Edited by C. G. Adams and D. V. Ager

8. The Soil Ecosystem (1969)[†]
   Edited by H. Sheals

9. Organisms and Continents through Time (1973)[*]
   Edited by N. F. Hughes

10. Cladistics: A Practical Course in Systematics (1992)[‡]
    P. L. Forey, C. J. Humphries, I. J. Kitching, R. W. Scotland, D. J. Siebert and D. M. Williams

11. Cladistics: The Theory and Practice of Parsimony Analysis, 2nd edition (1998)[‡]
    I. J. Kitching, P. L. Forey, C. J. Humphries and D. M. Williams

---

[†] Published by the Systematics Association (out of print)
[*] Published by the Palaeontological Association in conjunction with the Systematics Association
[‡] Published by Oxford University Press for the Systematics Association

Systematics Association Special Volumes

55. Arthropod Relationships (1998)**
Edited by R. A. Fortey and R. H. Thomas

56. Evolutionary Relationships among Protozoa (1998)**
Edited by G. H. Coombs, K. Vickerman, M. A. Sleigh and A. Warren

57. Molecular Systematics and Plant Evolution (1999)‡‡
Edited by P. M. Hollingsworth, R. M. Bateman and R. J. Gornall

58. Homology and Systematics (2000)‡‡
Edited by R. Scotland and R. T. Pennington

59. The Flagellates: Unity, Diversity and Evolution (2000)‡‡
Edited by B. S. C. Leadbeater and J. C. Green

60. Interrelationships of the Platyhelminthes (2001)‡‡
Edited by D. T. J. Littlewood and R. A. Bray

61. Major Events in Early Vertebrate Evolution (2001)‡‡
Edited by P. E. Ahlberg

62. The Changing Wildlife of Great Britain and Ireland (2001)‡‡
Edited by D. L. Hawksworth

63. Brachiopods Past and Present (2001)‡‡
Edited by H. Brunton, L. R. M. Cocks and S. L. Long

64. Morphology, Shape and Phylogeny (2002)‡‡
Edited by N. MacLeod and P. L. Forey

65. Developmental Genetics and Plant Evolution (2002)‡‡
Edited by Q. C. B. Cronk, R. M. Bateman and J. A. Hawkins

66. Telling the Evolutionary Time: Molecular Clocks and the Fossil Record (2003)‡‡
Edited by P. C. J. Donoghue and M. P. Smith

67. Milestones in Systematics (2004)‡‡
Edited by D. M. Williams and P. L. Forey

68. Organelles, Genomes and Eukaryote Phylogeny (2004)‡‡
Edited by R. P. Hirt and D. S. Horner

69. Neotropical Savannas and Seasonally Dry Forests: Plant Diversity, Biogeography and Conservation (2006)‡‡
Edited by R. T. Pennington, G. P. Lewis and J. A. Rattan

70. Biogeography in a Changing World (2006)‡‡
Edited by M. C. Ebach and R. S. Tangney

71. Pleurocarpous Mosses: Systematics & Evolution (2006)‡‡
Edited by A. E. Newton and R. S. Tangney

72. Reconstructing the Tree of Life: Taxonomy and Systematics of Species Rich Taxa (2006)‡‡
Edited by T. R. Hodkinson and J. A. N. Parnell

73. Biodiversity Databases: Techniques, Politics, and Applications (2007)‡‡
Edited by G. B. Curry and C. J. Humphries

74. Automated Taxon Identification in Systematics: Theory, Approaches and Applications (2007)[‡‡]
   Edited by N. MacLeod

75. Unravelling the Algae: The Past, Present, and Future of Algal Systematics (2008)[‡‡]
   Edited by J. Brodie and J. Lewis

76. The New Taxonomy (2008)[‡‡]
   Edited by Q. D. Wheeler

77. Palaeogeography and Palaeobiogeography: Biodiversity in Space and Time (in press)[‡‡]
   Edited by P. Upchurch, A. McGowan and C. Slater

[a] Published by Clarendon Press for the Systematics Association
[*] Published by Academic Press for the Systematics Association
[‡] Published by Oxford University Press for the Systematics Association
[**] Published by Chapman & Hall for the Systematics Association
[‡‡] Published by CRC Press for the Systematics Association

Printed in the United States
By Bookmasters